Introductory Algebra

AN APPLIED APPROACH

Second Edition

Richard N. Aufmann □ Vernon C. Barker

Palomar College California

HOUGHTON MIFFLIN COMPANY
BOSTON Dallas Geneva, Illinois
Lawrenceville, New Jersey Palo Alto

Printed in the U.S.A.

ISBN Numbers:
Text: 0-395-38094-4
Instructor's Annotated Edition: 0-395-42787-8
Solutions Manual: 0-395-42788-6
Alternate Testing Program, Forms A and B: 0-395-42789-4
Alternate Testing Program, Forms C and D: 0-395-42790-8
Instructor's Computerized Test Generator: 0-395-37794-3
COMPUTER TUTOR™: 0-395-37798-6
COMPUTER TUTOR™ Backup Disks: 0-395-37816-8

Cover Design by Jill Haber.

Title Page and Chapter Opener Designs by Margaret Ong Tsao.

EFGHIJ-M-89

Contents

Overview

Purpose

INTRODUCTORY ALGEBRA: AN APPLIED APPROACH, SECOND EDITION is a newly-developed text which covers all of the topics considered essential in a first-year algebra course, emphasizing applications of algebra throughout. The only mathematical prerequisite is a working knowledge of the basic computational skills. The text has been specifically developed to meet not only the needs of the traditional college student, but also the needs of the mature student whose mathematical proficiency may have declined during years away from formal schooling.

Contents

INTRODUCTORY ALGEBRA: AN APPLIED APPROACH is the second in a series of three texts by the authors:

BASIC COLLEGE MATHEMATICS: AN APPLIED APPROACH, THIRD EDITION
INTRODUCTORY ALGEBRA: AN APPLIED APPROACH, SECOND EDITION
INTERMEDIATE ALGEBRA: AN APPLIED APPROACH, SECOND EDITION

The first book, BASIC COLLEGE MATHEMATICS: AN APPLIED APPROACH, provides a comprehensive coverage of computational skills and their applications. INTRODUCTORY ALGEBRA: AN APPLIED APPROACH contains a complete development of the basic skills and applications typically found in a first-year algebra course. INTERMEDIATE ALGEBRA: AN APPLIED APPROACH covers the essentials of second-year algebra as well as certain more advanced pre-calculus topics. Since the three texts share several important pedagogical and organizational features, they may be used sequentially to reap the benefits of a smoothly-integrated series of learning materials. However, because the three texts have been written so that the content of each is independent of the other two, any one book in the series may be used independently of the other two.

Organization

INTRODUCTORY ALGEBRA: AN APPLIED APPROACH is organized into 12 chapters. Each chapter is divided into a varying number of sections, and each section contains several related objectives. Any one objective contains the exposition of a single skill or application. The exercise sets found at the end of each section are grouped by objective to establish a simple matching between exposition and related practice problems. The Review/Test found at the end of each chapter is also organized by objective, in order to define a clear correspondence between exposition and related testing.

Features

INTRODUCTORY ALGEBRA: AN APPLIED APPROACH is built around the three proven and effective teaching strategies that are characteristic of the entire series. First, an **applied approach** generates an awareness of the value of algebra as a practical tool. Second, an **interactive approach** encourages the student to practice each skill while it is being presented, thus avoiding needless confusion later when working practice assignments. Third, an **objective-specific approach** helps the student or the instructor manage instruction, improving both the efficiency and the effectiveness of the instruction. These three strategies are described pictorially on the next three pages.

New Features

This new edition is primarily a refinement and enhancement of the previous edition and contains the same organizational features and pedagogical approach that have made it so successful. Features new to this edition include **chapter summaries,** which call out the "Key Words" and "Essential Rules" in each chapter and assist students preparing for class testing; **cumulative chapter tests,** which help students evaluate their mastery of math skills; special **calculator and computer enrichment topics,** which provide students with valuable instruction and practice in using calculators or computer applications to solve selected types of exercises; and **historical notes,** which briefly examine mathematically-oriented topics of interest.

An Applied Approach

The traditional approach to teaching or reviewing algebra, which places major emphasis on problems requiring only manipulation of numbers and variables, is lacking in that it fails to teach the practical value of algebra. By contrast, INTRODUCTORY ALGEBRA: AN APPLIED APPROACH places a heavy stress on applications. Where applicable, the last objective of any section is an applications objective in which the skills covered in the section are used in the solution of practical problems. Also, an entire chapter of the text (Chapter 4-Solving Equations: Applications) and portions of several other chapters are devoted to certain standard types of applications. This carefully-integrated applied approach generates awareness on the student's part of the value of algebra as a real-life tool.

Chapter 4—Solving Equations: Applications **is devoted entirely to applications of algebra.**

A strategy which the student may use in solving application problems is stated and explained for each major type of problem.

This strategy is used in the solution of the worked example.

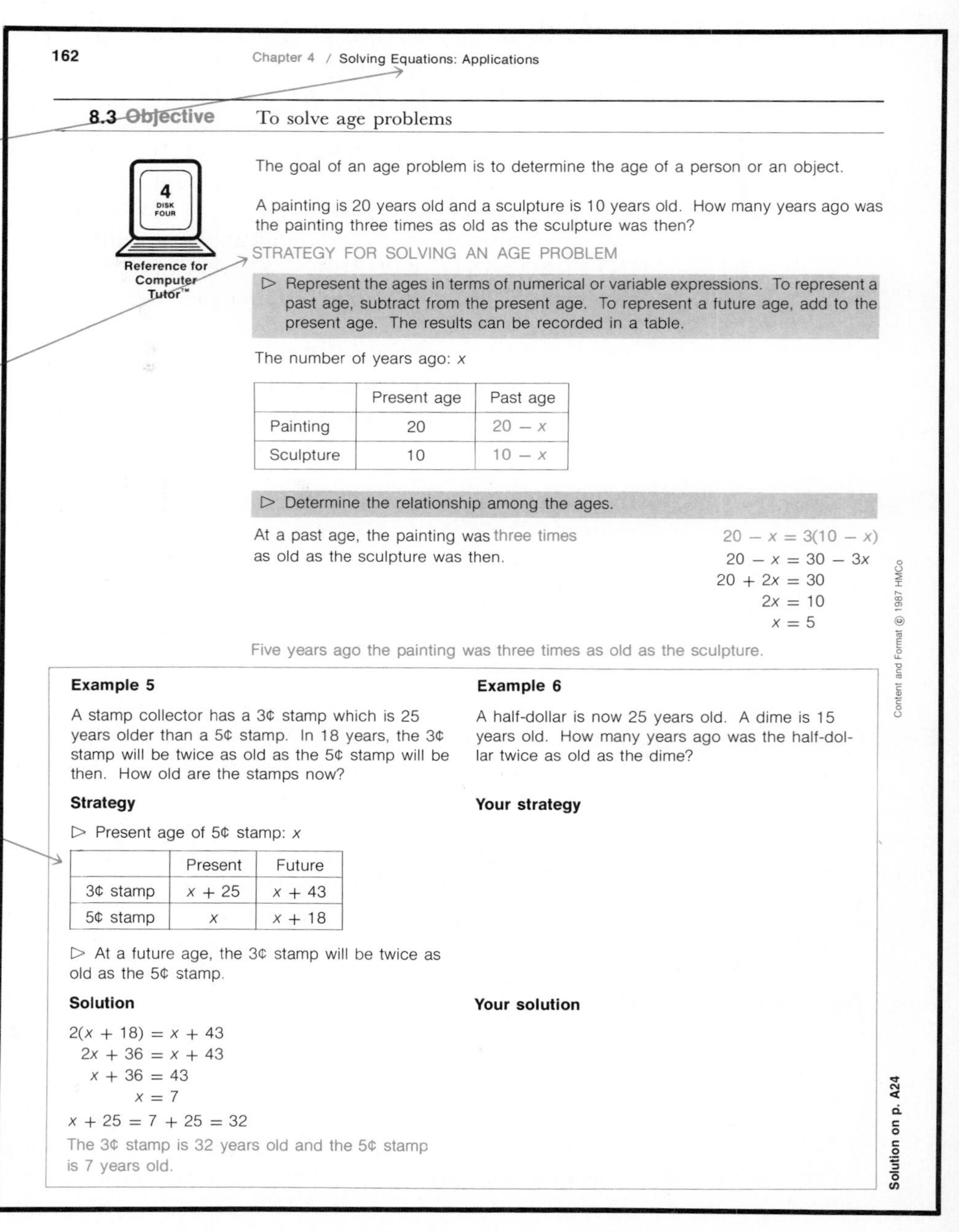

162 Chapter 4 / Solving Equations: Applications

8.3 Objective To solve age problems

4 DISK FOUR

Reference for Computer Tutor™

The goal of an age problem is to determine the age of a person or an object.

A painting is 20 years old and a sculpture is 10 years old. How many years ago was the painting three times as old as the sculpture was then?

STRATEGY FOR SOLVING AN AGE PROBLEM

▷ Represent the ages in terms of numerical or variable expressions. To represent a past age, subtract from the present age. To represent a future age, add to the present age. The results can be recorded in a table.

The number of years ago: x

	Present age	Past age
Painting	20	$20 - x$
Sculpture	10	$10 - x$

▷ Determine the relationship among the ages.

At a past age, the painting was three times as old as the sculpture was then.

$$20 - x = 3(10 - x)$$
$$20 - x = 30 - 3x$$
$$20 + 2x = 30$$
$$2x = 10$$
$$x = 5$$

Five years ago the painting was three times as old as the sculpture.

Example 5

A stamp collector has a 3¢ stamp which is 25 years older than a 5¢ stamp. In 18 years, the 3¢ stamp will be twice as old as the 5¢ stamp will be then. How old are the stamps now?

Strategy

▷ Present age of 5¢ stamp: x

	Present	Future
3¢ stamp	$x + 25$	$x + 43$
5¢ stamp	x	$x + 18$

▷ At a future age, the 3¢ stamp will be twice as old as the 5¢ stamp.

Solution

$$2(x + 18) = x + 43$$
$$2x + 36 = x + 43$$
$$x + 36 = 43$$
$$x = 7$$
$$x + 25 = 7 + 25 = 32$$

The 3¢ stamp is 32 years old and the 5¢ stamp is 7 years old.

Example 6

A half-dollar is now 25 years old. A dime is 15 years old. How many years ago was the half-dollar twice as old as the dime?

Your strategy

Your solution

Solution on p. A24

An Interactive Approach

Instructors have long realized the need for a text which requires the student to use a skill *as it is being taught.* INTRODUCTORY ALGEBRA: AN APPLIED APPROACH uses an interactive technique which meets this need. Every objective, including the one pictured below, contains at least one pair of examples in which one example is worked. The second example in the pair is not worked so that the student may "interact" with the text by solving it. In order to provide immediate feedback, complete solutions to these examples are provided in the Answer Section. The benefit of this interactive strategy is that the student can check that a new skill has been learned in advance of attempting a homework assignment.

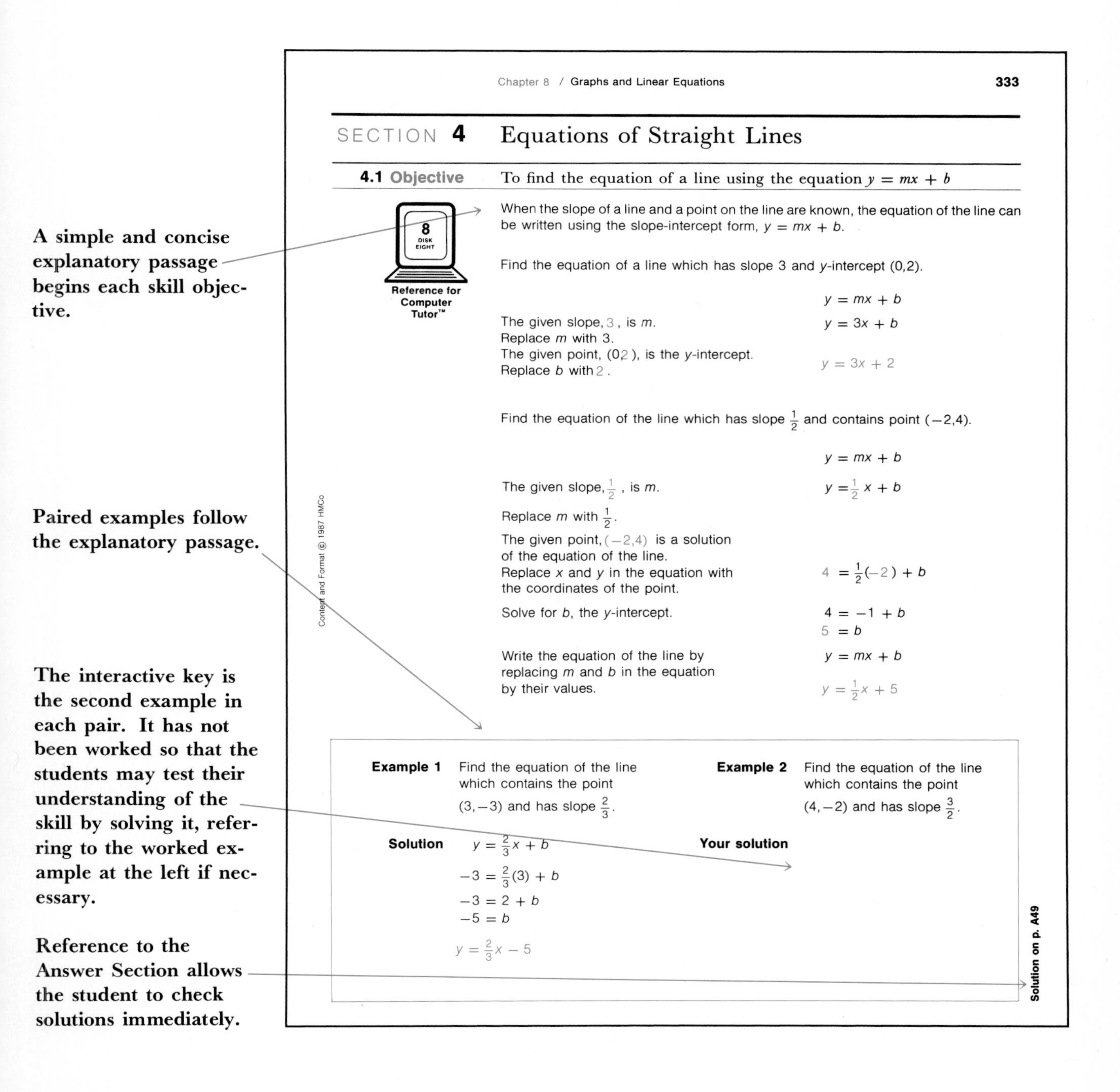

Chapter 8 / Graphs and Linear Equations 333

SECTION **4** Equations of Straight Lines

4.1 Objective To find the equation of a line using the equation $y = mx + b$

8 DISK EIGHT
Reference for Computer Tutor™

When the slope of a line and a point on the line are known, the equation of the line can be written using the slope-intercept form, $y = mx + b$.

Find the equation of a line which has slope 3 and y-intercept (0,2).

	$y = mx + b$
The given slope, 3, is m. Replace m with 3.	$y = 3x + b$
The given point, (0,2), is the y-intercept. Replace b with 2.	$y = 3x + 2$

Find the equation of the line which has slope $\frac{1}{2}$ and contains point (−2,4).

	$y = mx + b$
The given slope, $\frac{1}{2}$, is m. Replace m with $\frac{1}{2}$.	$y = \frac{1}{2}x + b$
The given point, (−2,4) is a solution of the equation of the line. Replace x and y in the equation with the coordinates of the point.	$4 = \frac{1}{2}(-2) + b$
Solve for b, the y-intercept.	$4 = -1 + b$ $5 = b$
Write the equation of the line by replacing m and b in the equation by their values.	$y = mx + b$ $y = \frac{1}{2}x + 5$

Example 1 Find the equation of the line which contains the point $(3, -3)$ and has slope $\frac{2}{3}$.

Solution

$y = \frac{2}{3}x + b$

$-3 = \frac{2}{3}(3) + b$

$-3 = 2 + b$

$-5 = b$

$y = \frac{2}{3}x - 5$

Example 2 Find the equation of the line which contains the point $(4, -2)$ and has slope $\frac{3}{2}$.

Your solution

Solution on p. A49

An Objective-Specific Approach

Many texts in mathematics are not organized in a manner which facilitates management of learning. Typically, students are left entirely to their own devices to wander through a maze of apparently unrelated lessons, practice sets, and tests. INTRODUCTORY ALGEBRA: AN APPLIED APPROACH solves this problem by organizing all lessons, practice sets, and tests around a carefully-constructed hierarchy of 122 objectives. The advantage of this objective-by-objective organization is that it enables the student who is uncertain at any step in the learning process to refer easily to the original presentation of a skill in order to review the skill or application involved.

A numbered objective statement names the skill taught in each lesson.

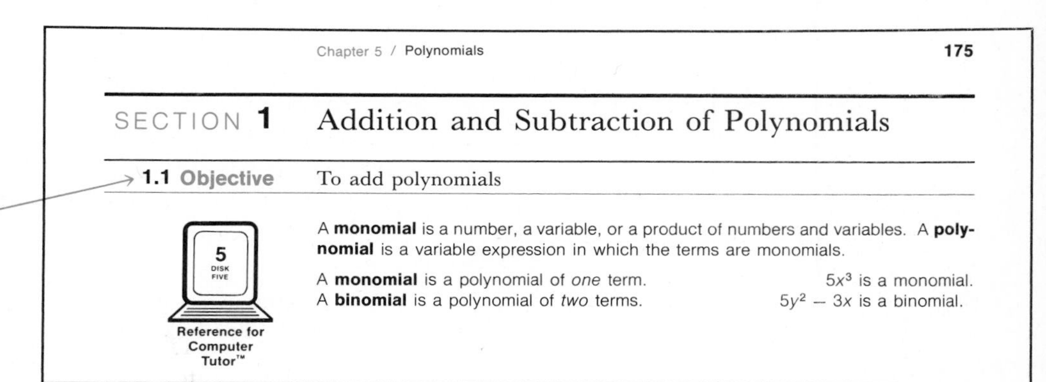

Chapter 5 / Polynomials 175

SECTION **1** Addition and Subtraction of Polynomials

1.1 Objective To add polynomials

A **monomial** is a number, a variable, or a product of numbers and variables. A **polynomial** is a variable expression in which the terms are monomials.

A **monomial** is a polynomial of *one* term. $5x^3$ is a monomial.
A **binomial** is a polynomial of *two* terms. $5y^2 - 3x$ is a binomial.

End-of-section Exercise Sets are referenced by number to objectives.

Chapter 5 / Polynomials 177

1.1 Exercises

Simplify. Use a vertical format.

1. $(x^2 + 7x) + (-3x^2 - 4x)$

2. $(3y^2 - 2y) + (5y^2 + 6y)$

3. $(y^2 + 4y) + (-4y - 8)$

4. $(3x^2 + 9x) + (6x - 24)$

The Review/Test at the end of each chapter is also referenced by number to objectives.

Chapter 5 / Polynomials 205

Review/Test

SECTION **1**

1.1 Simplify $(3x^3 - 2x^2 - 4) + (8x^2 - 8x + 7)$.

1.2 Simplify $(3a^2 - 2a - 7) - (5a^3 + 2a - 10)$.

Instructor's Computerized Test Generator

For the Instructor

Reference for Computerized Test Generator

The INSTRUCTOR'S COMPUTERIZED TEST GENERATOR is a test-making tool designed to produce an infinite variety of both multiple-choice and free-response objective-referenced tests for each chapter of the text. (Cumulative Tests and Final Exams may also be created.)

The INSTRUCTOR'S COMPUTERIZED TEST GENERATOR is *educationally sound.* The data base consists of 1588 customized test items that are organized around the same hierarchy of objectives that organize the lessons of the text. Thus, the "generator" is an instructional management tool that makes it possible to determine which objectives have been mastered and which objectives require the recycling of instruction for any individual student. The tests *directly* support the text!

As an aid to the Instructor, the Instructor's Annotated Edition (IAE) is cross-referenced to the appropriate disk in the INSTRUCTOR'S COMPUTERIZED TEST GENERATOR by virtue of a computer-referencing logo and annotation which is to be found at the end of the Historical Note for each chapter. (See example upper left.)

While there are other computer-based test generators available, the INSTRUCTOR'S COMPUTERIZED TEST GENERATOR for the Aufmann/Barker program clearly distinguishes itself as technically *superior.* Printouts of complex math symbolisms as well as graphic representations are faithful to the text and are of exceptionally high quality.

The INSTRUCTOR'S COMPUTERIZED TEST GENERATOR is currently available for the Apple® II family of microcomputers.

The Computer Tutor™

For the Student

Reference for Computer Tutor™

The COMPUTER TUTOR™ is an "interactive" instructional-delivery vehicle designed for student use. The objectives which organize the "tutor" are the same as those of the text. Thus, each lesson of the "tutor" directly supports a corresponding lesson in the text. Each lesson in the Student Text (as well as its replica in the Instructor's Annotated Edition) is now cross-referenced to the COMPUTER TUTOR™ by virtue of a computer-referencing logo which is found adjacent to the lesson objective. The COMPUTER TUTOR™ lessons, in turn, are cross-referenced to the corresponding text lessons.

The COMPUTER TUTOR™ can prove to be a useful adjunct to basal-text instruction for a variety of reasons:

- An individual student might require help with *initial instruction* because of class absence.
- An individual student might require the *recycling of instruction* because testing has revealed lack of mastery on a given skill or concept the first time around.
- An individual student might require *review instruction* as they prepare for competency exams or as they prepare for enrollment in higher-level courses.

The COMPUTER TUTOR™ is not only educationally sound (for the reasons listed above) but it is also technically well-executed. A special typeface has been especially created for this screen-only program to enhance readability, and complex math symbolisms and graphics are particularly well-executed.

The COMPUTER TUTOR™ is currently available for the Apple® II family of microcomputers.

Other Ancillaries

Instructor's Annotated Edition

In order to facilitate the Instructor's grading of exercise sets and Review/Tests, the ancillary package of INTRODUCTORY ALGEBRA: AN APPLIED APPROACH includes an Instructor's Annotated Edition (IAE). The IAE is an exact replica of the student text except that the answer to every problem in the text has been printed in red directly adjacent to the problem. An uncommon item in college mathematics packages, the IAE can serve as an invaluable timesaver to the Instructor.

Solutions Manual

The ancillary package which accompanies INTRODUCTORY ALGEBRA: AN APPLIED APPROACH includes a Solutions Manual which contains the *complete solution for every exercise in the text.* At the Instructor's discretion, students may be granted access to the Solutions Manual. Use of the Solutions Manual allows the student to check the answers *and* the solution to every exercise. In the event that an answer is found to be incorrect, the student's solution may be compared to the solution found in the Solutions Manual in order to find the exact nature of the error. Students who are permitted use of the Solutions Manual are often able to help themselves, reducing the demand on the Instructor's time for tutorial help.

Alternate Testing Program

Instructors frequently request testing materials which are not available to the student. For this reason, the ancillary package for INTRODUCTORY ALGEBRA: AN APPLIED APPROACH includes *two* printed Alternate Testing Booklets. In each booklet, the first half is a battery of free-response tests, one for each chapter. After every third chapter test, there is a cumulative test which covers three chapters. The first half of the booklet ends with a Final Exam covering all twelve chapters in free-response form. The second half of each booklet is identical to the first except that the tests are multiple choice instead of free response. Thus, the Instructor has four printed tests (in two formats) for every chapter including cumulative tests and final exams. All tests are on easy-to-copy, permission-to-reproduce pages.

Acknowledgements

The authors would like to thank the people who have reviewed this manuscript and provided many valuable suggestions:

Donald J. Albers
Menlo College, California

LaVerne Blagman
University of the District of Columbia, Washington, D.C.

Rhona Noll
New York City Technical College, New York

Ellen Casey
Massachusetts Bay Community College, Massachusetts

1

Real Numbers

Objectives

- To use the inequality symbols $<$ and $>$ with integers
- To use opposites and absolute value
- To add integers
- To subtract integers
- To multiply integers
- To divide integers
- To solve application problems
- To write a rational number as a decimal
- To add or subtract rational numbers
- To multiply or divide rational numbers
- To evaluate exponential expressions
- To use the Order of Operations Agreement to simplify expressions

Early Egyptian Fractions

One of the earliest written documents of mathematics is the Rhind Papyrus. This tablet was found in Egypt in 1858, but it is estimated that the writings date back to 1650 B.C.

The Rhind Papyrus contains over 80 problems. A study of these problems has enabled mathematicians and scientists to understand some of the methods by which the early Egyptians used mathematics.

Evidence gained from the Papyrus shows that the Egyptian method of calculating with fractions was much different from the methods used today. All fractions were represented in terms of what are called "unit fractions." A unit fraction is a fraction in which the numerator is one. This fraction was symbolized (using modern numbers) with a bar over the number.

For example, $\overline{3} = \frac{1}{3}$; $\overline{15} = \frac{1}{15}$

The early Egyptians also tended to deal with powers of two (2, 4, 8, 16, . . .). As a result, representing fractions with a two in the numerator in terms of unit fractions was an important matter. The Rhind Papyrus has a table giving the equivalent unit fractions for all odd denominators from 5 to 101 and 2 as the numerator. Some of these are listed below.

$$\frac{2}{5} = \overline{3}\ \overline{15} \qquad \left(\frac{2}{5} = \frac{1}{3} + \frac{1}{15}\right)$$

$$\frac{2}{7} = \overline{4}\ \overline{28}$$

$$\frac{2}{11} = \overline{6}\ \overline{66}$$

$$\frac{2}{19} = \overline{12}\ \overline{76}\ \overline{114}$$

SECTION 1 Introduction to Integers

1.1 Objective To use the inequality symbols $<$ and $>$ with integers

Reference for Computer Tutor™

The **natural numbers** are 1, 2, 3, 4, 5, 6, 7, 8, . . .

The three dots mean that the list continues on and on, and that there is no largest natural number.

The natural numbers can be shown on the **number line.**

The number line

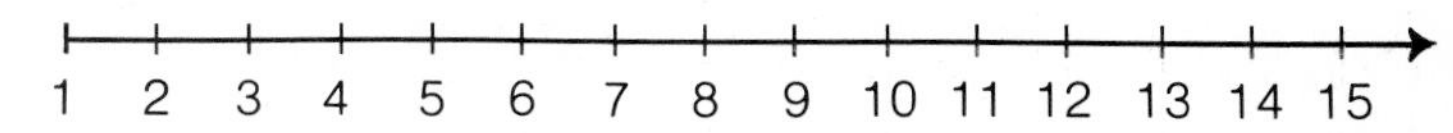

The **graph** of a natural number is shown by placing a heavy dot on the number line directly above the number.

The graph of 6 on the number line

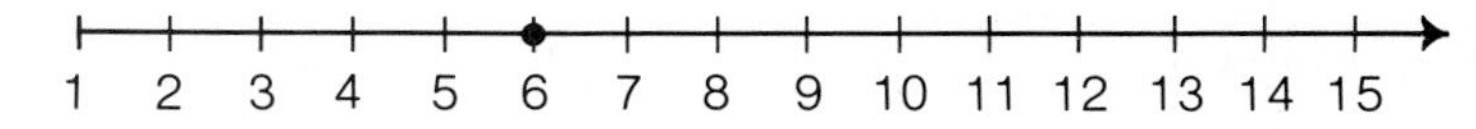

The **integers** are . . . −4, −3, −2, −1, 0, 1, 2, 3, 4, . . .

Each integer can be shown on the number line.

The integers to the left of zero on the number line are called **negative integers.** The integers to the right of zero are called **positive integers** or natural numbers. Zero is neither a positive nor a negative number.

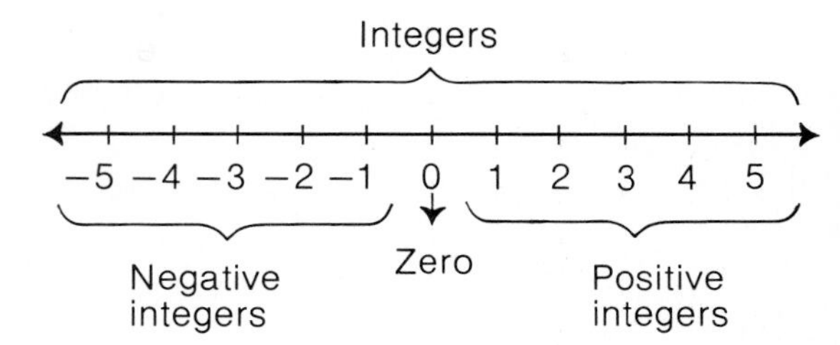

Just as the word "it" is used in language to stand for an object, a letter of the alphabet can be used in mathematics to stand for a number. Such a letter is called a **variable.**

A number line can be used to visualize the relative order of two integers. If a and b are two integers, and a is to the left of b on the number line, then a is **less than** b ($a < b$). If a is to the right of b on the number line, then a is **greater than** b ($a > b$).

Negative 4 is less than negative 1.

$-4 < -1$

−5 −4 −3 −2 −1 0 1 2 3 4 5

5 is greater than 0.

$5 > 0$

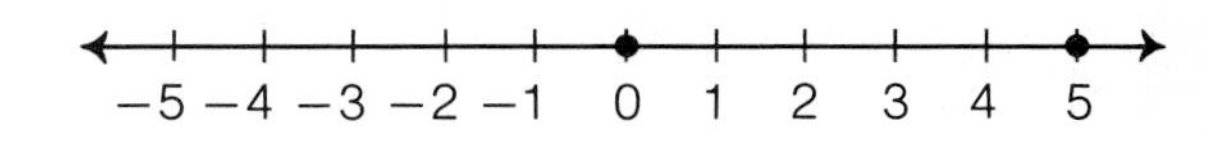

Example 1 Place the correct symbol, $<$ or $>$, between the two numbers.
a. −17 6 b. −30 −3

Solution a. $-17 < 6$ b. $-30 < -3$

Example 2 Place the correct symbol, $<$ or $>$, between the two numbers.
a. 5 −13 b. −8 −22

Your solution

Solution on p. A5

1.2 Objective To use opposites and absolute value

Reference for Computer Tutor™

Two numbers that are the same distance from zero on the number line but on opposite sides of zero are **opposite numbers,** or **opposites.**

The opposite of 5 is -5.

The opposite of -5 is 5.

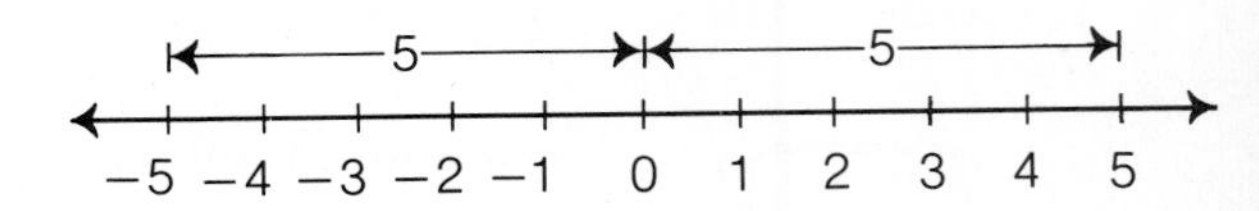

The negative sign can be read "the opposite of."

$-(2) = -2$ The opposite of 2 is -2.

$-(-2) = 2$ The opposite of -2 is 2.

The **absolute value** of a number is its distance from zero on the number line. Therefore, the absolute value of a number is a positive number or zero. The symbol for absolute value is $|\ |$.

The distance from 0 to 3 is 3.
Therefore, the absolute value of 3 is 3.

$|3| = 3$

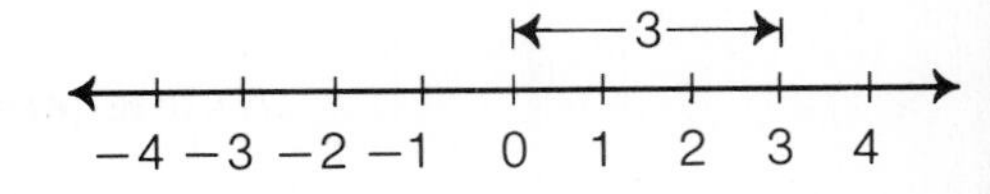

The distance from 0 to -3 is 3.
Therefore, the absolute value of -3 is 3.

$|-3| = 3$

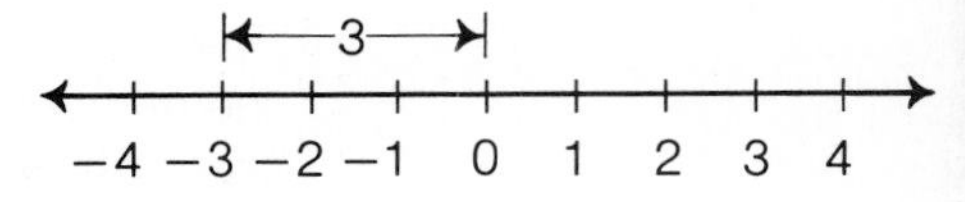

The absolute value of a positive number is the number itself. The absolute value of a negative number is the opposite of the negative number. The absolute value of zero is zero.

Evaluate $-|-7|$.

The absolute value sign does not affect the negative sign in front of the absolute value sign. $-|-7| = -7$

Example 3 Find the opposite number.
a. 6 b. -51

Solution a. -6 b. 51

Example 4 Find the opposite number.
a. -9 b. 62

Your solution

Example 5 Evaluate $|-4|$ and $-|-10|$.

Solution $|-4| = 4$

$-|-10| = -10$

Example 6 Evaluate $|-5|$ and $-|-9|$.

Your solution

Solutions on p. A5

1.1 Exercises

Place the correct symbol, < or >, between the two numbers.

1. 3 5 **2.** 7 4 **3.** −2 −5

4. −6 −1 **5.** −16 1 **6.** −2 13

7. 3 −7 **8.** 5 −6 **9.** −11 −8

10. −4 −10 **11.** −1 −6 **12.** −9 −4

13. 0 −3 **14.** 8 0 **15.** 6 −8

16. 8 −6 **17.** −14 16 **18.** −12 1

19. 35 28 **20.** 42 19 **21.** −42 27

22. −36 49 **23.** 21 −34 **24.** 53 −46

25. −27 −39 **26.** −51 −20 **27.** −87 63

28. −75 92 **29.** 68 −79 **30.** 95 −71

31. −62 −84 **32.** −91 −70 **33.** 94 83

34. 76 81 **35.** 59 −67 **36.** 48 −66

37. −93 −55 **38.** −64 −86 **39.** −88 57

40. −58 82 **41.** 0 129 **42.** −136 0

43. −131 101 **44.** 127 −150 **45.** −194 −180

1.2 Exercises

Find the opposite number.

46. 4 **47.** 16 **48.** -2 **49.** -3

50. 22 **51.** 45 **52.** -31 **53.** -88

Evaluate.

54. $|2|$ **55.** $|-2|$ **56.** $|-6|$ **57.** $|6|$

58. $|8|$ **59.** $|5|$ **60.** $|-9|$ **61.** $|-1|$

62. $-|-1|$ **63.** $-|-8|$ **64.** $-|-5|$ **65.** $-|0|$

66. $|16|$ **67.** $|19|$ **68.** $|-12|$ **69.** $|-22|$

70. $-|29|$ **71.** $-|20|$ **72.** $-|-14|$ **73.** $-|-18|$

74. $|-15|$ **75.** $|-23|$ **76.** $-|33|$ **77.** $-|27|$

78. $-|-36|$ **79.** $-|-41|$ **80.** $|32|$ **81.** $|25|$

82. $|-38|$ **83.** $|-30|$ **84.** $-|37|$ **85.** $-|34|$

86. $-|-42|$ **87.** $-|-45|$ **88.** $|44|$ **89.** $|36|$

90. $|-74|$ **91.** $|-61|$ **92.** $-|88|$ **93.** $-|52|$

94. $-|-81|$ **95.** $-|-93|$ **96.** $|-107|$ **97.** $|-119|$

SECTION 2 Addition and Subtraction of Integers

2.1 Objective To add integers

Reference for Computer Tutor™

A number can be represented anywhere along the number line by an arrow. A positive number is represented by an arrow pointing to the right and a negative number is represented by an arrow pointing to the left. The size of a number is represented by the length of the arrow.

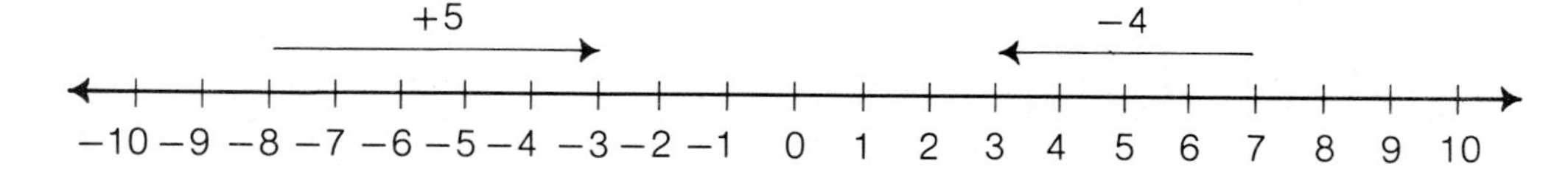

Addition of integers can be shown on the number line. To add integers, find the point on the number line corresponding to the first addend. At that point draw an arrow representing the second addend. The sum is the number directly below the tip of the arrow.

$4 + 2 = 6$

$-4 + (-2) = -6$

$-4 + 2 = -2$

$4 + (-2) = 2$

The pattern for addition shown on the number lines above is summarized in the following rules for adding integers.

Like Signs To add numbers with like signs, add the absolute values of the numbers. Then attach the sign of the addends.

$2 + 8 = 10$

$-2 + (-8) = -10$

Unlike Signs To add numbers with unlike signs, find the difference between the absolute values of the numbers. Then attach the sign of the number with the greater absolute value.

$-2 + 8 = 6$

$2 + (-8) = -6$

Add: $162 + (-247)$

Since the signs are unlike, find the difference between the absolute values of the numbers and attach the sign of the number with the greater absolute value.

$162 + (-247)$
-85

Add: $-14 + (-47)$

Since the signs are like, add the absolute values of the numbers and attach the sign of the addends.

$-14 + (-47)$
-61

Add: $-4 + (-6) + (-8) + 9$

To add more than two numbers, add the first two numbers. Then add the sum to the third number. Continue until all the numbers have been added.

$\underbrace{-4 + (-6)} + (-8) + 9$

$\underbrace{-10 \quad + (-8)} + 9$

$\underbrace{-18 \quad + 9}$

-9

Example 1 Add: $-162 + 98$

Solution $-162 + 98$
-64

Example 2 Add: $-154 + (-37)$

Your solution

Example 3 Add: $42 + (-12) + (-30)$

Solution $42 + (-12) + (-30)$
$30 + (-30)$
0

Example 4 Add: $-36 + 17 + (-21)$

Your solution

Example 5 Add: $-2 + (-7) + 4 + (-6)$

Solution $-2 + (-7) + 4 + (-6)$
$-9 + 4 + (-6)$
$-5 + (-6)$
-11

Example 6 Add: $-5 + (-2) + 9 + (-3)$

Your solution

Solutions on p. A5

2.2 Objective To subtract integers

Reference for Computer Tutor™

Subtraction of an integer is defined as addition of the opposite integer.

Subtract 8 − 3 by using addition of the opposite.

Subtraction ⟶ Addition of the Opposite

$8 - (+3) = 8 + (-3) = 5$

Opposites

To subtract one integer from another, add the opposite of the second integer to the first number.

First number	−	second number	=	first number	+	the opposite of the second number	
40	−	60	=	40	+	(−60)	= −20
−40	−	60	=	−40	+	(−60)	= −100
−40	−	(−60)	=	−40	+	60	= 20
40	−	(−60)	=	40	+	60	= 100

When subtraction occurs several times in a problem, rewrite each subtraction as addition of the opposite. Then add.

Subtract: $-12 - 4 - (-15)$

Rewrite each subtraction as addition of the opposite. Add.

$-12 - 4 - (-15)$
$-12 + (-4) + 15$
$-16 + 15$
-1

Subtract:
$-12 - 4 - (-6) - 7 - (-8)$

Rewrite each subtraction as addition of the opposite. Add.

$-12 - 4 - (-6) - 7 - (-8)$
$-12 + (-4) + 6 + (-7) + 8$
$-16 + 6 + (-7) + 8$
$-10 + (-7) + 8$
$-17 + 8$
-9

Example 7 Subtract: $-12 - 8$

Solution
$-12 - 8$
$-12 + (-8)$
-20

Example 8 Subtract: $-8 - 14$

Your solution

Example 9 Subtract: $8 - 6 - (-20)$

Solution
$8 - 6 - (-20)$
$8 + (-6) + 20$
$2 + 20$
22

Example 10 Subtract: $3 - (-4) - 15$

Your solution

Example 11 Subtract:
$-8 - 30 - (-12) - 7 - (-14)$

Solution
$-8 - 30 - (-12) - 7 - (-14)$
$-8 + (-30) + 12 + (-7) + 14$
$-38 + 12 + (-7) + 14$
$-26 + (-7) + 14$
$-33 + 14$
-19

Example 12 Subtract:
$4 - (-3) - 12 - (-7) - 20$

Your solution

Example 13 Subtract:
$12 - 12 - (-3) - 5 - 7$

Solution
$12 - 12 - (-3) - 5 - 7$
$12 + (-12) + 3 + (-5) + (-7)$
$0 + 3 + (-5) + (-7)$
$3 + (-5) + (-7)$
$-2 + (-7)$
-9

Example 14 Subtract:
$17 - 10 - 2 - (-6) - 9$

Your solution

Solutions on p. A5

2.1 Exercises

Add.

1. $3 + (-5)$
2. $-4 + 2$
3. $8 + 12$
4. $16 + 23$
5. $-3 + (-8)$
6. $-12 + (-1)$
7. $-4 + (-5)$
8. $-12 + (-12)$
9. $6 + (-9)$
10. $4 + (-9)$
11. $-6 + 7$
12. $-12 + 6$
13. $2 + (-3) + (-4)$
14. $7 + (-2) + (-8)$
15. $-3 + (-12) + (-15)$
16. $9 + (-6) + (-16)$
17. $-17 + (-3) + 29$
18. $13 + 62 + (-38)$
19. $-3 + (-8) + 12$
20. $-27 + (-42) + (-18)$
21. $13 + (-22) + 4 + (-5)$
22. $-14 + (-3) + 7 + (-6)$
23. $-22 + 10 + 2 + (-18)$
24. $-6 + (-8) + 13 + (-4)$

25. $-16 + (-17) + (-18) + 10$
26. $-25 + (-31) + 24 + 19$
27. $-126 + (-247) + (-358) + 339$
28. $-651 + (-239) + 524 + 487$

2.2 Exercises

Subtract.

29. $16 - 8$

30. $12 - 3$

31. $7 - 14$

32. $6 - 9$

33. $-7 - 2$

34. $-9 - 4$

35. $7 - (-2)$

36. $3 - (-4)$

37. $-6 - (-3)$

38. $-4 - (-2)$

39. $6 - (-12)$

40. $-12 - 16$

41. $-4 - 3 - 2$

42. $4 - 5 - 12$

43. $12 - (-7) - 8$

44. $-12 - (-3) - (-15)$

45. $4 - 12 - (-8)$

46. $13 - 7 - 15$

47. $-6 - (-8) - (-9)$

48. $7 - 8 - (-1)$

49. $-30 - (-65) - 29 - 4$

50. $42 - (-82) - 65 - 7$

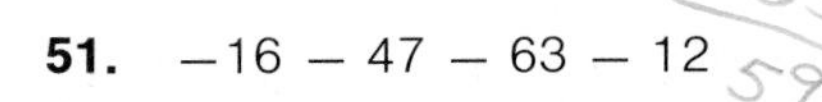

51. $-16 - 47 - 63 - 12$

52. $42 - (-30) - 65 - (-11)$

53. $-47 - (-67) - 13 - 15$

54. $-18 - 49 - (-84) - 27$

55. $167 - 432 - (-287) - 359$

56. $-521 - (-350) - 164 - (-299)$

SECTION 3 Multiplication and Division of Integers

3.1 Objective To multiply integers

Reference for Computer Tutor™

Multiplication is the repeated addition of the same number.

Several different symbols are used to indicate multiplication.

$3 \times 2 = 6$
$3 \cdot 2 = 6$
$(3)(2) = 6$

When 5 is multiplied by a sequence of decreasing integers, each product decreases by 5.

$5 \times 3 = 15$
$5 \times 2 = 10$
$5 \times 1 = 5$
$5 \times 0 = 0$

The pattern developed can be continued so that 5 is multiplied by a sequence of negative numbers. The resulting products must be negative in order to maintain the pattern of decreasing by 5.

$5 \times (-1) = -5$
$5 \times (-2) = -10$
$5 \times (-3) = -15$
$5 \times (-4) = -20$

This illustrates that the product of a positive number and a negative number is negative.

When -5 is multiplied by a sequence of decreasing integers, each product increases by 5.

$-5 \times 3 = -15$
$-5 \times 2 = -10$
$-5 \times 1 = -5$
$-5 \times 0 = 0$

The pattern developed can be continued so that -5 is multiplied by a sequence of negative numbers. The resulting products must be positive in order to maintain the pattern of increasing by 5.

$-5 \times (-1) = 5$
$-5 \times (-2) = 10$
$-5 \times (-3) = 15$
$-5 \times (-4) = 20$

This illustrates that the product of two negative numbers is positive.

The pattern for multiplication shown above is summarized in the following rules for multiplying integers.

Like Signs To multiply numbers with like signs, multiply the absolute values of the factors. The product is positive.

$4 \cdot 8 = 32$

$(-4)(-8) = 32$

Unlike Signs To multiply numbers with unlike signs, multiply the absolute values of the factors. The product is negative.

$-4 \cdot 8 = -32$

$(4)(-8) = -32$

Multiply: $2(-3)(-5)(-7)$

To multiply more than two numbers, multiply the first two numbers. Then multiply the product by the third number. Continue until all the numbers have been multiplied.

$2(-3)(-5)(-7)$

$-6 \cdot (-5)(-7)$

$30 \cdot (-7)$

-210

Example 1 Multiply: $(-2) \cdot 3 \cdot 6$

Solution $(-2) \cdot 3 \cdot 6$

$-6 \cdot 6$

-36

Example 2 Multiply: $(-3) \cdot 4 \cdot (-5)$

Your solution

Example 3 Multiply: $-42 \cdot 62$

Solution $-42 \cdot 62$

-2604

Example 4 Multiply: $-38 \cdot 51$

Your solution

Example 5 Multiply: $-2 \cdot (-10) \cdot 7 \cdot 12$

Solution $-2 \cdot (-10) \cdot 7 \cdot 12$

$20 \cdot 7 \cdot 12$

$140 \cdot 12$

1680

Example 6 Multiply: $-6 \cdot 8 \cdot (-11) \cdot 3$

Your solution

Example 7 Multiply: $-5(-4)(6)(-3)$

Solution $-5(-4)(6)(-3)$

$20(6)(-3)$

$120(-3)$

-360

Example 8 Multiply: $-7(-8)(9)(-2)$

Your solution

Solutions on pp. A5–A6

3.2 Objective To divide integers

Reference for Computer Tutor™

For every division problem there is a related multiplication problem.

Division: $\frac{8}{2} = 4$ Related multiplication: $4 \cdot 2 = 8$

This fact can be used to illustrate the rules for dividing signed numbers.

Like Signs The quotient of two numbers with like signs is positive.

$\frac{12}{3} = 4$ because $4 \cdot 3 = 12$

$\frac{-12}{-3} = 4$ because $4(-3) = -12$

Unlike Signs The quotient of two numbers with unlike signs is negative.

$\frac{12}{-3} = -4$ because $-4(-3) = 12$

$\frac{-12}{3} = -4$ because $-4 \cdot 3 = -12$

Note that $\frac{12}{-3}$, $\frac{-12}{3}$, and $-\frac{12}{3}$ are all equal to -4.

If a and b are two integers, then $\frac{a}{-b} = \frac{-a}{b} = -\frac{a}{b}$.

Properties of Zero and One in Division

Zero divided by any number other than zero is zero. $\frac{0}{a} = 0$ because $0 \cdot a = 0$

Division by zero is not defined. $\frac{4}{0} = ?$ $? \times 0 = 4$ There is no number whose product with zero is 4.

Any number other than zero divided by itself is one. $\frac{a}{a} = 1$ because $1 \cdot a = a$

Example 9 Divide: $(-120) \div (-8)$

Solution $(-120) \div (-8)$

15

Example 10 Divide: $(-135) \div (-9)$

Your solution

Example 11 Divide: $-81 \div 3$

Solution $-81 \div 3$

-27

Example 12 Divide: $-72 \div 4$

Your solution

Example 13 Divide: $95 \div (-5)$

Solution $95 \div (-5)$

-19

Example 14 Divide: $84 \div (-6)$

Your solution

Solutions on p. A6

3.3 Objective To solve application problems

Use COMPUTER TUTOR™ DISK 1

To solve an application problem, first read the problem carefully. The *Strategy* involves identifying the quantity to be found and planning the steps which are necessary to find that quantity. The *Solution* involves performing each operation stated in the Strategy and writing the answer.

Example 15 The average temperature on the sunlit side of the moon is approximately $215°F$. On the dark side, it is approximately $-250°F$. Find the difference between these average temperatures.

Strategy To find the difference, subtract the average temperature on the dark side of the moon $(-250°)$ from the average temperature on the sunlit side $(215°)$.

Solution

$215 - (-250)$

$215 + 250$

465

The difference is $465°F$.

Example 16 The average temperature throughout the earth's stratosphere is $-70°F$. The average temperature on the earth's surface is $57°F$. Find the difference between these average temperatures.

Your strategy

Your solution

Example 17 The daily high temperatures during one week were recorded as follows: $-9°$, $3°$, $0°$, $-8°$, $2°$, $1°$, $4°$. Find the average daily high temperature for the week.

Strategy To find the average daily high temperature:

▷ Add the seven temperature readings.

▷ Divide by 7.

Solution

$-9 + 3 + 0 + (-8) + 2 + 1 + 4$

$-6 + 0 + (-8) + 2 + 1 + 4$

$-6 + (-8) + 2 + 1 + 4$

$-14 + 2 + 1 + 4$

$-12 + 1 + 4$

$-11 + 4$

-7

$-7 \div 7 = -1$

The average daily high temperature was $-1°$.

Example 18 The daily low temperatures during one week were recorded as follows: $-6°$, $-7°$, $1°$, $0°$, $-5°$, $-10°$, $-1°$. Find the average daily low temperature for the week.

Your strategy

Your solution

3.1 Exercises

Multiply.

1. 14×3
2. 62×9
3. $-4 \cdot 6$
4. $-7 \cdot 3$
5. $-2 \cdot (-3)$
6. $-5 \cdot (-1)$
7. $(9)(2)$
8. $(3)(8)$
9. $5(-4)$
10. $4(-7)$
11. $-8(2)$
12. $-9(3)$
13. $(-5)(-5)$
14. $(-3)(-6)$
15. $(-7)(0)$
16. -32×4
17. -24×3
18. $19 \cdot (-7)$
19. $6(-17)$
20. $-8(-26)$
21. $-4(-35)$
22. $-5 \cdot (23)$
23. $-6 \cdot (38)$
24. $9(-27)$
25. $8(-40)$
26. $-7(-34)$
27. $-4(39)$
28. $4 \cdot (-8) \cdot 3$
29. $5 \times 7 \times (-2)$
30. $8 \cdot (-6) \cdot (-1)$
31. $(-9)(-9)(2)$
32. $-8(-7)(-4)$
33. $-5(8)(-3)$
34. $(-6)(5)(7)$
35. $-1(4)(-9)$
36. $6(-3)(-2)$

Multiply.

37. $4(-4) \cdot 6(-2)$

38. $-5 \cdot 9(-7) \cdot 3$

39. $-9(4) \cdot 3(1)$

40. $8(8)(-5)(-4)$

41. $(-6) \cdot 7 \cdot (-10)(-5)$

42. $-9(-6)(11)(-2)$

43. $-6(-5)(12)(0)$

44. $7(9) \cdot 10 \cdot (-1)$

45. $-19(28)(-43)(-11)$

46. $-65(13)(-47)(-92)$

3.2 Exercises

Divide.

47. $12 \div (-6)$

48. $18 \div (-3)$

49. $(-72) \div (-9)$

50. $(-64) \div (-8)$

51. $0 \div (-6)$

52. $-49 \div 7$

53. $45 \div (-5)$

54. $-24 \div 4$

55. $-36 \div 4$

56. $-56 \div 7$

57. $-81 \div (-9)$

58. $-40 \div (-5)$

59. $72 \div (-3)$

60. $44 \div (-4)$

61. $-60 \div 5$

62. $-66 \div 6$

63. $-93 \div (-3)$

64. $-98 \div (-7)$

Divide.

65. $(-85) \div (-5)$

66. $(-60) \div (-4)$

67. $120 \div 8$

68. $144 \div 9$

69. $78 \div (-6)$

70. $84 \div (-7)$

71. $-72 \div 4$

72. $-80 \div 5$

73. $-114 \div (-6)$

74. $-91 \div (-7)$

75. $-104 \div (-8)$

76. $-126 \div (-9)$

77. $57 \div (-3)$

78. $162 \div (-9)$

79. $-136 \div (-8)$

80. $-128 \div 4$

81. $-130 \div (-5)$

82. $(-280) \div 8$

83. $(-92) \div (-4)$

84. $-196 \div (-7)$

85. $-150 \div (-6)$

86. $(-261) \div 9$

87. $204 \div (-6)$

88. $165 \div (-5)$

89. $-132 \div (-12)$

90. $-156 \div (-13)$

91. $-182 \div 14$

92. $-144 \div 12$

93. $143 \div 11$

94. $168 \div 14$

95. $-180 \div (-15)$

96. $-169 \div (-13)$

97. $154 \div (-11)$

98. $274{,}883 \div 367$

99. $398{,}750 \div 1375$

100. $841{,}662 \div 2461$

3.3 Application Problems

Solve.

1. Find the temperature after a rise of 7°C from −8°C.

2. Find the temperature after a rise of 5°C from −19°C.

3. During a card game of Hearts, you had a score of 11 points before your opponent "shot the moon," subtracting a score of 26 from your total. What was your score after your opponent "shot the moon"?

4. In a card game of Hearts, you had a score of −19 before you "shot the moon," entitling you to add 26 points to your score. What was your score after you "shot the moon"?

The elevation, or height, of places on earth is measured in relation to sea level, or the average level of the ocean's surface. The table below shows height above sea level as a positive number, depth below sea level as a negative number.

Place	Elevation (in feet)
Mt. Everest	29,028
Mt. Aconcagua	23,035
Mt. McKinley	20,320
Mt. Kilimanjaro	19,340
Salinas Grandes	−131
Death Valley	−282
Qattara Depression	−436
Dead Sea	−1286

5. Use the table to find the difference in elevation between Mt. McKinley and Death Valley (the highest and lowest points in North America).

6. Use the table to find the difference in elevation between Mt. Kilimanjaro and the Qattara Depression (the highest and lowest points in Africa).

7. Use the table to find the difference in elevation between Mt. Everest and the Dead Sea (the highest and lowest points in Asia).

8. Use the table to find the difference in elevation between Mt. Aconcagua and Salinas Grandes (the highest and lowest points in South America).

9. The daily low temperatures during one week were recorded as follows: 5°, −4°, 9°, 0°, −11°, −13°, −7°. Find the average daily low temperature for the week.

10. The daily high temperatures during one week were recorded as follows: −7°, −10°, 4°, 6°, −2°, −8°, −4°. Find the average daily high temperature for the week.

SECTION 4 Rational Numbers

4.1 Objective To write a rational number as a decimal

Reference for Computer Tutor™

A **rational number** is the quotient of two integers. Therefore, a rational number is a number which can be written in the form $\frac{a}{b}$, where a and b are integers and b is not zero. A rational number written in this way is commonly called a fraction.

a ← an integer
b ← a non-zero integer

$\frac{2}{3}, \frac{-4}{9}, \frac{18}{-5}, \frac{4}{1}$ } Rational numbers

Since an integer can be written as the quotient of the integer and 1, every integer is a rational number.

$5 = \frac{5}{1}$ $\qquad -3 = \frac{-3}{1}$

A number written in **decimal notation** is also a rational number.

three-tenths	0.3	$= \frac{3}{10}$
thirty-five hundredths	0.35	$= \frac{35}{100}$
negative four-tenths	−0.4	$= \frac{-4}{10}$

A rational number written as a fraction can be written in decimal notation.

Write $\frac{5}{8}$ as a decimal.

The fraction bar can be read "÷".

$\frac{5}{8} = 5 \div 8$

$$\begin{array}{r} 0.625 \\ 8\overline{)5.000} \\ -4\,8 \\ \hline 20 \\ -16 \\ \hline 40 \\ -40 \\ \hline 0 \end{array}$$

0.625 ← This is called a **terminating decimal.**

0 ← The remainder is zero.

$\frac{5}{8} = 0.625$

Write $\frac{4}{11}$ as a decimal.

$$\begin{array}{r} 0.3636... \\ 11\overline{)4.0000} \\ -3\,3 \\ \hline 70 \\ -66 \\ \hline 40 \\ -33 \\ \hline 70 \\ -66 \\ \hline 4 \end{array}$$

0.3636. . . ← This is called a **repeating decimal.**

4 ← The remainder is never zero.

$\frac{4}{11} = 0.\overline{36}$ The bar over the digits 36 is used to show that these digits repeat.

Every rational number can be written as a terminating or a repeating decimal.

Some numbers for example, $\sqrt{2}$ and π, have decimal representations which never terminate or repeat. These numbers are called **irrational numbers.**

$\sqrt{2} = 1.414213502\ldots$

$\pi = 3.141592604\ldots$

The rational numbers and the irrational numbers taken together are called the **real numbers.**

Example 1 Write $\frac{3}{20}$ as a decimal.

Solution

$$\begin{array}{r} 0.15 \\ 20\overline{)3.00} \\ \underline{-2\,0} \\ 1\,00 \\ \underline{-1\,00} \\ 0 \end{array}$$

$\frac{3}{20} = 0.15$

Example 2 Write $\frac{4}{25}$ as a decimal.

Your solution

Example 3 Write $\frac{3}{22}$ as a decimal. Place a bar over the repeating digits of the decimal.

Solution

$$\begin{array}{r} 0.13636 \\ 22\overline{)3.00000} \\ \underline{-2\,2} \\ 80 \\ \underline{-66} \\ 140 \\ \underline{-132} \\ 80 \\ \underline{-66} \\ 140 \\ \underline{-132} \\ 8 \end{array}$$

$\frac{3}{22} = 0.1\overline{36}$

Example 4 Write $\frac{4}{9}$ as a decimal. Place a bar over the repeating digits of the decimal.

Your solution

Solutions on p. A6

4.2 Objective To add or subtract rational numbers

Reference for Computer Tutor™

To add or subtract fractions, first rewrite the fractions as equivalent fractions with a common denominator, using the least common multiple (LCM) of the denominators as the common denominator. Then add the numerators and place the sum over the common denominator. Reduce the answer to simplest form.

Simplify $-\frac{5}{6} + \frac{3}{10}$.

Step 1

Find the LCM of the denominators 6 and 10. The LCM of denominators is sometimes called the **least common denominator** (LCD).

Prime factorizations of 6 and 10: $6 = 2 \cdot 3$

$10 = 2 \cdot 5$

$\text{LCM} = 2 \cdot 3 \cdot 5 = 30$

Step 2

Rewrite the fractions as equivalent fractions using the LCM of the denominators as the common denominator.

Step 3

Add the numerators and place the sum over the common denominator.

Step 4

Reduce the answer to simplest form.

$$-\frac{5}{6} + \frac{3}{10} = -\frac{25}{30} + \frac{9}{30} = \frac{-25+9}{30} = \frac{-16}{30} = -\frac{8}{15}$$

Simplify $-\frac{5}{9} - \frac{7}{12}$.

$9 = 3 \cdot 3$

$12 = 2 \cdot 2 \cdot 3$

$\text{LCM} = 2 \cdot 2 \cdot 3 \cdot 3 = 36$

$$-\frac{5}{9} - \frac{7}{12} = -\frac{20}{36} - \frac{21}{36} = \frac{-20}{36} + \frac{-21}{36} = \frac{-20+(-21)}{36} = \frac{-41}{36} = -1\frac{5}{36}$$

To add or subtract decimals, write the numbers so that the decimal points are in a vertical line. Then proceed as in the addition or subtraction of integers. Write the decimal point in the answer directly below the decimal points in the problem.

Simplify 14.02 + 137.6 + 9.852.

Write the decimals so that the decimal points are in a vertical line.

Write the decimal point in the sum directly below the decimal points in the problem.

$$\begin{array}{r} 14.02 \\ 137.6 \\ +\ \ 9.852 \\ \hline 161.472 \end{array}$$

Simplify −114.039 + 84.76.

Since the signs are unlike, find the difference between the absolute values of the numbers.

$$\begin{array}{r} {\scriptstyle 10\ 13\ 9} \\ {\scriptstyle 0\ \ \not{0}\ \ \not{3}\ \ 10\ 13} \\ \not{1}\,\not{1}\,\not{4}.\not{0}\,\not{3}\,9 \\ -\quad 8\,4.7\,6 \\ \hline 2\,9.2\,7\,9 \end{array}$$

Attach the sign of the number with the greater absolute value.

$-114.039 + 84.76 = -29.279$

Example 5 Simplify $\frac{5}{16} - \frac{7}{40}$.

Solution The LCM of 16 and 40 is 80.

$$\frac{5}{16} - \frac{7}{40} = \frac{25}{80} - \frac{14}{80} =$$

$$\frac{25}{80} + \frac{-14}{80} = \frac{25 + (-14)}{80} = \frac{11}{80}$$

Example 6 Simplify $\frac{5}{9} - \frac{11}{12}$.

Your solution

Example 7 Simplify $-\frac{3}{4} + \frac{1}{6} - \frac{5}{8}$.

Solution The LCM of 4, 6, and 8 is 24.

$$-\frac{3}{4} + \frac{1}{6} - \frac{5}{8} =$$

$$-\frac{18}{24} + \frac{4}{24} - \frac{15}{24} =$$

$$\frac{-18}{24} + \frac{4}{24} + \frac{-15}{24} =$$

$$\frac{-18 + 4 + (-15)}{24} = \frac{-29}{24} =$$

$$-1\frac{5}{24}$$

Example 8 Simplify $-\frac{7}{8} - \frac{5}{6} + \frac{1}{2}$.

Your solution

Example 9 Simplify 4.027 + 19.66 + 3.09.

Solution

$$\begin{array}{r} 4.027 \\ 19.66 \\ +\ \ 3.09 \\ \hline 26.777 \end{array}$$

Example 10 Simplify 3.907 + 4.9 + 6.63.

Your solution

Example 11 Simplify 42.987 − 98.61.

Solution

$$\begin{array}{r} 98.61 \\ -\ 42.987 \\ \hline 55.623 \end{array}$$

Find the difference between the absolute values.

42.987 − 98.61 = −55.623

Example 12 Simplify 16.127 − 67.91.

Your solution

Example 13 Simplify 1.02 + (−3.6) + 9.24.

Solution 1.02 + (−3.6) + 9.24
−2.58 + 9.24
6.66

Example 14 Simplify 2.7 + (−9.44) + 6.2.

Your solution

4.3 Objective — To multiply or divide rational numbers

Reference for Computer Tutor™

The product of two fractions is the product of the numerators over the product of the denominators.

Simplify $\frac{3}{8} \times \frac{12}{17}$.

Step 1	Step 2	Step 3
Multiply the numerators. Multiply the denominators.	Write the prime factorization of each factor. Cancel the common factors.	Multiply the numbers remaining in the numerator. Multiply the numbers remaining in the denominator.
$\frac{3}{8} \times \frac{12}{17} = \frac{3 \cdot 12}{8 \cdot 17} =$	$\frac{3 \cdot \overset{1}{\cancel{2}} \cdot \overset{1}{\cancel{2}} \cdot 3}{2 \cdot \underset{1}{\cancel{2}} \cdot \underset{1}{\cancel{2}} \cdot 17} =$	$\frac{9}{34}$

To divide fractions, invert the divisor. Then proceed as in the multiplication of fractions.

Simplify $\frac{3}{10} \div \left(-\frac{18}{25}\right)$.

The signs are unlike. The quotient is negative.

$$\frac{3}{10} \div \left(-\frac{18}{25}\right) = -\left(\frac{3}{10} \div \frac{18}{25}\right) = -\left(\frac{3}{10} \times \frac{25}{18}\right) = -\left(\frac{3 \cdot 25}{10 \cdot 18}\right) = -\left(\frac{\overset{1}{\cancel{3}} \cdot \overset{1}{\cancel{5}} \cdot 5}{2 \cdot \underset{1}{\cancel{5}} \cdot 2 \cdot \underset{1}{\cancel{3}} \cdot 3}\right) = -\frac{5}{12}$$

To multiply decimals, multiply as in integers. Write the decimal point in the product so that the number of decimal places in the product equals the sum of the decimal places in the factors.

Simplify -6.89×0.00035.

The signs are unlike.
Multiply the absolute values.

```
      6.89     2 decimal places
 × 0.00035     5 decimal places
      3445
     2067
 0.0024115     7 decimal places
```

The product is negative. $-6.89 \times 0.00035 = -0.0024115$

To divide decimals, move the decimal point in the divisor to make it a whole number. Move the decimal point in the dividend the same number of places to the right. Place the decimal point in the quotient directly over the decimal point in the dividend. Then divide as in whole numbers.

Simplify $1.32 \div 0.27$. Round to the nearest tenth.

$0.27.\overline{)1.32.}$

Move the decimal point 2 places to the right in the divisor and then in the dividend. Place the decimal point in the quotient.

```
       4.88 ≈ 4.9
27.)132.00
   -108
     240
    -216
      240
     -216
       24
```

The symbol ≈ is used to indicate that the quotient is an approximate value after being rounded off.

Example 15 Simplify $\frac{2}{3} \times \left(-\frac{9}{10}\right)$.

Solution The product is negative.

$\frac{2}{3} \times \left(-\frac{9}{10}\right) = -\left(\frac{2}{3} \times \frac{9}{10}\right) =$

$-\frac{2 \cdot 9}{3 \cdot 10} = -\frac{\overset{1}{\cancel{2}} \cdot \overset{1}{\cancel{3}} \cdot 3}{\underset{1}{\cancel{3}} \cdot \underset{1}{\cancel{2}} \cdot 5} = -\frac{3}{5}$

Example 16 Simplify $-\frac{7}{12} \times \frac{9}{14}$.

Your solution

Example 17 Simplify $-\frac{5}{8} \div \left(-\frac{5}{40}\right)$.

Solution The quotient is positive.

$-\frac{5}{8} \div \left(-\frac{5}{40}\right) = \frac{5}{8} \div \frac{5}{40} =$

$\frac{5}{8} \times \frac{40}{5} = \frac{5 \cdot 40}{8 \cdot 5} =$

$\frac{\overset{1}{\cancel{5}} \cdot \overset{1}{\cancel{2}} \cdot \overset{1}{\cancel{2}} \cdot \overset{1}{\cancel{2}} \cdot 5}{\underset{1}{\cancel{2}} \cdot \underset{1}{\cancel{2}} \cdot \underset{1}{\cancel{2}} \cdot \underset{1}{\cancel{5}}} = \frac{5}{1} = 5$

Example 18 Simplify $-\frac{3}{8} \div -\frac{5}{12}$.

Your solution

Example 19 Simplify -4.29×8.2.

Solution The product is negative.

```
   4.29
 × 8.2
   858
 3432
 35.178
```

$-4.29 \times 8.2 = -35.178$

Example 20 Simplify -5.44×3.8.

Your solution

Example 21 Simplify $-3.2 \times (-0.4) \times 6.9$.

Solution $-3.2 \times (-0.4) \times 6.9$

1.28×6.9

8.832

Example 22 Simplify $3.44 \times (-1.7) \times 0.6$.

Your solution

Example 23 Simplify $-0.0792 \div (-0.42)$. Round to the nearest hundredth.

Solution

```
         0.188 ≈ 0.19
0.42.)0.07.920
        -4 2
         3 72
        -3 36
           360
          -336
            24
```

$-0.0792 \div (-0.42) \approx 0.19$

Example 24 Simplify $-0.394 \div 1.7$. Round to the nearest hundredth.

Your solution

4.1 Exercises

Write as a decimal. Place a bar over the repeating digits of a repeating decimal.

1. $\frac{1}{3}$ **2.** $\frac{2}{3}$ **3.** $\frac{1}{4}$ **4.** $\frac{3}{4}$

5. $\frac{2}{5}$ **6.** $\frac{4}{5}$ **7.** $\frac{1}{6}$ **8.** $\frac{5}{6}$

9. $\frac{1}{8}$ **10.** $\frac{7}{8}$ **11.** $\frac{2}{9}$ **12.** $\frac{8}{9}$

13. $\frac{5}{11}$ **14.** $\frac{10}{11}$ **15.** $\frac{7}{12}$ **16.** $\frac{11}{12}$

17. $\frac{4}{15}$ **18.** $\frac{8}{15}$ **19.** $\frac{9}{16}$ **20.** $\frac{15}{16}$

21. $\frac{7}{18}$ **22.** $\frac{17}{18}$ **23.** $\frac{1}{20}$ **24.** $\frac{13}{20}$

25. $\frac{6}{25}$ **26.** $\frac{14}{25}$ **27.** $\frac{7}{30}$ **28.** $\frac{19}{30}$

29. $\frac{9}{40}$ **30.** $\frac{21}{40}$ **31.** $\frac{13}{36}$ **32.** $\frac{29}{36}$

33. $\frac{15}{22}$ **34.** $\frac{19}{22}$ **35.** $\frac{11}{24}$ **36.** $\frac{19}{24}$

37. $\frac{5}{33}$ **38.** $\frac{25}{33}$ **39.** $\frac{3}{37}$ **40.** $\frac{14}{37}$

4.2 Exercises

Simplify.

41. $\frac{2}{3} + \frac{5}{12}$

42. $\frac{1}{2} + \frac{3}{8}$

43. $\frac{5}{8} - \frac{5}{6}$

44. $\frac{1}{9} - \frac{5}{27}$

45. $-\frac{5}{12} - \frac{3}{8}$

46. $-\frac{5}{6} - \frac{5}{9}$

47. $-\frac{6}{13} + \frac{17}{26}$

48. $-\frac{7}{12} + \frac{5}{8}$

49. $-\frac{5}{8} - \left(-\frac{11}{12}\right)$

50. $\frac{1}{3} + \frac{5}{6} - \frac{2}{9}$

51. $\frac{1}{2} - \frac{2}{3} + \frac{1}{6}$

52. $-\frac{3}{8} - \frac{5}{12} - \frac{3}{16}$

53. $-\frac{5}{16} + \frac{3}{4} - \frac{7}{8}$

54. $\frac{1}{2} - \frac{3}{8} - \left(-\frac{1}{4}\right)$

55. $\frac{3}{4} - \left(-\frac{7}{12}\right) - \frac{7}{8}$

56. $\frac{1}{3} - \frac{1}{4} - \frac{1}{5}$

57. $\frac{2}{3} - \frac{1}{2} + \frac{5}{6}$

58. $\frac{5}{16} + \frac{1}{8} - \frac{1}{2}$

59. $\frac{5}{8} - \left(-\frac{5}{12}\right) + \frac{1}{3}$

60. $\frac{1}{8} - \frac{11}{12} + \frac{1}{2}$

61. $-\frac{7}{9} + \frac{14}{15} + \frac{8}{21}$

62. $1.09 + 6.2$

63. $-32.1 - 6.7$

64. $5.13 - 8.179$

65. $-13.092 + 6.9$

66. $2.54 - 3.6$

67. $5.43 + 7.925$

68. $-16.92 - 6.925$

69. $-3.87 + 8.546$

70. $6.9027 - 17.692$

Simplify.

71. $2.09 - 6.72 - 5.4$

72. $16.4 + 3.09 - 7.93$

73. $-18.39 + 4.9 - 23.7$

74. $19 - (-3.72) - 82.75$

75. $-3.07 - (-2.97) - 17.4$

76. $-3.09 - 4.6 - 27.3$

77. $317.09 - 46.902 + 583.0714$

78. $71.0235 - 86.0974 + 254.309$

4.3 Exercises

Simplify.

79. $\frac{1}{2} \times \left(-\frac{3}{4}\right)$

80. $-\frac{2}{9} \times \left(-\frac{3}{14}\right)$

81. $\left(-\frac{3}{8}\right)\left(-\frac{4}{15}\right)$

82. $\left(-\frac{3}{4}\right)\left(-\frac{8}{27}\right)$

83. $-\frac{1}{2} \times \frac{8}{9}$

84. $\frac{5}{12} \times \left(-\frac{8}{15}\right)$

85. $\frac{5}{8} \times \left(-\frac{7}{12}\right) \times \frac{16}{25}$

86. $\left(\frac{1}{2}\right)\left(-\frac{3}{4}\right)\left(-\frac{5}{8}\right)$

87. $\left(\frac{5}{12}\right)\left(-\frac{8}{15}\right)\left(-\frac{1}{3}\right)$

88. $\frac{3}{8} \div \frac{1}{4}$

89. $\frac{5}{6} \div \left(-\frac{3}{4}\right)$

90. $-\frac{5}{12} \div \frac{15}{32}$

91. $-\frac{7}{8} \div \frac{4}{21}$

92. $\frac{7}{10} \div \frac{2}{5}$

93. $-\frac{15}{64} \div \left(-\frac{3}{40}\right)$

Simplify.

94. $\frac{1}{8} \div \left(-\frac{5}{12}\right)$ **95.** $-\frac{4}{9} \div \left(-\frac{2}{3}\right)$ **96.** $-\frac{6}{11} \div \frac{4}{9}$

97. 1.2×3.47 **98.** -0.8×6.2 **99.** $(-1.89)(-2.3)$

100. $(6.9)(-4.2)$ **101.** $1.06 \times (-3.8)$ **102.** $-2.7 \times (-3.5)$

103. $1.2 \times (-0.5) \times 3.7$ **104.** $-2.4 \times 6.1 \times 0.9$

105. $(-0.8)(3.006)(-5.1)$ **106.** $(-3.4)(-0.08)(1.06)$

Simplify. Round to the nearest hundredth.

107. $-24.7 \div 0.09$ **108.** $-1.27 \div (-1.7)$ **109.** $9.07 \div (-3.5)$

110. $0.0976 \div 0.042$ **111.** $-6.904 \div 1.35$ **112.** $-7.894 \div (-2.06)$

113. $-354.2086 \div 0.1719$ **114.** $-2658.3109 \div (-0.0473)$

115. $(-3.92)(-27.1)(45.008)$ **116.** $-0.461 \times 0.087 \times (-9.675)$

SECTION 5 Exponents and the Order of Operations Agreement

5.1 Objective To evaluate exponential expressions

Reference for Computer Tutor™

Repeated multiplication of the same factor can be written using an exponent.

$2 \cdot 2 \cdot 2 \cdot 2 \cdot 2 = 2^5$ ← **exponent** (2 is the **base**)

$a \cdot a \cdot a \cdot a = a^4$ ← **exponent** (a is the **base**)

The **exponent** indicates how many times the factor, called the **base,** occurs in the multiplication. The multiplication $2 \cdot 2 \cdot 2 \cdot 2 \cdot 2$ is in **factored form.** The exponential expression 2^5 is in **exponential form.**

2^1 is read "the first power of two" or just "two." Usually the exponent 1 is not written.

2^2 is read "the second power of two" or "two squared."

2^3 is read "the third power of two" or "two cubed."

2^4 is read "the fourth power of two."

2^5 is read "the fifth power of two."

a^5 is read "the fifth power of a."

To evaluate an exponential expression, write each factor as many times as indicated by the exponent. Then multiply.

$3^5 = 3 \cdot 3 \cdot 3 \cdot 3 \cdot 3 = 243$

$2^3 \cdot 3^2 = (2 \cdot 2 \cdot 2) \cdot (3 \cdot 3) = 8 \cdot 9 = 72$

Evaluate $(-4)^2$ and -4^2.

The -4 is squared only when the negative sign is *inside* parentheses.

$(-4)^2 = (-4)(-4) = 16$

$-4^2 = -(4 \times 4) = -16$

Evaluate $(-2)^4$ and $(-2)^5$.

The product of an even number of negative factors is positive.

$(-2)^4 = (-2)(-2)(-2)(-2) = 4(-2)(-2) = -8(-2) = 16$

The product of an odd number of negative factors is negative.

$(-2)^5 = (-2)(-2)(-2)(-2)(-2) = 4(-2)(-2)(-2) = -8(-2)(-2) = 16(-2) = -32$

Example 1 Evaluate -5^3.

Solution $-5^3 = -(5 \cdot 5 \cdot 5) = -125$

Example 2 Evaluate -6^3.

Your solution

Example 3 Evaluate $(-4)^4$.

Solution $(-4)^4 = (-4)(-4)(-4)(-4) = 256$

Example 4 Evaluate $(-3)^4$.

Your solution

Example 5 Evaluate $(-3)^2 \cdot 2^3$.

Solution $(-3)^2 \cdot 2^3 = (-3)(-3) \cdot (2)(2)(2) = 9 \cdot 8 = 72$

Example 6 Evaluate $(3^3)(-2)^3$.

Your solution

Example 7 Evaluate $\left(-\frac{2}{3}\right)^3$.

Solution $\left(-\frac{2}{3}\right)^3 = \left(-\frac{2}{3}\right)\left(-\frac{2}{3}\right)\left(-\frac{2}{3}\right) = -\frac{2 \cdot 2 \cdot 2}{3 \cdot 3 \cdot 3} = -\frac{8}{27}$

Example 8 Evaluate $\left(-\frac{2}{5}\right)^2$.

Your solution

Example 9 Evaluate $-4(0.7)^2$.

Solution $-4(0.7)^2 = -4(0.7)(0.7) = -2.8(0.7) = -1.96$

Example 10 Evaluate $-3(0.3)^3$.

Your solution

Solutions on p. A8

5.2 Objective — To use the Order of Operations Agreement to simplify expressions

Reference for Computer Tutor™

Evaluate $2 + 3 \cdot 5$.

There are two arithmetic operations, addition and multiplication, in this problem. The operations could be performed in different orders.

Add first.	$\underbrace{2 + 3} \cdot 5$	Multiply first.	$2 + \underbrace{3 \cdot 5}$
Then multiply.	$\underbrace{5 \cdot 5}_{25}$	Then add.	$\underbrace{2 + 15}_{17}$

In order to prevent more than one answer to the same problem, an Order of Operations Agreement is followed.

The Order of Operations Agreement

Step 1 Perform operations inside grouping symbols. Grouping symbols include parentheses ( ), brackets [ ], and the fraction bar.
Step 2 Simplify exponential expressions.
Step 3 Do multiplication and division as they occur from left to right.
Step 4 Do addition and subtraction as they occur from left to right.

Simplify $12 - 24(8 - 5) \div 2^2$.

1) Perform operations inside grouping symbols. $12 - 24(8 - 5) \div 2^2$
2) Simplify exponential expressions. $12 - 24(3) \div 2^2$
3) Do multiplication and division as they occur from left to right. $12 - 24(3) \div 4$
$12 - 72 \div 4$
4) Do addition and subtraction as they occur from left to right. $12 - 18$
$12 + (-18)$ Do this step mentally.
-6

One or more of the above steps may not be needed to simplify an expression. In that case, proceed to the next step in the Order of Operations Agreement.

Simplify $\frac{4 + 8}{2 + 1} - (3 - 1) + 2$.

1) Perform operations inside grouping symbols. $\frac{4 + 8}{2 + 1} - (3 - 1) + 2$
3) Do multiplication and division as they occur from left to right. $\frac{12}{3} - 2 + 2$
4) Do addition and subtraction as they occur from left to right. $4 - 2 + 2$
$4 + (-2) + 2$ Do this step mentally.
$2 + 2$
4

When an expression has grouping symbols inside grouping symbols, perform the operations inside the inner grouping symbols first.

Simplify $6 \div [4 - (6 - 8)] + 2^2$.

1) Perform operations inside grouping symbols. $6 \div [4 - (6 - 8)] + 2^2$
$6 \div [4 - (-2)] + 2^2$
$6 \div [4 + 2] + 2^2$ Do this step mentally.
2) Simplify exponential expressions. $6 \div 6 + 2^2$
3) Do multiplication and division as they occur from left to right. $6 \div 6 + 4$
4) Do addition and subtraction as they occur from left to right. $1 + 4$
5

Example 11 Simplify
$4 - 3[4 - 2(6 - 3)] \div 2$.

Solution
$4 - 3[4 - 2(6 - 3)] \div 2$
$4 - 3[4 - 2 \cdot 3] \div 2$
$4 - 3[4 - 6] \div 2$
$4 - 3[-2] \div 2$
$4 + 6 \div 2$
$4 + 3$
7

Example 12 Simplify
$18 - 5[8 - 2(2 - 5)] \div 10$.

Your solution

Example 13 Simplify
$27 \div (5 - 2)^2 + (-3)^2 \cdot 4$.

Solution
$27 \div (5 - 2)^2 + (-3)^2 \cdot 4$
$27 \div 3^2 + (-3)^2 \cdot 4$
$27 \div 9 + 9 \cdot 4$
$3 + 9 \cdot 4$
$3 + 36$
39

Example 14 Simplify
$36 \div (8 - 5)^2 - (-3)^2 \cdot 2$.

Your solution

Example 15 Simplify
$(1.75 - 1.3)^2 \div 0.025 + 6.1$.

Solution
$(1.75 - 1.3)^2 \div 0.025 + 6.1$
$(0.45)^2 \div 0.025 + 6.1$
$0.2025 \div 0.025 + 6.1$
$8.1 + 6.1$
14.2

Example 16 Simplify
$(6.97 - 4.72)^2 \times 4.5 \div 0.05$.

Your solution

Example 17 Simplify $\frac{5}{8} - \left(\frac{2}{5} - \frac{1}{2}\right) \div \left(\frac{2}{3}\right)^2$.

Solution
$\frac{5}{8} - \left(\frac{2}{5} - \frac{1}{2}\right) \div \left(\frac{2}{3}\right)^2$
$\frac{5}{8} - \left(-\frac{1}{10}\right) \div \left(\frac{2}{3}\right)^2$
$\frac{5}{8} + \frac{1}{10} \div \frac{4}{9}$
$\frac{5}{8} + \frac{1}{10} \cdot \frac{9}{4}$
$\frac{5}{8} + \frac{9}{40}$
$\frac{25}{40} + \frac{9}{40}$
$\frac{34}{40} = \frac{17}{20}$

Example 18 Simplify $\frac{5}{8} \div \left(\frac{1}{3} - \frac{3}{4}\right) + \frac{7}{12}$.

Your solution

Solutions on p. A8

5.1 Exercises

Evaluate.

1. 6^2
2. 7^4
3. -7^2
4. -4^3
5. $(-3)^2$
6. $(-2)^3$
7. $(-3)^4$
8. $(-5)^3$
9. $\left(\frac{1}{2}\right)^2$
10. $\left(-\frac{3}{4}\right)^3$
11. $(0.3)^2$
12. $(1.5)^3$
13. $\left(\frac{2}{3}\right)^2 \cdot 3^3$
14. $\left(-\frac{1}{2}\right)^3 \cdot 8$
15. $(0.3)^3 \cdot 2^3$
16. $(0.5)^2 \cdot 3^3$
17. $(-3) \cdot 2^2$
18. $(-5) \cdot 3^4$
19. $(-2) \cdot (-2)^3$
20. $(-2) \cdot (-2)^2$
21. $2^3 \cdot 3^3 \cdot (-4)$
22. $(-3)^3 \cdot 5^2 \cdot 10$
23. $(-7) \cdot 4^2 \cdot 3^2$
24. $(-2) \cdot 2^3 \cdot (-3)^2$
25. $\left(\frac{2}{3}\right)^2 \cdot \frac{1}{4} \cdot 3^3$
26. $\left(\frac{3}{4}\right)^2 \cdot (-4) \cdot 2^3$
27. $8^2 \cdot (-3)^5 \cdot 5$

5.2 Exercises

Simplify by using the Order of Operations Agreement.

28. $4 - 8 \div 2$
29. $2^2 \cdot 3 - 3$
30. $2(3 - 4) - (-3)^2$
31. $16 - 32 \div 2^3$
32. $24 - 18 \div 3 + 2$
33. $8 - (-3)^2 - (-2)$

Simplify by using the Order of Operations Agreement.

34. $16 + 15 \div (-5) - 2$

35. $14 - 2^2 - (4 - 7)$

36. $3 - 2[8 - (3 - 2)]$

37. $-2^2 + 4[16 \div (3 - 5)]$

38. $6 + \dfrac{16 - 4}{2^2 + 2} - 2$

39. $24 \div \dfrac{3^2}{8 - 5} - (-5)$

40. $96 \div 2[12 + (6 - 2)] - 3^3$

41. $4 \cdot [16 - (7 - 1)] \div 10$

42. $16 \div 2 - 4^2 - (-3)^2$

43. $18 \div (9 - 2^3) + (-3)$

44. $16 - 3(8 - 3)^2 \div 5$

45. $4(-8) \div [2(7 - 3)^2]$

46. $\dfrac{(-10) + (-2)}{6^2 - 30} \div (2 - 4)$

47. $16 - 4 \cdot \dfrac{3^3 - 7}{2^3 + 2} - (-2)^2$

48. $(0.2)^2 \cdot (-0.5) + 1.72$

49. $0.3(1.7 - 4.8) + (1.2)^2$

50. $(1.8)^2 - 2.52 \div (1.8)$

51. $(1.65 - 1.05)^2 \div 0.4 + 0.8$

52. $\left(\frac{1}{2}\right)^2 - \left(\frac{3}{4} - \frac{1}{2}\right)$

53. $\frac{3}{8} \div \left(\frac{5}{6} + \frac{2}{3}\right)$

54. $\left(\frac{5}{12} - \frac{9}{16}\right) \cdot \frac{3}{7}$

55. $\left(\frac{3}{4}\right)^2 - \left(\frac{1}{2}\right)^3 \div \frac{3}{5}$

Calculators and Computers

Extended Precision on an Electronic Calculator

Consider the decimal equivalents of the following fractions:

$$\frac{7}{33} = 0.\overline{21} \qquad \text{and} \qquad \frac{15}{37} = 0.\overline{405}$$

These decimal equivalents were calculated using an electronic calculator which displays 7 decimal places. Some fractions, however, do not have decimal equivalents which repeat until well after 7 places.

Some examples of fractions with repeating cycles which are longer than seven places are:

$$\frac{4}{17} = 0.\overline{2352941176470588} \qquad \text{and} \qquad \frac{9}{23} = 0.\overline{3913043478260869565217}$$

A calculator which displays 7 decimal places was used to determine each of the above decimal equivalents. The procedure for these calculations is illustrated below.

Find the repeating decimal expression for $\frac{8}{17}$.

1) Divide the numerator by the denominator. $\frac{8}{17} = 0.4705882$

   The decimal approximation is the first four digits.

   $\frac{8}{17} = 0.4705$

2) Take the last three digits (as a decimal) and form the product with the denominator.

   $0.882 \times 17 = 14.994 \approx 15$
   Round this to the nearest integer and divide by the denominator.

   $\frac{15}{17} = 0.8823529$

   The new decimal approximation is the approximation from Step 1 and the first four digits from Step 2.

   $\frac{8}{17} = 0.47058823$

3) Repeat Step 2. Continue to repeat Step 2 until the decimal representation repeats.

   $0.529 \times 17 = 8.993 \approx 9$ $\quad$ $0.117 \times 17 = 1.989 \approx 2$ $\quad$ $0.470 \times 17 = 7.99 \approx 8$

   $\frac{9}{17} \approx 0.5294117$ $\quad$ $\frac{2}{17} \approx 0.1176470$ $\quad$ $\frac{8}{17} \approx 0.4705882$

   $\frac{8}{17} \approx 0.470588235294$ $\quad$ $\frac{8}{17} \approx 0.4705882352941176$

   The decimal begins to repeat with the digits 4705.

   The decimal equivalent of $\frac{8}{17}$ is $0.\overline{4705882352941176}$.

Exercises: Find the decimal equivalents of each of the following:

1. $\frac{3}{17}$ **2.** $\frac{10}{19}$ **3.** $\frac{10}{23}$

Chapter Summary

KEY WORDS

The **natural numbers** are 1, 2, 3, 4, 5, 6, 7 . . .

The **integers** are −4, −3, −2, −1, 0, 1, 2, 3, 4 . . .

The **absolute value** of a number is its distance from zero on a number line.

A **rational number** is a number of the form $\frac{a}{b}$, where a and b are integers and b is not equal to zero A rational number written in this form is commonly called a **fraction.**

An **irrational number** is a number which has a decimal representation which never terminates or repeats.

An expression of the form a^n is in **exponential form,** where a is the base and n is the exponent.

ESSENTIAL RULES

Addition of Integers with Like Signs To add numbers with like signs, add the absolute values of the numbers. Then attach the sign of the addends.

Addition of Integers with Unlike Signs To add numbers with unlike signs, find the difference between the absolute values of the numbers. Then attach the sign of the number with the greater absolute value.

Subtraction of Integers To subtract one integer from another, add the opposite of the second integer to the first integer.

Multiplication of Integers with Like Signs To multiply numbers with like signs, multiply the absolute values of the numbers. The product is positive.

Multiplication of Integers with Unlike Signs To multiply numbers with unlike signs, multiply the absolute values of the numbers. The product is negative.

Division of Integers with Like Signs The quotient of two numbers with like signs is positive.

Division of Integers with Unlike Signs The quotient of two numbers with unlike signs is negative.

Order of Operations Agreement

Step 1 Perform operations inside grouping symbols.
Step 2 Simplify exponential expressions.
Step 3 Do multiplication and division as they occur from left to right.
Step 4 Do addition and subtraction as they occur from left to right.

Test will be like review

Review/Test

SECTION 1

1.1a Place the correct symbol, $<$ or $>$, between the two numbers.

-2 > -40

1.1b Place the correct symbol, $<$ or $>$, between the two numbers.

-1 < 0

1.2a Find the opposite of -4.

4

1.2b Evaluate $-|-4|$.

-4

SECTION 2

2.1a Add: $13 + (-16)$

2.1b Add: $-22 + 14 + (-8)$

2.2a Subtract: $16 - 30$

2.2b Subtract: $16 - (-30) - 42$

SECTION 3

3.1a Multiply: -4×12

3.1b Multiply: $-5 \times (-6) \times 3$

3.2a Divide: $-72 \div 8$

3.2b Divide: $-561 \div (-33)$

Review/Test

3.3a Find the temperature after a rise of 11 °C from −4 °C.

7°C

3.3b The daily low temperature readings for a three-day period were as follows: −7°, 9°, −8°. Find the average low temperature for the three-day period.

(−7)+(+9)+(−8)
2+(−8) = −6
3)−6 = −2°

SECTION **4**

4.1a Write $\frac{17}{20}$ as a decimal.

4.1b Write $\frac{7}{9}$ as a decimal. Place a bar over the repeating digits of the decimal.

4.2a Simplify $-\frac{2}{5} + \frac{7}{15}$.

4.2b Simplify $6.039 - 12.92$.

4.3a Simplify $\frac{5}{12} \div \left(-\frac{5}{6}\right)$.

4.3b Simplify $6.02 \times (-0.89)$.

SECTION **5**

5.1a Evaluate $(-3^3) \cdot 2^2$.

5.1b Evaluate $\frac{3}{4} \cdot (4)^2$.

5.2a Simplify $3^2 - 4 + 20 \div 5$.

5.2b Simplify $\frac{-10 + 2}{2 + (-4)} \div 2 + 6$.

Cumulative Review/Test

1. Place the correct symbol, $<$ or $>$, between the two numbers.
   $-8 \quad -4$

2. Place the correct symbol, $<$ or $>$, between the two numbers.
   $1 \quad -1$

3. Find the opposite of -2.

4. Evaluate $-|-3|$.

5. Add: $12 + (-14)$

6. Add: $-12 + 8 + (-4)$

7. Add: $-8 + (-4) + 3 + (-2)$

8. Subtract: $7 - 21$

9. Subtract: $12 - (-10) - 4$

10. Subtract:
    $-8 - (-3) - 12 - (-4)$

11. Multiply: -3×18

12. Multiply: $2(-3)(-12)$

13. Divide: $96 \div (-12)$

14. Divide: $(-204) \div (-17)$

15. Find the temperature after a rise of 5°C from −12°C.

16. The daily high temperature readings for a four-day period were recorded as follows: −19°, −7°, 1°, and 9°. Find the average high temperature for the four-day period.

Cumulative Review/Test

17. Write $\frac{7}{20}$ as a decimal.

18. Write $\frac{7}{11}$ as a decimal. Place a bar over the repeating digits of the decimal.

19. Simplify $-\frac{5}{8} + \frac{1}{6}$.

20. Simplify $\frac{2}{3} - \frac{5}{8}$.

21. Simplify $-1.329 + 4.89$.

22. Simplify $-\frac{7}{16} \div \frac{3}{8}$.

23. Simplify $\left(\frac{3}{4}\right)\left(-\frac{8}{9}\right)\left(-\frac{3}{16}\right)$.

24. Simplify $0.2654 \div (-0.023)$. Round to the nearest tenth.

25. Evaluate $-4^2 \cdot \left(\frac{1}{2}\right)^2$.

26. Evaluate $(-3)^3 \cdot \left(\frac{1}{6}\right)^2$.

27. Evaluate $-2^3 \cdot \left(-\frac{3}{8}\right)^2$.

28. Evaluate $(0.5)^2 \cdot (-8)^2$.

29. Simplify $2^3 \div 4 - 2(2 - 7)$.

30. Simplify $(7 - 2)^2 - 5 - 3 \cdot 4$.

31. Simplify $\left(\frac{5}{8}\right)^2 \div \left(\frac{7}{12} - \frac{3}{2}\right)$.

32. Simplify $\left(\frac{7}{16} \div \frac{5}{8}\right) - \left(\frac{2}{5}\right)^2$.

2

Variable Expressions

Objectives

- To evaluate a variable expression
- To simplify a variable expression using the Properties of Addition
- To simplify a variable expression using the Properties of Multiplication
- To simplify a variable expression using the Distribution Property
- To simplify general variable expressions
- To translate a verbal expression into a variable expression given the variable
- To translate a verbal expression into a variable expression by assigning the variable
- To translate a verbal expression into a variable expression and then simplify the resulting expression

History of Variables

Prior to the 16th century, unknown quantities were represented by words. In Latin, the language in which most scholarly works were written, the word 'res', meaning thing, was used. In Germany, the word 'zahl', meaning number, was used. In Italy, the word 'cosa', also meaning thing, was used.

Then in 1637 René Descartes, a French mathematician, began using the letters x, y, and z to represent variables. It is interesting to note, upon examining Descartes' work that toward the end of the book the letters y and z were no longer used and x became the choice for a variable.

One explanation as to why the letters y and z appeared less frequently has to do with the nature of printing presses during Descartes' time. A printer had a large tray which contained all the letters of the alphabet. There were many copies of each letter, especially those letters which are used frequently. For example, there were more e's than q's. Since the letters y and z do not occur frequently in French, a printer would have few of these letters on hand. Consequently when Descartes started using these letters as variables, it quickly depleted the printer's supply and x's had to be used instead.

Today, x is used by most nations as the standard letter for a single unknown. In fact, x-rays were so named because the scientists who discovered them did not know what they were and thus labeled them the "unknown rays" or x-rays.

SECTION 1 Evaluating Variable Expressions

1.1 Objective To evaluate a variable expression

Reference for Computer Tutor™

Often we discuss a quantity without knowing its exact value, for example, the price of gold next month, the cost of a new automobile next year, or the tuition cost for next semester. In algebra, a letter of the alphabet is used to stand for a quantity which is unknown, or which can change or *vary*. The letter is called a **variable.** An expression which contains one or more variables is called a **variable expression.**

A variable expression is shown at the right. The expression can be rewritten by writing subtraction as the addition of the opposite.

$3x^2 - 5y + 2xy - x - 7$

$3x^2 + (-5y) + 2xy + (-x) + (-7)$

Note that the expression has 5 addends. The **terms** of a variable expression are the addends of the expression. The expression has 5 terms.

5 terms

$3x^2 \quad - \quad 5y \quad + \quad 2xy \quad - \quad x \quad - \quad 7$

variable terms   constant term

The terms $3x^2$, $-5y$, $2xy$, and $-x$ are **variable terms.**

The term -7 is a **constant term,** or simply a **constant.**

Each variable term is composed of a **numerical coefficient** and a **variable part** (the variable or variables and their exponents).

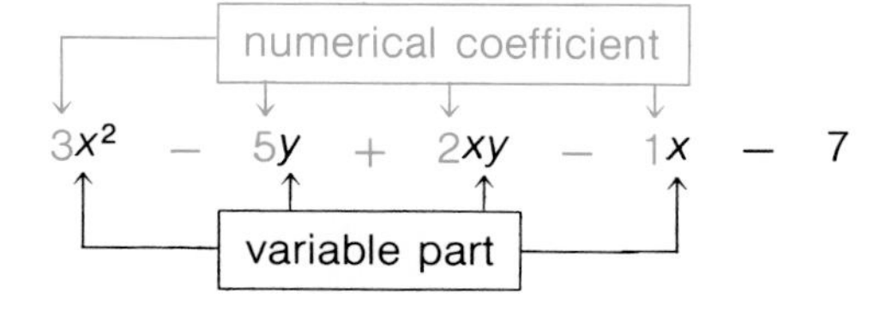

When the numerical coefficient is 1 or -1, the 1 is usually not written ($x = 1x$ and $-x = -1x$).

Variable expressions occur naturally in science. In a physics lab, a student may discover that a weight of one pound will stretch a spring $\frac{1}{2}$ inch. Two pounds will stretch the spring 1 inch. By experimenting, the student can discover that the distance the spring will stretch is found by multiplying the weight by $\frac{1}{2}$. By letting W represent the weight attached to the spring, the distance the spring stretches can be represented by the variable expression $\frac{1}{2}W$.

With a weight of W pounds, the spring will stretch $\frac{1}{2} \cdot W = \frac{1}{2}W$ inches.

With a weight of 10 pounds, the spring will stretch $\frac{1}{2} \cdot 10 = 5$ inches.

With a weight of 3 pounds, the spring will stretch $\frac{1}{2} \cdot 3 = 1\frac{1}{2}$ inches.

Replacing the variable or variables in a variable expression and then simplifying the resulting numerical expression is called **evaluating the variable expression.**

Evaluate $ab - b^2$ when $a = 2$ and $b = -3$.

Step 1 Replace each variable in the expression with the number it stands for.

$ab - b^2$
$2(-3) - (-3)^2$

Step 2 Use the Order of Operations Agreement to simplify the resulting numerical expression.

$-6 - 9$
-15

Example 1 Name the variable terms of the expression $2a^2 - 5a + 7$.

Solution $2a^2$
$-5a$

Example 2 Name the constant term of the expression $6n^2 + 3n - 4$.

Your solution

Example 3 Evaluate $x^2 - 3xy$ when $x = 3$ and $y = -4$.

Solution
$x^2 - 3xy$
$3^2 - 3(3)(-4)$
$9 - 3(3)(-4)$
$9 - 9(-4)$
$9 - (-36)$
45

Example 4 Evaluate $2xy + y^2$ when $x = -4$ and $y = 2$.

Your solution

Example 5 Evaluate $\frac{a^2 - b^2}{a - b}$ when $a = 3$ and $b = -4$.

Solution
$\frac{a^2 - b^2}{a - b}$
$\frac{3^2 - (-4)^2}{3 - (-4)}$
$\frac{9 - 16}{3 - (-4)}$
$\frac{-7}{7} = -1$

Example 6 Evaluate $\frac{a^2 + b^2}{a + b}$ when $a = 5$ and $b = -3$.

Your solution

Example 7 Evaluate $x^2 - 3(x - y) - z^2$ when $x = 2$, $y = -1$, and $z = 3$.

Solution
$x^2 - 3(x - y) - z^2$
$2^2 - 3[2 - (-1)] - 3^2$
$2^2 - 3(3) - 3^2$
$4 - 3(3) - 9$
$4 - 9 - 9$
$-5 - 9$
-14

Example 8 Evaluate $x^3 - 2(x + y) + z^2$ when $x = 2$, $y = -4$, and $z = -3$.

Your solution

Solutions on p. A10

1.1 Exercises

Name the terms of the variable expression. Then underline the constant term.

1. $2x^2 + 5x - 8$ **2.** $-3n^2 - 4n + 7$ **3.** $6 - a^4$

Name the variable terms of the expression. Then underline the variable part of each term.

4. $9b^2 - 4ab + a^2$ **5.** $7x^2y + 6xy^2 + 10$ **6.** $5 - 8n - 3n^2$

Name the coefficients of the variable terms.

7. $x^2 - 9x + 2$ **8.** $12a^2 - 8ab - b^2$ **9.** $n^3 - 4n^2 - n + 9$

Evaluate the variable expression when $a = 2$, $b = 3$, and $c = -4$.

10. $3a + 2b$ **11.** $a - 2c$ **12.** $-a^2$

13. $2c^2$ **14.** $-3a + 4b$ **15.** $3b - 3c$

16. $b^2 - 3$ **17.** $-3c + 4$ **18.** $16 \div (2c)$

19. $6b \div (-a)$ **20.** $bc \div (2a)$ **21.** $-2ab \div c$

22. $a^2 - b^2$ **23.** $b^2 - c^2$ **24.** $(a + b)^2$

25. $a^2 + b^2$ **26.** $2a - (c + a)^2$ **27.** $(b - a)^2 + 4c$

28. $b^2 - \frac{ac}{8}$ **29.** $\frac{5ab}{6} - 3cb$ **30.** $(b - 2a)^2 + bc$

Evaluate the variable expression when $a = -2$, $b = 4$, $c = -1$, and $d = 3$.

31. $\frac{b + c}{d}$ 32. $\frac{d - b}{c}$ 33. $\frac{2d + b}{-a}$

34. $\frac{b + 2d}{b}$ 35. $\frac{b - d}{c - a}$ 36. $\frac{2c - d}{-ad}$

37. $(b + d)^2 - 4a$ 38. $(d - a)^2 - 3c$ 39. $(d - a)^2 \div 5$

40. $(b - c)^2 \div 5$ 41. $b^2 - 2b + 4$ 42. $a^2 - 5a - 6$

43. $\frac{bd}{a} \div c$ 44. $\frac{2ac}{b} \div (-c)$ 45. $2(b + c) - 2a$

46. $3(b - a) - bc$ 47. $\frac{b - 2a}{bc^2 - d}$ 48. $\frac{b^2 - a}{ad + 3c}$

49. $\frac{1}{3}d^2 - \frac{3}{8}b^2$ 50. $\frac{5}{8}a^4 - c^2$ 51. $\frac{-4bc}{2a - b}$

52. $\frac{abc}{b - d}$ 53. $a^3 - 3a^2 + a$ 54. $d^3 - 3d - 9$

55. $-\frac{3}{4}b + \frac{1}{2}(ac + bd)$ 56. $-\frac{2}{3}d - \frac{1}{5}(bd - ac)$ 57. $(b - a)^2 - (d - c)^2$

58. $(b + c)^2 + (a + d)^2$ 59. $4ac + (2a)^2$ 60. $3dc - (4c)^2$

Evaluate the variable expression when $a = 2.7$, $b = -1.6$, and $c = -0.8$.

61. $c^2 - ab$ 62. $(a + b)^2 - c$ 63. $\frac{b^3}{c} - 4a$

SECTION 2 Simplifying Variable Expressions

2.1 Objective To simplify a variable expression using the Properties of Addition

Reference for Computer Tutor™

Like terms of a variable expression are the terms with the same variable part. (Since $x^2 = x \cdot x$, x^2 and x are not like terms.)

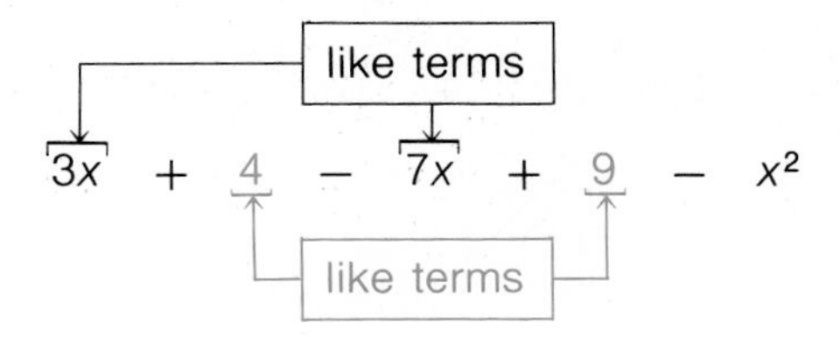

Constant terms are like terms. 4 and 9 are like terms.

To **simplify a variable expression,** add, or **combine,** like terms by adding the numerical coefficients. The variable part of the terms remains unchanged.

Simplify $2x + 3x$.

Add the numerical coefficients of the like variable terms.

$2x + 3x$
$(x + x) + (x + x + x)$
$5x$

Simplify $5x - 11x$.

Add the numerical coefficients of the like variable terms.

$5x - 11x$
$5x + (-11x)$ — Do this step mentally.
$-6x$

In simplifying more complicated expressions, the following Properties of Addition are used.

The Associative Property of Addition

If a, b, and c are real numbers, then $a + b + c = (a + b) + c = a + (b + c)$.

When adding three or more terms, the terms can be grouped in any order. The sum is the same.

$3x + 5x + 9x = (3x + 5x) + 9x = 3x + (5x + 9x)$
$8x + 9x = 3x + 14x$
$17x = 17x$

The Commutative Property of Addition

If a and b are real numbers, then $a + b = b + a$.

When adding two like terms, the terms can be added in either order. The sum is the same.

$2x + (-4x) = -4x + 2x$
$-2x = -2x$

The Addition Property of Zero

If a is a real number, then $a + 0 = 0 + a = a$.

The sum of a term and zero is the term. $5x + 0 = 0 + 5x = 5x$

The Inverse Property of Addition

If a is a real number, then $a + (-a) = (-a) + a = 0$.

The sum of a term and its opposite is zero. The opposite of a number is called its **additive inverse.** $7x + (-7x) = -7x + 7x = 0$

Simplify $8x + 3y - 8x$.

Use the Commutative and Associative Properties of Addition to rearrange and group like terms. $8x + 3y - 8x$

$3y + (8x - 8x)$ Do these steps mentally.

Combine like terms. $3y + 0$

$3y$

Simplify $4x^2 + 5x - 6x^2 - 2x$.

Use the Commutative and Associative Properties of Addition to rearrange and group like terms. $4x^2 + 5x - 6x^2 - 2x$

$(4x^2 - 6x^2) + (5x - 2x)$ Do these steps mentally.

Combine like terms. $-2x^2 + 3x$

Example 1 Simplify $3x + 4y - 10x + 7y$.

Solution $3x + 4y - 10x + 7y$

$-7x + 11y$

Example 2 Simplify $3a - 2b - 5a + 6b$.

Your solution

Example 3 Simplify $x^2 - 7 + 4x^2 - 16$.

Solution $x^2 - 7 + 4x^2 - 16$

$5x^2 - 23$

Example 4 Simplify $-3y^2 + 7 + 8y^2 - 14$.

Your solution

2.2 Objective To simplify a variable expression using the Properties of Multiplication

Reference for Computer Tutor™

In simplifying variable expressions, the following Properties of Multiplication are used.

The Associative Property of Multiplication

If a, b, and c are real numbers, then $a \cdot b \cdot c = (a \cdot b) \cdot c = a \cdot (b \cdot c)$.

When multiplying three or more factors, the factors can be grouped in any order. The product is the same.

$2(3x) = (2 \cdot 3)x = 6x$

The Commutative Property of Multiplication

If a and b are real numbers, then $a \cdot b = b \cdot a$.

When multiplying two factors, the factors can be multiplied in either order. The product is the same.

$(2x) \cdot 3 = 3 \cdot (2x) = (3 \cdot 2)x = 6x$

The Multiplication Property of One

If a is a real number, then $a \cdot 1 = 1 \cdot a = a$.

The product of a term and one is the term.

$(8x)(1) = (1)(8x) = 8x$

The Inverse Property of Multiplication

If a is a real number, and a is not equal to zero, then $a \cdot \frac{1}{a} = \frac{1}{a} \cdot a = 1$.

$\frac{1}{a}$ is called the **reciprocal** of a. $\frac{1}{a}$ is also called the **multiplicative inverse** of a.

The product of a number and its reciprocal is one.

$7 \cdot \frac{1}{7} = \frac{1}{7} \cdot 7 = 1$

Simplify $2(-x)$.

Use the Associative Property of Multiplication to group factors.

$2(-x)$

$2(-1 \cdot x)$
$[2 \cdot (-1)]x$

Do these steps mentally.

$-2x$

Simplify $\frac{3}{2}\left(\frac{2x}{3}\right)$.

Note that $\frac{2x}{3} = \frac{2}{3} \cdot \frac{x}{1} = \frac{2}{3}x$.

$\frac{3}{2}\left(\frac{2x}{3}\right)$

$\frac{3}{2}\left(\frac{2}{3}x\right)$

Use the Associative Property of Multiplication to group factors. Use the Inverse Property of Multiplication and the Multiplication Property of One.

$\left(\frac{3}{2} \cdot \frac{2}{3}\right)x$
$1x$

Do these steps mentally.

x

Simplify $(16x)2$.

Use the Commutative and Associative Properties of Multiplication to rearrange and group factors.

$(16x)2$

$2(16x)$
$(2 \cdot 16)x$

Do these steps mentally.

$32x$

Example 5 Simplify $-2(3x^2)$.

Solution $-2(3x^2)$
$-6x^2$

Example 6 Simplify $-5(4y^2)$.

Your solution

Example 7 Simplify $-5(-10x)$.

Solution $-5(-10x)$
$50x$

Example 8 Simplify $-7(-2a)$.

Your solution

Example 9 Simplify $(6x)(-4)$.

Solution $(6x)(-4)$
$-24x$

Example 10 Simplify $(-5x)(-2)$.

Your solution

Solutions on p. A10

2.3 Objective To simplify a variable expression using the Distributive Property

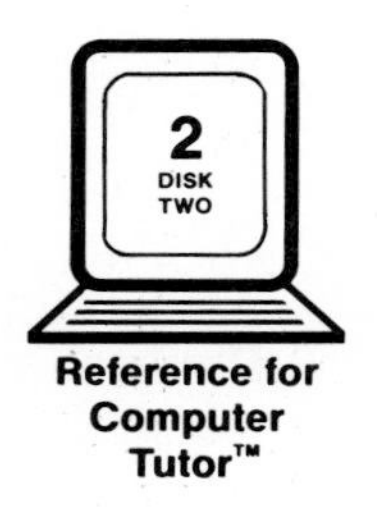

Reference for Computer Tutor™

A student works 3 hours on Friday and 5 hours on Saturday. The hourly rate of pay is \$4 an hour. Find the total wages received for the two days.

The total income can be found in two ways.

1. Multiply the hourly wage by the total number of hours worked.

   $\$4(3 + 5) = \$4 \cdot 8 = \$32$

2. Find the income for each day and add.

   $\$4(3) + \$4(5) = \$12 + \$20 = \$32$

Note that $4(3 + 5) = 4(3) + 4(5)$. This is an example of the Distributive Property of Multiplication over Addition.

The Distributive Property

If a, b, and c are real numbers, then $a(b + c) = ab + ac$ or
$(b + c)a = ba + ca$.

The Distributive Property is used to remove parentheses from a variable expression.

$3(2x - 5)$

$3(2x) + 3(-5)$

Do this step mentally.

$6x - 15$

Simplify $-3(5 + x)$.

Use the Distributive Property to remove parentheses from the variable expression.

$-3(5 + x)$

$-3(5) + (-3)x$ Do this step mentally.

$-15 - 3x$

Simplify $-(2x - 4)$.

Use the Distributive Property to remove parentheses from the variable expression.

$-(2x - 4)$

$-1(2x - 4)$
$-1(2x) - (-1)(4)$ Do these steps mentally.

$-2x + 4$

Notice: when a negative sign immediately precedes the parentheses, the sign of *each* term inside the parentheses is changed.

Example 11 Simplify $7(4 + 2x)$.

Solution $7(4 + 2x)$
$28 + 14x$

Example 12 Simplify $5(3 + 7b)$.

Your solution

Example 13 Simplify $(2x - 6)2$

Solution $(2x - 6)2$
$4x - 12$

Example 14 Simplify $(3a - 1)5$.

Your solution

Example 15 Simplify $-3(-5a + 7b)$.

Solution $-3(-5a + 7b)$
$15a - 21b$

Example 16 Simplify $-8(-2a + 7b)$.

Your solution

Example 17 Simplify $-(3a - 2)$.

Solution $-(3a - 2)$
$-3a + 2$

Example 18 Simplify $-(5x - 12)$.

Your solution

Example 19 Simplify $-2(x^2 + 5x - 4)$.

Solution $-2(x^2 + 5x - 4)$
$-2x^2 - 10x + 8$

Example 20 Simplify $3(-a^2 - 6a + 7)$.

Your solution

Solutions on p. A10

2.4 Objective To simplify general variable expressions

Reference for Computer Tutor™

When simplifying variable expressions, use the Distributive Property to remove parentheses and brackets used as grouping symbols.

Simplify $4(x - y) - 2(-3x + 6y)$.

Use the Distributive Property to remove parentheses. $4(x - y) - 2(-3x + 6y)$

$4x - 4y + 6x - 12y$

Combine like terms. $10x - 16y$

Example 21 Simplify $2x - 3(2x - 7y)$.

Solution $2x - 3(2x - 7y)$
$2x - 6x + 21y$
$-4x + 21y$

Example 22 Simplify $3y - 2(y - 7x)$.

Your solution

Example 23 Simplify $7(x - 2y) - 3(-x - 2y)$.

Solution $7(x - 2y) - 3(-x - 2y)$
$7x - 14y + 3x + 6y$
$10x - 8y$

Example 24 Simplify $-2(x - 2y) + 4(x - 3y)$.

Your solution

Example 25 Simplify $-2(-3x + 7y) - 14x$.

Solution $-2(-3x + 7y) - 14x$
$6x - 14y - 14x$
$-8x - 14y$

Example 26 Simplify $-5(-2y - 3x) + 4y$.

Your solution

Example 27 Simplify $2x - 3[2x - 3(x + 7)]$.

Solution $2x - 3[2x - 3(x + 7)]$
$2x - 3[2x - 3x - 21]$
$2x - 3[-x - 21]$
$2x + 3x + 63$
$5x + 63$

Example 28 Simplify $3y - 2[x - 4(2 - 3y)]$.

Your solution

Solutions on p. A10

2.1 Exercises

Simplify.

1. $6x + 8x$
2. $12x + 13x$
3. $9a - 4a$
4. $12a - 3a$
5. $4y + (-10y)$
6. $8y + (-6y)$
7. $-3b - 7$
8. $-12y - 3$
9. $-12a + 17a$
10. $-3a + 12a$
11. $5ab - 7ab$
12. $9ab - 3ab$
13. $-12xy + 17xy$
14. $-15xy + 3xy$
15. $-3ab + 3ab$
16. $-7ab + 7ab$
17. $-\frac{1}{2}x - \frac{1}{3}x$
18. $-\frac{2}{5}y + \frac{3}{10}y$
19. $\frac{3}{8}x^2 - \frac{5}{12}x^2$
20. $\frac{2}{3}y^2 - \frac{4}{9}y^2$
21. $3x + 5x + 3x$
22. $8x + 5x + 7x$
23. $5a - 3a + 5a$
24. $10a - 17a + 3a$
25. $-5x^2 - 12x^2 + 3x^2$
26. $-y^2 - 8y^2 + 7y^2$
27. $7x + (-8x) + 3y$
28. $8y + (-10x) + 8x$
29. $7x - 3y + 10x$
30. $8y + 8x - 8y$
31. $3a + (-7b) - 5a + b$
32. $-5b + 7a - 7b + 12a$
33. $3x + (-8y) - 10x + 4x$
34. $3y + (-12x) - 7y + 2y$
35. $x^2 - 7x + (-5x^2) + 5x$
36. $3x^2 + 5x - 10x^2 - 10x$

2.2 Exercises

Simplify.

37. $4(3x)$

38. $12(5x)$

39. $-3(7a)$

40. $-2(5a)$

41. $-2(-3y)$

42. $-5(-6y)$

43. $(4x)2$

44. $(6x)12$

45. $(3a)(-2)$

46. $(7a)(-4)$

47. $(-3b)(-4)$

48. $(-12b)(-9)$

49. $-5(3x^2)$

50. $-8(7x^2)$

51. $\frac{1}{3}(3x^2)$

52. $\frac{1}{6}(6x^2)$

53. $\frac{1}{5}(5a)$

54. $\frac{1}{8}(8x)$

55. $-\frac{1}{2}(-2x)$

56. $-\frac{1}{4}(-4a)$

57. $-\frac{1}{7}(-7n)$

58. $-\frac{1}{9}(-9b)$

59. $(3x)\left(\frac{1}{3}\right)$

60. $(12x)\left(\frac{1}{12}\right)$

61. $(-6y)\left(-\frac{1}{6}\right)$

62. $(-10n)\left(-\frac{1}{10}\right)$

63. $\frac{1}{3}(9x)$

64. $\frac{1}{7}(14x)$

65. $-\frac{1}{5}(10x)$

66. $-\frac{1}{8}(16x)$

67. $-\frac{2}{3}(12a^2)$

68. $-\frac{5}{8}(24a^2)$

69. $-\frac{1}{2}(-16y)$

70. $-\frac{3}{4}(-8y)$

71. $(16y)\left(\frac{1}{4}\right)$

72. $(33y)\left(\frac{1}{11}\right)$

73. $(-6x)\left(\frac{1}{3}\right)$

74. $(-10x)\left(\frac{1}{5}\right)$

75. $(-8a)\left(-\frac{3}{4}\right)$

2.3 Exercises

Simplify.

76. $-(x + 2)$ **77.** $-(x + 7)$ **78.** $2(4x - 3)$

79. $5(2x - 7)$ **80.** $-2(a + 7)$ **81.** $-5(a + 16)$

82. $-3(2y - 8)$ **83.** $-5(3y - 7)$ **84.** $(5 - 3b)7$

85. $(10 - 7b)2$ **86.** $-3(3 - 5x)$ **87.** $-5(7 - 10x)$

88. $3(5x^2 + 2x)$ **89.** $6(3x^2 + 2x)$ **90.** $-2(-y + 9)$

91. $-5(-2x + 7)$ **92.** $(-3x - 6)5$ **93.** $(-2x + 7)7$

94. $2(-3x^2 - 14)$ **95.** $5(-6x^2 - 3)$ **96.** $-3(2y^2 - 7)$

97. $-8(3y^2 - 12)$ **98.** $3(x^2 - y^2)$ **99.** $5(x^2 + y^2)$

100. $-2(x^2 - 3y^2)$ **101.** $-4(x^2 - 5y^2)$ **102.** $-(6a^2 - 7b^2)$

103. $3(x^2 + 2x - 6)$ **104.** $4(x^2 - 3x + 5)$ **105.** $-2(y^2 - 2y + 4)$

106. $-3(y^2 - 3y - 7)$ **107.** $2(-a^2 - 2a + 3)$ **108.** $4(-3a^2 - 5a + 7)$

109. $-5(-2x^2 - 3x + 7)$ **110.** $-3(-4x^2 + 3x - 4)$ **111.** $3(2x^2 + xy - 3y^2)$

112. $5(2x^2 - 4xy - y^2)$ **113.** $-(3a^2 + 5a - 4)$ **114.** $-(8b^2 - 6b + 9)$

2.4 Exercises

Simplify.

115. $4x - 2(3x + 8)$

116. $6a - (5a + 7)$

117. $9 - 3(4y + 6)$

118. $10 - (11x - 3)$

119. $5n - (7 - 2n)$

120. $8 - (12 + 4y)$

121. $3(x + 2) - 5(x - 7)$

122. $2(x - 4) - 4(x + 2)$

123. $12(y - 2) + 3(7 - 3y)$

124. $6(2y - 7) - 3(3 - 2y)$

125. $3(a - b) - 4(a + b)$

126. $2(a + 2b) - (a - 3b)$

127. $4[x - 2(x - 3)]$

128. $2[x + 2(x + 7)]$

129. $-2[3x + 2(4 - x)]$

130. $-5[2x + 3(5 - x)]$

131. $-3[2x - (x + 7)]$

132. $-2[3x - (5x - 2)]$

133. $2x - 3[x - 2(4 - x)]$

134. $-7x + 3[x - 7(3 - 2x)]$

135. $-5x - 2[2x - 4(x + 7)] - 6$

136. $4a - 2[2b - (b - 2a)] + 3b$

137. $2x + 3(x - 2y) + 5(3x - 7y)$

138. $5y - 2(y - 3x) + 2(7x - y)$

SECTION 3 Translating Verbal Expressions into Variable Expressions

3.1 Objective To translate a verbal expression into a variable expression given the variable

Reference for Computer Tutor™

One of the major skills required in applied mathematics is to translate a verbal expression into a variable expression. This requires recognizing the verbal phrases that translate into mathematical operations. A partial list of the verbal phrases used to indicate the different mathematical operations is given below.

Addition	added to	6 added to y	$y + 6$
	more than	8 more than x	$x + 8$
	the sum of	the sum of x and z	$x + z$
	increased by	t increased by 9	$t + 9$
	the total of	the total of 5 and y	$5 + y$
Subtraction	minus	x minus 2	$x - 2$
	less than	7 less than t	$t - 7$
	decreased by	m decreased by 3	$m - 3$
	the difference between	the difference between y and 4	$y - 4$
Multiplication	times	10 times t	$10t$
	of	one half of x	$\frac{1}{2}x$
	the product of	the product of y and z	yz
	multiplied by	y multiplied by 11	$11y$
Division	divided by	x divided by 12	$\frac{x}{12}$
	the quotient of	the quotient of y and z	$\frac{y}{z}$
	the ratio of	the ratio of t to 9	$\frac{t}{9}$
Power	the square of	the square of x	x^2
	the cube of	the cube of a	a^3

Translate "14 less than the cube of x" into a variable expression.

Step 1 Identify the words which indicate the mathematical operations. — 14 less than the cube of x

Step 2 Use the identified operations to write the variable expression. — $x^3 - 14$

Translate "the difference between the square of x and the sum of y and z" into a variable expression.

Step 1 Identify the words which indicate the mathematical operations. — the difference between the square of x and the sum of y and z

Step 2 Use the identified operations to write the variable expression. — $x^2 - (y + z)$

Example 1 Translate "the total of 3 times *n* and *n*" into a variable expression.

Solution the total of 3 times *n* and *n*

$3n + n$

m - (n+12)

Example 2 Translate "the difference between twice *n* and one third of *n*" into a variable expression.

Your solution

Example 3 Translate "*m* decreased by the sum of *n* and 12" into a variable expression.

Solution *m* decreased by the sum of *n* and 12

$m - (n + 12)$

Example 4 Translate "the quotient of 7 less than *b* and 15" into a variable expression.

Your solution

Solutions on p. A11

3.2 Objective To translate a verbal expression into a variable expression by assigning the variable

Reference for Computer Tutor™

In most applications which involve translating phrases into variable expressions, the variable to be used is not given. To translate these phrases, a variable must be assigned to an unknown quantity before the variable expression can be written.

Translate "the sum of two consecutive integers" into a variable expression.

Step 1 Assign a variable to one of the unknown quantities. — the first integer: n

Step 2 Use the assigned variable to write an expression for any other unknown quantity. — the next consecutive integer: $n + 1$

Step 3 Use the assigned variable to write the variable expression. — $n + (n + 1)$

Translate "the quotient of twice a number and the difference between the number and twelve" into a variable expression.

Step 1 Assign a variable to one of the unknown quantities. — the unknown number: n

Step 2 Use the assigned variable to write an expression for any other unknown quantity. — twice a number: $2n$; the difference between the number and twelve: $n - 12$

Step 3 Use the assigned variable to write the variable expression. — $\frac{2n}{n - 12}$

Example 5 Translate "a number added to the product of four and the square of the number" into a variable expression.

Solution the unknown number: n
the square of the number: n^2
the product of 4 and the square of the number: $4n^2$

$n + 4n^2$

Example 6 Translate "a number multiplied by the total of ten and the cube of the number" into a variable expression.

Your solution

Example 7 Translate "the sum of the squares of two consecutive integers" into a variable expression.

Solution the first integer: x
the second integer: $x + 1$

$x^2 + (x + 1)^2$

Example 8 Translate "the sum of three consecutive integers" into a variable expression.

Your solution

Example 9 Translate "four times the sum of one half of a number and fourteen" into a variable expression.

Solution the unknown number: n
one half of the number: $\frac{1}{2}n$
the sum of one half of the number and fourteen: $\frac{1}{2}n + 14$

$4\left(\frac{1}{2}n + 14\right)$

Example 10 Translate "five times the difference between a number and sixty" into a variable expression.

Your solution

Solutions on p. A11

3.3 Objective To translate a verbal expression into a variable expression and then simplify the resulting expression

Reference for Computer Tutor™

After translating a verbal expression into a variable expression, simplify the variable expression by using the Addition, Multiplication, and Distributive Properties.

Translate and simplify "a number minus the difference between twice the number and eleven."

Step 1 Assign a variable to one of the unknown quantities. — the unknown number: n

Step 2 Use the assigned variable to write an expression for any other unknown quantity. — the difference between twice the number and eleven: $2n - 11$

Step 3 Use the assigned variable to write the variable expression. — $n - (2n - 11)$

Step 4 Simplify the variable expression. — $n - 2n + 11$

$-n + 11$

Example 11 Translate and simplify "the sum of one fourth of a number and one eighth of the same number."

Solution the unknown number: n

one fourth of the number: $\frac{1}{4}n$

one eighth of the number: $\frac{1}{8}n$

$\frac{1}{4}n + \frac{1}{8}n$

$\frac{2}{8}n + \frac{1}{8}n$

$\frac{3}{8}n$

Example 12 Translate and simplify "the difference between three eighths of a number and five sixths of the same number."

Your solution

Example 13 Translate and simplify "the total of five times an unknown number and twice the difference between the number and nine."

Solution the unknown number: n

five times the unknown number: $5n$

twice the difference between the number and nine: $2(n - 9)$

$5n + 2(n - 9)$

$5n + 2n - 18$

$7n - 18$

Example 14 Translate and simplify "the sum of three consecutive integers."

Your solution

Solutions on p. A11

3.1 Exercises

Translate into a variable expression.

1. the sum of 8 and y
2. a less than 16
3. t increased by 10
4. p decreased by 7
5. z added to 14
6. q multiplied by 13
7. 20 less than the square of x
8. 6 times the difference between m and 7
9. the sum of three fourths of n and 12
10. b decreased by the product of 2 and b
11. 8 increased by the quotient of n and 4
12. the product of -8 and y
13. the product of 3 and the total of y and 7
14. 8 divided by the difference between x and 6
15. the product of t and the sum of t and 16
16. the quotient of 6 less than n and twice n
17. 15 more than one half of the square of x
18. 19 less than the product of n and -2
19. the total of 5 times the cube of n and the square of n
20. the ratio of 9 more than m to m
21. r decreased by the quotient of r and 3
22. four fifths of the sum of w and 10
23. the difference between the square of x and the total of x and 17
24. s increased by the quotient of 4 and s
25. the product of 9 and the total of z and 4
26. n increased by the difference between 10 times n and 9

3.2 Exercises

Translate into a variable expression.

27. twelve minus a number

28. a number divided by eighteen

29. two thirds of a number

30. twenty more than a number

31. the quotient of twice a number and nine

32. ten times the difference between a number and fifty

33. eight less than the product of eleven and a number

34. the sum of five eighths of a number and six

35. nine less than the total of a number and two

36. the product of a number and three more than the number

37. the quotient of seven and the total of five and a number

38. four times the sum of a number and nineteen

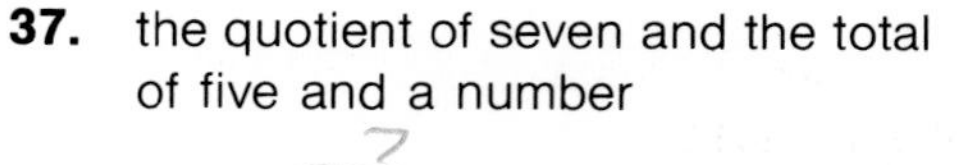

39. five increased by one half of the sum of a number and three

40. the quotient of fifteen and the sum of a number and twelve

41. a number multiplied by the difference between twice the number and four

42. the product of two thirds and the sum of a number and seven

43. the product of five less than a number and the number

44. the difference between forty and the quotient of a number and twenty

45. the quotient of five more than twice a number and the number

46. the sum of the square of a number and twice the number

47. a number decreased by the difference between three times the number and eight

48. the sum of eight more than a number and one third of the number

3.3 Exercises

Translate into a variable expression. Then simplify.

49. a number added to the product of three and the number

50. a number increased by the total of the number and nine

51. five more than the sum of a number and six

52. a number decreased by the difference between eight and the number

53. a number minus the sum of the number and ten

54. the difference between one third of a number and five eighths of the number

55. the sum of one sixth of a number and four ninths of the number

56. two more than the total of a number and five

57. the sum of a number divided by three and the number

58. twice the sum of six times a number and seven

59. fourteen multiplied by one seventh of a number

60. four times the product of six and a number

61. the difference between ten times a number and twice the number

62. the total of twelve times a number and twice the number

63. sixteen more than the difference between a number and six

64. a number plus the product of the number and nineteen

65. a number subtracted from the product of the number and four

66. eight times the sum of the square of a number and three

67. the difference between fifteen times a number and the product of the number and five

68. two thirds of the sum of nine times a number and three

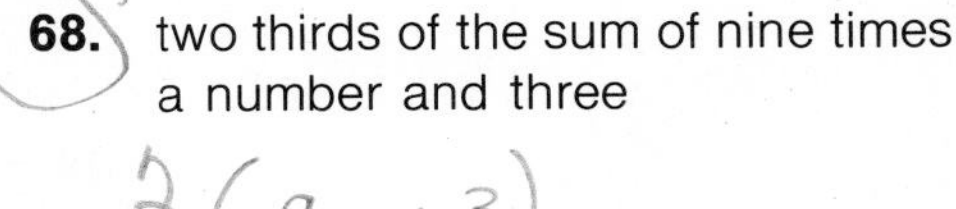

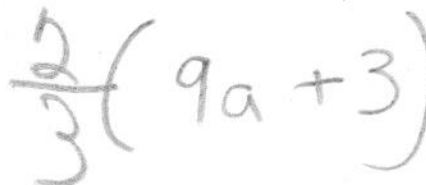

Translate into a variable expression. Then simplify.

69. twelve less than the difference between nine and a number

70. thirteen decreased by the sum of a number and fifteen

71. ten minus the sum of two thirds of a number and four

72. nine more than the quotient of eight times a number and four

73. five times the sum of two consecutive integers

74. seven times the total of two consecutive integers

75. six more than the sum of two consecutive integers

76. five minus the sum of two consecutive integers

77. twice the sum of three consecutive integers

78. one third of the sum of three consecutive integers

79. the sum of three more than the square of a number and twice the square of the number

80. the total of five increased by the cube of a number and twice the cube of the number

81. a number plus seven added to the difference between two and twice the number

82. ten more than a number added to the difference between the number and eleven

83. six increased by a number added to twice the difference between the number and three

84. the sum of a number and ten added to the difference between the number and twelve

85. a number plus the product of the number minus nine and four

86. eighteen minus the product of two less than a number and eight

Calculators and Computers

Evaluating Variable Expressions

Evaluating variable expressions with your calculator will at times require the use of the [+/−] key, the [x^2] key, and the parentheses keys.

The [+/−] key changes the sign of the number currently in the display. The parentheses keys are used as grouping symbols and the [x^2] key is used to square the number in the display. For the examples below, $a = 2$, $b = -3$, and $c = 4$.

Evaluate $a + bc$.

Replace a, b, and c by their values. $2 + (-3)(4)$

Use the Order of Operations Agreement (multiply before adding) and enter the following on your calculator:

3 [+/−] [×] 4 [+] 2 [=]

The answer in the display should be −10.

Evaluate: $ac^2 - b$.

Replace a, b, and c by their values. $(2)(4)^2 - (-3)$

Use the Order of Operations Agreement (exponents, multiply, subtract) and enter the following on your calculator:

4 [x^2] [×] 2 [−] 3 [+/−] [=]

The answer in the display should be 35.

Evaluate $5a^2 - 6bc$.

This example is a little more complicated and requires the use of the parentheses keys. Because of the Order of Operations Agreement, the expression $5a^2 - 6bc$ is evaluated as if there were parentheses around each of the terms: $(5a^2) - (6bc)$. (Do all multiplications before subtraction.)

Replace a, b, and c by their values. $5(2)^2 - 6(-3)(4)$

Use the Order of Operations Agreement and enter the following on your calculator:

[(] 2 [x^2] [×] 5 [)] [−] [(] 6 [×] 3 [+/−] [×] 4 [)] [=]

The answer in the display should be 92.

Practice evaluating variable expressions by trying some of the problems in your text. Remember that a term with more than one factor must have parentheses around it.

Chapter Summary

KEY WORDS

A **variable** is a letter which is used to stand for a quantity which is unknown.

A **variable expression** is an expression which contains one or more variables.

The **terms** of a variable expression are the addends of the expression.

A **variable term** is composed of a numerical coefficient and a variable part.

Like terms of a variable expression are the terms with the same variable part.

The **additive inverse** of a number is the opposite of the number.

The **multiplicative inverse** of a number is the reciprocal of the number.

ESSENTIAL RULES

The Associative Property of Addition — If a, b, and c are real numbers, then $(a + b) + c = a + (b + c)$.

The Commutative Property of Addition — If a and b are real numbers, then $a + b = b + a$.

The Addition Property of Zero — If a is a real number, then $a + 0 = 0 + a = a$.

The Additive Inverse Property — If a is a real number, then $a + (-a) = (-a) + a = 0$.

The Associative Property of Multiplication — If a, b, and c are real numbers, then $(ab)c = a(bc)$.

The Commutative Property of Multiplication — If a and b are real numbers, then $ab = ba$.

The Multiplication Property of One — If a is a real number, then $1 \cdot a = a \cdot 1 = a$.

The Inverse Property of Multiplication — If a is a non-zero real number, then $a\left(\frac{1}{a}\right) = \left(\frac{1}{a}\right)a = 1$.

The Distributive Property — If a, b, and c are real numbers, then $a(b + c) = ab + ac$.

Review/Test

SECTION **1**

1.1a Evaluate $b^2 - 3ab$ when $a = 3$ and $b = -2$.

1.1b Evaluate $\frac{-2ab}{2b - a}$ when $a = -4$ and $b = 6$.

SECTION **2**

2.1a Simplify $3x - 5x + 7x$.

2.1b Simplify $-7y^2 + 6y^2 - (-2y^2)$.

2.1c Simplify $3x - 7y - 12x$.

2.1d Simplify $3x + (-12y) - 5x - (-7y)$.

2.2a Simplify $\frac{1}{5}(10x)$.

2.2b Simplify $(12x)\left(\frac{1}{4}\right)$.

2.2c Simplify $(-3)(-12y)$.

2.2d Simplify $\frac{2}{3}(-15a)$.

2.3a Simplify $5(3 - 7b)$.

2.3b Simplify $-2(2x - 4)$.

Review/Test

2.3c Simplify $-3(2x^2 - 7y^2)$.

2.3d Simplify $-5(2x^2 - 3x + 6)$.

2.4a Simplify $2x - 3(x - 2)$.

2.4b Simplify $5(2x + 4) - 3(x - 6)$.

2.4c Simplify $2x + 3[4 - (3x - 7)]$.

2.4d Simplify $-2[x - 2(x - y)] + 5y$.

SECTION **3**

3.1a Translate "b decreased by the product of b and 7" into a variable expression.

3.1b Translate "10 times the difference between x and 3" into a variable expression.

3.2a Translate "the sum of a number and twice the square of the number" into a variable expression.

3.2b Translate "three less than the quotient of six and a number" into a variable expression.

3.3a Translate and simplify "eight times the sum of two consecutive integers."

3.3b Translate and simplify "eleven added to twice the sum of a number and four."

Cumulative Review/Test

1. Add: $-4 + 7 + (-10)$

2. Subtract: $-16 - (-25) - 4$

3. Multiply: $(-2)(3)(-4)$

4. Divide: $(-60) \div 12$

5. Write $1\frac{1}{4}$ as a decimal.

6. Simplify $\frac{7}{12} - \frac{11}{16} - \left(-\frac{1}{3}\right)$.

7. Simplify $\frac{5}{12} \div \left(2\frac{1}{2}\right)$.

8. Simplify $\left(-\frac{9}{16}\right) \cdot \left(\frac{8}{27}\right) \cdot \left(-\frac{3}{2}\right)$.

9. Simplify $-3^2 \cdot \left(-\frac{2}{3}\right)^3$.

10. Simplify $-2^5 \div (3 - 5)^2 - (-3)$.

11. Simplify $\left(-\frac{3}{4}\right)^2 - \left(\frac{3}{8} - \frac{11}{12}\right)$.

12. Evaluate $a^2 - 3b$ when $a = 2$ and $b = -4$.

13. Simplify $-2x^2 - (-3x^2) + 4x^2$.

14. Simplify $5a - 10b - 12a$.

15. Simplify $\frac{1}{2}(12a)$.

16. Simplify $\left(-\frac{5}{6}\right)(-36b)$.

Cumulative Review/Test

17. Simplify $3(8 - 2x)$.

18. Simplify $-2(-3y + 9)$.

19. Simplify $-4(2x^2 - 3y^2)$.

20. Simplify $-3(3y^2 - 3y - 7)$.

21. Simplify $-3x - 2(2x - 7)$.

22. Simplify $4(3x - 2) - 7(x + 5)$.

23. Simplify
$2x + 3[x - 2(4 - 2x)]$.

24. Simplify
$3[2x - 3(x - 2y)] + 3y$.

25. Translate "the sum of one half of *b* and *b*" into a variable expression.

26. Translate "10 divided by the difference between *y* and 2" into a variable expression.

27. Translate "the difference between eight and the quotient of a number and twelve" into a variable expression.

28. Translate "the product of a number and two more than the number" into a variable expression.

29. Translate and simplify "the product of four and the sum of two consecutive integers."

30. Translate and simplify "twelve more than the product of three plus a number and five."

3

Solving Equations

Objectives

- To determine if a given number is a solution of an equation
- To solve an equation of the form $x + a = b$
- To solve an equation of the form $ax = b$
- To solve an equation of the form $ax + b = c$
- To solve application problems
- To solve an equation of the form $ax + b = cx + d$
- To solve an equation containing parentheses
- To solve application problems
- To translate a sentence into an equation and solve
- To solve application problems

Mersenne Primes

A prime number which can be written in the form $2^n - 1$, where n is also prime, is called a Mersenne prime. The table below shows some Mersenne primes.

$$3 = 2^2 - 1$$
$$7 = 2^3 - 1$$
$$31 = 2^5 - 1$$
$$127 = 2^7 - 1$$

You might notice that not every prime number is a Mersenne prime. For example, 5 is a prime number but not a Mersenne prime. Also, not all prime numbers will yield a prime number. For example, $2^{11} - 1 = 2047$ is not a prime number.

The search for Mersenne primes has been quite extensive, especially since the advent of the computer. One reason for the extensive research into large prime numbers (not only Mersenne primes) has to do with cryptology.

Cryptology is the study of making or breaking secret codes. One method for making a code which is difficult to break is called public key cryptology. For this method to work, it is necessary to use very large prime numbers. To keep anyone from breaking the code, each prime should have at least 200 digits.

Today the largest known Mersenne prime is $2^{216091} - 1$. This number has 65,050 digits in its representation.

Another Mersenne prime got special recognition in a postage-meter stamp. It is the number $2^{11213} - 1$. This number has 3276 digits in its representation.

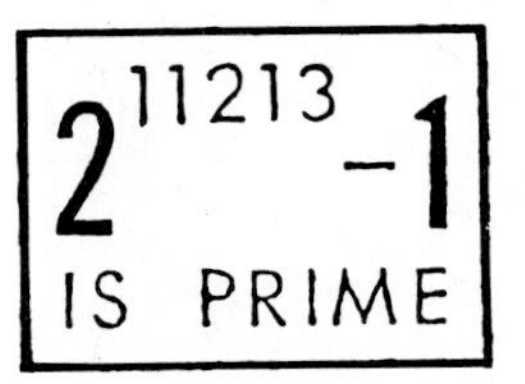

SECTION 1 Introduction to Equations

1.1 Objective To determine if a given number is a solution of an equation

Reference for Computer Tutor™

An **equation** expresses the equality of two mathematical expressions. The expressions can be either numerical or variable expressions.

$$\left.\begin{aligned} 9 + 3 &= 12 \\ 3x - 2 &= 10 \\ y^2 + 4 &= 2y - 1 \\ z &= 2 \end{aligned}\right\} \text{Equations}$$

The equation at the right is true if the variable is replaced by 5.

$x + 8 = 13$

$5 + 8 = 13$ A true equation

The equation is false if the variable is replaced by 7.

$7 + 8 = 13$ A false equation

A **solution** of an equation is a number which, when substituted for the variable, results in a true equation. 5 is a solution of the equation $x + 8 = 13$. 7 is not a solution of the equation $x + 8 = 13$.

Is -3 a solution of the equation $4x + 18 = x^2 - 3$?

Step 1 Replace the variable by the given number, -3.

Step 2 Evaluate the numerical expressions using the Order of Operations Agreement.

$$\begin{array}{r|l} \multicolumn{2}{c}{4x + 18 = x^2 - 3} \\ \hline 4(-3) + 18 & (-3)^2 - 3 \\ -12 + 18 & 9 - 3 \\ \multicolumn{2}{c}{6 = 6} \end{array}$$

Step 3 Compare the results. If the results are equal, the given number is a solution. If the results are not equal, the given number is not a solution.

Yes, -3 is a solution of the equation $4x + 18 = x^2 - 3$.

Example 1 Is $\frac{2}{3}$ a solution of $12x - 2 = 6x + 2$?

Solution

$$\begin{array}{r|l} \multicolumn{2}{c}{12x - 2 = 6x + 2} \\ \hline 12\left(\frac{2}{3}\right) - 2 & 6\left(\frac{2}{3}\right) + 2 \\ 8 - 2 & 4 + 2 \\ \multicolumn{2}{c}{6 = 6} \end{array}$$

Yes, $\frac{2}{3}$ is a solution.

Example 2 Is $\frac{1}{4}$ a solution of $5 - 4x = 8x + 2$?

Your solution

Solution on p. A13

Example 3 Is -4 a solution of $4 + 5x = x^2 - 2x$?

Solution

$$4 + 5x = x^2 - 2x$$

$4 + 5(-4)$	$(-4)^2 - 2(-4)$
$4 + (-20)$	$16 - (-8)$

$-16 \neq 24$ ($\neq$ means is not equal to)

No, -4 is not a solution.

Example 4 Is 5 a solution of $10x - x^2 = 3x - 10$?

Your solution

Solution on p. A13

1.2 Objective To solve an equation of the form $x + a = b$

Reference for Computer Tutor™

To **solve** an equation means to find a solution of the equation. The simplest equation to solve is an equation of the form **variable = constant,** since the constant is the solution.

If $x = 5$, then 5 is the solution of the equation, since $5 = 5$ is a true equation.

The solution of the equation shown at the right is 7.

$$x + 2 = 9 \qquad 7 + 2 = 9$$

Note that if 4 is added to each side of the equation, the solution is still 7.

$$x + 2 + 4 = 9 + 4$$
$$x + 6 = 13 \qquad 7 + 6 = 13$$

If -5 is added to each side of the equation, the solution is still 7.

$$x + 2 + (-5) = 9 + (-5)$$
$$x - 3 = 4 \qquad 7 - 3 = 4$$

This illustrates the Addition Property of Equations.

Addition Property of Equations

The same number can be added to each side of an equation without changing the solution of the equation.

If $a = b$, then $a + c = b + c$.

This property is used in solving equations. Note the effect of adding, to each side of the equation shown above, the opposite of the constant term 2. After simplifying each side of the equation, the equation is in the form *variable = constant*. The solution is the constant.

$$x + 2 = 9$$
$$x + 2 + (-2) = 9 + (-2)$$
$$x + 0 = 7$$
$$x = 7$$

variable = constant

The solution is 7.

In solving an equation, the goal is to rewrite the given equation in the form *variable = constant*. The Addition Property of Equations can be used to rewrite an equation in this form. The Addition Property of Equations is used to **remove a term** from one side of an equation **by adding the opposite of that term** to each side of the equation.

Solve: $y - 6 = 9$

The goal is to rewrite the equation in the form *variable = constant.*

Step 1 Add the opposite of the constant term -6 to each side of the equation. (Addition Property of Equations)

$$y - 6 = 9$$
$$y - 6 + 6 = 9 + 6$$

Step 2 Simplify using the Addition Property of Inverses and the Addition Property of Zero. Now the equation is in the form *variable = constant.*

$$y + 0 = 15$$ Do this step mentally.

$$y = 15$$

Step 3 Write the solution.

The solution is 15.

Check: $y - 6 = 9$
$15 - 6 = 9$
$9 = 9$ A true equation

Example 5 Solve: $x + \frac{1}{3} = -\frac{3}{4}$

Solution

$$x + \frac{1}{3} = -\frac{3}{4}$$
$$x + \frac{1}{3} + \left(-\frac{1}{3}\right) = -\frac{3}{4} + \left(-\frac{1}{3}\right)$$
$$x = -\frac{13}{12}$$

The solution is $-\frac{13}{12}$.

Example 6 Solve: $\frac{1}{2} = x - \frac{2}{3}$

Your solution

1.3 Objective To solve an equation of the form $ax = b$

Reference for Computer Tutor™

The solution of the equation shown at the right is 3. Note that if each side of the equation is multiplied by 5, the solution is still 3.

$2x = 6$ $2 \cdot 3 = 6$
$5 \cdot 2x = 5 \cdot 6$
$10x = 30$ $10 \cdot 3 = 30$

If each side of the equation is multiplied by -4, the solution is still 3.

$(-4) \cdot 2x = (-4) \cdot 6$
$-8x = -24$ $-8 \cdot 3 = -24$

This illustrates the Multiplication Property of Equations.

Multiplication Property of Equations

Each side of an equation can be multiplied by the same non-zero number without changing the solution of the equation.

If $a = b$ and $c \neq 0$, then $ac = bc$.

This property is used in solving equations. Note the effect of multiplying each side of the equation shown above by the reciprocal of the coefficient 2. After simplifying each side of the equation, the equation is in the form *variable* = *constant*. The solution is the constant.

$$2x = 6$$
$$\frac{1}{2} \cdot 2x = \frac{1}{2} \cdot 6$$
$$1x = 3$$
$$x = 3$$

variable = constant.

The solution is 3.

In solving an equation, the goal is to rewrite the given equation in the form *variable* = *constant*. The Multiplication Property of Equations can be used to rewrite an equation in this form. The Multiplication Property of Equations is used to **remove a coefficient** from one side of an equation **by multiplying each side of the equation by the reciprocal of the coefficient.**

Solve: $4x = 6$

The goal is to rewrite the equation in the form *variable* = *constant*.

Step 1 Multiply each side of the equation by the reciprocal of the coefficient 4. (Multiplication Property of Equations)

$$4x = 6$$
$$\frac{1}{4} \cdot 4x = \frac{1}{4} \cdot 6$$

Step 2 Simplify using the Multiplication Property of Reciprocals and the Multiplication Property of One. Now the equation is in the form *variable* = *constant*.

$$1x = \frac{6}{4}$$ Do this step mentally.

$$x = \frac{6}{4} = \frac{3}{2}$$

Step 3 Write the solution.

The solution is $\frac{3}{2}$.

Check:
$$4x = 6$$
$$4\left(\frac{3}{2}\right) = 6$$
$$6 = 6$$ A true equation

Example 7 Solve: $\frac{3x}{4} = -9$

Solution

$$\frac{3x}{4} = -9$$
$$\frac{4}{3} \cdot \frac{3}{4}x = \frac{4}{3}(-9) \qquad \left[\frac{3x}{4} = \frac{3}{4}x\right]$$
$$x = -12$$

The solution is -12.

Example 8 Solve: $-\frac{2x}{5} = 6$

Your solution

Example 9 Solve: $5x - 9x = 12$

Solution

$$5x - 9x = 12$$ Combine like terms.
$$-4x = 12$$
$$\left(-\frac{1}{4}\right)(-4x) = \left(-\frac{1}{4}\right)(12)$$
$$x = -3$$

The solution is -3.

Example 10 Solve: $4x - 8x = 16$

Your solution

Solutions on p. A13

1.1 Exercises

1. Is 4 a solution of $2x = 8$?

2. Is 3 a solution of $y + 4 = 7$?

3. Is -1 a solution of $2b - 1 = 3$?

4. Is -2 a solution of $3a - 4 = 10$?

5. Is 1 a solution of $4 - 2m = 3$?

6. Is 2 a solution of $7 - 3n = 2$?

7. Is 5 a solution of $2x + 5 = 3x$?

8. Is 4 a solution of $3y - 4 = 2y$?

9. Is 0 a solution of $4a + 5 = 3a + 5$?

10. Is 0 a solution of $4 - 3b = 4 - 5b$?

11. Is -2 a solution of $4 - 2n = n + 10$?

12. Is -3 a solution of $5 - m = 2 - 2m$?

13. Is 3 a solution of $z^2 + 1 = 4 + 3z$?

14. Is 2 a solution of $2x^2 - 1 = 4x - 1$?

15. Is -1 a solution of $y^2 - 1 = 4y + 3$?

16. Is -2 a solution of $m^2 - 4 = m + 3$?

17. Is 5 a solution of $x^2 + 2x + 1 = (x + 1)^2$?

18. Is -6 a solution of $(n - 2)^2 = n^2 - 4n + 4$?

19. Is 4 a solution of $x(x + 1) = x^2 + 5$?

20. Is 3 a solution of $2a(a - 1) = 3a + 3$?

21. Is $-\frac{1}{4}$ a solution of $8t + 1 = -1$?

22. Is $\frac{1}{2}$ a solution of $4y + 1 = 3$?

23. Is $\frac{2}{5}$ a solution of $5m + 1 = 10m - 3$?

24. Is $\frac{3}{4}$ a solution of $8x - 1 = 12x + 3$?

25. Is $\frac{1}{3}$ a solution of $2n + 2 = 5n - 1$?

26. Is $\frac{1}{4}$ a solution of $5x + 4 = x + 5$?

27. Is $-\frac{1}{3}$ a solution of $3b(b - 1) = 2b + 2$?

28. Is 2.1 a solution of $x^2 - 4x = x + 1.89$?

29. Is 1.56 a solution of $c^2 - 3c = 4c - 8.4864$?

30. Is -1.8 a solution of $a(a + 1) = 2.6 - 2a$?

1.2 Exercises

Solve and check.

31. $x + 5 = 7$

32. $y + 3 = 9$

33. $b - 4 = 11$

34. $z - 6 = 10$

35. $2 + a = 8$

36. $5 + x = 12$

37. $m + 9 = 3$

38. $t + 12 = 10$

39. $n - 5 = -2$

40. $x - 6 = -5$

41. $b + 7 = 7$

42. $y - 5 = -5$

43. $a - 3 = -5$

44. $x - 6 = -3$

45. $z + 9 = 2$

46. $n + 11 = 1$

47. $10 + m = 3$

48. $8 + x = 5$

49. $9 + x = -3$

50. $10 + y = -4$

51. $b - 5 = -3$

52. $t - 6 = -4$

53. $4 + x = 10$

54. $9 + a = 20$

55. $2 = x + 7$

56. $-8 = n + 1$

57. $4 = m - 11$

58. $-6 = y - 5$

59. $12 = 3 + w$

60. $-9 = 5 + x$

61. $4 = -10 + b$

62. $-7 = -2 + x$

63. $13 = -6 + a$

64. $m + \frac{2}{3} = -\frac{1}{3}$

65. $c + \frac{3}{4} = -\frac{1}{4}$

66. $x - \frac{1}{2} = \frac{1}{2}$

Solve and check.

67. $x - \frac{2}{5} = \frac{3}{5}$ **68.** $\frac{5}{8} + y = \frac{1}{8}$ **69.** $\frac{4}{9} + a = -\frac{2}{9}$

70. $m + \frac{1}{2} = -\frac{1}{4}$ **71.** $b + \frac{1}{6} = -\frac{1}{3}$ **72.** $x + \frac{2}{3} = \frac{3}{4}$

73. $n + \frac{2}{5} = \frac{2}{3}$ **74.** $-\frac{5}{6} = x - \frac{1}{4}$ **75.** $-\frac{1}{4} = c - \frac{2}{3}$

76. $d + 1.3619 = 2.0148$ **77.** $w + 2.932 = 4.801$

78. $-0.813 + x = -1.096$ **79.** $-1.926 + t = -1.042$

80. $6.149 = -3.108 + z$ **81.** $5.237 = -2.014 + x$

1.3 Exercises

Solve and check.

82. $5x = 15$ **83.** $4y = 28$ **84.** $3b = -12$

85. $2a = -14$ **86.** $-3x = 6$ **87.** $-5m = 20$

88. $-3x = -27$ **89.** $-6n = -30$ **90.** $20 = 4c$

91. $18 = 2t$ **92.** $-32 = 8w$ **93.** $-56 = 7x$

94. $8d = 0$ **95.** $-5x = 0$ **96.** $36 = 9z$

Solve and check.

97. $35 = -5x$

98. $-64 = 8a$

99. $-32 = -4y$

100. $-54 = 6c$

101. $49 = -7t$

102. $\frac{x}{3} = 2$

103. $\frac{x}{4} = 3$

104. $-\frac{y}{2} = 5$

105. $-\frac{b}{3} = 6$

106. $\frac{3}{4}y = 9$

107. $\frac{2}{5}x = 6$

108. $-\frac{2}{3}d = 8$

109. $-\frac{3}{5}m = 12$

110. $\frac{2n}{3} = 2$

111. $\frac{5x}{6} = -10$

112. $\frac{-3z}{8} = 9$

113. $\frac{-4x}{5} = -12$

114. $-6 = -\frac{2}{3}y$

115. $-15 = -\frac{3}{5}x$

116. $\frac{2}{5}a = 3$

117. $\frac{3x}{4} = 2$

118. $\frac{3}{4}c = \frac{3}{5}$

119. $\frac{2}{9} = \frac{2}{3}y$

120. $-\frac{6}{7} = -\frac{3}{4}b$

121. $-\frac{2}{5}m = -\frac{6}{7}$

122. $5x + 2x = 14$

123. $3n + 2n = 20$

124. $7d - 4d = 9$

125. $10y - 3y = 21$

126. $2x - 5x = 9$

127. $\frac{x}{1.46} = 3.25$

128. $\frac{z}{2.95} = -7.88$

129. $3.47a = 7.1482$

130. $2.31m = 2.4255$

131. $-3.7x = 7.881$

132. $\frac{n}{2.65} = 9.08$

SECTION 2 General Equations—Part I

2.1 Objective To solve an equation of the form $ax + b = c$

Reference for Computer Tutor™

In solving an equation of the form $ax + b = c$, the goal is to rewrite the equation in the form *variable = constant*. This requires the application of both the Addition and Multiplication Properties of Equations.

Solve: $\frac{2}{5}x - 3 = -7$

The goal is to rewrite the equation in the form *variable = constant*.

Step 1 Add the opposite of the constant term -3 to each side of the equation. Then simplify.

$$\frac{2}{5}x - 3 = -7$$
$$\frac{2}{5}x - 3 + 3 = -7 + 3$$
$$\frac{2}{5}x = -4$$

Step 2 Multiply each side of the equation by the reciprocal of the coefficient $\frac{2}{5}$. Simplify. Now the equation is in the form *variable = constant*.

$$\frac{5}{2} \cdot \frac{2}{5}x = \frac{5}{2}(-4)$$
$$x = -10$$

Step 3 Write the solution.

The solution is -10.

-10 checks as the solution.

Example 1 Solve: $3x - 7 = -5$

Solution

$$3x - 7 = -5$$
$$3x - 7 + 7 = -5 + 7$$
$$3x = 2$$
$$\frac{1}{3} \cdot 3x = \frac{1}{3} \cdot 2$$
$$x = \frac{2}{3}$$

The solution is $\frac{2}{3}$.

Example 2 Solve: $5x + 7 = 10$

Your solution

Example 3 Solve: $5 = 9 - 2x$

Solution

$$5 = 9 - 2x$$
$$5 + (-9) = 9 + (-9) - 2x$$
$$-4 = -2x$$
$$\left(-\frac{1}{2}\right)(-4) = \left(-\frac{1}{2}\right)(-2x)$$
$$2 = x$$

The solution is 2.

Example 4 Solve: $2 = 11 + 3x$

Your solution

Solutions on p. A14

12
5
60

Example 5 Solve: $2x + 4 - 5x = 10$

Solution

$$2x + 4 - 5x = 10$$
$$-3x + 4 = 10$$

Combine like terms.

$$-3x + 4 + (-4) = 10 + (-4)$$
$$-3x = 6$$
$$-\frac{1}{3}(-3x) = -\frac{1}{3}(6)$$
$$x = -2$$

The solution is -2.

Example 6 Solve: $x - 5 + 4x = 25$

Your solution

Solution on p. A14

2.2 Objective To solve application problems

Use COMPUTER TUTOR™ DISK 3

Example 7 To determine the total cost of production, an economist uses the equation $T = U \cdot N + F$, where T is the total cost, U is the unit cost, N is the number of units made, and F is the fixed cost. Use this equation to find the number of units made during a month when the total cost was \$9000, the unit cost was \$25, and the fixed costs were \$3000.

Strategy To find the number of units made, replace each of the variables by their given value and solve for N.

Solution

$$T = U \cdot N + F$$
$$9000 = 25N + 3000$$
$$9000 + (-3000) = 25N + 3000 + (-3000)$$
$$6000 = 25N$$
$$\frac{1}{25} \cdot 6000 = \frac{1}{25} \cdot 25N$$
$$240 = N$$

The number of units made was 240.

Example 8 The pressure at a certain depth in the ocean can be approximated by the equation $P = 15 + \frac{1}{2}D$, where P is the pressure in pounds per square inch and D is the depth in feet. Use this equation to find the depth when the pressure is 45 pounds per square inch.

Your strategy

Your solution

Solution on p. A14

2.1 Exercises

Solve and check.

1. $3x + 1 = 10$ **2.** $4y + 3 = 11$ **3.** $2a - 5 = 7$

4. $5m - 6 = 9$ **5.** $5 = 4x + 9$ **6.** $2 = 5b + 12$

7. $2x - 5 = -11$ **8.** $3n - 7 = -19$ **9.** $10 = 4 + 3d$

10. $13 = 9 + 4z$ **11.** $7 - c = 9$ **12.** $2 - x = 11$

13. $4 - 3w = -2$ **14.** $5 - 6x = -13$ **15.** $8 - 3t = 2$

16. $12 - 5x = 7$ **17.** $4a - 20 = 0$ **18.** $3y - 9 = 0$

19. $6 + 2b = 0$ **20.** $10 + 5m = 0$ **21.** $-2x + 5 = -7$

22. $-5d + 3 = -12$ **23.** $-8x - 3 = -19$ **24.** $-7n - 4 = -25$

25. $-12x + 30 = -6$ **26.** $-13 = -11y + 9$ **27.** $2 = 7 - 5a$

28. $3 = 11 - 4n$ **29.** $-35 = -6b + 1$ **30.** $-8x + 3 = -29$

31. $-3m - 21 = 0$ **32.** $-5x - 30 = 0$ **33.** $-4y + 15 = 15$

34. $-3x + 19 = 19$ **35.** $0 = 14 - 7c$ **36.** $0 = 24 - 6z$

Solve and check.

37. $7x - 3 = 3$ **38.** $8y + 3 = 7$ **39.** $6a + 5 = 9$

40. $3m + 4 = 11$ **41.** $14 = 9x + 1$ **42.** $4 = 5b + 6$

43. $11 = 15 + 4n$ **44.** $4 = 2 - 3c$ **45.** $9 - 4x = 6$

46. $3t - 2 = 0$ **47.** $9x - 4 = 0$ **48.** $7 - 8z = 0$

49. $1 - 3x = 0$ **50.** $9d + 10 = 7$ **51.** $12w + 11 = 5$

52. $6y - 5 = -7$ **53.** $8b - 3 = -9$ **54.** $7 + 12x = 3$

55. $9 + 14x = 7$ **56.** $5 - 6m = 2$ **57.** $7 - 9a = 4$

58. $9 = -12c + 5$ **59.** $10 = -18x + 7$ **60.** $2y + \frac{1}{3} = \frac{7}{3}$

61. $4a + \frac{3}{4} = \frac{19}{4}$ **62.** $2n - \frac{3}{4} = \frac{13}{4}$ **63.** $3x - \frac{5}{6} = \frac{13}{6}$

64. $5y + \frac{3}{7} = \frac{3}{7}$ **65.** $9x + \frac{4}{5} = \frac{4}{5}$ **66.** $8 = 7d - 1$

67. $8 = 10x - 5$ **68.** $4 = 7 - 2w$ **69.** $7 = 9 - 5a$

70. $8t + 13 = 3$ **71.** $12x + 19 = 3$ **72.** $-6y + 5 = 13$

Solve and check.

73. $-4x + 3 = 9$ **74.** $\frac{1}{2}a - 3 = 1$ **75.** $\frac{1}{3}m - 1 = 5$

76. $\frac{2}{5}y + 4 = 6$ **77.** $\frac{3}{4}n + 7 = 13$ **78.** $-\frac{2}{3}x + 1 = 7$

79. $-\frac{3}{8}b + 4 = 10$ **80.** $\frac{x}{4} - 6 = 1$ **81.** $\frac{y}{5} - 2 = 3$

82. $\frac{2x}{3} - 1 = 5$ **83.** $\frac{3c}{7} - 1 = 8$ **84.** $4 - \frac{3}{4}z = -2$

85. $3 - \frac{4}{5}w = -9$ **86.** $5 + \frac{2}{3}y = 3$ **87.** $17 + \frac{5}{8}x = 7$

88. $17 = 7 - \frac{5}{6}t$ **89.** $9 = 3 - \frac{2x}{7}$ **90.** $3 = \frac{3a}{4} + 1$

91. $7 = \frac{2x}{5} + 4$ **92.** $5 - \frac{4c}{7} = 8$ **93.** $7 - \frac{5}{9}y = 9$

94. $6a + 3 + 2a = 11$ **95.** $5y + 9 + 2y = 23$ **96.** $7x - 4 - 2x = 6$

97. $11z - 3 - 7z = 9$ **98.** $2x - 6x + 1 = 9$ **99.** $b - 8b + 1 = -6$

100. $3 = 7x + 9 - 4x$ **101.** $-1 = 5m + 7 - m$ **102.** $8 = 4n - 6 + 3n$

103. $0.135y + 0.0257 = -0.0742$ **104.** $1.42x - 3.449 = 1.308$

105. $3.58 = 3.5686 + 0.076x$ **106.** $-6.54 = 1.48y - 3.062$

2.2 Application Problems

Solve.

Use the equation $v = v_0 + 32t$, where v is the final velocity of a falling object, v_0 is the initial velocity, and t is the time it takes for the object to fall.

1. Find the time it takes for the velocity of a falling object to increase from 12 ft/s to 76 ft/s.

2. Find the time it takes for the velocity of an object to increase from rest to 96 ft/s. (Hint: initial velocity is zero.)

Use the equation $P = 2l + 2w$, where P is the perimeter of a rectangle, l is the length of the rectangle, and w is the width of the rectangle.

3. Find the length of a rectangle when the perimeter is 80 ft and the width is 10 ft.

4. Find the width of a rectangle when the perimeter is 100 ft and the length is 35 ft.

Use the equation $A = \frac{1}{2} \cdot h(b_1 + b_2)$, where A is the area of a trapezoid, h is the height of the trapezoid, b_1 is the length of one of the bases, and b_2 is the length of the other base.

5. Find the height of a trapezoid when the area is 35 in.^{2}, b_1 is 4 in., and b_2 is 10 in.

6. Find the height of a trapezoid when b_1 is 5 m, b_2 is 9 m, and the area is 21 m^2.

Use the equation $M = \frac{Q}{f} + 1$, where M is the magnification, Q is the distance of an object from the lens, and f is the focal length of the lens.

7. Find the magnification when the distance of the object from the lens is 25 cm and the focal length of the lens is 2.5 cm.

8. Find the distance an object is from the lens when the magnification is 10 and the focal length of the lens is 2 cm.

Use the equation $v^2 = v_0^2 + 64s$, where v is the final velocity of a falling object, v_0 is the initial velocity, and s is the height from which the object falls.

9. Find the height from which an object falls when its initial velocity is 24 ft/s and its final velocity is 32 ft/s.

10. Find the height from which an object falls when its final velocity is 48 ft/s and the object falls from rest.

SECTION 3 General Equations—Part II

3.1 Objective To solve an equation of the form $ax + b = cx + d$

Reference for Computer Tutor™

In solving an equation of the form $ax + b = cx + d$, the goal is to rewrite the equation in the form *variable = constant*. Begin by rewriting the equation so that there is only one variable term in the equation. Then rewrite the equation so that there is only one constant term.

Solve: $4x - 5 = 6x + 11$

The goal is to rewrite the equation in the form *variable = constant*.

Step 1 Add the opposite of the variable term $6x$ to each side of the equation. Simplify. Now there is only one variable term in the equation.

$$4x - 5 = 6x + 11$$
$$4x + (-6x) - 5 = 6x + (-6x) + 11$$
$$-2x - 5 = 11$$

Step 2 Add the opposite of the constant term -5 to each side of the equation. Simplify. Now there is only one constant term in the equation.

$$-2x - 5 + 5 = 11 + 5$$
$$-2x = 16$$

Step 3 Multiply each side of the equation by the reciprocal of the coefficient -2. Simplify. Now the equation is in the form *variable = constant*.

$$\left(-\frac{1}{2}\right)(-2x) = \left(-\frac{1}{2}\right)(16)$$
$$x = -8$$

Step 4 Write the solution.

The solution is -8.

-8 checks as the solution.

Example 1 Solve: $4x - 5 = 8x - 7$

Solution

$$4x - 5 = 8x - 7$$
$$4x + (-8x) - 5 = 8x + (-8x) - 7$$
$$-4x - 5 = -7$$
$$-4x - 5 + 5 = -7 + 5$$
$$-4x = -2$$
$$\left(-\frac{1}{4}\right)(-4x) = \left(-\frac{1}{4}\right)(-2)$$
$$x = \frac{1}{2}$$

The solution is $\frac{1}{2}$.

Example 2 Solve: $5x + 4 = 6 + 10x$

Your solution

Solution on p. A15

Example 3 Solve: $3x + 4 - 5x = 2 - 4x$

Solution

$$3x + 4 - 5x = 2 - 4x$$
$$-2x + 4 = 2 - 4x$$
$$-2x + 4x + 4 = 2 - 4x + 4x$$
$$2x + 4 = 2$$
$$2x + 4 + (-4) = 2 + (-4)$$
$$2x = -2$$
$$\frac{1}{2} \cdot 2x = \frac{1}{2}(-2)$$
$$x = -1$$

The solution is -1.

Example 4 Solve: $5x - 10 - 3x = 6 - 4x$

Your solution

Solution on p. A15

3.2 Objective To solve an equation containing parentheses

Reference for Computer Tutor™

When an equation contains parentheses, one of the steps in solving the equation requires the use of the Distributive Property.

The Distributive Property is used to remove parentheses from a variable expression. $a(b + c) = ab + ac$

Solve: $4 + 5(2x - 3) = 3(4x - 1)$

The goal is to rewrite the equation in the form *variable* = *constant*.

Step 1 Use the Distributive Property to remove parentheses. Simplify.

$$4 + 5(2x - 3) = 3(4x - 1)$$
$$4 + 10x - 15 = 12x - 3$$
$$10x - 11 = 12x - 3$$

Step 2 Add the opposite of the variable term $12x$ to each side of the equation. Simplify. Now there is only one variable term in the equation.

$$10x + (-12x) - 11 = 12x + (-12x) - 3$$
$$-2x - 11 = -3$$

Step 3 Add the opposite of the constant term -11 to each side of the equation. Simplify. Now there is only one constant term in the equation.

$$-2x - 11 + 11 = -3 + 11$$
$$-2x = 8$$

Step 4 Multiply each side of the equation by the reciprocal of the coefficient -2. Simplify. Now the equation is in the form *variable* = *constant*.

$$\left(-\frac{1}{2}\right)(-2x) = \left(-\frac{1}{2}\right)(8)$$
$$x = -4$$

Step 5 Write the solution.

The solution is -4.

-4 checks as the solution.

Example 5 Solve:
$3x - 4(2 - x) = 3(x - 2) - 4$

Solution

$$\begin{aligned} 3x - 4(2 - x) &= 3(x - 2) - 4 \\ 3x - 8 + 4x &= 3x - 6 - 4 \\ 7x - 8 &= 3x - 10 \\ 7x + (-3x) - 8 &= 3x + (-3x) - 10 \\ 4x - 8 &= -10 \\ 4x - 8 + 8 &= -10 + 8 \\ 4x &= -2 \\ \frac{1}{4} \cdot 4x &= \frac{1}{4} \cdot (-2) \\ x &= -\frac{1}{2} \end{aligned}$$

The solution is $-\frac{1}{2}$.

Example 6 Solve:
$5x - 4(3 - 2x) = 2(3x - 2) + 6$

Your solution

Example 7 Solve:
$3[2 - 4(2x - 1)] = 4x - 10$

Solution

$$\begin{aligned} 3[2 - 4(2x - 1)] &= 4x - 10 \\ 3[2 - 8x + 4] &= 4x - 10 \\ 3[6 - 8x] &= 4x - 10 \\ 18 - 24x &= 4x - 10 \\ 18 - 24x + (-4x) &= 4x + (-4x) - 10 \\ 18 - 28x &= -10 \\ 18 + (-18) - 28x &= -10 + (-18) \\ -28x &= -28 \\ -\frac{1}{28} \cdot (-28x) &= -\frac{1}{28} \cdot (-28) \\ x &= 1 \end{aligned}$$

The solution is 1.

Example 8 Solve:
$-2[3x - 5(2x - 3)] = 3x - 8$

Your solution

Solutions on p. A15

3.3 Objective To solve application problems

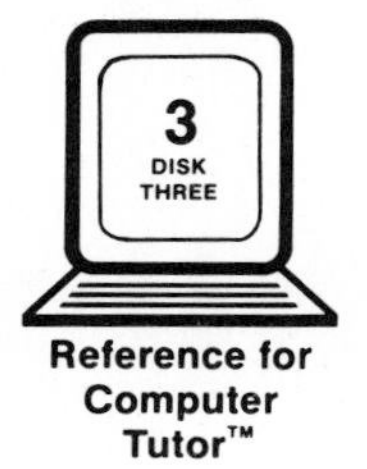

Reference for Computer Tutor™

A lever system is shown at the right. It consists of a lever, or bar; a fulcrum; and two forces, F_1 and F_2. The distance d represents the length of the lever, x represents the distance from F_1 to the fulcrum, and $d - x$ represents the distance from F_2 to the fulcrum.

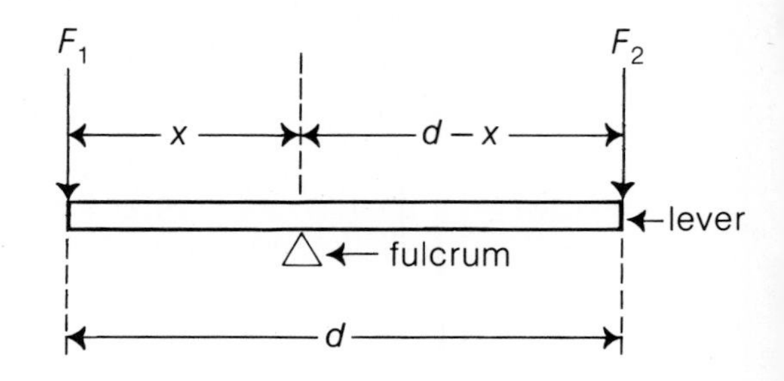

A principle of physics states that when the lever system balances,
$F_1 \cdot x = F_2 \cdot (d - x)$.

Example 9 A lever is 15 ft long. A force of 50 lb is applied to one end of the lever and a force of 100 lb is applied to the other end. Where is the fulcrum located when the system balances?

Strategy To find the location of the fulcrum when the system balances, replace the variables F_1, F_2, and d in the lever system equation by the given values and solve for x.

Solution

$$
\begin{aligned}
F_1 \cdot x &= F_2 \cdot (d - x) \\
50x &= 100(15 - x) \\
50x &= 1500 - 100x \\
50x + 100x &= 1500 - 100x + 100x \\
150x &= 1500 \\
\frac{1}{150} \cdot 150x &= \frac{1}{150} \cdot 1500 \\
x &= 10
\end{aligned}
$$

The fulcrum is 10 ft from the 50-pound force.

Example 10 A lever is 25 ft long. A force of 45 lb is applied to one end of the lever and a force of 80 lb is applied to the other end. What is the location of the fulcrum when the system balances?

Your strategy

Your solution

Solution on p. A15

3.1 Exercises

Solve and check.

1. $8x + 5 = 4x + 13$
2. $6y + 2 = y + 17$
3. $7m + 4 = 6m + 7$
4. $11n + 3 = 10n + 11$
5. $5x - 4 = 2x + 5$
6. $9a - 10 = 3a + 2$
7. $12y - 4 = 9y - 7$
8. $13b - 1 = 4b - 19$
9. $15x - 2 = 4x - 13$
10. $7a - 5 = 2a - 20$
11. $3x + 1 = 11 - 2x$
12. $n - 2 = 6 - 3n$
13. $2x - 3 = -11 - 2x$
14. $4y - 2 = -16 - 3y$
15. $2b + 3 = 5b + 12$
16. $m + 4 = 3m + 8$
17. $4x - 7 = 5x + 1$
18. $6d - 2 = 7d + 5$
19. $4y - 8 = y - 8$
20. $5a + 7 = 2a + 7$
21. $6 - 5x = 8 - 3x$
22. $10 - 4n = 16 - n$
23. $5 + 7x = 11 + 9x$
24. $3 - 2y = 15 + 4y$
25. $2x - 4 = 6x$
26. $2b - 10 = 7b$
27. $8m = 3m + 20$
28. $9y = 5y + 16$
29. $-3x - 4 = 2x + 6$
30. $-5a - 3 = 2a + 18$
31. $-8n + 3 = -5n - 6$
32. $-10x + 4 = -x - 14$
33. $-x - 4 = -3x - 16$

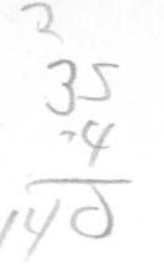

Solve and check.

34. $8 - 4x = 18 - 5x$ **35.** $6 - 10a = 8 - 9a$ **36.** $5 - 7m = 2 - 6m$

37. $8b + 5 = 5b + 7$ **38.** $6y - 1 = 2y + 2$ **39.** $7x - 8 = x - 3$

40. $10x - 3 = 3x - 1$ **41.** $5n + 3 = 2n + 1$ **42.** $8a - 2 = 4a - 5$

43. $8.7y = 3.9y + 16.8$ **44.** $4.5x + 6.03 = 2.7x$ **45.** $5.6x = 7.2x - 3.92$

3.2 Exercises

Solve and check.

46. $5x + 2(x + 1) = 23$ **47.** $6y + 2(2y + 3) = 16$

48. $9n - 3(2n - 1) = 15$ **49.** $12x - 2(4x - 6) = 28$

50. $7a - (3a - 4) = 12$ **51.** $9m - 4(2m - 3) = 11$

52. $2(3x + 1) - 4 = 16$ **53.** $4(2b + 3) - 9 = 11$

54. $5(3 - 2y) + 4y = 3$ **55.** $4(1 - 3x) + 7x = 9$

Solve and check.

56. $10x + 1 = 2(3x + 5) - 1$

57. $5y - 3 = 7 + 4(y - 2)$

58. $4 - 3a = 7 - 2(2a + 5)$

59. $9 - 5x = 12 - (6x + 7)$

60. $2x - 5 = 3(4x + 5)$

61. $3n - 7 = 5(2n + 7)$

62. $5b + 2(3b + 1) = 3b + 5$

63. $x + 3(4x - 2) = 7x - 1$

64. $3y - 7 = 5(2y - 3) + 4$

65. $2a - 5 = 4(3a + 1) - 2$

66. $5 - (9 - 6x) = 2x - 2$

67. $7 - (5 - 8x) = 4x + 3$

68. $3[2 - 4(y - 1)] = 3(2y + 8)$

69. $5[2 - (2x - 4)] = 2(5 - 3x)$

70. $3a + 2[2 + 3(a - 1)] = 2(3a + 4)$

71. $5 + 3[1 + 2(2x - 3)] = 6(x + 5)$

72. $-2[4 - (3b + 2)] = 5 - 2(3b + 6)$

73. $-4[x - 2(2x - 3)] + 1 = 2x - 3$

74. $0.36x - 2(1.63x) - 8.03 = 3(1.96x - 4.14)$

75. $0.593 - 0.14(2.1y - 3.15) = 0.26(2.9y - 6.1)$

3.3 Application Problems

Solve.

Use the lever system equation $F_1 \cdot x = F_2 \cdot (d - x)$.

1. A lever is 25 ft long. A force of 26 lb is applied to one end of the lever and a force of 24 lb is applied to the other end. Find the location of the fulcrum when the system balances.

2. A lever is 14 ft long. A force of 40 lb is applied to one end of the lever and a force of 30 lb is applied to the other end. Find the location of the fulcrum when the system balances.

3. A lever is 10 ft long. At a distance of 8 ft from the fulcrum, a force of 100 lb is applied. How large a force must be applied to the other end of the lever so that the system will balance?

4. A lever is 8 ft long. At a distance of 7 ft from the fulcrum, a force of 80 lb is applied. How large a force must be applied to the other end of the lever so that the system will balance?

To determine the break-even point, or the number of units which must be sold so that no profit or loss occurs, an economist uses the equation $Px = Cx + F$, where P is the selling price per unit, x is the number of units sold, C is the cost to make each unit, and F is the fixed cost.

5. An economist has determined that the selling price per unit for a power saw is \$50. The cost to make each unit is \$32 and the fixed cost is \$3600. Find the break-even point.

6. A business analyst has determined that the selling price per unit for a sweater is \$25. The cost to make each unit is \$10 and the fixed cost is \$1500. Find the break-even point.

7. An economist has determined that the selling price for a single popcorn popper is \$30. The cost to make each popper is \$8. The fixed cost is \$880. Find the break-even point.

8. A financial manager has determined that the selling price for a single screen door is \$48. The cost to make one screen door is \$16. The fixed cost is \$1600. Find the break-even point.

9. A business analyst has determined that the cost to make a cassette tape is \$6 and that the fixed cost is \$1000. The selling price per unit is \$10. How many units must be sold to break even?

10. A marketing analyst has determined that the cost to make a small table is \$7 and that the fixed cost is \$1500. The selling price per unit is \$19. How many units must be sold to break even?

SECTION 4 Translating Sentences into Equations

4.1 Objective To translate a sentence into an equation and solve

Reference for Computer Tutor™

An equation states that two mathematical expressions are equal. Therefore, to translate a sentence into an equation requires recognition of the words or phrases which mean "equals." Some of these phrases are listed below.

equals
is
is equal to
amounts to
represents

} translate to =

Once the sentence is translated into an equation, the equation can be solved by rewriting the equation in the form *variable = constant*.

Translate "five less than a number is thirteen" into an equation and solve.

Step 1 Assign a variable to the unknown quantity. — The unknown number: n

Step 2 Find two verbal expressions for the same value. — | Five less than a number | is | thirteen |

Step 3 Write a mathematical expression for each verbal expression. Write the equals sign. — $n - 5 \quad = \quad 13$

Step 4 Solve the equation.

$$n - 5 + 5 = 13 + 5$$
$$n = 18$$

The number is 18.

Example 1 Translate "three more than twice a number is the number plus six" into an equation and solve.

Solution The unknown number: n

Three more than twice a number	is	the number plus six

$$2n + 3 = n + 6$$
$$2n + (-n) + 3 = n + (-n) + 6$$
$$n + 3 = 6$$
$$n + 3 + (-3) = 6 + (-3)$$
$$n = 3$$

The number is 3.

Example 2 Translate "four less than one third of a number equals five minus two thirds of the number" into an equation and solve.

Your solution

Solution on p. A16

Example 3 Translate "four times the sum of a number and three is six less than twice the number" into an equation and solve.

Solution The unknown number: n

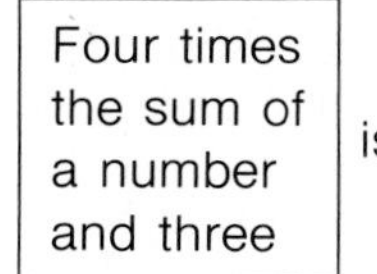

is

six less than twice the number

$$4(n + 3) = 2n - 6$$
$$4n + 12 = 2n - 6$$
$$4n + (-2n) + 12 = 2n + (-2n) - 6$$
$$2n + 12 = -6$$
$$2n + 12 + (-12) = -6 + (-12)$$
$$2n = -18$$
$$\frac{1}{2} \cdot 2n = \frac{1}{2}(-18)$$
$$n = -9$$

The number is -9.

Example 4 Translate "two times the difference between a number and eight is equal to the sum of six times the number and eight" into an equation and solve.

Your solution

Example 5 The sum of two numbers is sixteen. The difference between four times the smaller number and two is two more than twice the larger number. Find the two numbers.

Solution The smaller number: n
The larger number: $16 - n$

The difference between four times the smaller and two	is	two more than twice the larger

$$4n - 2 = 2(16 - n) + 2$$
$$4n - 2 = 32 - 2n + 2$$
$$4n - 2 = 34 - 2n$$
$$4n + 2n - 2 = 34 - 2n + 2n$$
$$6n - 2 = 34$$
$$6n - 2 + 2 = 34 + 2$$
$$6n = 36$$
$$\frac{1}{6} \cdot 6n = \frac{1}{6} \cdot 36$$
$$n = 6 \qquad 16 - n = 10$$

The smaller number is 6
The larger number is 10.

Example 6 The sum of two numbers is twelve. The total of three times the smaller number and six amounts to seven less than the product of four and the larger number. Find the two numbers.

Your solution

Solutions on p. A16

4.2 Objective To solve application problems

Use COMPUTER TUTOR™ DISK 3

Example 7 The temperature of the sun on the Kelvin scale is 6500°. This is 454° more than the temperature on the Fahrenheit scale. Find the Fahrenheit temperature.

Strategy To find the Fahrenheit temperature, write and solve an equation using F to represent the Fahrenheit temperature.

Solution

6500	is	454° more than Fahrenheit temperature

$$6500 = F + 454$$
$$6500 + (-454) = F + 454 + (-454)$$
$$6046 = F$$

The Fahrenheit temperature of the sun is 6046°.

Example 8 A molecule of octane gas has 8 carbon atoms. This represents twice the number of carbon atoms in a butane gas molecule. Find the number of carbon atoms in a butane molecule.

Your strategy

Your solution

Example 9 The Fahrenheit temperature is 68°. This is 32° more than $\frac{9}{5}$ the Celsius temperature. Find the Celsius temperature.

Strategy To find the Celsius temperature, write and solve an equation using C to represent the Celsius temperature.

Solution

68	is	32 degrees more than $\frac{9}{5}$ the Celsius temperature

$$68 = \frac{9}{5}C + 32$$
$$68 + (-32) = \frac{9}{5}C + 32 + (-32)$$
$$36 = \frac{9}{5}C$$
$$\frac{5}{9} \cdot 36 = \frac{5}{9} \cdot \frac{9}{5}C$$
$$20 = C$$

The Celsius temperature is 20°.

Example 10 The Celsius temperature is 20°. This is equal to $\frac{5}{9}$ of the difference between the Fahrenheit temperature and 32. Find the Fahrenheit temperature.

Your strategy

Your solution

Solutions on p. A17

Example 11 Primary and secondary waves are two types of waves which occur after an earthquake. The speed of a secondary wave is 8 mi/s. This is 12 mi/s less than twice the speed of a primary wave. Find the speed of a primary wave.

Strategy To find the speed, write and solve an equation using s to represent the unknown speed.

Solution

8	is	12 mi/s less than twice the speed of a primary wave

$$8 = 2s - 12$$
$$8 + 12 = 2s - 12 + 12$$
$$20 = 2s$$
$$\frac{1}{2} \cdot 20 = \frac{1}{2} \cdot 2s$$
$$10 = s$$

The speed is 10 mi/s.

Example 12 The length of a wire which produces a C note on a piano is 10 in. This represents 6 in. less than $\frac{1}{2}$ the length of the wire which produces an A note. Find the length of the wire which produces an A note.

Your strategy

Your solution

Example 13 A board 20 ft long is cut into two pieces. Five times the length of the smaller piece is two feet more than twice the length of the longer piece. Find the length of each piece.

Strategy To find the length of each piece, write and solve an equation using x to represent the length of the shorter piece and $20 - x$ to represent the length of the longer piece.

Solution

Five times the smaller piece	is	two feet more than twice the longer

$$5x = 2(20 - x) + 2$$
$$5x = 40 - 2x + 2$$
$$5x = 42 - 2x$$
$$5x + 2x = 42 - 2x + 2x$$
$$7x = 42$$
$$\frac{1}{7} \cdot 7x = \frac{1}{7} \cdot 42$$
$$x = 6 \qquad 20 - x = 14$$

The shorter piece is 6 ft.
The longer piece is 14 ft.

Example 14 A company makes 140 televisions per day. Three times the number of black and white TV's made equals 20 less than the number of color TV's made. Find the number of color TV's made each day.

Your strategy

Your solution

Solutions on p. A17

4.1 Exercises

Translate into an equation and solve.

opposite means if (-3) is (+3) or 5 is -5

1. The sum of a number and twelve is ten. Find the number.

2. The difference between a number and five is two. Find the number.

$a - 5 = 2$

3. Two-thirds of a number is six. Find the number.

4. The quotient of a number and seven is the opposite of three. Find the number.

$\frac{a}{7} = -3$

5. Three times the sum of a number and four is fifteen. Find the number.

6. Seven less than twice a number is equal to eleven. Find the number.

$2x - 7 = 11$

7. The difference between twelve and the product of five and a number equals seven. Find the number.

8. The sum of six times a number and five is the opposite of thirteen. Find the number.

$6x + 5 = -13$

9. The sum of three times a number and the number is twelve. Find the number.

10. The difference between five times a number and twice the number is nine. Find the number.

$5x - 2x = 9$

11. The total of twice a number and the sum of the number and three equals fifteen. Find the number.

12. The difference between twice a number and three times the sum of the number and three is equal to fourteen. Find the number.

$2x - 3(x + 3) = 14$

13. Fifteen is one less than the product of four and a number. Find the number.

14. Twenty is two minus the product of six and a number. Find the number.

Translate into an equation and solve.

15. Three times the sum of a number and two is fifteen. Find the number.

16. Five times the difference between a number and four equals thirty. Find the number.

17. Four times a number is equal to ten more than twice the number. Find the number.

18. Twice a number is twenty less than the product of seven and the number. Find the number.

19. The sum of a number and four equals eight less than the product of three and the number. Find the number.

20. A number minus three is equal to the difference between six times the number and eighteen. Find the number.

21. Four times the difference between three times a number and one is equal to six more than twice the number. Find the number.

22. Five times the sum of a number and one equals four less than twice the number. Find the number.

23. The sum of two numbers is twenty-five. Twice the larger number is three times the smaller number. Find the two numbers.

24. The sum of two numbers is thirty-five. Four times the smaller number is three times the larger number. Find the two numbers.

25. The sum of two numbers is ten. The total of three times the smaller number and one is equal to the difference between twice the larger number and four. Find the two numbers.

26. The sum of two numbers is thirteen. The difference between four times the larger number and five equals the sum of five times the smaller number and two. Find the two numbers.

27. The sum of two numbers is sixteen. Twelve more than the larger number is equal to the sum of four times the smaller number and three. Find the two numbers.

28. The sum of two numbers is twenty-four. The difference between twice the larger number and eleven equals the difference between four times the smaller number and seventeen. Find the two numbers.

4.2 Application Problems

Write an equation and solve.

1. The retail selling price of a radio is \$52. This price is \$23 more than the cost of the radio. Find the cost of the radio.

2. The sale price of a raincoat is \$45. This is \$19 less than the original price. Find the original price.

3. Due to depreciation, the value of a car is now \$3600. This is three-fifths of its original value. Find the original value.

4. An executive's salary is \$60,000. This is three times the executive's salary fifteen years ago. Find the executive's salary fifteen years ago.

5. An advertising agency spent three times as much for television advertising as it spent for radio advertising. The total advertising budget for radio and television was \$36,000. How much was spent for each?

6. An attorney spent 60 hours on two cases. Four times as many hours were spent on one case as on the second case. Find the number of hours spent on each case.

7. During a holiday season a company employs 2100 people. There are twice as many part-time employees as full-time employees. How many part-time employees are employed during the holiday season?

8. A management consultant estimates that a company should have seven times as many non-management employees as management employees. Using this estimate, how many management employees should a company with 480 employees have?

9. A ribose sugar molecule in RNA contains oxygen, carbon, and hydrogen atoms. The number of oxygen atoms equals the number of carbon atoms. There are twice as many hydrogen atoms as carbon atoms. If the total number of atoms in the molecule is 20, find the number of each type of atom.

10. A vitamin supplement contains 25 mg of three different B vitamins. The number of milligrams of B-1 is twice the number of milligrams of B-6. There are also twice the number of milligrams of B-2 as B-6. Find the number of milligrams of each B vitamin.

11. The purchase price of a car, including finance charges, was \$9536. A down payment of \$600 was made. The remainder was paid in 48 equal monthly payments. Find the monthly payment. Round to the nearest cent.

12. The total purchase price, including finance charges, for a microwave oven was \$618. A down payment of \$100 was made. The remainder was paid in 12 equal monthly payments. Find the monthly payment. Round to the nearest cent.

Write an equation and solve.

13. The pressure in the power stroke of an engine is 570 lb/in.^{2}. This is 30 lb/in.^{2} more than three times the pressure in the compression stroke. Find the pressure in the compression stroke.

14. It requires 540 Calories of heat to produce 1 g of steam from boiling water. This is 60 Calories more than 6 times the number of Calories required to make 1 g of water from ice. How many Calories are required to make 1 g of water from ice?

15. The engine speed of a car is 2000 rpm (revolutions per minute). This is equal to 400 rpm less than twice the drive shaft speed. Find the drive shaft speed.

16. The area of one piston head in a hydraulic lift is 50 in.^{2}. This equals 2 in.^{2} more than four times the area of a second piston head. Find the area of the second piston head.

17. A plumbing repair bill was $155. This included $80 for parts and $25 for each hour of labor. Find the number of hours of labor.

18. An auto repair bill was $75. This included $30 for parts and $15 for each hour of labor. Find the number of hours of labor.

19. The sum of the angles of a triangle is 180°. The measure of the first angle is 90°. The measure of the second angle is twice the measure of the third angle. Find the measures of the second and third angles.

20. A cement patio slab is in the shape of a rectangle. The length of the slab is 20 ft. This is 4 ft more than twice the width of the slab. Find the width of the slab.

21. A piano wire is 42 in. long. A fourth cord can be produced by dividing this wire into two parts so that three times the length of the longer piece is four times the length of the shorter piece. Find the length of each piece.

22. A piano wire is 36 in. long. A major third chord can be produced by dividing this wire into two parts so that four times the length of the longer piece is five times the length of the shorter piece. Find the length of each piece.

23. A board 12 ft long is cut into two pieces. Three times the length of the shorter piece is 1 ft more than twice the longer piece. Find the length of each piece.

24. An investor deposits $10,000 into two accounts. Two times the larger deposit is equal to the difference between four times the smaller deposit and $4000. Find the amount deposited in each account.

Calculators and Computers

Solving a First-Degree Equation

Chapter 3, Solving Equations, is a core chapter of this text. You will be using the equation-solving skills taught in this chapter throughout the remainder of your studies in algebra.

The order in which the skills used to solve equations are used, is very important. The order in which these skills are used to solve an equation is as follows:

1) Remove parentheses.

2) Get all variable terms on one side of the equation and simplify.

3) Get all constant terms on the other side of the equation and simplify.

4) Multiply both sides of the equation by the reciprocal of the coefficient of the variable term.

Solving equations is a learned skill and requires practice. To provide you with additional practice, the program SOLVE A FIRST-DEGREE EQUATION on the Student Disk will allow you to practice solving the following three types of equations:

1) $ax + b = c$

2) $ax + b = cx + d$

3) Equations with parentheses

1) After you select the type of equation you want to practice, a problem will be displayed on the screen.

2) Using paper and pencil, solve the problem.

3) When you are ready, press the RETURN key and the complete solution will be displayed.

4) Compare your solution with the displayed solution.

5) All answers are rounded to the nearest hundredth.

6) When you finish a problem, you may continue practicing the type of problem you have selected, return to the main menu and select a different type, or quit the program.

Chapter Summary

KEY WORDS

An **equation** expresses the equality of two mathematical expressions.

A **solution** of an equation is a number which, when substituted for the variable, results in a true equation.

To **solve** an equation means to find a solution of the equation. The goal is to rewrite the equation in the form **variable = constant.**

To **translate** a sentence into an equation requires recognition of the words or phrases which mean "equals." Some of these phrases are "equals," "is," "is equal to," "amounts to," and "represents."

ESSENTIAL RULES

Addition Property of Equations

The same number or variable term can be added to each side of an equation without changing the solution of the equation.

If $a = b$, then $\boldsymbol{a + c = b + c}$.

Multiplication Property of Equations

Each side of an equation can be multiplied by the same non-zero number without changing the solution of the equation.

If $a = b$ and $c \neq 0$, then $\boldsymbol{ac = bc}$.

Review/Test

SECTION **1**

1.1a Is -2 a solution of $x^2 - 3x = 2x - 6$?

1.1b Is $\frac{2}{3}$ a solution of $6x - 7 = 3 - 9x$?

1.2 Solve: $x - 3 = -8$

1.3 Solve: $\frac{3}{4}x = -9$

SECTION **2**

2.1a Solve: $3x - 5 = -14$

2.1b Solve: $7 - 4x = -13$

2.2 A financial manager has determined that the cost per unit for a calculator is $15 and that the fixed costs per month are $2000. Find the number of calculators produced during a month in which the total cost was $5000. Use the equation $T = U \cdot N + F$, where T is the total cost, U is the cost per unit, N is the number of units produced, and F is the fixed costs.

SECTION **3**

3.1a Solve: $3x - 2 = 5x + 8$

3.1b Solve: $6 - 5x = 5x + 11$

3.2a Solve: $5x - 2(4x - 3) = 6x + 9$

3.2b Solve: $9 - 3(2x - 5) = 12 + 5x$

Review/Test

3.3 A chemist mixes 100 g of water at 80°C with 50 g of water at 20°C. Use the equation $m_1 \cdot (T_1 - T) = m_2 \cdot (T - T_2)$ to find the final temperature of the water after mixing. m_1 is the quantity of water at the hotter temperature, T_1 is the temperature of the hotter water, m_2 is the quantity of water at the cooler temperature, T_2 is the temperature of the cooler water, and T is the final temperature of the water after mixing.

SECTION **4**

4.1a Translate "the difference between three times a number and fifteen is twenty-seven" into an equation and solve.

4.1b Translate "the sum of five times a number and six equals the product of the number plus twelve and three" into an equation and solve.

4.1c The sum of two numbers is 18. The difference between four times the smaller number and seven is equal to the sum of two times the larger number and five. Find the two numbers.

4.2a A train travels between two cities in 26 h. This is 5 h more than the product of three and the time required for a plane to fly between the two cities. Find the number of hours required for the plane to fly between the two cities.

4.2b A board 18 ft long is cut into two pieces. Two feet less than the product of five and the length of the smaller piece is equal to the difference between three times the length of the longer piece and eight. Find the length of each piece.

Cumulative Review/Test

1. Subtract: $-6-(-20)-8$

2. Multiply: $(-2)(-6)(-4)$

3. Simplify $-\frac{5}{6}-\left(-\frac{7}{16}\right)$.

4. Simplify $-2\frac{1}{3}\div 1\frac{1}{6}$.

5. Simplify $-4^2\cdot\left(-\frac{3}{2}\right)^3$.

6. Simplify $25-3\frac{(5-2)^2}{2^3+1}-(-2)$.

7. Evaluate $3(a-c)-2ab$ when $a=2$, $b=3$, and $c=-4$.

8. Simplify $3x-8x+(-12x)$.

9. Simplify $2a-(-3b)-7a-5b$.

10. Simplify $(16x)\left(\frac{1}{8}\right)$.

11. Simplify $-4(-9y)$.

12. Simplify $-2(-x^2-3x+2)$.

13. Simplify $-2(x-3)+2(4-x)$.

14. Simplify $-3[2x-4(x-3)]+2$.

15. Is -3 a solution of $x^2+6x+9=x+3$?

16. Is $\frac{1}{2}$ a solution of $3-8x=12x-2$?

Cumulative Review/Test

17. Solve: $x - 4 = -9$

18. Solve: $\frac{3}{5}x = -15$

19. Solve: $7x - 8 = -29$

20. Solve: $13 - 9x = -14$

21. Solve:
$8x - 3(4x - 5) = -2x - 11$

22. Solve:
$6 - 2(5x - 8) = 3x - 4$

23. Solve: $5x - 8 = 12x + 13$

24. Solve: $11 - 4x = 2x + 8$

25. A business manager has determined that the cost per unit for a camera is \$70 and that the fixed costs per month are \$3500. Find the number of cameras that are produced during a month in which the total cost was \$21,000. Use the equation $T = U \cdot N + F$, where T is the total cost, U is the cost per unit, N is the number of units produced, and F is the fixed cost.

26. A chemist mixes 300 g of water at 75°C with 100 g of water at 15°C. Use the equation $m_1 \cdot (T_1 - T) = m_2 \cdot (T - T_2)$ to find the final temperature of the water. m_1 is the quantity of water at the hotter temperature, T_1 is the temperature of the hotter water, m_2 is the quantity of water at the cooler temperature, T_2 is the temperature of the cooler water, and T is the final temperature of the water after mixing.

27. Translate "the difference between twelve and the product of five and a number is negative eighteen" into an equation and solve.

28. Translate "the sum of six times a number and thirteen is five less than the product of three and the number" into an equation and solve.

29. The area of the cement foundation of a house is 2000 ft². This is 200 ft² more than three times the area of the garage. Find the area of the garage.

30. A board 16 ft long is cut into two pieces. Four feet more than the product of three and the length of the shorter piece is equal to three feet less than twice the length of the longer piece. Find the length of each piece.

4

Solving Equations: Applications

Objectives

- To write a percent as a fraction or a decimal
- To write a fraction or a decimal as a percent
- To solve the basic percent equation
- To solve application problems
- To solve markup problems
- To solve discount problems
- To solve investment problems
- To solve value mixture problems
- To solve percent mixture problems
- To solve uniform motion problems
- To solve perimeter problems
- To solve problems involving the angles of a triangle
- To solve consecutive integer problems
- To solve coin and stamp problems
- To solve age problems

Word Problems

Word problems have been challenging students of mathematics for a long time. Here are two types of problems you may have seen before:

A number added to $\frac{1}{7}$ of the number is 19. What is the number?

A dog is chasing a rabbit which has a head start of 150 feet.
The dog jumps 9 feet every time the rabbit jumps 7 feet.
In how many jumps will the dog catch up with the rabbit?

The unusual facts about these problems is that the first one is around 4000 years old and occurred as Problem 1 in the Rhind Papyrus. The second problem is around 1500 years old and comes from a Latin algebra book written in A.D. 450.

These examples illustrate that word problems have been around for a long time. The long history of word problems also recognizes the importance that each generation has put on solving these problems. It is through word problems that the initial steps of applying mathematics are taken.

The answer to the first problem is $16\frac{5}{8}$.

The answer to the second problem is 75 jumps.

SECTION 1 Introduction to Percent

1.1 Objective To write a percent as a fraction or a decimal

Reference for Computer Tutor™

"A population growth rate of 3%," "a manufacturer's discount of 25%," and "an 8% increase in pay" are typical examples of the many ways in which percent is used in applied problems. **Percent** means "parts of 100." Thus, 27% means 27 parts of 100.

In applied problems involving a percent, it is usually necessary either to rewrite the percent as a fraction or a decimal, or to rewrite a fraction or a decimal as a percent.

To write a percent as a fraction, drop the percent sign and multiply by $\frac{1}{100}$.

Write 27% as a fraction.

Drop the percent sign and multiply by $\frac{1}{100}$. $27\% = 27\left(\frac{1}{100}\right) = \frac{27}{100}$

To write a percent as a decimal, drop the percent sign and multiply by 0.01.

Write 33% as a decimal.

Drop the percent sign. Then multiply by 0.01. $33\% = 33(0.01) = 0.33$

Move the decimal point two places to the left. Then drop the percent sign.

Example 1 Write 130% as a fraction and as a decimal.

Solution $130\% = 130\left(\frac{1}{100}\right) = \frac{130}{100} = 1\frac{3}{10}$

$130\% = 130(0.01) = 1.30$

Example 2 Write 125% as a fraction and as a decimal.

Your solution

Example 3 Write $33\frac{1}{3}\%$ as a fraction.

Solution $33\frac{1}{3}\% = 33\frac{1}{3}\left(\frac{1}{100}\right) = \frac{100}{3}\left(\frac{1}{100}\right) = \frac{1}{3}$

Example 4 Write $16\frac{2}{3}\%$ as a fraction.

Your solution

Example 5 Write 0.25% as a decimal.

Solution $0.25\% = 0.25(0.01) = 0.0025$

Example 6 Write 0.5% as a decimal.

Your solution

Solutions on p. A19

1.2 Objective To write a fraction or a decimal as a percent

Reference for Computer Tutor™

A fraction or decimal can be written as a percent by multiplying by 100%.

Write $\frac{5}{8}$ as a percent.

Multiply by 100% $\frac{5}{8} = \frac{5}{8}(100\%) = \frac{500}{8}\% = 62.5\%$ or $62\frac{1}{2}\%$

Write 0.82 as a percent.

Multiply by 100% $0.82 = 0.82(100\%) = 82\%$

Move the decimal point two places to the right. Then write the percent sign.

Example 7 Write 0.027 as a percent.

Solution $0.027 = 0.027(100\%) = 2.7\%$

Example 8 Write 0.043 as a percent.

Your solution

Example 9 Write 1.34 as a percent.

Solution $1.34 = 1.34(100\%) = 134\%$

Example 10 Write 2.57 as a percent.

Your solution

Example 11 Write $\frac{5}{6}$ as a percent. Round to the nearest tenth of a percent.

Solution $\frac{5}{6} = \frac{5}{6}(100\%) = \frac{500}{6}\% \approx 83.3\%$

Example 12 Write $\frac{5}{9}$ as a percent. Round to the nearest tenth of a percent.

Your solution

Example 13 Write $\frac{7}{16}$ as a percent. Write the remainder in fractional form.

Solution $\frac{7}{16} = \frac{7}{16}(100\%) = \frac{700}{16}\% = 43\frac{3}{4}\%$

Example 14 Write $\frac{9}{16}$ as a percent. Write the remainder in fractional form.

Your solution

1.1 Exercises

Write as a fraction and a decimal.

1. 75% **2.** 40% **3.** 50% **4.** 10%

5. 64% **6.** 88% **7.** 125% **8.** 160%

9. 19% **10.** 87% **11.** 5% **12.** 2%

13. 450% **14.** 380% **15.** 8% **16.** 4%

Write as a fraction.

17. $11\frac{1}{9}\%$ **18.** $4\frac{2}{7}\%$ **19.** $12\frac{1}{2}\%$ **20.** $37\frac{1}{2}\%$

21. $31\frac{1}{4}\%$ **22.** $66\frac{2}{3}\%$ **23.** $\frac{1}{4}\%$ **24.** $\frac{1}{2}\%$

25. $5\frac{3}{4}\%$ **26.** $68\frac{3}{4}\%$ **27.** $6\frac{1}{4}\%$ **28.** $83\frac{1}{3}\%$

Write as a decimal.

29. 7.3% **30.** 9.1% **31.** 15.8% **32.** 16.7%

33. 0.3% **34.** 0.9% **35.** 9.15% **36.** 121.2%

37. 18.23% **38.** 62.14% **39.** 0.15% **40.** 0.27%

1.2 Exercises

Write as a percent.

41. 0.15
42. 0.37
43. 0.05
44. 0.02
45. 0.175
46. 0.125
47. 1.15
48. 1.36
49. 0.62
50. 0.96
51. 3.165
52. 2.142
53. 0.008
54. 0.004
55. 0.065
56. 0.083

Write as a percent. Round to the nearest tenth of a percent.

57. $\frac{27}{50}$
58. $\frac{83}{100}$
59. $\frac{1}{3}$
60. $\frac{3}{8}$
61. $\frac{5}{11}$
62. $\frac{4}{9}$
63. $\frac{7}{8}$
64. $\frac{9}{20}$
65. $1\frac{2}{3}$
66. $2\frac{1}{2}$
67. $1\frac{2}{7}$
68. $1\frac{11}{12}$

Write as a percent. Write the remainder in fractional form.

69. $\frac{17}{50}$
70. $\frac{17}{25}$
71. $\frac{3}{8}$
72. $\frac{7}{16}$
73. $\frac{5}{14}$
74. $\frac{3}{19}$
75. $\frac{3}{16}$
76. $\frac{4}{7}$
77. $1\frac{1}{4}$
78. $2\frac{5}{8}$
79. $1\frac{5}{9}$
80. $1\frac{13}{16}$

SECTION 2 The Percent Equation

2.1 Objective To solve the basic percent equation

Reference for Computer Tutor™

The solution of a problem which involves a percent requires solving the basic percent equation shown at the right.

BASIC PERCENT EQUATION

Percent × Base = Amount

$P \times B = A$

In any percent problem, two parts of the equation are given, and one is unknown.

To translate a problem involving a percent into an equation, remember that the word "of" translates to "multiply" and the word "is" translates to "=". The base usually follows the word "of."

20% of what number is 30?

Given: $P = 20\% = 0.20$
$A = 30$
Unknown: Base

$$P \times B = A$$
$$(0.20)B = 30$$
$$\frac{1}{0.20}(0.20)B = \frac{1}{0.20}(30)$$
$$B = 150$$

The number is 150.

Find 25% of 200.

Given: $P = 25\% = 0.25$
$B = 200$
Unknown: Amount

$$P \times B = A$$
$$0.25(200) = A$$
$$50 = A$$

25% of 200 is 50.

In most cases, the percent is written as a decimal before solving the basic percent equation. However, some percents are more easily written as a fraction. For example,

$33\frac{1}{3}\% = \frac{1}{3}$ $\qquad 66\frac{2}{3}\% = \frac{2}{3}$ $\qquad 16\frac{2}{3}\% = \frac{1}{6}$ $\qquad 83\frac{1}{3}\% = \frac{5}{6}$

Example 1 12 is $33\frac{1}{3}\%$ of what number?

Solution

$$12 = \frac{1}{3}B \qquad \left(33\frac{1}{3}\% = \frac{1}{3}\right)$$
$$3 \cdot 12 = 3 \cdot \frac{1}{3}B$$
$$36 = B$$

The number is 36.

Example 2 27 is what percent of 60?

Your solution

Solution on p. A19

2.2 Objective To solve application problems

Use COMPUTER TUTOR™ DISK 4

The key to solving a percent problem is identifying the percent, the base, and the amount. The base usually follows the word "of."

Example 3 A student answered 76 of the 80 questions on a test correctly. What percent of the questions were answered correctly?

Strategy To find the percent of the questions answered correctly, solve the basic percent equation using $B = 80$ and $A = 76$. The percent is unknown.

Solution

$$P \times B = A$$
$$P(80) = 76$$
$$P(80)\left(\frac{1}{80}\right) = 76\left(\frac{1}{80}\right)$$
$$P = 0.95$$

95% of the questions were answered correctly.

Example 4 A quality control inspector found that 6 out of 200 wheel bearings inspected were defective. What percent of the wheel bearings were defective?

Your strategy

Your solution

Example 5 A new labor contract increased an employee's hourly wage by 5%. What is the amount of increase for an employee who was making \$9.60 an hour?

Strategy To find the amount of increase, solve the basic percent equation using $B = 9.60$ and $P = 5\% = 0.05$. The amount is unknown.

Solution

$$P \times B = A$$
$$(0.05)(9.60) = A$$
$$0.48 = A$$

The amount of increase is \$.48.

Example 6 A company was producing 2500 gal of paint each week. Due to a decrease in demand for the paint, the company reduced its weekly production by 500 gal. What percent decrease does this represent?

Your strategy

Your solution

Solutions on p. A19

2.1 Exercises

Solve.

1. 12 is what percent of 50?

2. What percent of 125 is 50?

3. Find 18% of 40.

4. What is 25% of 60?

5. 12% of what is 48?

6. 45% of what is 9?

7. What is $33\frac{1}{3}$% of 27?

8. Find $16\frac{2}{3}$% of 30.

9. What percent of 12 is 3?

10. 10 is what percent of 15?

11. 60% of what is 3?

12. 75% of what is 6?

13. 12 is what percent of 6?

14. 20 is what percent of 16?

15. $5\frac{1}{4}$% of what is 21?

16. $37\frac{1}{2}$% of what is 15?

17. Find 15.4% of 50.

18. What is 18.5% of 46?

19. 1 is 0.5% of what?

20. 3 is 1.5% of what?

21. $\frac{3}{4}$% of what is 3?

22. $\frac{1}{2}$% of what is 3?

23. Find 125% of 16.

24. What is 250% of 12?

25. 16.43 is what percent of 20.45? Round to the nearest hundredth of a percent.

26. Find 18.37% of 625.43. Round to the nearest hundredth.

2.2 Application Problems

Solve.

1. A company's budget for the development of a new product is \$250,000. Of this amount, \$50,000 is for materials. What percent of the total budget is for materials?

2. Fifteen years ago, a painting was priced at \$6000. Today the painting has a value of \$15,000. What percent of the price 15 years ago is its value today?

3. An engineer estimates that 32% of the gasoline used by a car is used efficiently. Using this estimate, how many gallons out of 20 gal of gasoline are used efficiently?

4. Approximately 80% of the air in the atmosphere is nitrogen. Using this estimate, how many liters of nitrogen are in a room which contains 500 L of air?

5. The value of a car today is $66\frac{2}{3}\%$ of its value two years ago. The value of the car two years ago was \$6000. What is the value of the car today?

6. An appliance store estimates that 15% of the washing machines it sells will require service within one year. Using this estimate, how many washing machines were sold in a year in which 27 new machines were serviced?

7. The normal underwater visibility off the coast of an island is 30 ft. Unusual turbulence reduced the visibility by 12 ft. What percent decrease does this represent?

8. The seating capacity of a baseball stadium, which had seated 50,000, was expanded by 6250 seats. What percent increase does this represent?

9. The number of take-offs and landings at a municipal airport this year was 303,750. This represents 112.5% of last year's take-offs and landings. How many take-offs and landings were there last year?

10. The annual license fee on a car is 1.4% of the value of the car. What is the value of a car during a year in which the license fee was \$91.00?

11. In a recent city election, 14,375 people out of 50,000 registered voters voted. What percent of the people voted in the election?

12. There are approximately 8760 h in one year. An adult sleeps approximately 2700 h during a year. What percent of the year does an adult spend sleeping? Round to the nearest hundredth of a percent.

SECTION 3 Markup and Discount

3.1 Objective To solve markup problems

Reference for Computer Tutor™

Cost is the price which a business pays for a product. **Selling price** is the price for which a business sells a product to a customer. The difference between selling price and cost is called **markup.** Markup is added to a retailer's cost to cover the expenses of operating a business. Markup is usually expressed as a percent of the retailer's cost. This percent is called the **markup rate.**

The basic markup equations used by a business are:

Selling price = cost + markup
$S = C + M$

Markup = markup rate × cost
$M = r \times C$

Substituting $r \times C$ for M in the first equation, selling price can also be written as $S = C + (r \times C) = C + rC$

The manager of a clothing store buys a suit for \$80 and sells the suit for \$116. Find the markup rate.

Given: $C = \$80$
$S = \$116$
Unknown: r
Use the equation $S = C + rC$.

$$S = C + rC$$
$$116 = 80 + 80r$$
$$116 - 80 = 80 - 80 + 80r$$
Do this step mentally.
$$36 = 80r$$
$$\frac{1}{80} \cdot 36 = \frac{1}{80} \cdot 80r$$
Do this step mentally.
$$\frac{36}{80} = r$$
$$0.45 = r$$

The markup rate is 45%.

Example 1 A hardware store employee uses a markup rate of 40% on all items. The selling price of a lawn mower is \$105. Find the cost.

Strategy Given: $r = 40\% = 0.40$
$S = \$105$
Unknown: C
Use the equation $S = C + rC$.

Solution
$$S = C + rC$$
$$105 = C + 0.40C$$
$$105 = 1.40C$$
$$75 = C$$

The cost is \$75.

Example 2 The cost to the manager of a sporting goods store for a tennis racket is \$40. The selling price of the racket is \$60. Find the markup rate.

Your strategy

Your solution

Solution on p. A20

3.2 Objective To solve discount problems

Reference for Computer Tutor™

Discount is the amount by which a retailer reduces the regular price of a product for a promotional sale. Discount is usually expressed as a percent of the regular price. This percent is called the **discount rate.**

The basic discount equations used by a business are:

Sale Price = regular price − discount: $S = R - D$

Discount = discount rate × regular price: $D = r \times R$

Substituting $r \times R$ for D in the first equation, sale price can also be written as $S = R - (r \times R) = R - rR$

In a garden supply store, the regular price of a 100-foot garden hose is \$32. During an "after-summer sale," the hose is being sold for \$24. Find the discount rate.

Given: $R = \$32$
$S = \$24$
Unknown: r
Use the equation $S = R - rR$.

$$S = R - rR$$
$$24 = 32 - 32r$$
$$24 - 32 = 32 - 32 - 32r \quad \text{Do this step mentally.}$$
$$-8 = -32r$$
$$\left(-\frac{1}{32}\right)(-8) = \left(-\frac{1}{32}\right)(-32r) \quad \text{Do this step mentally.}$$
$$\frac{1}{4} = r$$
$$0.25 = r$$

The discount rate is 25%.

Example 3 The sale price for a chemical sprayer is \$23.40. This price is 35% off the regular price. Find the regular price.

Strategy Given: $S = \$23.40$
$r = 35\% = 0.35$
Unknown: R
Use the equation $S = R - rR$.

Solution
$$S = R - rR$$
$$23.40 = R - 0.35R$$
$$23.40 = 0.65R$$
$$36 = R$$

The regular price is \$36.

Example 4 A case of motor oil which regularly sells for \$27.60 is on sale for \$20.70. What is the discount rate?

Your strategy

Your solution

3.1 Application Problems

Solve.

1. The manager of a garden supply store buys a 50-foot rubber hose for $12. A markup rate of 35% is used. What is the selling price?

2. A bookstore buys a paperback book for $3 and uses a markup rate of 20%. Find the selling price of the book.

3. A shoe store uses a markup rate of 45%. One pair of shoes has a selling price of $72.50. Find the cost of the shoes.

4. The selling price for a gallon of ice cream is $6.57. The markup rate used by the ice cream store is 46%. Find the cost of the ice cream.

5. The cost to a furniture manufacturer for making a kitchen chair is $15. The manufacturer then sells the chair for $24. What is the markup rate?

6. The manufacturer's cost for a car air conditioner is $300. The manufacturer then sells the air conditioner for $450. What is the markup rate?

7. The meat manager at a market uses a markup rate of 22%. What is the selling price for a steak which costs $2.50?

8. A pen and pencil set costs a retailer $15. Find the selling price when the markup rate is $66\frac{2}{3}\%$.

9. A restaurant serves a breakfast for $1.75. The restaurant's cost to make the breakfast is $1.00. What is the markup rate?

10. The selling price for a silk jacket is $200. The cost to the clothing store for the jacket is $125. What is the markup rate?

11. The cost to a car dealer for a car is $5696. The selling price of the car is $8900. What is the markup rate?

12. The cost to an appliance dealer for a refrigerator is $483. The dealer sells the refrigerator for $920. What is the markup rate? Round to the nearest hundredth of a percent.

3.2 Application Problems

Solve.

13. A hair dryer which regularly sells for $28 is on sale for 20% off the regular price. Find the sale price.

14. A briefcase is on sale for 25% off the regular price of $85. Find the sale price.

15. The sale price of a desk lamp is $26, which is 20% off the regular price. Find the regular price.

16. The sale price for a car tire is $35.75, which is 45% off the regular price. Find the regular price.

17. A box of stationery which regularly sells for $4 is on sale for $3. Find the discount rate.

18. An art print which regularly sells for $50 is on sale for $40. Find the discount rate.

19. A gas barbecue is on sale for $200. This is $33\frac{1}{3}$% off the regular price. Find the regular price.

20. During a "Going-Out-of-Business" sale, a clothing store reduced all items 60%. What was the regular price of an umbrella which was on sale for $7.20?

21. A children's gym set, which regularly sells for $130, is on sale for $96.20. Find the discount rate.

22. A radio with cassette recorder, which regularly sells for $60, is on sale for $37.80. Find the discount rate.

23. A cabin tent, which regularly sells for $129, is on sale for $99. Find the discount rate to the nearest tenth of a percent.

24. A portable television set, which regularly sells for $139, is on sale for $89.90. Find the discount rate to the nearest tenth of a percent.

SECTION 4 Investment Problems

4.1 Objective To solve investment problems

Reference for Computer Tutor™

The annual simple interest which an investment earns is given by the equation $I = Pr$, where I is the simple interest, P is the principal, or the amount invested, and r is the simple interest rate.

The annual interest rate on a \$2500 investment is 14%. Find the annual simple interest earned on the investment.

Given: $P = \$2500$
$r = 14\% = 0.14$
Unknown: I

$I = Pr$
$I = 2500(0.14)$
$I = 350$

The annual simple interest is \$350.

An investor has a total of \$10,000 to deposit into two simple interest accounts. On one account, the annual simple interest rate is 7%. On the second account, the annual simple interest rate is 11%. How much should be invested in each account so that the total annual interest earned is \$1000?

STRATEGY FOR SOLVING A PROBLEM INVOLVING MONEY DEPOSITED IN TWO SIMPLE INTEREST ACCOUNTS

▷ For each amount invested, write a numerical or variable expression for the principal, the interest rate, and the interest earned. The results can be recorded in a table.

The sum of the amounts at each interest rate is \$10,000.

Amount invested at 7%: x
Amount invested 11%: $\$10,000 - x$

	Principal, P	·	Interest rate, r	=	Interest earned, I
Amount at 7%	x	·	0.07	=	$0.07x$
Amount at 11%	$10,000 - x$	·	0.11	=	$0.11(10,000 - x)$

▷ Determine how the amounts of interest earned on each amount are related. For example, the total interest earned by both accounts may be known or it may be known that the interest earned on one account is equal to the interest earned by the other account.

The total annual interest earned is \$1000.

$$0.07x + 0.11(10,000 - x) = 1000$$
$$0.07x + 1100 - 0.11x = 1000$$
$$-0.04x + 1100 = 1000$$
$$-0.04x = -100$$
$$x = 2500$$
$$10,000 - x = 10,000 - 2500 = 7500$$

The amount invested at 7% is \$2500.
The amount invested at 11% is \$7500.

Example 1

An investment counselor invested 75% of a client's money into a 13% annual simple interest money market fund. The remainder was invested in 9% annual simple interest government securities. Find the amount invested in each if the total annual interest earned is \$5400.

Strategy

▷ Amount invested: x
Amount invested at 9%: $0.25x$
Amount invested at 13%: $0.75x$

	Principal	Rate	Interest
Amount at 9%	$0.25x$	0.09	$0.0225x$
Amount at 13%	$0.75x$	0.13	$0.0975x$

▷ The sum of the interest earned by the two investments equals the total annual interest earned (\$5400).

Solution

$$0.0225x + 0.0975x = 5400$$
$$0.12x = 5400$$
$$x = 45{,}000$$

$$0.25x = 0.25(45{,}000) = 11{,}250$$

$$0.75x = 0.75(45{,}000) = 33{,}750$$

The amount invested at 9% is \$11,250.
The amount invested at 13% is \$33,750.

Example 2

An investment of \$5000 is made at an annual simple interest rate of 8%. How much additional money must be invested at 14% so that the total interest earned will be 11% of the total investment?

Your strategy

Your solution

Solution on p. 420

4.1 Application Problems

Solve.

1. A total of $5000 is deposited into two simple interest accounts. On one account the annual simple interest rate is 8%, while on the second account the annual simple interest rate is 12%. How much should be invested in each account so that the total annual interest earned is $520?

2. An investment club invested a part of $10,000 in a 9.5% annual simple interest account and the remainder in a 14% annual simple interest account. The amount of interest earned for one year was $1085. How much was invested in each account?

3. An investment of $4000 is made at an annual simple interest rate of 12%. How much additional money must be invested at an annual simple interest rate of 8% so that the total interest earned is 10% of the total investment?

4. An investment of $3500 is made at an annual simple interest rate of 9%. How much additional money must be invested at an annual simple interest rate of 13% so that the total interest earned is 11% of the total investment?

5. A total of $4000 is invested into two simple interest accounts. On one account the annual simple interest rate is 12%, while on the second account the annual simple interest rate is 8%. How much should be invested in each account so that the interest earned by each account is the same?

6. An investment advisor deposited a total of $6000 into two money market funds. One fund earns 14% annual simple interest, while a second tax-free fund earns 7% annual simple interest. How much must be invested in each fund so that the interest earned by each is the same?

7. An accountant deposited an amount of money into a 12% annual simple interest account. Another deposit, $2000 more than the first, was placed in a 9.8% annual simple interest account. The total interest earned on both investments for one year was $741. How much money was deposited in the 12% account?

8. A deposit was made into a 6% annual simple interest savings account. Another deposit, $3500 less than the first, was placed in a 10% annual simple interest bond market account. The total interest earned on both accounts for one year was $450. How much money was deposited in the 6% account?

Solve.

9. An investment of $12,000 is made into a 10.5% simple interest account. How much additional money must be deposited into an 8% simple interest account so that the total interest earned on both accounts is 9.5% of the total investment?

10. To provide for retirement income, an engineer purchases a $10,000 bond. The simple interest rate on the bond is 8.5%. How much money must be invested in additional bonds which have an interest rate of 9.25% so that the total interest earned each year is $2700?

11. A stock broker's client has $25,000 to invest. The broker recommends that part of the $25,000 be placed in 7.5% tax-free municipal bonds and the remainder in 11.25% commercial bonds. How much should be invested in each type of bond so that the total interest earned is $2250?

12. A corporation gave a university $250,000 to support research assistants in science. The university deposited some of the money in a 12% simple interest account and the remainder in a 10.5% simple interest account. How much should be deposited in each so that the total interest earned is $28,875?

13. The manager of a trust decided to invest 60% of a client's account in stocks which earn 6% simple interest. The remainder was invested in a tax shelter which earns 8% simple interest. The annual interest earned from the investments was $1632. What was the total amount invested?

14. The portfolio manager for a corporation invested 75% of the company's investment account in 12.5% short term certificates. The remainder was invested in 10% corporate bonds. The annual interest earned from the two investments was $47,500. What was the total amount invested?

15. A financial manager recommended an investment plan in which 20% of a client's cash be placed in a 7% simple interest account, 40% be placed in 10% high grade bonds, and the remainder in a 13% high-risk investment. The total interest earned from the investments would be $5300. What is the total amount to be invested?

16. An investment company deposited 70% of its investment capital in a 9.8% simple interest account. The remainder was deposited in a 11.5% simple interest account. The total interest earned from the investments was $46,395. How much was invested in each account?

SECTION 5 Mixture Problems

5.1 Objective To solve value mixture problems

Reference for Computer Tutor™

A value mixture problem involves combining two ingredients which have different prices into a single blend. For example, a coffee merchant may blend two types of coffee into a single blend, or a candy manufacturer may combine two types of candy to sell as a "variety pack."

The solution of a value mixture problem is based upon the equation $V = AC$, where V is the value of an ingredient, A is the amount of the ingredient, and C is the cost per unit of the ingredient.

A coffee merchant wants to make 6 lb of a blend of coffee to sell for $5 per pound. The blend is made using a $6 grade and a $3 grade of coffee. How many pounds of each of these grades should be used?

STRATEGY FOR SOLVING A VALUE MIXTURE PROBLEM

▷ For each ingredient in the mixture, write a numerical or variable expression for the amount of the ingredient used, the unit cost of the ingredient, and the value of the amount used. For the blend, write a numerical or variable expression for the amount, the unit cost of the blend, and the value of the amount. The results can be recorded in a table.

The sum of the amounts is 6 lb.

Amount of $6 coffee: x
Amount of $3 coffee: $6 - x$

	Amount, A	$\cdot$	Unit cost, C	=	Value, V
$6 grade	x	$\cdot$	$6	=	$6x$
$3 grade	$6 - x$	$\cdot$	$3	=	$3(6 - x)$
$5 blend	6	$\cdot$	$5	=	$5(6)$

▷ Determine how the values of each ingredient are related. Use the fact that the sum of the values of each ingredient is equal to the value of the blend.

The sum of the values of the $6 grade and the $3 grade is equal to the value of the $5 blend.

$$6x + 3(6 - x) = 5(6)$$
$$6x + 18 - 3x = 30$$
$$3x + 18 = 30$$
$$3x = 12$$
$$x = 4$$

$$6 - x = 6 - 4 = 2$$

The merchant must use 4 lb of the $6 coffee and 2 lb of the $3 coffee.

Example 1

How many ounces of a silver alloy which costs \$4 an ounce must be mixed with 10 oz of an alloy which costs \$6 an ounce to make a mixture which costs \$4.32 an ounce?

Strategy

▷ Ounces of \$4 alloy: x

	Amount	Cost	Value
\$4 alloy	x	\$4	$4x$
\$6 alloy	10	\$6	$6(10)$
\$4.32 mixture	$10 + x$	\$4.32	$4.32(10 + x)$

▷ The sum of the values before mixing equals the value after mixing.

Solution

$$\begin{aligned} 4x + 6(10) &= 4.32(10 + x) \\ 4x + 60 &= 43.2 + 4.32x \\ -0.32x + 60 &= 43.2 \\ -0.32x &= -16.8 \\ x &= 52.5 \end{aligned}$$

52.5 oz of the \$4 silver alloy must be used.

Example 2

A gardener has 20 lb of a lawn fertilizer which costs \$.80 per pound. How many pounds of a fertilizer which costs \$.55 per pound should be mixed with the 20 lb of lawn fertilizer to produce a mixture which costs \$.75 per pound?

Your strategy

Your solution

Solution on p. A21

5.2 Objective To solve percent mixture problems

Reference for Computer Tutor™

The amount of a substance in a solution can be given as a percent of the total solution. For example, a 5% salt water solution means that 5% of the total solution is salt. The remaining 95% is water.

The solution of a percent mixture problem is based upon the equation $Q = Ar$, where Q is the quantity of a substance in the solution, r is the percent of concentration, and A is the amount of solution.

A 500-milliliter bottle contains a 4% solution of hydrogen peroxide. Find the amount of hydrogen peroxide in the solution.

Given: $A = 500$
$r = 4\% = 0.04$
Unknown: Q

$Q = Ar$
$Q = 500(0.04)$
$Q = 20$

The bottle contains 20 ml of hydrogen peroxide.

How many gallons of a 20% salt solution must be mixed with 6 gal of a 30% salt solution to make a 22% salt solution?

STRATEGY FOR SOLVING A PERCENT MIXTURE PROBLEM

▷ For each solution, write a numerical or variable expression for the amount of solution, percent of concentration, and the quantity of the substance in the solution. The results can be recorded in a table.

The unknown quantity of 20% solution: x

	Amount of solution, A	$\cdot$	Percent of concentration, r	$=$	Quantity of substance, Q
20% solution	x	$\cdot$	0.20	$=$	$0.20x$
30% solution	6	$\cdot$	0.30	$=$	$0.30(6)$
22% solution	$x + 6$	$\cdot$	0.22	$=$	$0.22(x + 6)$

▷ Determine how the quantities of the substance in each solution are related. Use the fact that the sum of the quantities of the substances being mixed is equal to the quantity of the substance after mixing.

The sum of the quantities of the substances in the 20% solution and the 30% solution is equal to the quantity of the substance in the 22% solution.

$$0.20x + 0.30(6) = 0.22(x + 6)$$
$$0.20x + 1.80 = 0.22x + 1.32$$
$$-0.02x + 1.80 = 1.32$$
$$-0.02x = -0.48$$
$$x = 24$$

24 gal of the 20% solution are required.

Example 3

A chemist wishes to make 2 L of an 8% acid solution by mixing a 10% acid solution and a 5% acid solution. How many liters of each solution should the chemist use?

Strategy

▷ Liters of 10% solution: x
Liters of 5% solution: $2 - x$

	Amount	Percent	Quantity
10%	x	0.10	$0.10x$
5%	$2 - x$	0.05	$0.05(2 - x)$
8%	2	0.08	$0.08(2)$

▷ The sum of the quantities before mixing is equal to the quantity after mixing.

Solution

$$0.10x + 0.05(2 - x) = 0.08(2)$$
$$0.10x + 0.10 - 0.05x = 0.16$$
$$0.05x + 0.10 = 0.16$$
$$0.05x = 0.06$$
$$x = 1.2$$
$$2 - x = 2 - 1.2 = 0.8$$

The chemist needs 1.2 L of 10% solution and 0.8 L of the 5% solution.

Example 4

A pharmacist dilutes 5 L of a 12% solution by adding water. How many liters of water are added to make an 8% solution?

Your strategy

Your solution

5.1 Application Problems

Solve.

1. A butcher combined hamburger which cost $2.50 per pound with hamburger which cost $3.10 per pound. How many pounds of each were used to make an 80-pound mixture to sell for $2.65 per pound?

2. A butcher combined hamburger which cost $4.40 per kilogram with hamburger which cost $8.40 per kilogram. How many kilograms of each were used to make a mixture of 50 kg to sell for $6.00 per kilogram?

3. How many ounces of pure gold which cost $400 an ounce must be mixed with 20 oz of an alloy which cost $220 an ounce to make an alloy which would cost $300 an ounce?

4. A goldsmith combined pure gold which cost $675 per ounce with an alloy costing $325 per ounce. How many ounces of each should be used to make 5 oz of a gold alloy which sells for $465 per ounce?

5. Find the selling price per pound of a mixture made from 12 lb of chocolate which cost $4.00 per pound and 30 lb of chocolate which cost $2.25 per pound.

6. How many kilograms of chocolates which cost $7.00 per kilogram must be mixed with 20 kg of chocolates which cost $3.50 per kilogram to make a box of mixed chocolates to sell for $4.50 per kilogram?

7. A grocer combined cranberry juice which cost $3.25 per gallon with apple juice which cost $2.25 per gallon. How many gallons of each should be used to make 100 gal of cranapple juice to sell for $2.50 per gallon?

8. Find the selling price per liter of a mixture made from 40 L of cranberry juice which cost $1.00 per liter and 120 L of apple juice which cost $.60 per liter.

9. How many pounds of walnuts which cost $1.60 per pound must be mixed with 18 lb of cashews which cost $2.50 per pound to make a mixture which costs $1.90 per pound?

10. A grocer combined peanuts which cost $2.50 per kilogram with walnuts which cost $4.50 per kilogram. How many kilograms of each were used to make a 100-kilogram mixture to sell for $3.24 per kilogram?

Solve.

11. Find the selling price per pound of a mixture of coffee made from 25 lb of coffee which cost $4.82 per pound and 40 lb of coffee which cost $3.00 per pound.

12. How many kilograms of coffee which cost $9 per kilogram must be mixed with 16 kg of coffee which cost $5 per kilogram to make a mixture which costs $6.50 per kilogram?

13. How many pounds of cheese which cost $4.20 per pound must be mixed with 12 lb of cheese which cost $2.25 per pound to make a grated cheese topping which costs $3.40 per pound?

14. Find the selling price per kilogram of a grated cheese mixture made from 8 kg of cheese which cost $9.20 per kilogram and 12 kg of cheese which cost $5.50 per kilogram.

15. To make a feed for cattle, a feed store operation combined soybeans which cost $8 per bushel with corn which cost $3 per bushel. How many bushels of each were used to make a mixture of 5000 bushels to sell for $4.50 per bushel?

16. How many bushels of soybeans which cost $7.50 per bushel must be mixed with 2400 bushels of corn which cost $3.25 per bushel to make a mixture which costs $4.50 per bushel?

17. Find the selling price per ounce of a mixture of 200 oz of silver which cost $5.50 per ounce and 500 oz of an alloy which cost $2.00 per ounce.

18. A silversmith combined a silver alloy which cost $4.30 per ounce with an alloy which cost $1.80 per ounce. How many ounces of each were used to make a mixture of 200 oz to sell for $2.50 per ounce?

19. How many liters of face cream which cost $80 per liter must be mixed with 6 L of face cream which cost $25 per liter to make a face cream which sells for $36 per liter?

20. Find the selling price per ounce of a face cream mixture made from 40 oz of face cream which cost $4.40 per ounce and 100 oz of face cream which cost $2.30 per ounce.

5.2 Application Problems

Solve.

21. A farmer has some cream which is 21% butterfat and some which is 15% butterfat. How many gallons of each must be mixed to produce 60 gal of cream which is 19% butterfat?

22. A chemist has some 8% hydrogen peroxide solution and some 5% hydrogen peroxide solution. How many milliliters of each should be mixed to make a 300-milliliter solution which is 6% hydrogen peroxide?

23. How many grams of pure acid must be added to 40 g of a 20% acid solution to make a solution which is 36% acid?

24. How many ounces of pure water must be added to 60 oz of a 15% salt solution to make a salt solution which is 10% salt?

25. A hygienist mixed 50 L of a 36% disinfectant solution with 40 L of water. What is the percent concentration of the resulting solution?

26. A researcher mixed 80 lb of a 30% aluminum alloy with 120 lb of a 25% aluminum alloy. What is the percent concentration of the resulting alloy?

27. A syrup manufacturer has some pure maple syrup and some which is 85% maple syrup. How many liters of each should be mixed to make 150 L which is 96% maple syrup?

28. A butcher has some hamburger which is 20% fat and some which is 15% fat. How many pounds of each should be mixed to make 50 lb of hamburger which is 18% fat?

29. A 100-pound bag of animal feed is 40% oats. How many pounds of oats must be added to this feed to produce a mixture which is 50% oats?

30. A goldsmith has 10 g of a 50% gold alloy. How many grams of pure gold should be added to the alloy to make an alloy which is 75% gold?

Solve.

31. Ten grams of sugar are added to a 40-gram serving of a breakfast cereal which is 30% sugar. What is the percent concentration of sugar in the resulting mixture?

32. Thirty ounces of pure grapefruit juice are added to 50 oz of a fruit punch which is 20% grapefruit juice. What is the percent concentration of grapefruit juice in the resulting mixture?

33. A clothing manufacturer has some fiber which is 20% polyester and some which is 50% polyester. How many pounds of each fiber should be woven together to produce 600 lb of a fabric which is 35% polyester?

34. A nurse wants to make 50 ml of a 16% salt solution. How many milliliters each of a 13% salt solution and an 18% salt solution should be mixed to produce the desired solution?

35. A baker mixed some flour which was 40% wheat with 80 lb of flour which was 30% wheat to make a mixture which is 32% wheat. How many pounds of the 40% wheat flour were used?

36. A manufacturer mixed a chemical which was 60% fire retardant with 70 lb of a chemical which was 80% fire retardant to make a mixture which is 74% fire retardant. How much of the 60% mixture was used?

37. A 200-pound alloy of tin, which is 35% tin, is mixed with 300 lb of another tin alloy. The resulting alloy is 20% tin. Find the percent of tin in the 300-pound alloy.

38. A silversmith mixes 50 g of one alloy which is 50% silver with 150 g of another silver alloy. The resulting alloy is 68% silver. Find the percent of silver in the 150-gram alloy.

39. A chemist mixes a 5% silver nitrate solution with an 8% silver nitrate solution. How many ounces of each should be used to make 50 oz of a 7% silver nitrate solution? Round to the nearest hundredth.

40. How many grams of salt must be added to 180 g of a solution which is 15% salt to make a solution which is 23% salt? Round to the nearest tenth.

SECTION 6 Uniform Motion Problems

6.1 Objective To solve uniform motion problems

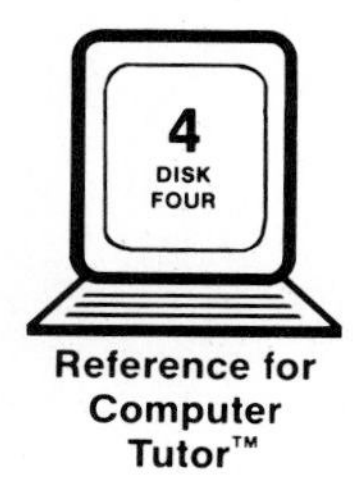

Reference for Computer Tutor™

A train which travels constantly in a straight line at 50 mph is in *uniform motion*. **Uniform motion** means that the speed of an object does not change.

The solution of a uniform motion problem is based upon the equation $d = rt$, where d is the distance traveled, r is the rate of travel, and t is the time spent traveling.

A car leaves a town traveling at 40 mph. Two hours later, a second car leaves the same town, on the same road, traveling at 60 mph. In how many hours will the second car be passing the first car?

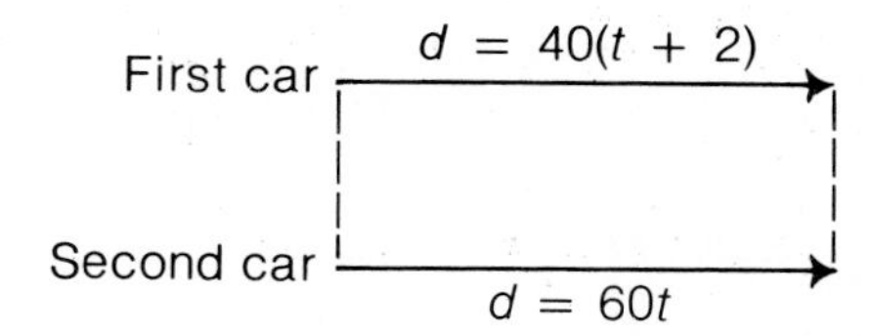

STRATEGY FOR SOLVING A UNIFORM MOTION PROBLEM

▷ For each object, write a numerical or variable expression for the distance, rate, and time. The results can be recorded in a table.

The first car traveled 2 h longer than the second car.

Unknown time for the second car: t
Time for the first car: $t + 2$

	Rate, r	·	Time, t	=	Distance, d
First car	40	·	$t + 2$	=	$40(t + 2)$
Second car	60	·	t	=	$60t$

▷ Determine how the distances traveled by each object are related. For example, the total distance traveled by both objects may be known or it may be known that the two objects traveled the same distance.

The two cars travel the same distance.

$$40(t + 2) = 60t$$
$$40t + 80 = 60t$$
$$80 = 20t$$
$$4 = t$$

The second car will be passing the first car in 4 h.

Example 1

Two cars, one traveling 10 mph faster than the second car, start at the same time from the same point and travel in opposite directions. In 3 h they are 300 mi apart. Find the rate of each car.

Strategy

▷ Rate of 1st car: r
Rate of 2nd car: $r + 10$

	Rate	Time	Distance
1st car	r	3	$3r$
2nd car	$r + 10$	3	$3(r + 10)$

▷ The total distance traveled by the two cars is 300 mi.

Solution

$$\begin{aligned} 3r + 3(r + 10) &= 300 \\ 3r + 3r + 30 &= 300 \\ 6r + 30 &= 300 \\ 6r &= 270 \\ r &= 45 \end{aligned}$$

$r + 10 = 45 + 10 = 55$

The first car is traveling 45 mph.
The second car is traveling 55 mph.

Example 2

Two trains, one traveling at twice the speed of the other, start at the same time from stations which are 288 mi apart and travel toward each other. In 3 h, the trains pass each other. Find the rate of each train.

Your strategy

Your solution

Example 3

How far can a bicycling club ride out into the country at a speed of 12 mph and return over the same road at 8 mph if they travel a total of 10 h?

Strategy

▷ Time spent riding out: t
Time spent riding back: $10 - t$

	Rate	Time	Distance
Out	12	t	$12t$
Back	8	$10 - t$	$8(10 - t)$

▷ The distance out equals the distance back.

Solution

$$\begin{aligned} 12t &= 8(10 - t) \\ 12t &= 80 - 8t \\ 20t &= 80 \\ t &= 4 \text{ (The time is 4 h.)} \end{aligned}$$

The distance out $= 12t = 12(4) = 48$ mi.

The club can ride 48 mi into the country.

Example 4

On a survey mission, a pilot flew out to a parcel of land and back in 5 h. The rate out was 150 mph and the rate returning was 100 mph. How far away was the parcel of land?

Your strategy

Your solution

Solutions on p. A22

6.1 Application Problems

Solve.

1. Two cyclists start from the same point and ride in opposite directions. One cyclist rides twice as fast as the other. In three hours they are 72 mi apart. Find the rate of each cyclist.

2. Two small planes start from the same point and fly in opposite directions. The first plane is flying 25 mph slower than the second plane. In two hours the planes are 430 mi apart. Find the rate of each plane.

3. A motorboat leaves a harbor and travels at an average speed of 8 mph toward a small island. Two hours later a cabin cruiser leaves the same harbor and travels at an average speed of 16 mph toward the same island. In how many hours after the cabin cruiser leaves will the cabin cruiser be alongside the motorboat?

4. A long-distance runner started on a course running at an average speed of 6 mph. One hour later, a second runner began the same course at an average speed of 8 mph. How long after the second runner started will the second runner overtake the first runner?

5. A family drove to a resort at an average speed of 30 mph and later returned over the same road at an average speed of 50 mph. Find the distance to the resort if the total driving time was 8 h.

6. As part of flight training, a student pilot was required to fly to an airport and then return. The average speed to the airport was 90 mph, and the average speed returning was 120 mph. Find the distance between the two airports if the total flying time was 7 h.

7. Running at an average rate of 8 m/s, a sprinter ran to the end of a track and then jogged back to the starting point at an average rate of 3 m/s. The sprinter took 55 s to run to the end of the track and jog back. Find the length of the track.

8. Three campers left their campsite by canoe and paddled downstream at an average rate of 8 mph. They then turned around and paddled back upstream at an average rate of 4 mph to return to their campsite. How long did it take the campers to canoe downstream if the total trip took 1 h?

9. A car traveling at 48 mph overtakes a cyclist who, riding at 12 mph, has had a 3 h head start. How far from the starting point does the car overtake the cyclist?

10. A jet plane traveling at 600 mph overtakes a propeller-driven plane which has had a 2 h head start. The propeller-driven plane is traveling at 200 mph. How far from the starting point does the jet overtake the propeller-driven plane?

Solve.

11. On a 195-mile trip, a car traveled at an average speed of 45 mph and then reduced the speed to 30 mph for the remainder of the trip. The trip took a total of 5 h. For how long did the car travel at each speed?

12. A 555-mile, 5-hour plane trip was flown at two speeds. For the first part of the trip, the average speed was 105 mph. For the remainder of the trip, the average speed was 115 mph. For how long did the plane fly at each speed?

13. A bus traveled on a level road for 2 h at an average speed of 20 mph faster than it traveled on a winding road. The time spent on the winding road was 3 h. Find the average speed on the winding road if the total trip was 200 mi.

14. After a sailboat had been on the water for 3 h, a change in wind direction reduced the average speed of the boat by 5 mph. The entire distance sailed was 57 mi. The total time spent sailing was 6 h. How far did the sailboat travel in the first 3 h?

15. An executive drove from home at an average speed of 30 mph to an airport where a helicopter was waiting. The executive boarded the helicopter and flew to the corporate offices at an average speed of 60 mph. The entire distance was 150 mi. The entire trip took 3 h. Find the distance from the airport to the corporate offices.

16. A passenger train leaves a train depot two hours after a freight train leaves the same depot. The freight train is traveling 20 mph slower than the passenger train. Find the rate of each train if the passenger train overtakes the freight train in 3 h.

17. A car and a bus set out at 2 p.m. from the same point headed in the same direction. The average speed of the car is 30 mph slower than twice the speed of the bus. In two hours, the car is 20 mi ahead of the bus. Find the rate of the car.

18. A cyclist and a jogger set out at 11 a.m. from the same point headed in the same direction. The average speed of the cyclist is twice the speed of the jogger. In one hour, the cyclist is 8 mi ahead of the jogger. Find the rate of the cyclist.

19. Two cyclists start at the same time from opposite ends of a course which is 45 mi long. One cyclist is riding at 14 mph and the second cyclist is riding at 16 mph. How long after they begin will they meet?

20. Two joggers start at the same time from opposite ends of a 10-mile course. One jogger is running at 4 mph and the other is running at 6 mph. How long after they begin will they meet?

SECTION 7 Geometry Problems

7.1 Objective To solve perimeter problems

Reference for Computer Tutor™

The **perimeter** of a geometric figure is a measure of the distance around the figure. The equations for the perimeters of a rectangle and a triangle are shown at the right.

Rectangle: length l, width w. Perimeter $= 2l + 2w$

Triangle: sides a, b, c. Perimeter $= a + b + c$

The perimeter of a rectangle is 26 ft. The length of the rectangle is 1 ft more than twice the width. Find the width of the rectangle.

STRATEGY FOR SOLVING A PERIMETER PROBLEM

▷ Let a variable represent the measure of one of the unknown sides of the figure. Express the measures of the remaining sides in terms of that variable.

Width: w
Length: $2w + 1$

▷ Determine which perimeter equation to use.

Use the equation for the perimeter of a rectangle.

$$\begin{aligned} 2l + 2w &= P \\ 2(2w + 1) + 2w &= 26 \\ 4w + 2 + 2w &= 26 \\ 6w + 2 &= 26 \\ 6w &= 24 \\ w &= 4 \end{aligned}$$

The width is 4 ft.

Example 1

The perimeter of a triangle is 25 ft. Two sides of the triangle are equal. The third side is 2 ft less than the length of one of the equal sides. Find the measure of one of the equal sides.

Strategy

▷ Each equal side: x
The third side: $x - 2$
▷ Use the equation for the perimeter of a triangle.

Solution

$$\begin{aligned} a + b + c &= P \\ x + x + (x - 2) &= 25 \\ 3x - 2 &= 25 \\ 3x &= 27 \\ x &= 9 \end{aligned}$$

Each of the equal sides measures 9 ft.

Example 2

The perimeter of a rectangle is 34 m. The width of the rectangle is 3 m less than the length. Find the measure of the width.

Your strategy

Your solution

Solution on p. A22

7.2 Objective To solve problems involving the angles of a triangle

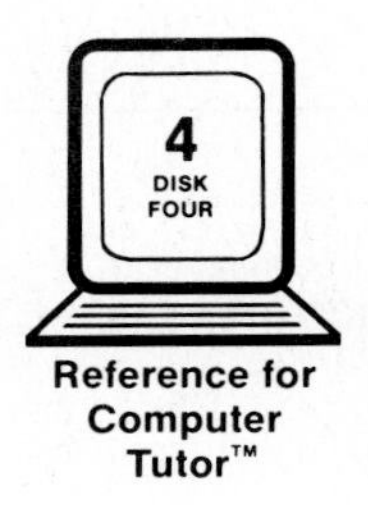

In a triangle, the sum of the measures of all the angles is 180°.

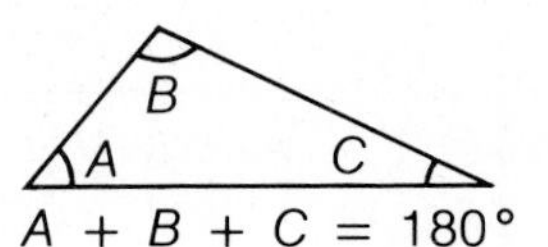

Two special types of triangles are shown at the right. A **right triangle** has one right angle (90°). An **isosceles triangle** has two equal angles.

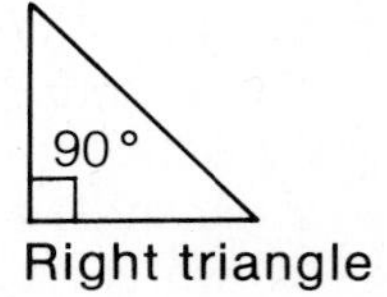

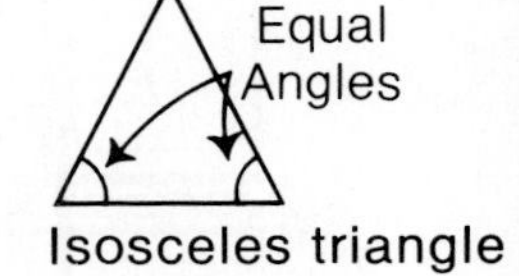

In a right triangle, the measure of one angle is twice the measure of the smallest angle. Find the measure of the smallest angle.

STRATEGY FOR SOLVING A PROBLEM INVOLVING THE ANGLES OF A TRIANGLE

▷ Let a variable represent one of the unknown angles. Express the other angles in terms of that variable.

Measure of smallest angle: x
Measure of second angle: $2x$
Measure of right angle: 90°

▷ Use the equation $A + B + C = 180°$

$$x + 2x + 90 = 180$$
$$3x + 90 = 180$$
$$3x = 90$$
$$x = 30$$

The measure of the smallest angle is 30°.

Example 3

In an isosceles triangle, the measure of one angle is 20° more than twice the measure of one of the equal angles. Find the measure of one of the equal angles.

Strategy

▷ Measure of one of the equal angles: x
Measure of the second equal angle: x
Measure of the third angle: $2x + 20$
▷ Use the equation $A + B + C = 180°$

Solution

$$x + x + (2x + 20) = 180$$
$$4x + 20 = 180$$
$$4x = 160$$
$$x = 40$$

The measure of one of the equal angles is 40°.

Example 4

In a triangle, the measure of one angle is twice the measure of the second angle. The measure of the third angle is 4° less than the measure of the second angle. Find the measure of each angle.

Your strategy

Your solution

Solution on p. A22

7.1 Application Problems

Solve.

1. The perimeter of a rectangle is 50 m. The width of the rectangle is 5 m less than the length. Find the length and width of the rectangle.

2. The perimeter of a rectangle is 120 ft. The length of the rectangle is twice the width. Find the length and width of the rectangle.

3. The width of a rectangle is 25% of the length. The perimeter is 250 cm. Find the length and width of the rectangle.

4. The width of a rectangle is 30% of the length. The perimeter of the rectangle is 338 ft. Find the length and width of the rectangle.

5. In an isosceles triangle, two sides are equal. The third side is 2 m less than one of the equal sides. The perimeter is 10 m. Find the length of each side.

6. The perimeter of a triangle is 33 ft. One side of the triangle is 1 ft longer than the second side. The third side is 2 ft longer than the second side. Find the measure of each side.

7. The perimeter of a rectangle is 42 m. The length of the rectangle is 3 m less than twice the width. Find the length and width of the rectangle.

8. The perimeter of a triangle is 20 ft. The first side is 1 ft less than twice the second side. The third side is 1 ft more than twice the second side. Find the measure of each side.

9. The perimeter of a triangle is 110 cm. One side is twice the second side. The third side is 30 cm more than the second side. Find the measure of each side.

10. In an isosceles triangle, two sides are equal. The third side is 50% of the length of one of the equal sides. Find the length of each side when the perimeter is 125 ft.

11. The perimeter of a rectangle is 56.24 m. The width of the rectangle is 0.84 m less than the length. Find the length and width of the rectangle.

12. In an isosceles triangle, two sides are equal. The length of one of the equal sides is 3.52 times the length of the third side. The perimeter is 10.43 m. Find the length of each side. Round to the nearest hundredth.

7.2 Application Problems

Solve.

13. In an equiangular triangle, all three angles are equal. Find the measures of the equal angles.

14. In an isosceles triangle, one angle is three times the measure of one of the equal angles. Find the measure of each angle.

15. In an isosceles right triangle, two angles are equal and the third angle is 90°. Find the measures of the equal angles.

16. One angle of a right triangle is 3° less than twice the measure of the smallest angle. Find the measure of each angle.

17. In an isosceles triangle, one angle is 12° more than twice the measure of one of the equal angles. Find the measure of each angle.

18. In an isosceles triangle, one angle is 5° less than three times the measure of one of the equal angles. Find the measure of each angle.

19. In a triangle, one angle is twice the measure of the second angle. The third angle is three times the measure of the second angle. Find the measure of each angle.

20. In a triangle, one angle is 5° more than the measure of the second angle. The third angle is 10° more than the measure of the second angle. Find the measure of each angle.

21. One angle of a triangle is three times the measure of the third angle. The second angle is 5° less than the measure of the third angle. Find the measure of each angle.

22. One angle of a triangle is twice the measure of the second angle. The third angle is three times the measure of the first angle. Find the measure of each angle.

23. The first angle of a triangle is twice the measure of the second angle. The third angle is 10° less than the measure of the first angle. Find the measure of each angle.

24. The first angle of a triangle is three times the measure of the second angle. The third angle is 33° more than the measure of the first angle. Find the measure of each angle.

SECTION 8 Puzzle Problems

8.1 Objective To solve consecutive integer problems

Reference for Computer Tutor™

Recall that the integers are the numbers . . . $-4, -3, -2, -1, 0, 1, 2, 3, 4,$ . . . An **even integer** is an integer that is divisible by 2. Examples of even integers are -8, 0, and 22. An **odd integer** is an integer that is not divisible by 2. Examples of odd integers are -17, 1, and 39.

Consecutive integers are integers which follow one another in order. Examples of consecutive integers are shown at the right. (Assume that the variable n represents an integer.)

$11, 12, 13$
$-8, -7, -6$
$n, n + 1, n + 2$

Examples of **consecutive even integers** are shown at the right. (Assume that the variable n represents an even integer.)

$24, 26, 28$
$-10, -8, -6$
$n, n + 2, n + 4$

Examples of **consecutive odd integers** are shown at the right. (Assume that the variable n represents an odd integer.)

$19, 21, 23$
$-1, 1, 3$
$n, n + 2, n + 4$

The sum of three consecutive odd integers is 45. Find the integers.

STRATEGY FOR SOLVING A CONSECUTIVE INTEGER PROBLEM

▷ Let a variable represent one of the integers. Express each of the other integers in terms of that variable. Remember that consecutive integers will differ by 1. Consecutive even or consecutive odd integers differ by 2.

Represent three consecutive odd integers.

First odd integer: n
Second odd integer: $n + 2$
Third odd integer: $n + 4$

▷ Determine the relationship among the integers.

The sum of the three odd integers is 45.

$$n + (n + 2) + (n + 4) = 45$$
$$3n + 6 = 45$$
$$3n = 39$$
$$n = 13$$

$$n + 2 = 13 + 2 = 15$$

$$n + 4 = 13 + 4 = 17$$

The three consecutive odd integers are 13, 15, and 17.

Example 1

Find three consecutive even integers such that three times the second is four more than the sum of the first and third.

Strategy

▷ First even integer: n
Second even integer: $n + 2$
Third even integer: $n + 4$
▷ Three times the second equals four more than the sum of the first and third.

Solution

$$3(n + 2) = n + (n + 4) + 4$$
$$3n + 6 = 2n + 8$$
$$n + 6 = 8$$
$$n = 2$$

$n + 2 = 2 + 2 = 4$

$n + 4 = 2 + 4 = 6$

The three even integers are 2, 4, and 6.

Example 2

Find three consecutive integers whose sum is -6.

Your strategy

Your solution

8.2 Objective To solve coin and stamp problems

Reference for Computer Tutor™

In solving problems dealing with coins or stamps of different values, it is necessary to represent the value of the coins or stamps in the same unit of money. The unit of money is frequently cents. For example:

The value of 3 quarters in cents is $3 \cdot 25$, or 75 cents.
The value of 4 nickels in cents is $4 \cdot 5$, or 20 cents.
The value of d dimes in cents is $d \cdot 10$, or $10d$ cents.

A coin bank contains $1.35 in dimes and quarters. In all, there are nine coins in the bank. Find the number of dimes and the number of quarters in the bank.

STRATEGY FOR SOLVING A COIN PROBLEM

▷ For each denomination of coin, write a numerical or variable expression for the number of coins, the value of the coin, and the total value of the coins in cents. The results can be recorded in a table.

The total number of coins is 9.

Number of quarters: x
Number of dimes: $9 - x$

Coin	Number of coins	·	Value of coin in cents	=	Total value in cents
Quarter	x	·	25	=	$25x$
Dime	$9 - x$	·	10	=	$10(9 - x)$

▷ Determine the relationship between the total values of the coins. Use the fact that the sum of the total values of each denomination of coin is equal to the total value of all the coins.

The sum of the total values of each denomination of coin is equal to the total value of all the coins (135 cents).

$$25x + 10(9 - x) = 135$$
$$25x + 90 - 10x = 135$$
$$15x + 90 = 135$$
$$15x = 45$$
$$x = 3$$

$$9 - x = 9 - 3 = 6$$

There are 3 quarters and 6 dimes in the bank.

Example 3

A collection of stamps consists of 3¢ stamps and 8¢ stamps. The number of 8¢ stamps is five more than three times the number of 3¢ stamps. The total value of all the stamps is $1.48. Find the number of 3¢ stamps.

Strategy

▷ Number of 3¢ stamps: x
Number of 8¢ stamps: $3x + 5$

Stamp	Number	Value	Total value
3¢	x	3	$3x$
8¢	$3x + 5$	8	$8(3x + 5)$

▷ The sum of the total values of each type of stamp equals the total value of all the stamps (148 cents).

Solution

$$3x + 8(3x + 5) = 148$$
$$3x + 24x + 40 = 148$$
$$27x + 40 = 148$$
$$27x = 108$$
$$x = 4$$

There are four 3¢ stamps in the collection.

Example 4

A coin bank contains nickels, dimes, and quarters. There are four times as many nickels as dimes, and five more quarters than dimes. The total value of all the coins is $6.75. Find the number of each kind of coin in the bank.

Your strategy

Your solution

Solution on p. A23

8.3 Objective To solve age problems

Reference for Computer Tutor™

The goal of an age problem is to determine the age of a person or an object.

A painting is 20 years old and a sculpture is 10 years old. How many years ago was the painting three times as old as the sculpture was then?

STRATEGY FOR SOLVING AN AGE PROBLEM

▷ Represent the ages in terms of numerical or variable expressions. To represent a past age, subtract from the present age. To represent a future age, add to the present age. The results can be recorded in a table.

The number of years ago: x

	Present age	Past age
Painting	20	$20 - x$
Sculpture	10	$10 - x$

▷ Determine the relationship among the ages.

At a past age, the painting was three times as old as the sculpture was then.

$$20 - x = 3(10 - x)$$
$$20 - x = 30 - 3x$$
$$20 + 2x = 30$$
$$2x = 10$$
$$x = 5$$

Five years ago the painting was three times as old as the sculpture.

Example 5

A stamp collector has a 3¢ stamp which is 25 years older than a 5¢ stamp. In 18 years, the 3¢ stamp will be twice as old as the 5¢ stamp will be then. How old are the stamps now?

Strategy

▷ Present age of 5¢ stamp: x

	Present	Future
3¢ stamp	$x + 25$	$x + 43$
5¢ stamp	x	$x + 18$

▷ At a future age, the 3¢ stamp will be twice as old as the 5¢ stamp.

Solution

$$2(x + 18) = x + 43$$
$$2x + 36 = x + 43$$
$$x + 36 = 43$$
$$x = 7$$
$$x + 25 = 7 + 25 = 32$$

The 3¢ stamp is 32 years old and the 5¢ stamp is 7 years old.

Example 6

A half-dollar is now 25 years old. A dime is 15 years old. How many years ago was the half-dollar twice as old as the dime?

Your strategy

Your solution

8.1 Application Problems

Solve.

1. The sum of three consecutive integers is 48. Find the integers.

2. The sum of three consecutive integers is 60. Find the integers.

3. The sum of three consecutive even integers is 66. Find the integers.

4. The sum of three consecutive even integers is 42. Find the integers.

5. The sum of three consecutive odd integers is 51. Find the integers.

6. The sum of three consecutive odd integers is 75. Find the integers.

7. Find two consecutive even integers such that three times the first equals twice the second.

8. Find two consecutive even integers such that four times the first is three times the second.

9. Five times the first of two consecutive odd integers equals three times the second. Find the integers.

10. Seven times the first of two consecutive odd integers is five times the second. Find the integers.

11. Find three consecutive integers whose sum is negative twenty-one.

12. Find three consecutive even integers whose sum is negative eighteen.

13. Twice the smallest of three consecutive odd integers is seven more than the largest. Find the integers.

14. Three times the smallest of three consecutive even integers is four more than twice the largest. Find the integers.

Solve.

15. Find three consecutive odd integers such that three times the middle integer is one more than the sum of the first and third.

16. Find three consecutive even integers such that three times the middle integer is four more than the sum of the first and third.

8.2 Application Problems

Solve.

17. A coin purse contains 16 coins in nickels and dimes. The coins have a total value of $1. Find the number of nickels and dimes in the coin purse.

18. A bank contains 30 coins in dimes and quarters. The coins have a total value of $5.40. Find the number of dimes and quarters in the bank.

19. A postal clerk sold some 15¢ stamps and some 25¢ stamps. Altogether 10 stamps were sold for a total cost of $1.70. How many of each type of stamp were sold?

20. A business executive purchased 40 stamps for $7.70. The purchase included 15¢ stamps and 20¢ stamps. How many of each type of stamp were purchased?

21. The total value of the dimes and quarters in a bank is $6.50. There are five more quarters than dimes. Find the number of each type of coin in the bank.

22. A drawer contains 15¢ stamps and 18¢ stamps. The number of 15¢ stamps is two less than three times the number of 18¢ stamps. The total value of all the stamps is $.96. How many 15¢ stamps are in the drawer?

23. A bank teller cashed a check for $150 using twenty-dollar bills and ten-dollar bills. In all, nine bills were handed to the customer. Find the number of twenty-dollar bills and ten-dollar bills.

24. A total of 30 bills are in a cash box. Some of the bills are one-dollar bills and the rest are five-dollar bills. The total amount of cash in the box is $50. Find the number of each type of bill in the cash box.

Solve.

25. A coin bank contains pennies, nickels, and quarters. There are six times as many nickels as pennies and four times as many quarters as pennies. The total amount of money in the bank is $6.55. Find the number of pennies in the bank.

26. A coin bank contains pennies, nickels, and dimes. There are five times as many nickels as pennies and three times as many dimes as pennies. The total amount of money in the bank is $6.72. Find the number of pennies in the bank.

27. A collection of stamps consists of 2¢ stamps, 5¢ stamps, and 7¢ stamps. There are eight more 2¢ stamps than 5¢ stamps, and twice as many 7¢ stamps as 5¢ stamps. The total value of the stamps is $1.63. Find the number of each type of stamp in the collection.

28. A collection of stamps consists of 3¢ stamps, 7¢ stamps, and 12¢ stamps. The number of 3¢ stamps is six less than the number of 7¢ stamps. The number of 12¢ stamps is one half the number of 7¢ stamps. The total value of all the stamps is $3.02. Find the number of each type of stamp in the collection.

29. A child's piggy bank contains nickels, dimes, and quarters. There are twice as many nickels as dimes and three more quarters than nickels. The total value of all the coins is $11.25. Find the number of each type of coin.

30. A collection of stamps consists of 6¢ stamps, 8¢ stamps, and 15¢ stamps. The number of 6¢ stamps is three times the number of 8¢ stamps. There are seven more 15¢ stamps than there are 6¢ stamps. The total value of all the stamps is $4.60. Find the number of each type of stamp.

8.3 Application Problems

Solve.

31. A book dealer has an autographed, first-edition book which is 35 years old and a reprint of the book which is 7 years old. In how many years will the autographed first edition be three times as old as the reprint will be then?

32. A collector of hand woven rugs has an oval rug which is 42 years old and a circular rug which is 14 years old. How many years ago was the oval rug five times as old as the circular rug was then?

33. A coin collector has a dime which is 24 years older than a nickel. In 8 years the dime will be twice as old as the nickel will be then. Find the present age of the dime and nickel.

34. An oil painting is 10 years older than a lithograph. Five years ago the painting was twice as old as the lithograph was then. Find the present age of each.

Solve.

35. An art collector has a porcelain vase which is 15 years old and a crystal vase which is 95 years old. In how many years will the crystal vase be three times as old as the porcelain vase will be then?

36. A stamp collector has a 2¢ stamp which is 20 years old and a 3¢ stamp which is 16 years old. How many years ago was the 2¢ stamp twice as old as the 3¢ stamp was then?

37. An antique butterchurn is 85 years old and an antique ice box is 75 years old. How many years ago was the butterchurn twice the age the ice box was then?

38. A diamond ring is 2 years old and a ruby ring is 22 years old. In how many years will the ruby ring be twice the age the diamond ring will be then?

39. An antique car is 45 years older than a replica of the car. In 13 years, the antique car will be four times as old as the replica will be then. Find the present ages of the two cars.

40. A gold coin is 84 years older than a silver coin. Twenty years ago, the gold coin was three times as old as the silver coin was then. Find the present ages of the two coins.

41. The sum of the ages of an oil painting and a watercolor is 20 years. The oil painting one year from now will be nine times the age of the watercolor one year ago. Find the present age of each painting.

42. The sum of the ages of two cars is 8 years. Two years ago the age of the older car was three times the age the younger car was then. Find the present age of each car.

43. The sum of the ages of a 5¢ coin and a 10¢ coin is 12. Two years from now the age of the 5¢ coin will equal the age of the 10¢ coin two years ago. Find the present age of each coin.

44. The sum of the ages of two children is 18. Six years from now the age of the older child will be twice the age of the younger child. Find the present ages of the two children.

Calculators and Computers

The EEX Key on a Calculator

Many application problems require the use of very large or very small numbers. The EEX key, or on some calculators, the EXP key, is used for entering these numbers. For example, the speed of light is approximately 29,800,000,000 cm/sec. This number cannot be directly entered on a calculator. The EEX key (which means exponent) is used.

To understand this key, consider the following table:

$$5489 = 548.9 \times 10^1$$
$$5489 = 54.89 \times 10^2$$
$$5489 = 5.489 \times 10^3$$

The number 5489 can be represented in various forms. Note that each time the decimal point is moved to the *left,* it is necessary to *multiply* by a power of ten.

Now study the following table:

$$0.004638 = 0.04638 \div 10^1$$
$$0.004638 = 0.4638 \div 10^2$$
$$0.004638 = 4.638 \div 10^3$$

The number 0.004638 can be represented in various forms. Note that each time the decimal point is moved to the *right,* it is necessary to *divide* by a power of ten. It is customary to express dividing by a power of ten by using multiplication and a negative exponent. Thus the above table could be written as follows:

$$0.004638 = 0.04638 \times 10^{-1}$$
$$0.004638 = 0.4638 \times 10^{-2}$$
$$0.004638 = 4.638 \times 10^{-3}$$

To enter the speed of light on a calculator, rewrite the number by moving the decimal point 10 places to the left and then multiplying by 10^{10}.

Now enter the number. 2.98 EEX 10

The EEX key is used to enter the exponent on 10.

The wave length of an x-ray is approximately 0.00000000537 cm. To enter this number on your calculator, rewrite the number by moving the decimal point 9 places to the right and then dividing by 10^9 (or multiplying by 10^{-9}).

$$0.00000000537 = 5.37 \div 10^9 = 5.37 \times 10^{-9}$$

Now enter the number. 5.37 EEX 9 +/−

The +/− key is used to change the sign of the exponent and thus make it negative.

Chapter Summary

KEY WORDS

Percent means "parts of 100."

Cost is the price which a business pays for a product.

Selling price is the price for which a business sells a product to a customer.

Markup is the difference between selling price and cost.

Discount is the amount by which a retailer reduces the regular price of a product.

Uniform motion means that an object at a constant speed moves in a straight line.

Perimeter of a geometric figure is a measure of the distance around the figure.

A **right angle** is an angle whose measure is 90 degrees.

A **right triangle** has one right angle.

An **isosceles triangle** has two equal angles and two equal sides.

Consecutive integers are integers which follow one another in order.

ESSENTIAL RULES

Basic Percent Equation: Percent × Base = Amount
$P \times B = A$

Basic Markup Equation: Selling Price = cost + markup
$S = C + M$

Markup = markup rate × cost
$M = r \times C$

Basic Discount Equations: Sale price = regular price − discount
$S = R - D$

Discount = discount rate × regular price
$D = r \times R$

Annual Simple Interest Equation: Simple Interest = Principal × simple interest rate
$I = P \times r$

Value Mixture Equation: Value = amount × unit cost
$V = A \times C$

Percent Mixture Equation: Quantity = amount × percent concentration
$Q = A \times r$

Uniform Motion Equation: Distance = rate × time
$D = r \times t$

Triangle Equation: $\angle A + \angle B + \angle C = 180°$

Review/Test

SECTION 1

1.1a Write 60% as a fraction and a decimal.

1.1b Write $62\frac{1}{2}\%$ as a fraction.

1.2a Write 0.375 as a percent.

1.2b Write $\frac{7}{8}$ as a percent. Write the remainder in fractional form.

SECTION 2

2.1a Find 16% of 40.

2.1b 20 is what percent of 16?

2.2 The value of a personal computer today is $2400. This is 80% of the computer's value last year. Find the value of the computer last year.

SECTION 3

3.1 The manager of a sports shop uses a markup rate of 50%. The selling price for a set of golf clubs is $300. Find the cost of the golf clubs.

3.2 A portable typewriter which regularly sells for $100 is on sale for $80. Find the discount rate.

SECTION 4

4.1 A total of $7000 is deposited into two simple interest accounts. On one account, the annual simple interest rate is 7%, while on the second account the simple interest rate is 9%. How much should be invested in each account so that the total annual interest earned is $600?

Review/Test

SECTION 5

5.1 A coffee merchant wants to make 12 lb of a blend of coffee to sell for \$6 per pound. The blend is made using a \$7 grade and a \$4 grade of coffee. How many pounds of each of these grades should be used?

5.2 How many gallons of a 15% acid solution must be mixed with 5 gal of a 20% acid solution to make a 16% acid solution?

SECTION 6

6.1 Two planes start at the same time from the same point and fly in opposite directions. The first plane is flying 100 mph faster than the second plane. In three hours the two planes are 1050 mi apart. Find the rate of each plane.

SECTION 7

7.1 The perimeter of a rectangle is 38 m. The length of the rectangle is 1 m less than three times the width. Find the length and width of the rectangle.

7.2 In a triangle, the first angle is 15° more than the second angle. The third angle is three times the second angle. Find the measure of each angle.

SECTION 8

8.1 Find three consecutive odd integers such that three times the first integer is one less than the sum of the second and third integers.

8.2 A coin bank contains 50 coins in nickels and quarters. The total amount of money in the bank is \$9.50. Find the number of nickels and the number of quarters in the bank.

8.3 The age of a 20¢ stamp is 5 years, and the age of a 5¢ stamp is 35 years. In how many years will the 5¢ stamp be three times the age the 20¢ stamp will be then?

Cumulative Review/Test

1. Simplify $-2 + (-8) - (-16)$.

2. Simplify $\left(-\frac{2}{3}\right)^3\left(-\frac{3}{4}\right)^2$.

3. Simplify $\frac{5}{6} - \left(\frac{2}{3}\right)^2 \div \left(\frac{1}{2} - \frac{1}{3}\right)$.

4. Evaluate $b^2 - (a - b)^2$ when $a = 3$ and $b = -2$.

5. Simplify $2x - 3y - (-4x) + 7y$.

6. Simplify $-5(3 - 2x - 5x^3)$.

7. Simplify $-3[x - 2(x - 3) - 4]$.

8. Is 2 a solution of $4 - 2x - x^2 = 2 - 4x$?

9. Solve: $5 - x = 2$

10. Solve: $-\frac{5}{8}x = 10$

11. Solve: $5 - 8x = -3$

12. Solve: $-2x - 3(4 - 2x) = 5x + 4$

13. Write 80% as a fraction.

14. Write $16\frac{2}{3}\%$ as a fraction.

15. The sum of two numbers is fifteen. The sum of five times the smaller number and eight equals five less than the product of three and the larger number. Find the two numbers.

16. An auto repair bill was $213. This includes $88 for parts and $25 for each hour of labor. Find the number of hours of labor.

Cumulative Review/Test

17. Write 0.075 as a percent.

18. Write $\frac{2}{25}$ as a percent.

19. Find $83\frac{1}{3}\%$ of 24.

20. 30% of what is 12?

21. A survey of 250 librarians showed that 50 of the libraries had a particular reference book on their shelves. What percent of the libraries had the reference book?

22. A deposit of $4000 is made into an account which earns 11% simple interest. How much additional money must be deposited into an account which pays 14% simple interest so that the total interest earned is 12% of the total investment?

23. A department store buys a chain necklace for $8 and sells it for $14. Find the markup rate.

24. A file cabinet which normally sells for $99 is on sale for 20% off. Find the sale price.

25. How many grams of a gold alloy which cost $4 a gram must be mixed with 30 g of a gold alloy which costs $7 a gram to make an alloy which sells for $5 a gram?

26. How many ounces of pure water must be added to 70 oz of a 10% salt solution to make a 7% salt solution?

27. The perimeter of a triangle is 49 ft. The first side is twice the length of the third side, and the second side is 5 ft more than the length of the third side. Find the measure of the first side.

28. In an isosceles triangle, two angles are equal. The third angle is 8° less than twice the measure of one of the equal angles. Find the measure of one of the equal angles.

29. Three times the second of three consecutive even integers is 14 more than the sum of the first and third integers. Find the middle even integer.

30. A coin bank contains dimes and quarters. The number of quarters is five less than four times the number of dimes. The total amount in the bank is $6.45. Find the number of dimes in the bank.

5

Polynomials

Objectives

- To add polynomials
- To subtract polynomials
- To multiply monomials
- To simplify powers of monomials
- To multiply a polynomial by a monomial
- To multiply two polynomials
- To multiply two binomials
- To multiply binomials which have special products
- To solve application problems
- To divide monomials
- To divide a polynomial by a monomial
- To divide polynomials
- To simplify expressions containing negative and zero exponents

Early Egyptian Arithmetic Operations

The early Egyptian arithmetic processes are recorded on the early Rhind Papyrus without showing the underlying principles. Scholars of today can only guess as to how these early developments were discovered.

Egyptian hieroglyphics used a base ten system of numbers in which a vertical line represented 1, a heel bone, ∩, represented 10, and a scroll, 9, represented 100.

The symbols at the right represent the number 237.
There are 7 vertical lines, 3 heel bones and 2 scrolls.
Thus, the symbols at the right represent 7 + 30 + 200 or 237.

|||| ||| ∩∩∩ 99

Addition in hieroglyphic notation does not require memorization of addition facts. Addition is done just by counting symbols.

Addition is a simple grouping operation.

Write down the total of each kind of symbol.

Group the 10 straight lines into one heel bone.

Hieroglyphic Notation	Modern Notation
|||| ∩∩ 999	324
+ |||| |||| ∩∩∩ 9	+ 138
|||| |||| |||| ∩∩∩∩ ∩ 9999	
12 + 50 + 400 =	462
|| ∩∩∩∩ ∩∩ 9999	

Subtraction in the hieroglyphic system is similar to making change. For example, what change do you get from a $1.00 bill when buying a $.55 item?

5 cannot be subtracted from 3, so a 10 is "borrowed" and 10 ones are added.

Notice that no zero is provided in this number system. That place value symbol is just not used. As shown at the right, the heel bone is not used as there are no 10's necessary in 208.

Hieroglyphic Notation	Modern Notation
||| ∩∩∩ 9999	433
− |||| | ∩∩ 99	−225
|||| |||| |||| | ∩∩ 9999	4 ~~3~~ 3 (2, 13)
− |||| | ∩∩ 99	−225
|||| |||| 99	208
8 + 200 =	208

SECTION 1 Addition and Subtraction of Polynomials

1.1 Objective To add polynomials

Reference for Computer Tutor™

A **monomial** is a number, a variable, or a product of numbers and variables. A **polynomial** is a variable expression in which the terms are monomials.

A **monomial** is a polynomial of *one* term. $5x^3$ is a monomial.
A **binomial** is a polynomial of *two* terms. $5y^2 - 3x$ is a binomial.
A **trinomial** is a polynomial of *three* terms. $6xy - 2r^2s + 4r$ is a trinomial.

The terms of a polynomial in one variable are usually arranged so that the exponents of the variable decrease from left to right. This is called **descending order.** The **degree of a polynomial** in one variable is its largest exponent. The degree of $4x^3 - 3x^2 + 6x - 1$ is 3. The degree of $5y^4 - 2y^3 + y^2 - 7y + 8$ is 4.

$$4x^3 - 3x^2 + 6x - 1$$
$$5y^4 - 2y^3 + y^2 - 7y + 8$$

Polynomials can be added, using either a vertical or horizontal format, by combining like terms.

Simplify $(2x^2 + x - 1) + (3x^3 + 4x^2 - 5)$. Use a vertical format.

Arrange the terms of each polynomial in descending order with like terms in the same column.
Combine the terms in each column.

$$\begin{array}{r} 2x^2 + x - 1 \\ + \; 3x^3 + 4x^2 \qquad - 5 \\ \hline 3x^3 + 6x^2 + x - 6 \end{array}$$

Simplify $(3x^3 - 7x + 2) + (7x^2 + 2x - 7)$. Use a horizontal format.

Use the Commutative and Associative Properties of Addition to rearrange and group like terms.

$(3x^3 - 7x + 2) + (7x^2 + 2x - 7)$

$3x^3 + 7x^2 + (-7x + 2x) + (2 - 7)$ Do this step mentally.

Combine like terms.
Write the polynomial in descending order.

$3x^3 + 7x^2 - 5x - 5$

Example 1
Simplify $(7y^2 - 6y + 9) + (-8y^2 - 2)$.
Use a vertical format.

Solution

$$\begin{array}{r} 7y^2 - 6y + 9 \\ + \; -8y^2 \qquad - 2 \\ \hline -y^2 - 6y + 7 \end{array}$$

Example 2
Simplify $(2x^2 + 4x - 3) + (5x^2 - 6x)$.
Use a vertical format.

Your solution

$$\begin{array}{r} 2x^2 + 4x - 3 \\ + \; 5x^2 - 6x \qquad \\ \hline 7x^2 - 2x - 3 \end{array}$$

Example 3
Simplify $(-4x^2 - 3xy + 2y^2) + (3x^2 - 4y^2)$.
Use a horizontal format.

Solution
$(-4x^2 - 3xy + 2y^2) + (3x^2 - 4y^2) =$
$-x^2 - 3xy - 2y^2$

Example 4
Simplify $(-3x^2 + 2y^2) + (-8x^2 + 9xy)$.
Use a horizontal format.

Your solution

Solutions on p. A26

1.2 Objective To subtract polynomials

Reference for Computer Tutor™

The **opposite** of the polynomial $x^2 - 2x + 3$ is $-(x^2 - 2x + 3)$

To simplify the opposite of a polynomial, change the sign of every term inside the parentheses.

$$-(x^2 - 2x + 3) = -x^2 + 2x - 3$$

Polynomials can be subtracted using either a vertical or horizontal format. To subtract, add the opposite of the second polynomial to the first.

Simplify $(-3x^2 - 7) - (-8x^2 + 3x - 4)$. Use a vertical format.

Arrange the terms of each polynomial in descending order with like terms in the same column.
Rewrite subtraction as addition of the opposite.
Combine the terms in each column.

$$\begin{array}{r} -3x^2 \qquad\quad -7 \\ - \;\; -8x^2 + 3x - 4 \\ \hline \end{array} = \begin{array}{r} -3x^2 \qquad\quad -7 \\ + \;\; 8x^2 - 3x + 4 \\ \hline 5x^2 - 3x - 3 \end{array}$$

Simplify $(5x^2 - 3x + 4) - (-3x^3 - 2x + 8)$. Use a horizontal format.

Rewrite subtraction as addition of the opposite.
Combine like terms.
Write the polynomial in descending order.

$$(5x^2 - 3x + 4) - (-3x^3 - 2x + 8)$$
$$(5x^2 - 3x + 4) + (3x^3 + 2x - 8)$$
$$3x^3 + 5x^2 - x - 4$$

Example 5

Simplify $(6y^2 - 3y - 1) - (7y^2 - y)$.
Use a vertical format.

Solution

$$\begin{array}{r} 6y^2 - 3y - 1 \\ - \;\; 7y^2 - \;\; y \qquad \\ \hline \end{array} = \begin{array}{r} 6y^2 - 3y - 1 \\ + \;\; -7y^2 + \;\; y \qquad \\ \hline -y^2 - 2y - 1 \end{array}$$

Example 6

Simplify $(8y^2 - 4xy + x^2) - (2y^2 - xy + 5x^2)$.
Use a vertical format.

Your solution

Example 7

Simplify $(4x^3 - 3x - 7) - (7x^2 - 4x - 2)$.
Use a horizontal format.

Solution

$(4x^3 - 3x - 7) - (7x^2 - 4x - 2)$
$(4x^3 - 3x - 7) + (-7x^2 + 4x + 2)$
$4x^3 - 7x^2 + x - 5$

Example 8

Simplify $(-3a^2 - 4a + 2) - (5a^3 + 2a - 6)$.
Use a horizontal format.

Your solution

Solutions on p. A26

1.1 Exercises

Simplify. Use a vertical format.

1. $(x^2 + 7x) + (-3x^2 - 4x)$
2. $(3y^2 - 2y) + (5y^2 + 6y)$
3. $(y^2 + 4y) + (-4y - 8)$
4. $(3x^2 + 9x) + (6x - 24)$
5. $(2x^2 + 6x + 12) + (3x^2 + x + 8)$
6. $(x^2 + x + 5) + (3x^2 - 10x + 4)$
7. $(x^3 - 7x + 4) + (2x^2 + x - 10)$
8. $(3y^3 + y^2 + 1) + (-4y^3 - 6y - 3)$
9. $(2a^3 - 7a + 1) + (-3a^2 - 4a + 1)$
10. $(5r^3 - 6r^2 + 3r) + (r^2 - 2r - 3)$

Simplify. Use a horizontal format.

11. $(4x^2 + 2x) + (x^2 + 6x)$
12. $(-3y^2 + y) + (4y^2 + 6y)$
13. $(4x^2 - 5xy) + (3x^2 + 6xy - 4y^2)$
14. $(2x^2 - 4y^2) + (6x^2 - 2xy + 4y^2)$
15. $(2a^2 - 7a + 10) + (a^2 + 4a + 7)$
16. $(-6x^2 + 7x + 3) + (3x^2 + x + 3)$
17. $(5x^3 + 7x - 7) + (10x^2 - 8x + 3)$
18. $(3y^3 + 4y + 9) + (2y^2 + 4y - 21)$
19. $(2r^2 - 5r + 7) + (3r^3 - 6r)$
20. $(3y^3 + 4y + 14) + (-4y^2 + 21)$
21. $(3x^2 + 7x + 10) + (-2x^3 + 3x + 1)$
22. $(7x^3 + 4x - 1) + (2x^2 - 6x + 2)$

1.2 Exercises

Simplify. Use a vertical format.

23. $(x^2 - 6x) - (x^2 - 10x)$

24. $(y^2 + 4y) - (y^2 + 10y)$

25. $(2y^2 - 4y) - (-y^2 + 2)$

26. $(-3a^2 - 2a) - (4a^2 - 4)$

27. $(x^2 - 2x + 1) - (x^2 + 5x + 8)$

28. $(3x^2 + 2x - 2) - (5x^2 - 5x + 6)$

29. $(4x^3 + 5x + 2) - (-3x^2 + 2x + 1)$

30. $(5y^2 - y + 2) - (-2y^3 + 3y - 3)$

31. $(2y^3 + 6y - 2) - (y^3 + y^2 + 4)$

32. $(-2x^2 - x + 4) - (-x^3 + 3x - 2)$

Simplify. Use a horizontal format.

33. $(y^2 - 10xy) - (2y^2 + 3xy)$

34. $(x^2 - 3xy) - (-2x^2 + xy)$

35. $(3x^2 + x - 3) - (x^2 + 4x - 2)$

36. $(5y^2 - 2y + 1) - (-3y^2 - y - 2)$

37. $(-2x^3 + x - 1) - (-x^2 + x - 3)$

38. $(2x^2 + 5x - 3) - (3x^3 + 2x - 5)$

39. $(4a^3 - 2a + 1) - (a^3 - 2a + 3)$

40. $(b^2 - 8b + 7) - (4b^3 - 7b - 8)$

41. $(4y^3 - y - 1) - (2y^2 - 3y + 3)$

42. $(3x^2 - 2x - 3) - (2x^3 - 2x^2 + 4)$

SECTION 2 Multiplication of Monomials

2.1 Objective To multiply monomials

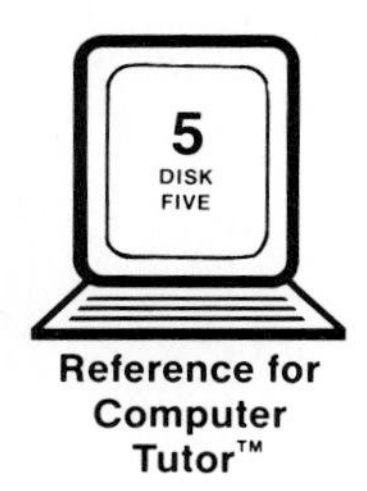

Reference for Computer Tutor™

Recall that in the exponential expression x^5, x is the base and 5 is the exponent. The exponent indicates the number of times the base occurs as a factor.

The product of exponential expressions with the *same* base can be simplified by writing each expression in factored form and writing the result with an exponent.

$$x^3 \cdot x^2 = \underbrace{\overbrace{(x \cdot x \cdot x)}^{3 \text{ factors}} \cdot \overbrace{(x \cdot x)}^{2 \text{ factors}}}_{5 \text{ factors}}$$
$$= x \cdot x \cdot x \cdot x \cdot x$$
$$= x^5$$

Note that adding the exponents results in the same product.

$$x^3 \cdot x^2 = x^{3+2} = x^5$$

Rule for Multiplying Exponential Expressions

If m and n are integers, then $x^m \cdot x^n = x^{m+n}$.

Simplify $a^2 \cdot a^6 \cdot a$.

The bases are the same. Add the exponents.

$$a^2 \cdot a^6 \cdot a = a^{2+6+1}$$ Do this step mentally.
$$= a^9$$

Simplify $(2xy)(3x^2y)$.

Use the Commutative and Associative Properties of Multiplication to rearrange and group factors. Multiply variables with like bases by adding the exponents.

$$(2xy)(3x^2y) = (2 \cdot 3)(x \cdot x^2)(y \cdot y)$$
$$= 6x^{1+2}y^{1+1}$$ Do this step mentally.
$$= 6x^3y^2$$

Example 1

Simplify $(-4y)(5y^3)$.

Solution

$(-4y)(5y^3) = (-4 \cdot 5)(y \cdot y^3) = -20y^4$

Example 2

Simplify $(3x^2)(6x^3)$.

Your solution

Example 3

Simplify $(2x^2y)(-5xy^4)$.

Solution

$(2x^2y)(-5xy^4) = [2(-5)](x^2 \cdot x)(y \cdot y^4) = -10x^3y^5$

Example 4

Simplify $(-3xy^2)(-4x^2y^3)$.

Your solution

Solutions on p. A26

2.2 Objective To simplify powers of monomials

Reference for Computer Tutor™

A power of a monomial can be simplified by rewriting the expression in factored form and then using the Rule for Multiplying Exponential Expressions.

$$(x^2)^3 = x^2 \cdot x^2 \cdot x^2 = x^{2+2+2} = x^6$$

$$(x^4y^3)^2 = (x^4y^3)(x^4y^3) = x^4 \cdot y^3 \cdot x^4 \cdot y^3 = (x^4 \cdot x^4)(y^3 \cdot y^3) = x^{4+4}y^{3+3} = x^8y^6$$

Note that multiplying each exponent inside the parentheses by the exponent outside the parentheses gives the same result.

$$(x^2)^3 = x^{2 \cdot 3} = x^6 \qquad (x^4y^3)^2 = x^{4 \cdot 2}y^{3 \cdot 2} = x^8y^6$$

Rule for Simplifying Powers of Exponential Expressions

If m and n are integers, then $(x^m)^n = x^{m \cdot n}$.

Rule for Simplifying Powers of Products

If m, n, and p are integers, then $(x^m \cdot y^n)^p = x^{m \cdot p}y^{n \cdot p}$.

Simplify $(x^5)^2$.

Multiply the exponents.

$$(x^5)^2 = x^{5 \cdot 2} = x^{10}$$

Do this step mentally.

Simplify $(3a^2b)^3$.

Multiply each exponent inside the parentheses by the exponent outside the parentheses.

$$(3a^2b)^3 = 3^{1 \cdot 3}a^{2 \cdot 3}b^{1 \cdot 3} = 3^3a^6b^3 = 27a^6b^3$$

Do this step mentally.

Example 5

Simplify $(2xy^3)^4$.

Solution

$(2xy^3)^4 = 2^4x^4y^{12} = 16x^4y^{12}$

Example 6

Simplify $(3x)(2x^2y)^3$.

Your solution

Example 7

Simplify $(-2x)(-3xy^2)^3$.

Solution

$(-2x)(-3xy^2)^3 = (-2x)(-3)^3x^3y^6 =$
$(-2x)(-27)x^3y^6 = [-2(-27)](x \cdot x^3)y^6 = 54x^4y^6$

Example 8

Simplify $(3x^2)^2(-2xy^2)^3$.

Your solution

Solutions on p. A26

2.1 Exercises

Simplify. write in descending order

1. $(x)(2x)$
2. $(-3y)(y)$
3. $(3x)(4x)$
4. $(7y^3)(7y^2)$
5. $(-2a^3)(-3a^4)$
6. $(5a^6)(-2a^5)$
7. $(x^2y)(xy^4)$
8. $(x^2y^4)(xy^7)$
9. $(-2x^4)(5x^5y)$
10. $(-3a^3)(2a^2b^4)$
11. $(x^2y^4)(x^5y^4)$
12. $(a^2b^4)(ab^3)$
13. $(2xy)(-3x^2y^4)$
14. $(-3a^2b)(-2ab^3)$
15. $(x^2yz)(x^2y^4)$
16. $(-ab^2c)(a^2b^5)$
17. $(a^2b^3)(ab^2c^4)$
18. $(x^2y^3z)(x^3y^4)$
19. $(-a^2b^2)(a^3b^6)$
20. $(xy^4)(-xy^3)$
21. $(-6a^3)(a^2b)$
22. $(2a^2b^3)(-4ab^2)$
23. $(-5y^4z)(-8y^6z^5)$
24. $(3x^2y)(-4xy^2)$
25. $(10ab^2)(-2ab)$
26. $(x^2y)(yz)(xyz)$
27. $(xy^2z)(x^2y)(z^2y^2)$
28. $(-2x^2y^3)(3xy)(-5x^3y^4)$
29. $(4a^2b)(-3a^3b^4)(a^5b^2)$
30. $(3ab^2)(-2abc)(4ac^2)$
31. $(3a^2b)(-6bc)(2ac^2)$

2.2 Exercises

Simplify.

32. $(2^2)^3$ **33.** $(3^2)^2$ **34.** $(-2)^2$ **35.** $(-3)^3$

36. $(-2^2)^3$ **37.** $(-2^3)^3$ **38.** $(x^3)^3$ **39.** $(y^4)^2$

40. $(x^7)^2$ **41.** $(y^5)^3$ **42.** $(-x^2)^2$ **43.** $(-x^2)^3$

44. $(2x)^2$ **45.** $(3y)^3$ **46.** $(-2x^2)^3$ **47.** $(-3y^3)^2$

48. $(x^2y^3)^2$ **49.** $(x^3y^4)^5$ **50.** $(3x^2y)^2$ **51.** $(-2ab^3)^4$

52. $(a^2)(3a^2)^3$ **53.** $(b^2)(2a^3)^4$ **54.** $(-2x)(2x^3)^2$

55. $(2y)(-3y^4)^3$ **56.** $(x^2y)(x^2y)^3$ **57.** $(a^3b)(ab)^3$

58. $(ab^2)^2(ab)^2$ **59.** $(x^2y)^2(x^3y)^3$ **60.** $(-2x)(-2x^3y)^3$

61. $(-3y)(-4x^2y^3)^3$ **62.** $(-2x)(-3xy^2)^2$ **63.** $(-3y)(-2x^2y)^3$

64. $(ab^2)(-2a^2b)^3$ **65.** $(a^2b^2)(-3ab^4)^2$ **66.** $(-2a^3)(3a^2b)^3$

67. $(-3b^2)(2ab^2)^3$ **68.** $(-3ab)^2(-2ab)^3$ **69.** $(-3a^2b)^3(-3ab)^3$

SECTION 3 Multiplication of Polynomials

3.1 Objective To multiply a polynomial by a monomial

Reference for Computer Tutor™

To multiply a polynomial by a monomial, use the Distributive Property and the Rule for Multiplying Exponential Expressions.

Simplify $-2x(x^2 - 4x - 3)$.

Use the Distributive Property. $-2x(x^2 - 4x - 3)$

Use the Rule for Multiplying Exponential Expressions. $-2x(x^2) - (-2x)(4x) - (-2x)(3)$ Do this step mentally.

$-2x^3 + 8x^2 + 6x$

Example 1

Simplify $(5x + 4)(-2x)$.

Solution

$(5x + 4)(-2x) = -10x^2 - 8x$

Example 2

Simplify $(-2y + 3)(-4y)$.

Your solution

Example 3

Simplify $x^3(2x^2 - 3x + 2)$.

Solution

$x^3(2x^2 - 3x + 2) = 2x^5 - 3x^4 + 2x^3$

Example 4

Simplify $-a^2(3a^2 + 2a - 7)$.

Your solution

Solutions on p. A27

3.2 Objective To multiply two polynomials

Reference for Computer Tutor™

Multiplication of two polynomials requires the repeated application of the Distributive Property.

$(y - 2)(y^2 + 3y + 1) =$
$(y - 2)(y^2) + (y - 2)(3y) + (y - 2)(1) =$
$y^3 - 2y^2 + 3y^2 - 6y + y - 2 =$
$y^3 + y^2 - 5y - 2$

A more convenient method of multiplying two polynomials is to use a vertical format, similar to that used for multiplication of whole numbers.

Multiply each term in the trinomial by -2.
Multiply each term in the trinomial by y.
Like terms must be in the same column.
Add the terms in each column.

$$\begin{array}{rrrr} & y^2 & + 3y & + 1 \\ \times & & y & - 2 \\ \hline & -2y^2 & - 6y & - 2 \\ y^3 + & 3y^2 & + y & \\ \hline y^3 + & y^2 & - 5y & - 2 \end{array}$$

Simplify $(a^2 - 3)(a + 5)$.

Multiply each term of $a^2 - 3$ by 5.
Mutliply each term of $a^2 - 3$ by a.
Arrange the terms in descending order.
Add the terms in each column.

$$\begin{array}{rrrr} & a^2 & & -\ 3 \\ \times & & a & +\ 5 \\ \hline & 5a^2 & & -15 \\ a^3 & & -\ 3a & \\ \hline a^3 & +\ 5a^2 & -\ 3a & -15 \end{array}$$

Example 5

Simplify $(2b^3 - b + 1)(2b + 3)$.

Solution

$$\begin{array}{rrrrr} & 2b^3 & & -\ b & +\ 1 \\ & \times & & 2b & +\ 3 \\ \hline & 6b^3 & & -\ 3b & +\ 3 \\ 4b^4\ + & & -\ 2b^2 & +\ 2b & \\ \hline 4b^4 & +\ 6b^3 & -\ 2b^2 & -\ b & +\ 3 \end{array}$$

Example 6

Simplify $(2y^3 + 2y^2 - 3)(3y - 1)$.

Your solution

Solution on p. A27

3.3 Objective To multiply two binomials

Reference for Computer Tutor™

It is frequently necessary to find the product of two binomials. The product can be found using a method called **FOIL,** which is based upon the Distributive Property. The letters of FOIL stand for **F**irst, **O**uter, **I**nner, and **L**ast.

Simplify $(2x + 3)(x + 5)$.

Multiply the First terms.	$(2x + 3)(x + 5)$	$2x \cdot x = 2x^2$
Multiply the Outer terms.	$(2x + 3)(x + 5)$	$2x \cdot 5 = 10x$
Multiply the Inner terms.	$(2x + 3)(x + 5)$	$3 \cdot x = 3x$
Multiply the Last terms.	$(2x + 3)(x + 5)$	$3 \cdot 5 = 15$

F O I L

Add the products. $(2x + 3)(x + 5)$ $= 2x^2 + 10x + 3x + 15$

Combine like terms. $= 2x^2 + 13x + 15$

Simplify $(4x - 3)(3x - 2)$.

$(4x - 3)(3x - 2) = 4x(3x) + 4x(-2) + (-3)(3x) + (-3)(-2)$ Do this step mentally.

$= 12x^2 - 8x - 9x + 6$

$= 12x^2 - 17x + 6$

Simplify $(3x - 2y)(x + 4y)$.

$(3x - 2y)(x + 4y) = 3x(x) + 3x(4y) + (-2y)(x) + (-2y)(4y)$ Do this step mentally.

$= 3x^2 + 12xy - 2xy - 8y^2$
$= 3x^2 + 10xy - 8y^2$

Example 7

Simplify $(2a - 1)(3a - 2)$.

Solution

$(2a - 1)(3a - 2) = 6a^2 - 4a - 3a + 2$
$= 6a^2 - 7a + 2$

Example 8

Simplify $(4y - 5)(2y - 3)$.

Your solution

Example 9

Simplify $(3x - 2)(4x + 3)$.

Solution

$(3x - 2)(4x + 3) = 12x^2 + 9x - 8x - 6$
$= 12x^2 + x - 6$

Example 10

Simplify $(3b + 2)(3b - 5)$.

Your solution

Solutions on p. A27

3.4 Objective To multiply binomials which have special products

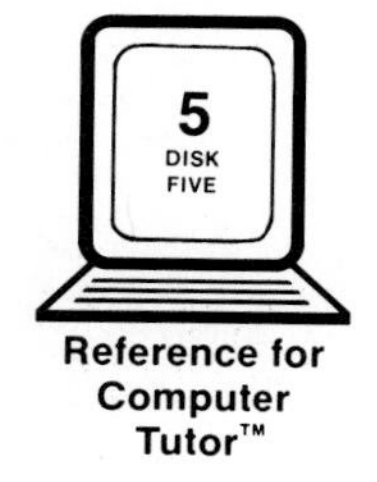

Reference for Computer Tutor™

Using FOIL, a pattern for the product of the sum and difference of two terms and for the square of a binomial can be found.

The Sum and Difference of Two Terms $(a + b)(a - b) = a^2 - ab + ab - b^2$
$= a^2 - b^2$

Square of first term (a^2)
Square of second term (b^2)

The Square of a Binomial $(a + b)^2 = (a + b)(a + b) = a^2 + ab + ab + b^2$
$= a^2 + 2ab + b^2$

Square of first term (a^2)
Twice the product of the two terms ($2ab$)
Square of the last term (b^2)

Simplify $(2x + 3)(2x - 3)$.

$(2x + 3)(2x - 3)$ is the sum and difference of two terms.

$(2x + 3)(2x - 3) = (2x)^2 - 3^2$ Do this step mentally.

$= 4x^2 - 9$

Simplify $(3x - 2)^2$.

$(3x - 2)^2$ is the square of a binomial.

$$(3x - 2)^2 = (3x)^2 + 2(3x)(-2) + (-2)^2$$

Do this step mentally.

$$= 9x^2 - 12x + 4$$

Example 11

Simplify $(4z - 2w)(4z + 2w)$.

Solution

$(4z - 2w)(4z + 2w) = 16z^2 - 4w^2$

Example 12

Simplify $(2a + 5c)(2a - 5c)$.

Your solution

Example 13

Simplify $(2r - 3s)^2$.

Solution

$(2r - 3s)^2 = 4r^2 - 12rs + 9s^2$

Example 14

Simplify $(3x + 2y)^2$.

Your solution

Solutions on p. A27

3.5 Objective To solve application problems

Use COMPUTER TUTOR™ DISK 5

Example 15

The length of a rectangle is $x + 7$. The width is $x - 4$. Find the area of the rectangle in terms of the variable x.

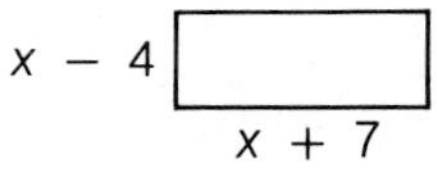

Strategy

To find the area, replace the variables l and w in the equation $A = l \cdot w$ by the given values and solve for A.

Solution

$A = l \cdot w$
$A = (x + 7)(x - 4)$
$A = x^2 - 4x + 7x - 28$
$A = x^2 + 3x - 28$

The area is $x^2 + 3x - 28$.

Example 16

The radius of a circle is $x - 4$. Use the equation $A = \pi r^2$, where r is the radius, to find the area of the circle in terms of x. Use 3.14 for π.

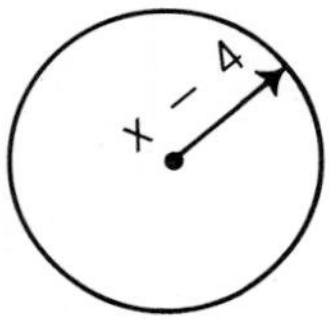

Your strategy

Your solution

Solution on p. A27

3.1 Exercises

Simplify.

1. $x(x - 2)$
2. $y(3 - y)$
3. $-x(x + 7)$
4. $-y(7 - y)$
5. $3a^2(a - 2)$
6. $4b^2(b + 8)$
7. $-5x^2(x^2 - x)$
8. $-6y^2(y + 2y^2)$
9. $-x^3(3x^2 - 7)$
10. $-y^4(2y^2 - y^6)$
11. $2x(6x^2 - 3x)$
12. $3y(4y - y^2)$
13. $(2x - 4)3x$
14. $(3y - 2)y$
15. $(3x + 4)x$
16. $(2x + 1)2x$
17. $-xy(x^2 - y^2)$
18. $-x^2y(2xy - y^2)$
19. $x(2x^3 - 3x + 2)$
20. $y(-3y^2 - 2y + 6)$
21. $-a(-2a^2 - 3a - 2)$
22. $-b(5b^2 + 7b - 35)$
23. $x^2(3x^4 - 3x^2 - 2)$
24. $y^3(-4y^3 - 6y + 7)$
25. $2y^2(-3y^2 - 6y + 7)$
26. $4x^2(3x^2 - 2x + 6)$
27. $(a^2 + 3a - 4)(-2a)$
28. $(b^3 - 2b + 2)(-5b)$
29. $-3y^2(-2y^2 + y - 2)$
30. $-5x^2(3x^2 - 3x - 7)$
31. $xy(x^2 - 3xy + y^2)$
32. $ab(2a^2 - 4ab - 6b^2)$

3.2 Exercises

Simplify.

33. $(x^2 + 3x + 2)(x + 1)$

34. $(x^2 - 2x + 7)(x - 2)$

35. $(a^2 - 3a + 4)(a - 3)$

36. $(x^2 - 3x + 5)(2x - 3)$

37. $(-2b^2 - 3b + 4)(b - 5)$

38. $(-a^2 + 3a - 2)(2a - 1)$

39. $(-2x^2 + 7x - 2)(3x - 5)$

40. $(-a^2 - 2a + 3)(2a - 1)$

41. $(x^2 + 5)(x - 3)$

42. $(y^2 - 2y)(2y + 5)$

43. $(x^3 - 3x + 2)(x - 4)$

44. $(y^3 + 4y^2 - 8)(2y - 1)$

45. $(5y^2 + 8y - 2)(3y - 8)$

46. $(3y^2 + 3y - 5)(4y - 3)$

47. $(5a^3 - 5a + 2)(a - 4)$

48. $(3b^3 - 5b^2 + 7)(6b - 1)$

49. $(y^3 + 2y^2 - 3y + 1)(y + 2)$

50. $(2a^3 - 3a^2 + 2a - 1)(2a - 3)$

3.3 Exercises

Simplify.

51. $(x + 1)(x + 3)$
52. $(y + 2)(y + 5)$
53. $(a - 3)(a + 4)$

54. $(b - 6)(b + 3)$
55. $(y + 3)(y - 8)$
56. $(x + 10)(x - 5)$

57. $(y - 7)(y - 3)$
58. $(a - 8)(a - 9)$
59. $(2x + 1)(x + 7)$

60. $(y + 2)(5y + 1)$
61. $(3x - 1)(x + 4)$
62. $(7x - 2)(x + 4)$

63. $(4x - 3)(x - 7)$
64. $(2x - 3)(4x - 7)$
65. $(3y - 8)(y + 2)$

66. $(5y - 9)(y + 5)$
67. $(3x + 7)(3x + 11)$
68. $(5a + 6)(6a + 5)$

69. $(7a - 16)(3a - 5)$
70. $(5a - 12)(3a - 7)$
71. $(3b + 13)(5b - 6)$

72. $(x + y)(2x + y)$
73. $(2a + b)(a + 3b)$
74. $(3x - 4y)(x - 2y)$

75. $(2a - b)(3a + 2b)$
76. $(5a - 3b)(2a + 4b)$
77. $(2x + y)(x - 2y)$

78. $(3x - 7y)(3x + 5y)$
79. $(2x + 3y)(5x + 7y)$
80. $(5x + 3y)(7x + 2y)$

81. $(3a - 2b)(2a - 7b)$
82. $(5a - b)(7a - b)$
83. $(a - 9b)(2a + 7b)$

84. $(2a + 5b)(7a - 2b)$
85. $(10a - 3b)(10a - 7b)$
86. $(12a - 5b)(3a - 4b)$

87. $(5x + 12y)(3x + 4y)$
88. $(11x + 2y)(3x + 7y)$
89. $(2x - 15y)(7x + 4y)$

90. $(5x + 2y)(2x - 5y)$
91. $(8x - 3y)(7x - 5y)$
92. $(2x - 9y)(8x - 3y)$

3.4 Exercises

Simplify.

93. $(y - 5)(y + 5)$

94. $(y + 6)(y - 6)$

95. $(2x + 3)(2x - 3)$

96. $(4x - 7)(4x + 7)$

97. $(x + 1)^2$

98. $(y - 3)^2$

99. $(3a - 5)^2$

100. $(6x - 5)^2$

101. $(3x - 7)(3x + 7)$

102. $(9x - 2)(9x + 2)$

103. $(2a + b)^2$

104. $(x + 3y)^2$

105. $(x - 2y)^2$

106. $(2x - 3y)^2$

107. $(4 - 3y)(4 + 3y)$

108. $(4x - 9y)(4x + 9y)$

109. $(5x + 2y)^2$

110. $(2a - 9b)^2$

3.5 Application Problems

Solve.

1. The length of a rectangle is $4x$. The width is $2x - 3$. Find the area of the rectangle in terms of the variable x.

2. The width of a rectangle is $x - 1$. The length is $2x + 1$. Find the area of the rectangle in terms of the variable x.

3. The length of a side of a square is $x + 2$. Use the equation $A = s^2$, where s is the length of the side of a square, to find the area of the square in terms of the variable x.

4. The length of a side of a square is $2x - 1$. Use the equation $A = s^2$, where s is the length of the side of a square, to find the area of the square in terms of the variable x.

5. The radius of a circle is $x + 3$. Use the equation $A = \pi r^2$, where r is the radius, to find the area of the circle in terms of the variable x. Use 3.14 for π.

6. The radius of a circle is $x - 2$. Use the equation $A = \pi r^2$, where r is the radius, to find the area of the circle in terms of the variable x. Use 3.14 for π.

SECTION 4 Division of Polynomials

4.1 Objective To divide monomials

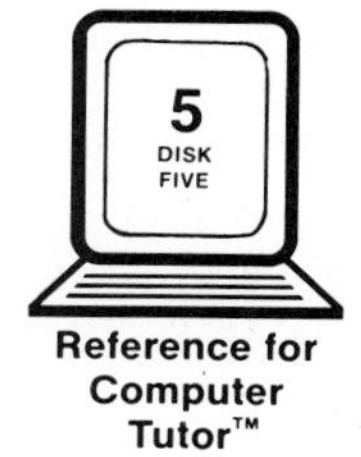

Reference for Computer Tutor™

The quotient of two exponential expressions with the *same* base can be simplified by writing each expression in factored form, canceling the common factors, and then writing the result with an exponent.

$$\frac{x^5}{x^2} = \frac{\overset{1}{\cancel{x}} \cdot \overset{1}{\cancel{x}} \cdot x \cdot x \cdot x}{\underset{1}{\cancel{x}} \cdot \underset{1}{\cancel{x}}} = x^3$$

$$\frac{x^4}{x^6} = \frac{\overset{1}{\cancel{x}} \cdot \overset{1}{\cancel{x}} \cdot \overset{1}{\cancel{x}} \cdot \overset{1}{\cancel{x}}}{\underset{1}{\cancel{x}} \cdot \underset{1}{\cancel{x}} \cdot \underset{1}{\cancel{x}} \cdot \underset{1}{\cancel{x}} \cdot x \cdot x} = \frac{1}{x^2}$$

Note that subtracting the smaller exponent from the larger exponent results in the same quotient.

$$\frac{x^5}{x^2} = \frac{x^{5-2}}{1} = x^3$$

$$\frac{x^4}{x^6} = \frac{1}{x^{6-4}} = \frac{1}{x^2}$$

Rule for Dividing Exponential Expressions

If m and n are integers and $x \neq 0$, then $\frac{x^m}{x^n} = x^{m-n}$ if $m > n$

and $\frac{x^m}{x^n} = \frac{1}{x^{n-m}}$ if $m < n$.

Simplify $\frac{x^7}{x^2}$.

$7 > 2$
The bases are the same.
Subtract the exponent in the denominator from the exponent in the numerator.

$$\frac{x^7}{x^2} \boxed{= x^{7-2}}$$ Do this step mentally.

$$= x^5$$

Simplify $\frac{a^3}{a^9}$.

$3 < 9$
The bases are the same.
Subtract the exponent in the numerator from the exponent in the denominator.

$$\frac{a^3}{a^9} \boxed{= \frac{1}{a^{9-3}}}$$ Do this step mentally.

$$= \frac{1}{a^6}$$

Simplify $\frac{x^4y^2}{-xy^3}$.

Negative signs are placed in front of a fraction.

$$\frac{x^4y^2}{-xy^3} = -\frac{x^4y^2}{xy^3}$$

Divide variables with like bases by subtracting the exponents.

$$\boxed{= -\frac{x^{4-1}}{y^{3-2}}}$$ Do this step mentally.

$$= -\frac{x^3}{y}$$

Simplify $\frac{10x^3y^5}{4x^6y^2}$.

Factor the coefficients.
Cancel the common factors.

$$\frac{10x^3y^5}{4x^6y^2} = \frac{\overset{1}{\cancel{2}} \cdot 5x^3y^5}{\underset{1}{\cancel{2}} \cdot 2x^6y^2}$$

Divide variables with like bases by subtracting the exponents.

$$= \frac{5y^{5-2}}{2x^{6-3}}$$

Do this step mentally.

$$= \frac{5y^3}{2x^3}$$

Example 1

Simplify $\frac{-16x^4}{4x^9}$.

Solution

$$\frac{-16x^4}{4x^9} = -\frac{\overset{1}{\cancel{2}} \cdot \overset{1}{\cancel{2}} \cdot 2 \cdot 2x^4}{\underset{1}{\cancel{2}} \cdot \underset{1}{\cancel{2}}x^9} = -\frac{4}{x^5}$$

Example 2

Simplify $\frac{42y^{12}}{-14y^{17}}$.

Your solution

Example 3

Simplify $\frac{-28x^4y^3}{6xy}$.

Solution

$$\frac{-28x^4y^3}{6xy} = -\frac{\overset{1}{\cancel{2}} \cdot 2 \cdot 7x^4y^3}{\underset{1}{\cancel{2}} \cdot 3xy} = -\frac{14x^3y^2}{3}$$

Example 4

Simplify $\frac{12r^4s^2}{-8r^3s}$.

Your solution

Example 5

Simplify $\frac{(-3ab)^2}{9a^3b}$.

Solution

$(-3)^2 = 9$

$$\frac{(-3ab)^2}{9a^3b} = \frac{(-3)^2a^2b^2}{9a^3b} = \frac{\overset{1}{\cancel{3}} \cdot \overset{1}{\cancel{3}}a^2b^2}{\underset{1}{\cancel{3}} \cdot \underset{1}{\cancel{3}}a^3b} = \frac{b}{a}$$

Example 6

Simplify $\frac{(2x^2y)^3}{-4xy^5}$.

Your solution

4.2 Objective — To divide a polynomial by a monomial

Reference for Computer Tutor™

Note that $\frac{8+4}{2}$ can be simplified by first adding the terms in the numerator and then dividing the result. It can also be simplified by first dividing each term in the numerator by the denominator and then adding the result.

$$\frac{8+4}{2} = \frac{12}{2} = 6$$

$$\frac{8+4}{2} = \frac{8}{2} + \frac{4}{2} = 4 + 2 = 6$$

To divide a polynomial by a monomial, divide each term in the numerator by the denominator, and write the sum of the quotients.

$$\frac{a+b}{c} = \frac{a}{c} + \frac{b}{c}$$

Simplify $\frac{6x^2 + 4x}{2x}$.

Divide each term of the polynomial by the monomial.
Simplify each expression.

$$\frac{6x^2 + 4x}{2x} = \frac{6x^2}{2x} + \frac{4x}{2x}$$
$$= 3x + 2$$

Check: $2x(3x + 2) = 6x^2 + 4x$

Example 7

Simplify $\frac{6x^3 - 3x^2 + 9x}{3x}$.

Solution

$$\frac{6x^3 - 3x^2 + 9x}{3x} = \frac{6x^3}{3x} - \frac{3x^2}{3x} + \frac{9x}{3x}$$
$$= 2x^2 - x + 3$$

Example 8

Simplify $\frac{4x^3y + 8x^2y^2 - 4xy^3}{2xy}$.

Your solution

Example 9

Simplify $\frac{12x^2y - 6xy + 4x^2}{2xy}$.

Solution

$$\frac{12x^2y - 6xy + 4x^2}{2xy} = \frac{12x^2y}{2xy} - \frac{6xy}{2xy} + \frac{4x^2}{2xy}$$
$$= 6x - 3 + \frac{2x}{y}$$

Example 10

Simplify $\frac{24x^2y^2 - 18xy + 6y}{6xy}$.

Your solution

Solutions on p. A28

4.3 Objective — To divide polynomials

Reference for Computer Tutor™

To divide polynomials, use a method similar to that used for division of whole numbers. The same equation used to check division of whole numbers is used to check an algebraic problem.

Dividend = (quotient × divisor) + remainder

Simplify $(x^2 - 5x + 8) \div (x - 3)$.

Step 1

$$\begin{array}{r} x \phantom{{}-5x+8} \\ x-3\overline{)x^2-5x+8} \\ \underline{x^2-3x} \phantom{{}+8} \\ -2x+8 \end{array}$$

Think: $x\overline{)x^2} = \dfrac{x^2}{x} = x$

Multiply: $x(x - 3) = x^2 - 3x$

Subtract: $(x^2 - 5x) - (x^2 - 3x) = -2x$

Step 2

$$\begin{array}{r} x-2 \phantom{{}+8} \\ x-3\overline{)x^2-5x+8} \\ \underline{x^2-3x} \phantom{{}+8} \\ -2x+8 \\ \underline{-2x+6} \\ 2 \end{array}$$

Think: $x\overline{)-2x} = \dfrac{-2x}{x} = -2$

Multiply: $-2(x - 3) = -2x + 6$

Subtract: $(-2x + 8) - (-2x + 6) = 2$

The remainder is 2.

Check: $(x - 2)(x - 3) + 2 = x^2 - 3x - 2x + 6 + 2 = x^2 - 5x + 8$

$(x^2 - 5x + 8) \div (x - 3) = x - 2 + \frac{2}{x - 3}$

Simplify $(6x + 2x^3 + 26) \div (x + 2)$.

Arrange the terms in descending order. There is no term of x^2 in $2x^3 + 6x + 26$. Insert a zero for the missing term so that like terms will be in columns.

$$\begin{array}{r} 2x^2-4x+14 \phantom{{}+26} \\ x+2\overline{)2x^3+0+6x+26} \\ \underline{2x^3+4x^2} \phantom{{}+6x+26} \\ -4x^2+6x \phantom{{}+26} \\ \underline{-4x^2-8x} \phantom{{}+26} \\ 14x+26 \\ \underline{14x+28} \\ -2 \end{array}$$

$(2x^3 + 6x + 28) \div (x + 2) = 2x^2 - 4x + 14 - \frac{2}{x + 2}$

Example 11

Simplify $(x^2 - 1) \div (x + 1)$.

Solution

Insert a zero for the missing term.

$$\begin{array}{r} x-1 \\ x+1\overline{)x^2+0-1} \\ \underline{x^2+x} \phantom{{}-1} \\ -x-1 \\ \underline{-x-1} \\ 0 \end{array}$$

$(x^2 - 1) \div (x + 1) = x - 1$

Example 12

Simplify $(2x^3 + x^2 - 8x - 3) \div (2x - 3)$.

Your solution

Solution on p. A28

4.1 Exercises

Simplify.

1. $\frac{3x^2}{x}$
2. $\frac{4y^2}{2y}$
3. $\frac{2x^2}{-2x}$

4. $\frac{-8y^2}{4y}$
5. $\frac{12x^4}{3x}$
6. $\frac{5x^2}{15x}$

7. $\frac{-16x}{4x^3}$
8. $\frac{27y}{-12y^3}$
9. $\frac{a^4b^5}{a^3b^9}$

10. $\frac{a^5b^7}{a^8b^2}$
11. $\frac{x^3y^4}{x^3y}$
12. $\frac{x^4y^5}{x^2y^5}$

13. $\frac{(3x)^2}{15x^2}$
14. $\frac{(4y)^2}{2y}$
15. $\frac{(6b)^3}{(-3b^2)^2}$

16. $\frac{(-3a^2)^3}{(9a)^2}$
17. $\frac{-36a^4b^7}{60a^5b^9}$
18. $\frac{-50a^2b^7}{45ab^2}$

19. $\frac{12a^2b^3}{-27a^2b^2}$
20. $\frac{-16xy^4}{96x^4y^4}$
21. $\frac{-8x^2y^4}{44y^2z^5}$

22. $\frac{22a^2b^4}{-132b^3c^2}$
23. $\frac{-(8a^2b^4)^3}{64a^3b^8}$
24. $\frac{-(14ab^4)^2}{28a^4b^2}$

25. $\frac{-20a^3b^4}{-45ab^7}$
26. $\frac{-14x^6y^4}{-70x^3y}$
27. $\frac{x^4y^7z}{x^2y^5z^4}$

28. $\frac{x^2y^4z^2}{x^3yz^5}$
29. $\frac{(-2ab^2)^3}{-8ab^7}$
30. $\frac{(-3x^3y)^2}{-12xy^5}$

31. $\frac{-a^8b^3c^7}{a^6b^3c^9}$
32. $\frac{x^5y^4z^6}{-x^3y^6}$
33. $\frac{x^3y^4z^6}{x^3z^6}$

4.2 Exercises

Simplify.

34. $\frac{2x + 2}{2}$

35. $\frac{5y + 5}{5}$

36. $\frac{10a - 25}{5}$

37. $\frac{16b - 40}{8}$

38. $\frac{3a^2 + 2a}{a}$

39. $\frac{6y^2 + 4y}{y}$

40. $\frac{4b^3 - 3b}{b}$

41. $\frac{12x^2 - 7x}{x}$

42. $\frac{3x^2 - 6x}{3x}$

43. $\frac{10y^2 - 6y}{2y}$

44. $\frac{5x^2 - 10x}{-5x}$

45. $\frac{3y^2 - 27y}{-3y}$

46. $\frac{x^3 + 3x^2 - 5x}{x}$

47. $\frac{a^3 - 5a^2 + 7a}{a}$

48. $\frac{x^6 - 3x^4 - x^2}{x^2}$

49. $\frac{a^8 - 5a^5 - 3a^3}{a^2}$

50. $\frac{5x^2y^2 + 10xy}{5xy}$

51. $\frac{8x^2y^2 - 24xy}{8xy}$

52. $\frac{9y^6 - 15y^3}{-3y^3}$

53. $\frac{4x^4 - 6x^2}{-2x^2}$

54. $\frac{3x^2 - 2x + 1}{x}$

55. $\frac{8y^2 + 2y - 3}{y}$

56. $\frac{-3x^2 + 7x - 6}{x}$

57. $\frac{2y^2 - 6y + 9}{y}$

58. $\frac{16a^2b - 20ab + 24ab^2}{4ab}$

59. $\frac{22a^2b + 11ab - 33ab^2}{11ab}$

60. $\frac{9x^2y + 6xy - 3xy^2}{xy}$

61. $\frac{5a^2b - 15ab + 30ab^2}{5ab}$

4.3 Exercises

Simplify.

62. $(x^2 + 2x + 1) \div (x + 1)$

63. $(x^2 + 10x + 25) \div (x + 5)$

64. $(a^2 - 6a + 9) \div (a - 3)$

65. $(b^2 - 14b + 49) \div (b - 7)$

66. $(x^2 - x - 6) \div (x - 3)$

67. $(y^2 + 2y - 35) \div (y + 7)$

68. $(2x^2 + 5x + 2) \div (x + 2)$

69. $(2y^2 - 13y + 21) \div (y - 3)$

70. $(4x^2 - 16) \div (2x + 4)$

71. $(2y^2 + 7) \div (y - 3)$

72. $(x^2 + 1) \div (x - 1)$

73. $(x^2 + 4) \div (x + 2)$

74. $(6x^2 - 7x) \div (3x - 2)$

75. $(6y^2 + 2y) \div (2y + 4)$

76. $(5x^2 + 7x) \div (x - 1)$

77. $(6x^2 - 5) \div (x + 2)$

78. $(a^2 + 5a + 10) \div (a + 2)$

79. $(b^2 - 8b - 9) \div (b - 3)$

Simplify.

80. $(2y^2 - 9y + 8) \div (2y + 3)$

81. $(3x^2 + 5x - 4) \div (x - 4)$

82. $(4x^2 + 8x + 3) \div (2x - 1)$

83. $(10y^2 + 21y + 10) \div (2y + 3)$

84. $(15a^2 - 8a - 8) \div (3a + 2)$

85. $(12a^2 - 25a - 7) \div (3a - 7)$

86. $(12x^2 - 23x + 5) \div (4x - 1)$

87. $(6a^2 + 25a + 24) \div (3a - 1)$

88. $(x^3 + 3x^2 + 5x + 3) \div (x + 1)$

89. $(x^3 - 6x^2 + 7x - 2) \div (x - 1)$

90. $(y^3 + 6y^2 + 4y - 5) \div (y + 3)$

91. $(4a^3 + 8a^2 + 5a + 9) \div (2a + 3)$

92. $(6a^3 + 5a^2 + 5) \div (3a - 2)$

93. $(4b^3 - b - 5) \div (2b + 3)$

94. $(2a^3 - 5a^2 + 2) \div (2a + 3)$

95. $(5x^3 - 7x^2 - 3x) \div (x - 2)$

96. $(x^4 - x^2 - 6) \div (x^2 + 2)$

97. $(x^4 + 3x^2 - 10) \div (x^2 - 2)$

SECTION 5 Negative and Zero Exponents

5.1 Objective To simplify expressions containing negative and zero exponents

Reference for Computer Tutor™

Note that when an exponential expression is divided by itself, the result contains a zero exponent.

$$\frac{x^3}{x^3} = x^{3-3} = x^0$$

When the same expression is simplified by factoring and canceling, the result is 1.

$$\frac{x^3}{x^3} = \frac{\overset{1}{\cancel{x}} \cdot \overset{1}{\cancel{x}} \cdot \overset{1}{\cancel{x}}}{\underset{1}{\cancel{x}} \cdot \underset{1}{\cancel{x}} \cdot \underset{1}{\cancel{x}}} = 1$$

To insure that the two answers are equal, a number or variable to the zero power must be equal to 1.

$$x^0 = 1, \ x \neq 0$$

Negative integers, as well as positive, can be used as exponents. The rules that have been developed for exponential expressions can be extended to include negative integers.

$x^m \cdot x^n = x^{m+n}$ $\qquad$ $a^{-2} \cdot a^3 = a^{-2+3} = a$

$(x^m)^n = x^{m \cdot n}$ $\qquad$ $(a^{-2})^3 = a^{-2 \cdot 3} = a^{-6}$

$(x^m \cdot y^n)^p = x^{m \cdot p}y^{n \cdot p}$ $\qquad$ $(a^{-2}b^4)^{-3} = a^{-2(-3)}b^{4(-3)} = a^6b^{-12}$

The Rule for Dividing Exponential Expressions can be stated as a single rule.

$\dfrac{x^m}{x^n} = x^{m-n}$ $\qquad$ $\dfrac{a^2}{a^5} = a^{2-5} = a^{-3}$

The meaning of a negative exponent can be developed using the rules presented in this unit.

The exponential expressions at the right are multiplied by adding the exponents. The product is a^0, or 1.

$$a^4 \cdot a^{-4} = a^{4+(-4)} = a^0 = 1$$

Recall that a number times its reciprocal is equal to 1.

$$a^4 \cdot \frac{1}{a^4} = 1$$

Since $a^4 \cdot a^{-4} = 1$, and $a^4 \cdot \dfrac{1}{a^4} = 1$, then a^{-4} must equal $\dfrac{1}{a^4}$.

Rule of Negative Exponents

If n is a positive integer and $x \neq 0$, then $x^{-n} = \dfrac{1}{x^n}$.

An exponential expression is considered in simplest form when written with a positive exponent.

A number with a negative exponent can be written with a positive exponent and then evaluated.

Write 2^{-3} with a positive exponent. Then evaluate.

Write the expression with a positive exponent. $2^{-3} = \frac{1}{2^3}$

Evaluate. $= \frac{1}{8}$

Simplify $\frac{x^{-4}y^6}{xy^2}$.

Divide variables with like bases by subtracting the exponents. $\frac{x^{-4}y^6}{xy^2} = x^{-4-1}y^{6-2}$ Do this step mentally.

Write the expression with positive exponents. $= x^{-5}y^4$

$= \frac{y^4}{x^5}$

Example 1

Write $\frac{3^{-3}}{3^2}$ with a positive exponent. Then evaluate.

Solution

$$\frac{3^{-3}}{3^2} = 3^{-5} = \frac{1}{3^5} = \frac{1}{243}$$

Example 2

Write $\frac{2^{-2}}{2^3}$ with a positive exponent. Then evaluate.

Your solution

Example 3

Simplify $(-2x)(3x^{-2})^{-3}$.

Solution

$$(-2x)(3x^{-2})^{-3} = (-2x)(3^{-3}x^6) = \frac{-2x \cdot x^6}{3^3} = -\frac{2x^7}{27}$$

Example 4

Simplify $(-2x^2)(x^{-3}y^{-4})^{-2}$.

Your solution

Example 5

Simplify $\frac{(2ab^{-2})^{-2}}{ab^2}$.

Solution

$$\frac{(2ab^{-2})^{-2}}{ab^2} = \frac{2^{-2}a^{-2}b^4}{ab^2} = 2^{-2}a^{-3}b^2 = \frac{b^2}{2^2a^3} = \frac{b^2}{4a^3}$$

Example 6

Simplify $\frac{(3x^{-2}y)^3}{9xy^0}$.

Your solution

Solutions on p. A28

5.1 Exercises

Write with a positive exponent. Then evaluate.

1. 5^{-2} **2.** 3^{-3} **3.** 7^{-1} **4.** 12^{-1}

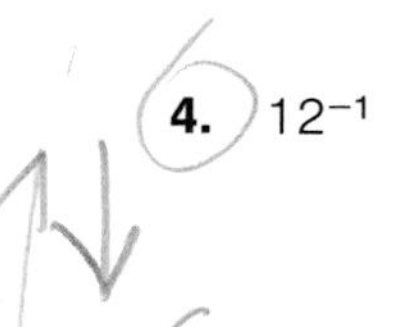

5. $\frac{3^{-2}}{3}$ **6.** $\frac{5^{-3}}{5}$ **7.** $\frac{2^{-3}}{2^3}$ **8.** $\frac{3^{-2}}{3^2}$

Simplify.

9. x^{-2} **10.** y^{-10} **11.** a^{-6} **12.** b^{-4}

13. x^2y^{-3} **14.** $a^{-2}b$ **15.** $x^{-1}y^{-2}$ **16.** $x^{-3}y^{-4}$

17. $x^{-3}x^4$ **18.** $x \cdot x^{-2}$ **19.** $a^{-3}a^{-4}$ **20.** $a^{-2}a^{-5}$

21. $\frac{x^{-2}}{x^2}$ **22.** $\frac{x^{-1}}{x}$ **23.** $\frac{a^{-3}}{a^5}$ **24.** $\frac{a^{-10}}{a^{10}}$

25. $\frac{x^{-2}y}{x}$ **26.** $\frac{x^4y^{-3}}{x^2}$ **27.** $\frac{a^{-2}b}{b^0}$ **28.** $\frac{a^2b^{-4}}{b^0}$

29. $(a^2)^{-2}$ **30.** $(b^3)^{-3}$ **31.** $(a^{-3})^2$ **32.** $(y^{-4})^3$

33. $(x^{-2})^{-3}$ **34.** $(y^{-3})^{-4}$ **35.** $(a^{-6})^{-3}$ **36.** $(a^{-1})^{-3}$

37. $(ab^{-1})^0$ **38.** $(a^2b)^0$ **39.** $(x^{-2}y^2)^2$ **40.** $(x^{-3}y^{-1})^2$

41. $\frac{x^2y^{-4}}{xy}$ **42.** $\frac{ab^{-5}}{a^7b}$ **43.** $\frac{x^{-2}y}{x^2y^4}$ **44.** $\frac{a^{-3}b}{ab^2}$

Simplify.

45. $(x^2y^{-1})^{-2}$

46. $(a^{-1}b^{-1})^{-4}$

47. $(x^{-3}y^{-3})^{-3}$

48. $(x^{-2}y^{-3})^{-4}$

49. $(-2xy^{-2})^3$

50. $(-3x^{-1}y^2)^2$

51. $(4x^2y^{-3})^2$

52. $(2a^{-3}b^2)^3$

53. $(3x^{-1}y^{-2})^2$

54. $(5xy^{-3})^{-2}$

55. $(2x^{-1})(x^{-3})$

56. $(-2x^{-5})x^7$

57. $(-5a^2)(a^{-5})^2$

58. $(2a^{-3})(a^7b^{-1})^3$

59. $(-2ab^{-2})(4a^{-2}b)^{-2}$

60. $(3ab^{-2})(2a^{-1}b)^{-3}$

61. $(-5x^{-2}y)(-2x^{-2}y^2)$

62. $\dfrac{a^{-1}b^{-1}}{ab}$

63. $\dfrac{a^{-3}b^{-4}}{a^2b^2}$

64. $\dfrac{3x^{-2}y^2}{6xy^2}$

65. $\dfrac{2x^{-2}y}{8xy}$

66. $\dfrac{3x^{-2}y}{xy}$

67. $\dfrac{16a^{-4}y}{3ay}$

68. $\dfrac{3x^{-2}y}{xy^2}$

69. $\dfrac{2x^{-1}y^4}{x^2y^3}$

70. $\dfrac{2x^{-1}y^{-4}}{4xy^2}$

71. $\dfrac{(x^{-1}y)^2}{xy^2}$

72. $\dfrac{(x^{-2}y)^2}{x^2y^3}$

73. $\dfrac{(x^{-3}y^{-2})^2}{x^6y^8}$

74. $\dfrac{(a^{-2}y^3)^{-3}}{a^2y}$

Calculators and Computers

Evaluating Polynomials

One way to evaluate a polynomial is to first express the polynomial in a form that suggests a sequence of steps on the calculator. To illustrate this method, consider the polynomial $4x^2 - 5x + 2$. First the polynomial is rewritten as:

$$4x^2 - 5x + 2 = (4x - 5)x + 2$$

To evaluate the polynomial, work through the rewritten expression from left to right, substituting the appropriate value for x.

Here are some examples:

Evaluate $5x^2 - 2x + 4$ when $x = 3$.

Rewrite the polynomial. $5x^2 - 2x + 4 = (5x - 2)x + 4$

Replace x in the rewritten expression by the given value. $(5 \cdot 3 - 2) \cdot 3 + 4$

Work through the expression from left to right. 5 [×] 3 [−] 2 [×] 3 [+] 4 [=]
The result in the display should be 43.

Evaluate $2x^3 - 4x^2 + 7x - 12$ when $x = 4$.

Rewrite the polynomial. $2x^3 - 4x^2 + 7x - 12 = [(2x - 4)x + 7]x - 12$

Replace x in the rewritten expression by the given value. $[(2 \cdot 4 - 4) \cdot 4 + 7] \cdot 4 - 12$

Work through the expression from left to right. 2 [×] 4 [−] 4 [×] 4 [+] 7 [×] 4 [−] 12 [=]
The result in the display should be 80.

Evaluate $4x^2 - 3x + 5$ when $x = -2$.

Rewrite the polynomial. $4x^2 - 3x + 5 = (4x - 3)x + 5$

Replace x in the given expression by the given value. $[4 \cdot (-2) - 3] \cdot (-2) + 5$

Work through the expression from left to right. 4 [×] 2 [+/−] [−] 3 [×] 2 [+/−] [+] 5
The result in the display should be 27.

Here are some practice exercises:
Evaluate for the given value.

1. $2x^2 - 3x + 7;\ x = 4$

2. $3x^2 + 7x - 12;\ x = -3$

3. $3x^3 - 2x^2 + 6x - 8;\ x = 3$

4. $2x^3 + 4x^2 - x - 2;\ x = -2$

5. $x^4 - 3x^3 + 6x^2 + 5x - 1;\ x = 2$

6. $2x^3 - 4x + 8;\ x = 2$
Hint: $2x^3 - 4x + 8 = 2x^3 + 0x^2 - 4x + 8$

Answers
1. 27 **2.** −6 **3.** 73 **4.** 0 **5.** 25 **5.** 16

Chapter Summary

KEY WORDS

A **monomial** is a number, a variable, or a product of numbers and variables.

A **polynomial** is a variable expression in which the terms are monomials.

A **monomial** is a polynomial of *one* term.

A **binomial** is a polynomial of *two* terms.

A **trinomial** is a polynomial of *three* terms.

The **degree of a polynomial** in one variable is the largest exponent on a variable.

ESSENTIAL RULES

Rule for Multiplying Exponential Expressions — If m and n are integers, then $x^m \cdot x^n = x^{m+n}$.

Rule for Simplifying Powers of Exponential Expressions — If m and n are integers, then $(x^m)^n = x^{m \cdot n}$.

Rule for Simplifying Powers of Products — If m, n, and p are integers, then $(x^m \cdot y^n)^p = x^{m \cdot p}y^{n \cdot p}$.

The Sum and Difference of Two Terms — $(a + b)(a - b) = a^2 - b^2$

The Square of a Binomial — $(a + b)^2 = a^2 + 2ab + b^2$
$(a - b)^2 = a^2 - 2ab + b^2$

Rule for Dividing Exponential Expressions — If m and n are integers and $x \neq 0$, then
$\frac{x^m}{x^n} = x^{m-n}$ **if** $m > n$ and
$\frac{x^m}{x^n} = \frac{1}{x^{n-m}}$ **if** $m < n$.

Rule for Negative Exponents — If n is a positive integer and $x \neq 0$, then $x^{-n} = \frac{1}{x^n}$.

Dividend = (quotient × divisor) + remainder

Review/Test

SECTION 1

1.1 Simplify $(3x^3 - 2x^2 - 4) + (8x^2 - 8x + 7)$.

1.2 Simplify $(3a^2 - 2a - 7) - (5a^3 + 2a - 10)$.

SECTION 2

2.1a Simplify $(ab^2)(a^3b^5)$.

2.1b Simplify $(-2xy^2)(3x^2y^4)$.

2.2a Simplify $(x^2y^3)^4$.

2.2b Simplify $(-2a^2b)^3$.

SECTION 3

3.1a Simplify $2x(2x^2 - 3x)$.

3.1b Simplify $-3y^2(-2y^2 + 3y - 6)$.

3.2a Simplify $(x - 3)(x^2 - 4x + 5)$.

3.2b Simplify $(-2x^3 + x^2 - 7)(2x - 3)$.

3.3a Simplify $(a - 2b)(a + 5b)$.

3.3b Simplify $(2x - 7y)(5x - 4y)$.

Review/Test

3.4a Simplify $(4y - 3)(4y + 3)$.

3.4b Simplify $(2x - 5)^2$.

3.5 The radius of a circle is $x - 5$. Use the equation $A = \pi r^2$, where r is the radius, to find the area of the circle in terms of the variable x. Use 3.14 for π.

SECTION **4**

4.1a Simplify $\frac{12x^2}{-3x^8}$.

4.1b Simplify $\frac{(3xy^3)^3}{3x^4y^3}$.

4.2a Simplify $\frac{16x^5 - 8x^3 + 20x}{4x}$.

4.2b Simplify $\frac{12x^3 - 3x^2 + 9}{3x^2}$.

4.3a Simplify
$(x^2 + 6x - 7) \div (x - 1)$.

4.3b Simplify
$(4x^2 - 7) \div (2x - 3)$.

SECTION **5**

5.1a Simplify $(a^2b^{-3})^2$.

5.1b Simplify $(-2ab^{-3})(3a^{-2}b^4)$.

Cumulative Review/Test

1. Simplify
$\frac{3}{16} - \left(-\frac{5}{8}\right) - \frac{7}{9}$.

2. Evaluate
$-3^2 \cdot \left(\frac{2}{3}\right)^3 \cdot \left(-\frac{5}{8}\right)$.

3. Simplify
$\left(-\frac{1}{2}\right)^3 \div \left(\frac{3}{8} - \frac{5}{6}\right) + 2$.

4. Evaluate $\frac{b - (a - b)^2}{b^2}$ when $a = -2$ and $b = 3$.

5. Simplify
$-2x - (-xy) + 7x - 4xy$.

6. Simplify $(12x)\left(-\frac{3}{4}\right)$.

7. Simplify
$-2[3x - 2(4 - 3x) + 2]$.

8. Solve: $12 = -\frac{3}{4}x$

9. Solve:
$2x - 9 = 3x + 7$

10. Solve:
$2 - 3(4 - x) = 2x + 5$

11. 35.2 is what percent of 160?

12. Simplify
$(4b^3 - 7b^2 - 7) +$
$(3b^2 - 8b + 3)$.

13. Simplify
$(3y^3 - 5y + 8) -$
$(-2y^2 + 5y + 8)$.

14. Simplify $(a^3b^5)^3$.

15. Simplify $(4xy^3)(-2x^2y^3)$.

16. Simplify
$-2y^2(-3y^2 - 4y + 8)$.

Cumulative Review/Test

17. Simplify $(2a - 7)(5a^2 - 2a + 3)$.

18. Simplify $(3b - 2)(5b - 7)$.

19. Simplify $(3b + 2)^2$.

20. Simplify $\frac{(-2a^2b^3)^2}{8a^4b^8}$.

21. Simplify $\frac{-18a^3 + 12a^2 - 6}{-3a^2}$.

22. Simplify $(a^2 - 4a - 21) \div (a + 3)$.

23. Simplify $(-2x^{-2}y)(-3x^{-4}y)$.

24. Translate "the difference between eight times a number and twice a number is eighteen" into an equation and solve.

25. A calculator costs a retailer \$24. Find the selling price when the markup rate is 80%.

26. Fifty ounces of pure orange juice are added to 200 oz of a fruit punch which is 10% orange juice. What is the percent concentration of orange juice in the resulting mixture?

27. A car traveling at 50 mph overtakes a cyclist who, riding at 10 mph, has had a 2 h head start. How far from the starting point does the car overtake the cyclist?

28. The width of a rectangle is 40% of the length. The perimeter of the rectangle is 42 m. Find the length and width of the rectangle.

29. The age of a gold coin is 60 years, and the age of a silver coin is 40 years. How many years ago was the gold coin twice the age the silver coin was then?

30. The length of a side of a square is $2x + 3$. Use the equation $A = s^2$, where s is the length of the side of a square, to find the area of the square in terms of the variable x.

6

Factoring

Objectives

- To find the greatest common factor (GCF) of two or more monomials
- To factor a monomial from a polynomial
- To factor a trinomial of the form $x^2 + bx + c$
- To factor completely
- To factor a trinomial of the form $ax^2 + bx + c$
- To factor completely
- To factor the difference of two perfect squares
- To factor a perfect square trinomial
- To factor a common binomial factor
- To factor completely
- To solve equations by factoring
- To solve application problems

Algebra from Geometry

The early Babylonians made substantial progress in both algebra and geometry. Often the progress they made in algebra was based on geometric concepts.

Here are some geometric proofs of algebraic identities.

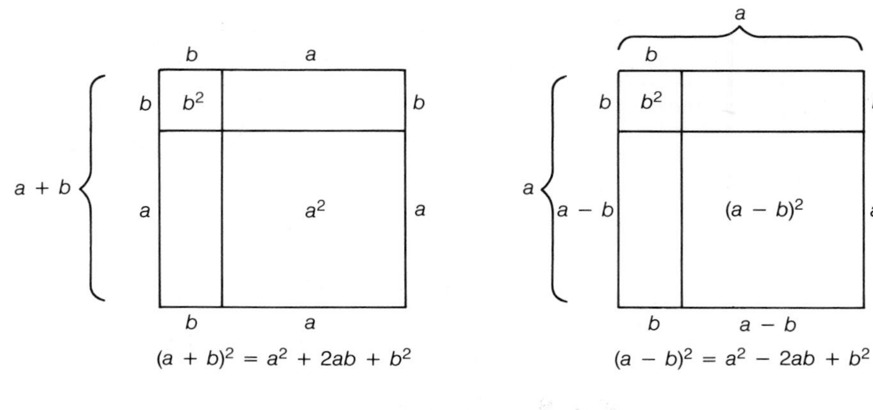

$(a + b)^2 = a^2 + 2ab + b^2$

$(a - b)^2 = a^2 - 2ab + b^2$

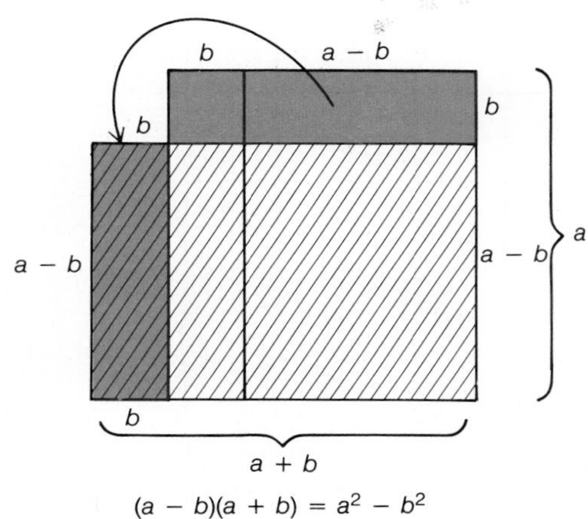

$(a - b)(a + b) = a^2 - b^2$

SECTION 1 Monomial Factors

1.1 Objective To find the greatest common factor (GCF) of two or more monomials

Reference for Computer Tutor™

The **greatest common factor (GCF)** of two or more integers is the greatest integer which is a factor of all of the integers.

$24 = 2 \cdot 2 \cdot 2 \cdot 3$
$60 = 2 \cdot 2 \cdot 3 \cdot 5$

$\text{GCF} = 2 \cdot 2 \cdot 3 = 12$

The GCF of two or more monomials is the product of the GCF of the coefficients and the common variable factors.

$6x^3y = 2 \cdot 3 \cdot x \cdot x \cdot x \cdot y$
$8x^2y^2 = 2 \cdot 2 \cdot 2 \cdot x \cdot x \cdot y \cdot y$

$\text{GCF} = 2 \cdot x \cdot x \cdot y = 2x^2y$

Note that the exponent of each variable in the GCF is the same as the *smallest* exponent of that variable in either of the monomials.

The GCF of $6x^3y$ and $8x^2y^2$ is $2x^2y$.

Find the GCF of $12a^4b$ and $18a^2b^2c$.

$12a^4b = 2 \cdot 2 \cdot 3 \cdot a^4 \cdot b$
$18a^2b^2c = 2 \cdot 3 \cdot 3 \cdot a^2 \cdot b^2 \cdot c$

The common variable factors are a^2 and b. c is not a common variable factor.

$\text{GCF} = 2 \cdot 3 \cdot a^2 \cdot b = 6a^2b$

Example 1

Find the GCF of $4x^4y$ and $18x^2y^6$.

Solution

$4x^4y = 2 \cdot 2 \cdot x^4 \cdot y$
$18x^2y^6 = 2 \cdot 3 \cdot 3 \cdot x^2 \cdot y^6$
The GCF is $2x^2y$.

Example 2

Find the GCF of $12x^3y^6$ and $15x^2y^3$.

Your Solution

Solution on p. A30

1.2 Objective To factor a monomial from a polynomial

Reference for Computer Tutor™

The Distributive Property is used to multiply factors of a polynomial. To **factor** a polynomial means to write the polynomial as a product of other polynomials.

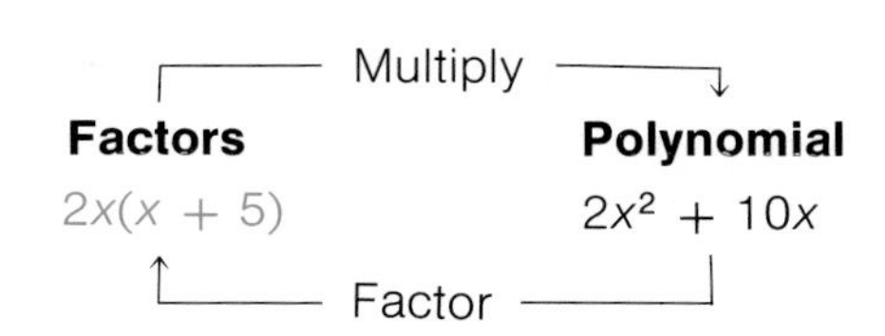

In the example above, $2x$ is the GCF of the terms $2x^2$ and $10x$. It is a **common monomial factor** of the terms. $x + 5$ is a **binomial factor** of $2x^2 + 10x$.

Factor $5x^3 - 35x^2 + 10x$.

Find the GCF of the terms of the polynomial.

$5x^3 = 5 \cdot x^3$
$35x^2 = 5 \cdot 7 \cdot x^2$
$10x = 2 \cdot 5 \cdot x$
The GCF is $5x$.

Divide each term of the polynomial by the GCF.

$$\frac{5x^3}{5x} = x^2 \quad \frac{-35x^2}{5x} = -7x \quad \frac{10x}{5x} = 2$$

Do this step mentally.

Use the quotients to rewrite the polynomial, expressing each term as a product with the GCF as one of the factors.

$5x^3 - 35x^2 + 10x = 5x(x^2) + 5x(-7x) + 5x(2)$

Use the Distributive Property to write the polynomial as a product of factors.

$= 5x(x^2 - 7x + 2)$

Example 3

Factor $8x^2 + 2xy$.

Solution

$8x^2 = 2 \cdot 2 \cdot 2 \cdot x^2$
$2xy = 2 \cdot x \cdot y$
The GCF is $2x$.

$8x^2 + 2xy = 2x(4x) + 2x(y) = 2x(4x + y)$

Example 4

Factor $14a^2 - 21a^4b$.

Your solution

Example 5

Factor $16x^2y + 8x^4y^2 - 12x^4y^5$.

Solution

$16x^2y = 2 \cdot 2 \cdot 2 \cdot 2 \cdot x^2 \cdot y$
$8x^4y^2 = 2 \cdot 2 \cdot 2 \cdot x^4 \cdot y^2$
$12x^4y^5 = 2 \cdot 2 \cdot 3 \cdot x^4 \cdot y^5$
The GCF is $4x^2y$.

$16x^2y + 8x^4y^2 - 12x^4y^5 =$
$4x^2y(4) + 4x^2y(2x^2y) + 4x^2y(-3x^2y^4) =$
$4x^2y(4 + 2x^2y - 3x^2y^4)$

Example 6

Factor $6x^4y^2 - 9x^3y^2 + 12x^2y^4$.

Your solution

Solutions on p. A30

1.1 Exercises

Find the greatest common factor.

1. x^7, x^3

2. y^6, y^{12}

3. x^2y^4, xy^6

4. a^5b^3, a^3b^8

5. $x^2y^4z^6, xy^8z^2$

6. ab^2c^3, a^3b^2c

7. $a^3b^2c^3, ab^4c^3$

8. x^3y^2z, x^4yz^5

9. $3x^4, 12x^2$

10. $12x, 30x^2$

11. $16a^3, 18a$

12. $8y^3, 12y^6$

13. $14a^3, 49a^7$

14. $12y^2, 27y^4$

15. $3x^2y^2, 5ab^2$

16. $8x^2y^3, 7ab^4$

17. $9a^2b^4, 24a^4b^2$

18. $15a^4b^2, 9ab^5$

19. $ab^3, 4a^2b, 12a^2b^3$

20. $12x^2y, x^4y, 16x$

21. $2x^2y, 4xy, 8x$

22. $16x^2, 8x^4y^2, 12xy$

23. $3x^2y^2, 6x, 9x^3y^3$

24. $4a^2b^3, 8a^3, 12ab^4$

1.2 Exercises

Factor.

25. $5a + 5$

26. $7b - 7$

27. $16 - 8a^2$

28. $12 + 12y^2$

29. $8x + 12$

30. $16a - 24$

31. $30a - 6$

32. $20b + 5$

33. $7x^2 - 3x$

34. $12y^2 - 5y$

35. $3a^2 + 5a^5$

36. $9x - 5x^2$

Factor.

37. $14y^2 + 11y$

38. $6b^3 - 5b^2$

39. $2x^4 - 4x$

40. $3y^4 - 9y$

41. $10x^4 - 12x^2$

42. $12a^5 - 32a^2$

43. $8a^8 - 4a^5$

44. $16y^4 - 8y^7$

45. $x^2y^2 - xy$

46. $a^2b^2 + ab$

47. $3x^2y^4 - 6xy$

48. $12a^2b^5 - 9ab$

49. $x^2y - xy^3$

50. $a^2b + a^4b^2$

51. $2a^5b + 3xy^3$

52. $5x^2y - 7ab^3$

53. $6a^2b^3 - 12b^2$

54. $8x^2y^3 - 4x^2$

55. $a^3 - 3a^2 + 5a$

56. $b^3 - 5b^2 - 7b$

57. $5x^2 - 15x + 35$

58. $8y^2 - 12y + 32$

59. $3x^3 + 6x^2 + 9x$

60. $5y^3 - 20y^2 + 10y$

61. $2x^4 - 4x^3 + 6x^2$

62. $3y^4 - 9y^3 - 6y^2$

63. $2x^3 + 6x^2 - 14x$

64. $3y^3 - 9y^2 + 24y$

65. $2y^5 - 3y^4 + 7y^3$

66. $6a^5 - 3a^3 - 2a^2$

67. $x^3y - 3x^2y^2 + 7xy^3$

68. $2a^2b - 5a^2b^2 + 7ab^2$

69. $5y^3 + 10y^2 - 25y$

70. $4b^5 + 6b^3 - 12b$

71. $3a^2b^2 - 9ab^2 + 15b^2$

72. $8x^2y^2 - 4x^2y + x^2$

SECTION 2 Factoring Polynomials of the Form $x^2 + bx + c$

2.1 Objective To factor a trinomial of the form $x^2 + bx + c$

Reference for Computer Tutor™

Trinomials of the form $x^2 + bx + c$, where b and c are integers, are shown at the right.

$x^2 + 8x + 12$, $b = 8$, $c = 12$
$x^2 - 7x + 12$, $b = -7$, $c = 12$
$x^2 - 2x - 15$, $b = -2$, $c = -15$

To **factor** a trinomial of this form means to express the trinomial as the product of two binomials.

Trinomials expressed as the product of binomials are shown at the right.

$x^2 + 8x + 12 = (x + 6)(x + 2)$
$x^2 - 7x + 12 = (x - 3)(x - 4)$
$x^2 - 2x - 15 = (x + 3)(x - 5)$

The method by which factors of a trinomial are found is based upon FOIL. Consider the following binomial products, noting the relationship between the constant terms of the binomials and the terms of the trinomials.

Signs in the binomials are the same

$(x + 6)(x + 2) = x^2 + 2x + 6x + (6)(2) = x^2 + 8x + 12$
sum of 6 and 2 → $8x$
product of 6 and 2 → 12

$(x - 3)(x - 4) = x^2 - 4x - 3x + (-3)(-4) = x^2 - 7x + 12$
sum of -3 and -4 → $-7x$
product of -3 and -4 → 12

Signs in the binomials are opposite

$(x + 3)(x - 5) = x^2 - 5x + 3x + (3)(-5) = x^2 - 2x - 15$
sum of 3 and -5 → $-2x$
product of 3 and -5 → -15

$(x - 4)(x + 6) = x^2 + 6x - 4x + (-4)(6) = x^2 + 2x - 24$
sum of -4 and 6 → $2x$
product of -4 and 6 → -24

Important Relationships

1. When the constant term of the trinomial is positive, the constant terms of the binomials have the same sign. They are both positive when the coefficient of the x term in the trinomial is positive. They are both negative when the coefficient of the x term in the trinomial is negative.

2. When the constant term of the trinomial is negative, the constant terms of the binomials have opposite signs.

3. In the trinomial, the coefficient of x is the sum of the constant terms of the binomials.

4. In the trinomial, the constant term is the product of the constant terms of the binomials.

The following trinomial factoring patterns help to summarize the relationships stated above.

Trinomial	Factoring Pattern
$x^2 + bx + c$	$(x + \square)(x + \square)$
$x^2 - bx + c$	$(x - \square)(x - \square)$
$x^2 + bx - c$	$(x + \square)(x - \square)$
$x^2 - bx - c$	$(x + \square)(x - \square)$

Factor $x^2 + 7x + 10$.

The constant term is positive.
The coefficient of x is positive.
The binomial constants will be positive.

$(x + \square)(x + \square)$

Find two positive factors of 10 whose sum is 7.

Factors	Sum
+1, +10	11
+2, +5	7

Write the factors of the trinomial.

$(x + 2)(x + 5)$

Check:
$$(x + 2)(x + 5) = x^2 + 5x + 2x + 10$$
$$= x^2 + 7x + 10$$

Factor $x^2 - 8x - 9$.

The constant term is negative.
The signs of the binomial constants will be opposites.

$(x + \square)(x - \square)$

Find two factors of 9, one of which is positive and one of which is negative, whose sum is -8.

Once the sum of -8 is found, other factors need not be tried.

Factors	Sum
−1, +9	8
+1, −9	−8
+3, −3	0

Write the factors of the trinomial.

$(x + 1)(x - 9)$

Check:
$$(x + 1)(x - 9) = x^2 - 9x + x - 9$$
$$= x^2 - 8x - 9$$

When only integers are used, some trinomials do not factor. For example, to factor $x^2 + 5x + 3$, it would be necessary to find two positive integers whose product is 3 and whose sum is 5. This is not possible, since the only positive factors of 3 are 1 and 3, and the sum of 1 and 3 is 4. This trinomial is **irreducible over the integers.** Binomials of the form $x + a$ or $x - a$ are also irreducible over the integers.

Example 1

Factor $x^2 - 8x + 15$.

Solution

$(x - \square)(x - \square)$

Factors	Sum
−1, −15	−16
−3, −5	−8

$(x - 3)(x - 5)$

$x^2 - 8x + 15 = (x - 3)(x - 5)$

Example 2

Factor $x^2 - 9x + 20$.

Your solution

Example 3

Factor $x^2 + 6x - 27$.

Solution

$(x + \square)(x - \square)$

Factors	Sum
+1, −27	−26
−1, +27	26
+3, −9	−6
−3, +9	6

$(x + 9)(x - 3)$

$x^2 + 6x - 27 = (x + 9)(x - 3)$

Example 4

Factor $x^2 + 3x - 18$.

Your solution

Solutions on p. A30

2.2 Objective To factor completely

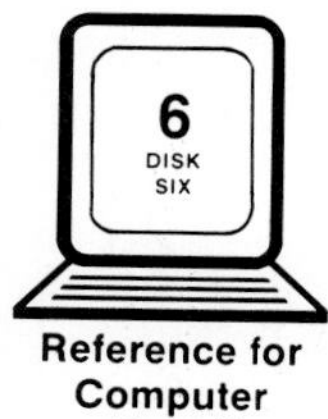

Reference for Computer Tutor™

A polynomial is factored completely when it is written as a product of factors which are irreducible over the integers.

Factor $3x^3 + 15x^2 + 18x$.

Find the GCF of the terms of the polynomial. — The GCF is $3x$.

Factor out the GCF.

$3x^3 + 15x^2 + 18x =$

$3x(x^2) + 3x(5x) + 3x(6) =$ (Do this step mentally.)

$3x(x^2 + 5x + 6)$

Factor the trinomial.
Find two positive factors of 6 whose sum is 5.

$3x(x + \square)(x + \square)$

Factors	Sum
+1, +6	7
+2 +3	5

Write the product of the GCF and the factors of the trinomial.

$3x(x + 2)(x + 3)$

Check: $3x(x + 2)(x + 3) =$
$3x(x^2 + 3x + 2x + 6) =$
$3x(x^2 + 5x + 6)$
$3x^3 + 15x^2 + 18x$

Factor $x^2 + 9xy + 20y^2$.

The terms have no common factor.
There are two variables.
Find two positive factors of 20 whose sum is 9.

$(x + \quad y)(x + \quad y)$

Factors	Sum
+1, +20	21
+2, +10	12
+4, +5	9

Write the factors of the trinomial.

$(x + 4y)(x + 5y)$

Check: $(x + 4y)(x + 5y) =$
$x^2 + 5xy + 4xy + 20y^2 =$
$x^2 + 9xy + 20y^2$

Example 5

Factor $2x^2y + 12xy - 14y$.

Solution

The GCF is $2y$.

$2x^2y + 12xy - 14y = 2y(x^2 + 6x - 7)$

Factor the trinomial.

$2y(x + \quad)(x - \quad)$

Factors	Sum
+1, −7	−6
−1, +7	6

$2y(x + 7)(x - 1)$

$2x^2y + 12xy - 14y = 2y(x + 7)(x - 1)$

Example 6

Factor $3a^2b - 18ab - 81b$.

Your solution

Example 7

Factor $4x^2 - 40xy + 84y^2$.

Solution

The GCF is 4.

$4x^2 - 40xy + 84y^2 = 4(x^2 - 10xy + 21y^2)$

Factor the trinomial.

$4(x - \quad y)(x - \quad y)$

Factors	Sum
−1, −21	−22
−3, −7	−10

$4(x - 3y)(x - 7y)$

$4x^2 - 40xy + 84y^2 = 4(x - 3y)(x - 7y)$

Example 8

Factor $3x^2 - 9xy - 12y^2$.

Your solution

Solutions on p. A31

2.1 Exercises

Factor.

1. $x^2 + 3x + 2$
2. $x^2 + 5x + 6$
3. $x^2 - x - 2$
4. $x^2 + x - 6$
5. $a^2 + a - 12$
6. $a^2 - 2a - 35$
7. $a^2 - 3a + 2$
8. $a^2 - 5a + 4$
9. $a^2 + a - 2$
10. $a^2 - 2a - 3$
11. $b^2 - 6b + 9$
12. $b^2 + 8b + 16$
13. $b^2 + 7b - 8$
14. $y^2 - y - 6$
15. $y^2 + 6y - 55$
16. $z^2 - 4z - 45$
17. $y^2 - 5y + 6$
18. $y^2 - 8y + 15$
19. $z^2 - 14z + 45$
20. $z^2 - 14z + 49$
21. $z^2 - 12z - 160$
22. $p^2 + 2p - 35$
23. $p^2 + 12p + 27$
24. $p^2 - 6p + 8$
25. $x^2 + 20x + 100$
26. $x^2 + 18x + 81$
27. $b^2 + 9b + 20$
28. $b^2 + 13b + 40$
29. $x^2 - 11x - 42$
30. $x^2 + 9x - 70$
31. $b^2 - b - 20$
32. $b^2 + 3b - 40$
33. $y^2 - 14y - 51$
34. $y^2 - y - 72$
35. $p^2 - 4p - 21$
36. $p^2 + 16p + 39$

Factor.

37. $y^2 - 8y + 32$

38. $y^2 - 9y + 81$

39. $x^2 - 20x + 75$

40. $p^2 + 24p + 63$

41. $x^2 - 15x + 56$

42. $x^2 + 21x + 38$

43. $x^2 + x - 56$

44. $x^2 + 5x - 36$

45. $a^2 - 21a - 72$

46. $a^2 - 7a - 44$

47. $a^2 - 15a + 36$

48. $a^2 - 21a + 54$

49. $z^2 - 9z - 136$

50. $z^2 + 14z - 147$

51. $c^2 - c - 90$

52. $c^2 - 3c - 180$

53. $z^2 + 15z + 44$

54. $p^2 + 24p + 135$

55. $c^2 + 19c + 34$

56. $c^2 + 11c + 18$

57. $x^2 - 4x - 96$

58. $x^2 + 10x - 75$

59. $x^2 - 22x + 112$

60. $x^2 + 21x - 100$

61. $b^2 + 8b - 105$

62. $b^2 - 22b + 72$

63. $a^2 - 9a - 36$

64. $a^2 + 42a - 135$

65. $b^2 - 23b + 102$

66. $b^2 - 25b + 126$

67. $a^2 + 27a + 72$

68. $z^2 + 24z + 144$

69. $x^2 + 25x + 156$

70. $x^2 - 29x + 100$

71. $x^2 - 10x - 96$

72. $x^2 + 9x - 112$

2.2 Exercises

Factor.

73. $2x^2 + 6x + 4$ **74.** $3x^2 + 15x + 18$ **75.** $3a^2 + 3a - 18$

76. $4x^2 - 4x - 8$ **77.** $ab^2 + 2ab - 15a$ **78.** $ab^2 + 7ab - 8a$

79. $xy^2 - 5xy + 6x$ **80.** $xy^2 + 8xy + 15x$ **81.** $z^3 - 7z^2 + 12z$

82. $2a^3 + 6a^2 + 4a$ **83.** $3y^3 - 15y^2 + 18y$ **84.** $4y^3 + 12y^2 - 72y$

85. $3x^2 + 3x - 36$ **86.** $2x^3 - 2x^2 + 4x$ **87.** $5z^2 - 15z - 140$

88. $6z^2 + 12z - 90$ **89.** $2a^3 + 8a^2 - 64a$ **90.** $3a^3 - 9a^2 - 54a$

91. $x^2 - 5xy + 6y^2$ **92.** $x^2 + 4xy - 21y^2$ **93.** $a^2 - 9ab + 20b^2$

94. $a^2 - 15ab + 50b^2$ **95.** $x^2 - 3xy - 28y^2$ **96.** $s^2 + 2st - 48t^2$

97. $y^2 - 15yz - 41z^2$ **98.** $y^2 + 85yz + 36z^2$ **99.** $z^4 - 12z^3 + 35z^2$

100. $z^4 + 2z^3 - 80z^2$ **101.** $b^4 - 22b^3 + 120b^2$ **102.** $b^4 - 3b^3 - 10b^2$

103. $2y^4 - 26y^3 - 96y^2$ **104.** $3y^4 + 54y^3 + 135y^2$ **105.** $x^4 + 7x^3 - 8x^2$

106. $x^4 - 11x^3 - 12x^2$ **107.** $4x^2y + 20xy - 56y$ **108.** $3x^2y - 6xy - 45y$

Factor.

109. $8y^2 - 32y + 24$

110. $10y^2 - 100y + 90$

111. $c^3 + 13c^2 + 30c$

112. $c^3 + 18c^2 - 40c$

113. $3x^3 - 36x^2 + 81x$

114. $4x^3 + 4x^2 - 24x$

115. $x^2 - 8xy + 15y^2$

116. $y^2 - 7xy - 8x^2$

117. $a^2 - 13ab + 42b^2$

118. $y^2 + 4yz - 21z^2$

119. $y^2 + 8yz + 7z^2$

120. $y^2 - 16yz + 15z^2$

121. $3x^2y + 60xy - 63y$

122. $4x^2y - 68xy - 72y$

123. $3x^3 + 3x^2 - 36x$

124. $4x^3 + 12x^2 - 160x$

125. $4z^3 + 32z^2 - 132z$

126. $5z^3 - 50z^2 - 120z$

127. $4x^3 + 8x^2 - 12x$

128. $5x^3 + 30x^2 + 40x$

129. $5p^2 + 25p - 420$

130. $4p^2 - 28p - 480$

131. $p^4 + 9p^3 - 36p^2$

132. $p^4 + p^3 - 56p^2$

133. $t^2 - 12ts + 35s^2$

134. $a^2 - 10ab + 25b^2$

135. $a^2 - 8ab - 33b^2$

136. $x^2 + 4xy - 60y^2$

137. $5x^4 - 30x^3 + 40x^2$

138. $6x^3 - 6x^2 - 120x$

139. $15ab^2 + 45ab - 60a$

140. $20a^2b - 100ab + 120b$

141. $3yx^2 + 36yx - 135y$

142. $4yz^2 - 52yz + 88y$

SECTION 3 Factoring Polynomials of the Form $ax^2 + bx + c$

3.1 Objective To factor a trinomial of the form $ax^2 + bx + c$

Reference for Computer Tutor™

Trinomials of the form $ax^2 + bx + c$, where a is a positive integer and b and c are integers, are shown at the right.

$3x^2 - x + 4$, $a = 3$, $b = -1$, $c = 4$
$6x^2 + 8x - 6$, $a = 6$, $b = 8$, $c = -6$

To factor a trinomial of this form, a trial-and-error method is used. Trial factors are written, using the factors of a and c to write the binomials. Then FOIL is used to check for b, the coefficient of the middle term.

To reduce the number of trial factors which must be considered, remember the following.

1. Use the signs of the constant and the coefficient of x in the trinomial to determine the signs of the terms in the binomial factors.

Trinomial	Factoring Pattern
$ax^2 + bx + c$	$(\ \ x + \ \ )(\ \ x + \ \ )$
$ax^2 - bx + c$	$(\ \ x - \ \ )(\ \ x - \ \ )$
$ax^2 - bx - c$	$(\ \ x + \ \ )(\ \ x - \ \ )$ or $(\ \ x - \ \ )(\ \ x + \ \ )$
$ax^2 + bx - c$	$(\ \ x + \ \ )(\ \ x - \ \ )$ or $(\ \ x - \ \ )(\ \ x + \ \ )$

2. If the terms of the trinomial do not have a common factor, then the two terms in either one of the binomial factors will not have a common factor.

Factor $2x^2 - 7x + 3$.

The terms have no common factor.
The constant term is positive.
The coefficient of x is negative.
The binomial constants will be negative.

$(\ \ x - \ \ )(\ \ x - \ \ )$

Write the factors of 2 (the coefficient of x^2). These factors will be the coefficients of the x terms in the binomial factors.

Factors of 2: 1, 2

Write the negative factors of 3 (the constant term). These factors will be the constants in the binomial factors.

Factors of 3: -1, -3

Write trial factors. Writing the 1 when it is the coefficient of x may be helpful. Use the Outer and Inner products of FOIL to determine the middle term of the trinomial.

Trial Factors	Middle Term
$(1x - 1)(2x - 3)$	$-3x - 2x = -5x$
$(1x - 3)(2x - 1)$	$-x - 6x = -7x$

Write the factors of the trinomial.

$(x - 3)(2x - 1)$

Check: $(x - 3)(2x - 1) =$
$2x^2 - x - 6x + 3 =$
$2x^2 - 7x + 3$

Factor $6x^2 - x - 2$.

The terms have no common factor. The constant term is negative. The signs of the binomial constants will be opposites.

$(\square x + \square)(\square x - \square)$

or

$(\square x - \square)(\square x + \square)$

Write the factors of 6. These factors will be the coefficients of the x terms in the binomial factors.

Factors of 6: 1, 6
2, 3

Write the factors of -2. These factors will be the constants in the binomial factors.

Factors of -2: $-1, +2$
$+1, -2$

Write the trial factors.
Use the Outer and Inner terms of FOIL to determine the middle term of the trinomial.
It is not necessary to test trial factors which have a common factor. For example, $6x + 2$ need not be tested because it has a common factor of 2.
Once a trial solution has the correct middle term, other trial factors need not be tried.

Trial Factors	Middle Term
$(1x - 1)(6x + 2)$	Common factor
$(1x + 2)(6x - 1)$	$-x + 12x = 11x$
$(1x + 1)(6x - 2)$	Common factor
$(1x - 2)(6x + 1)$	$x - 12x = -11x$
$(2x - 1)(3x + 2)$	$4x - 3x = x$
$(2x + 2)(3x - 1)$	Common factor
$(2x + 1)(3x - 2)$	$-4x + 3x = -x$
$(2x - 2)(3x + 1)$	Common factor

Write the factors of the trinomial.

$(2x + 1)(3x - 2)$

Check: $(2x + 1)(3x - 2) =$
$6x^2 - 4x + 3x - 2 =$
$6x^2 - x - 2$

Example 1

Factor $3x^2 + x - 2$.

Solution

$(\square x + \square)(\square x - \square)$ or $(\square x - \square)(\square x + \square)$

Factors of 3: 1, 3

Factors of -2: $+1, -2$
$-1, +2$

Trial Factors	Middle Term
$(1x + 1)(3x - 2)$	$-2x + 3x = x$
$(1x - 2)(3x + 1)$	$x - 6x = -5x$
$(1x - 1)(3x + 2)$	$2x - 3x = -x$
$(1x + 2)(3x - 1)$	$-x + 6x = 5x$

$(x + 1)(3x - 2)$

$3x^2 + x - 2 = (x + 1)(3x - 2)$

Example 2

Factor $2x^2 - x - 3$.

Your solution

Solution on p. A32

3.2 Objective To factor completely

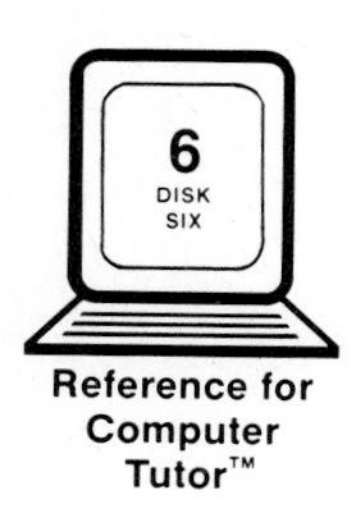

Reference for Computer Tutor™

Factor $3x^3 - 23x^2 + 14x$.

Find the GCF of the terms of the polynomial. — The GCF is x.

Factor out the GCF. — $3x^3 - 23x^2 + 14x = x(3x^2 - 23x + 14)$

Factor the trinomial. — $x(\;\;x - \;\;)(\;\;x - \;\;)$

Write the factors of 3. — Factors of 3: 1, 3

Write the negative factors of 14. — Factors of 14: $-1, -14$; $-2, -7$

Write trial factors. Writing the 1 when it is the coefficient of x may be helpful.
Determine the middle term of the trinomial.

Trial Factors	Middle Term
$(1x - 1)(3x - 14)$	$-14x - 3x = -17x$
$(1x - 14)(3x - 1)$	$-x - 42x = -43x$
$(1x - 2)(3x - 7)$	$-7x - 6x = -13x$
$(1x - 7)(3x - 2)$	$-2x - 21x = -23x$

Write the product of the GCF and the factors of the trinomial. — $x(x - 7)(3x - 2)$

Check: $x(x - 7)(3x - 2) =$
$x(3x^2 - 2x - 21x + 14) =$
$x(3x^2 - 23x + 14) =$
$3x^3 - 23x^2 + 14x$

Factor $15 - 2x - x^2$.

The terms have no common factor.
The coefficient of x^2 is -1.
The signs of the binomials will be opposites.

$(\;\; + \;\;x)(\;\; - \;\;x)$
or
$(\;\; - \;\;x)(\;\; + \;\;x)$

Write the factors of 15. — Factors of 15: 1, 15; 3, 5

Write the factors of -1. — Factors of -1: 1, -1

Write trial factors.
Determine the middle term of the trinomial.

Trial Factors	Middle Term
$(1 + 1x)(15 - 1x)$	$-x + 15x = 14x$
$(1 - 1x)(15 + 1x)$	$x - 15x = -14x$
$(3 + 1x)(5 - 1x)$	$-3x + 5x = 2x$
$(3 - 1x)(5 + 1x)$	$3x - 5x = -2x$

Write the factors of the trinomial. — $(3 - x)(5 + x)$

Check: $(3 - x)(5 + x) = 15 + 3x - 5x - x^2$
$= 15 - 2x - x^2$

Example 3

Factor $2x^2y + 19xy - 10y$.

Solution

The GCF is y.

$2x^2y + 19xy - 10y = y(2x^2 + 19x - 10)$

Factor the trinomial.

$y(\square x + \square)(\square x - \square)$ or $y(\square x - \square)(\square x + \square)$

Factors of 2: 1, 2 Factors of -10: $+1, -10$
$-1, +10$
$+2, -5$
$-2, +5$

Trial Factors	Middle Terms
$(1x + 1)(2x - 10)$	Common factor
$(1x - 10)(2x + 1)$	$x - 20x = -19x$
$(1x - 1)(2x + 10)$	Common factor
$(1x + 10)(2x - 1)$	$-x + 20x = 19x$
$(1x + 2)(2x - 5)$	$-5x + 4x = -x$
$(1x - 5)(2x + 2)$	Common factor
$(1x - 2)(2x + 5)$	$5x - 4x = x$
$(1x + 5)(2x - 2)$	Common factor

$y(x + 10)(2x - 1)$

$2x^2y + 19xy - 10y = y(x + 10)(2x - 1)$

Example 4

Factor $4a^2b^2 + 26a^2b - 14a^2$.

Your solution

Example 5

Factor $12x - 32x^2 - 12x^3$.

Solution

The GCF is $4x$.

$12x - 32x^2 - 12x^3 = 4x(3 - 8x - 3x^2)$

Factor the trinomial.

$4x(\square + \square x)(\square - \square x)$ or $4x(\square - \square x)(\square + \square x)$

Factors of 3: 1, 3 Factors of -3: $+1, -3$
$-1, +3$

Trial Factors	Middle Term
$(1 + 1x)(3 - 3x)$	Common factor
$(1 - 3x)(3 + 1x)$	$x - 9x = -8x$
$(1 - 1x)(3 + 3x)$	Common factor
$(1 + 3x)(3 - 1x)$	$-x + 9x = 8x$

$4x(1 - 3x)(3 + x)$

$12x - 32x^2 - 12x^3 = 4x(1 - 3x)(3 + x)$

Example 6

Factor $12y + 12y^2 - 45y^3$.

Your solution

Solutions on p. A32

3.1 Exercises

Factor.

1. $2x^2 + 3x + 1$
2. $5x^2 + 6x + 1$
3. $2y^2 + 7y + 3$
4. $3y^2 + 7y + 2$
5. $2a^2 - 3a + 1$
6. $3a^2 - 4a + 1$
7. $2b^2 - 11b + 5$
8. $3b^2 - 13b + 4$
9. $2x^2 + x - 1$
10. $4x^2 - 3x - 1$
11. $2x^2 - 5x - 3$
12. $3x^2 + 5x - 2$
13. $2t^2 - t - 10$
14. $2t^2 + 5t - 12$
15. $3p^2 - 16p + 5$
16. $6p^2 + 5p + 1$
17. $12y^2 - 7y + 1$
18. $6y^2 - 5y + 1$
19. $6z^2 - 7z + 3$
20. $9z^2 + 3z + 2$
21. $6t^2 - 11t + 4$
22. $10t^2 + 11t + 3$
23. $8x^2 + 33x + 4$
24. $7x^2 + 50x + 7$
25. $5x^2 - 62x - 7$
26. $9x^2 - 13x - 4$
27. $12y^2 + 19y + 5$
28. $5y^2 - 22y + 8$
29. $7a^2 + 47a - 14$
30. $11a^2 - 54a - 5$
31. $3b^2 - 16b + 16$
32. $6b^2 - 19b + 15$
33. $2z^2 - 27z - 14$
34. $4z^2 + 5z - 6$
35. $3p^2 + 22p - 16$
36. $7p^2 + 19p + 10$

Factor.

37. $6x^2 - 17x + 12$ **38.** $15x^2 - 19x + 6$ **39.** $5b^2 + 33b - 14$

40. $8x^2 - 30x + 25$ **41.** $6a^2 + 7a - 24$ **42.** $14a^2 + 15a - 9$

43. $4z^2 + 11z + 6$ **44.** $6z^2 - 25z + 14$ **45.** $22p^2 + 51p - 10$

46. $14p^2 - 41p + 15$ **47.** $8y^2 + 17y + 9$ **48.** $12y^2 - 145y + 12$

49. $18t^2 - 9t - 5$ **50.** $12t^2 + 28t - 5$ **51.** $6b^2 + 71b - 12$

52. $8b^2 + 65b + 8$ **53.** $9x^2 + 12x + 4$ **54.** $25x^2 - 30x + 9$

55. $6b^2 - 13b + 6$ **56.** $20b^2 + 37b + 15$ **57.** $33b^2 + 34b - 35$

58. $15b^2 - 43b + 22$ **59.** $18y^2 - 39y + 20$ **60.** $24y^2 + 41y + 12$

61. $15a^2 + 26a - 21$ **62.** $6a^2 + 23a + 21$ **63.** $8y^2 - 26y + 15$

64. $18y^2 - 27y + 4$ **65.** $8z^2 + 2z - 15$ **66.** $10z^2 + 3z - 4$

67. $15x^2 - 82x + 24$ **68.** $13z^2 + 49z - 8$ **69.** $10z^2 - 29z + 10$

70. $15z^2 - 44z + 32$ **71.** $36z^2 + 72z + 35$ **72.** $16z^2 + 8z - 35$

3.2 Exercises

Factor.

73. $4x^2 + 6x + 2$
74. $12x^2 + 33x - 9$
75. $15y^2 - 50y + 35$
76. $30y^2 + 10y - 20$
77. $2x^3 - 11x^2 + 5x$
78. $2x^3 - 3x^2 - 5x$
79. $3a^2b - 16ab + 16b$
80. $2a^2b - ab - 21b$
81. $3z^2 + 95z + 10$
82. $8z^2 - 36z + 1$
83. $3x^2 + xy - 2y^2$
84. $6x^2 + 10xy + 4y^2$
85. $3a^2 + 5ab - 2b^2$
86. $2a^2 - 9ab + 9b^2$
87. $4y^2 - 11yz + 6z^2$
88. $2y^2 + 7yz + 5z^2$
89. $12 - x - x^2$
90. $2 + x - x^2$
91. $28 + 3z - z^2$
92. $15 - 2z - z^2$
93. $8 - 7x - x^2$
94. $12 + 11x - x^2$
95. $9x^2 + 33x - 60$
96. $16x^2 - 16x - 12$
97. $80y^2 - 36y + 4$
98. $24y^2 - 24y - 18$
99. $8z^3 + 14z^2 + 3z$
100. $6z^3 - 23z^2 + 20z$
101. $6x^2y - 11xy - 10y$
102. $8x^2y - 27xy + 9y$
103. $24x^2 - 52x + 24$
104. $60x^2 + 95x + 20$
105. $35a^4 + 9a^3 - 2a^2$
106. $15a^4 + 26a^3 + 7a^2$
107. $15b^2 - 115b + 70$
108. $25b^2 + 35b - 30$

Factor.

109. $3x^2 - 26xy + 35y^2$

110. $4x^2 + 16xy + 15y^2$

111. $216y^2 - 3y - 3$

112. $360y^2 + 4y - 4$

113. $21 - 20x - x^2$

114. $18 + 17x - x^2$

115. $15a^2 + 11ab - 14b^2$

116. $15a^2 - 31ab + 10b^2$

117. $33z - 8z^2 - z^3$

118. $24z + 10z^2 - z^3$

119. $10x^3 + 12x^2 + 2x$

120. $9x^3 - 39x^2 + 12x$

121. $10t^2 - 5t - 50$

122. $16t^2 + 40t - 96$

123. $3p^3 - 16p^2 + 5p$

124. $6p^3 + 5p^2 + p$

125. $26z^2 + 98z - 24$

126. $30z^2 - 87z + 30$

127. $10y^3 - 44y^2 + 16y$

128. $14y^3 + 94y^2 - 28y$

129. $4yz^3 + 5yz^2 - 6yz$

130. $2yz^3 - 17yz^2 + 8yz$

131. $20b^4 + 41b^3 + 20b^2$

132. $6b^4 - 13b^3 + 6b^2$

133. $12a^3 + 14a^2 - 48a$

134. $42a^3 + 45a^2 - 27a$

135. $36p^2 - 9p^3 - p^4$

136. $9x^2y - 30xy^2 + 25y^3$

137. $8x^2y - 38xy^2 + 35y^3$

138. $9x^3y - 24x^2y^2 + 16xy^3$

139. $9x^3y + 12x^2y + 4xy$

140. $9a^3b - 9a^2b^2 - 10ab^3$

141. $2a^3b - 11a^2b^2 + 5ab^3$

SECTION 4 Special Factoring

4.1 Objective To factor the difference of two perfect squares

Reference for Computer Tutor™

The product of a term and itself is called a **perfect square.** The exponents of variables of perfect squares are always even numbers.

Term		Perfect Square
2	$2 \cdot 2 =$	4
x	$x \cdot x =$	x^2
$3y^3$	$3y^3 \cdot 3y^3 =$	$9y^6$

The **square root** of a perfect square is one of the two equal factors of the perfect square. "$\sqrt{\ }$", called a radical, is the symbol for square root. To find the exponent of the square root of a variable term, multiply the exponent by $\frac{1}{2}$.

$\sqrt{4} = 2$

$\sqrt{x^2} = x$

$\sqrt{9y^6} = 3y^3$

The difference of two perfect squares is the product of the sum and difference of two terms.

Sum and Difference of Two Terms		**Difference of Two Perfect Squares**
$(a + b)(a - b)$	$=$	$a^2 - b^2$

The factors of the difference of two perfect squares are the sum and difference of the square roots of the perfect squares.

$a^2 + b^2$ is the *sum* of two perfect squares. It is irreducible over the integers.

Factor $x^2 - 16$.

Write $x^2 - 16$ as the difference of two perfect squares.

$$x^2 - 16 = x^2 - 4^2$$

The factors are the sum and difference of the square roots of the perfect squares.

$$= (x + 4)(x - 4)$$

Check: $(x + 4)(x - 4) = x^2 - 4x + 4x - 16$
$= x^2 - 16$

Factor $x^2 - 10$.

Since 10 is not a perfect square, $x^2 - 10$ cannot be written as the difference of two perfect squares. $x^2 - 10$ is irreducible over the integers.

Example 1

Factor $16x^2 - y^2$.

Solution

$16x^2 - y^2 = (4x)^2 - y^2 = (4x + y)(4x - y)$

Example 2

Factor $25a^2 - b^2$.

Your solution

Example 3

Factor $z^6 - 25$.

Solution

$z^6 - 25 = (z^3)^2 - 5^2 = (z^3 + 5)(z^3 - 5)$

Example 4

Factor $n^8 - 36$.

Your solution

Solutions on p. A33

4.2 Objective To factor a perfect square trinomial

Reference for Computer Tutor™

A perfect square trinomial is the square of a binomial.

Square of a Binomial		Perfect Square Trinomial
$(a + b)^2$	$= (a + b)(a + b) =$	$a^2 + 2ab + b^2$
$(a - b)^2$	$= (a - b)(a - b) =$	$a^2 - 2ab + b^2$

In factoring a perfect square trinomial, remember that the terms of the binomial are the square roots of the perfect squares of the trinomial. The sign in the binomial is the sign of the middle term of the trinomial.

Factor $x^2 + 10x + 25$.

Check that the trinomial is a perfect square.

$\sqrt{x^2} = x$
$\sqrt{25} = 5$ $\quad 2(5x) = 10x$

The trinomial is a perfect square.

Write the factors as the square of a binomial.

$(x + 5)^2$

Check: $(x + 5)^2 = (x + 5)(x + 5)$
$= x^2 + 5x + 5x + 25$
$= x^2 + 10x + 25$

Factor $x^2 + 10x - 25$.

Since the constant term is negative, $x^2 + 10x - 25$ is not a perfect square trinomial. $x^2 + 10x - 25$ is irreducible over the integers.

Example 5

Factor $y^2 - 14y + 49$.

Solution

$\sqrt{y^2} = y$
$\sqrt{49} = 7$ $\quad 2(7y) = 14y$

The trinomial is a perfect square.

$y^2 - 14y + 49 = (y - 7)^2$

Example 6

Factor $a^2 + 20a + 100$.

Your solution

Example 7

Factor $9x^2 - 24xy + 16y^2$.

Solution

$\sqrt{9x^2} = 3x$
$\sqrt{16y^2} = 4y$ $\quad 2(3x \cdot 4y) = 24xy$

The trinomial is a perfect square.

$9x^2 - 24xy + 16y^2 = (3x - 4y)^2$

Example 8

Factor $25a^2 - 30ab + 9b^2$.

Your solution

...tions on p. A23

4.3 Objective To factor a common binomial factor

Reference for Computer Tutor™

In the examples at the right, the binomials in parentheses are called binomial factors.

$2a(a + b)^2$

$3xy(x - y)$

The Distributive Property is used to factor a common binomial factor from an expression.

Factor $6(x - 3) + y^2(x - 3)$.

The common binomial factor is $x - 3$. Use the Distributive property to write the expression as a product of factors.

$6(x - 3) + y^2(x - 3) =$

$(x - 3)(6 + y^2)$

Factor $2x(a - b) + 5(b - a)$.

Rewrite the expression as a difference of terms which have a common factor. Note that $(b - a) = (-a + b) = -(a - b)$

$2x(a - b) + 5(b - a)$

$2x(a - b) + 5[-(a - b)]$ — Do this step mentally.

$2x(a - b) - 5(a - b)$

Write the expression as a product of factors.

$(a - b)(2x - 5)$

Example 9

Factor $4x(3x - 2) - 7(3x - 2)$.

Solution

$4x(3x - 2) - 7(3x - 2) = (3x - 2)(4x - 7)$

Example 10

Factor $5x(2x + 3) - 4(2x + 3)$.

Your solution

Example 11

Factor $5a(2x - 7) + 2(7 - 2x)$.

Solution

$5a(2x - 7) + 2(7 - 2x) =$
$5a(2x - 7) - 2(2x - 7) = (2x - 7)(5a - 2)$

Example 12

Factor $2y(5x - 2) - 3(2 - 5x)$.

Your solution

Solutions on p. A33

4.4 Objective To factor completely

Reference for Computer Tutor™

When factoring a polynomial completely, ask the following questions about the polynomial.

1. Is there a common factor? If so, factor out the common factor.
2. Is the polynomial the difference of two perfect squares? If so, factor.
3. Is the polynomial a perfect square trinomial? If so, factor.
4. Is the polynomial a trinomial which is the product of two binomials? If so, factor.
5. Is each factor irreducible over the integers? If not, factor.

Example 13

Factor $3x^2 - 48$.

Solution

The GCF is 3.
$3x^2 - 48 = 3(x^2 - 16)$
Factor the difference of two perfect squares.
$3(x + 4)(x - 4)$

$3x^2 - 48 = 3(x + 4)(x - 4)$

Example 14

Factor $12x^3 - 75x$.

Your solution

Example 15

Factor $x^2(x - 3) + 4(3 - x)$.

Solution

The common binomial factor is $x - 3$.
$x^2(x - 3) + 4(3 - x) =$
$x^2(x - 3) - 4(x - 3) = (x - 3)(x^2 - 4)$
Factor the difference of two perfect squares.
$(x - 3)(x + 2)(x - 2)$

$x^2(x - 3) + 4(3 - x) = (x - 3)(x + 2)(x - 2)$

Example 16

Factor $a^2(b - 7) + (7 - b)$.

Your solution

Example 17

Factor $4x^2y^2 + 12xy^2 + 9y^2$.

Solution

The GCF is y^2.
$4x^2y^2 + 12xy^2 + 9y^2 = y^2(4x^2 + 12x + 9)$
Factor the perfect square trinomial.
$y^2(2x + 3)^2$

$4x^2y^2 + 12xy^2 + 9y^2 = y^2(2x + 3)^2$

Example 18

Factor $4x^3 + 28x^2 - 120x$.

Your solution

4.1 Exercises

Factor.

1. $x^2 - 4$
2. $x^2 - 9$
3. $a^2 - 81$
4. $a^2 - 49$
5. $4x^2 - 1$
6. $9x^2 - 16$
7. $x^6 - 9$
8. $y^{12} - 64$
9. $25x^2 - 1$
10. $4x^2 - 1$
11. $1 - 49x^2$
12. $1 - 64x^2$
13. $t^2 + 36$
14. $x^2 + 64$
15. $x^4 - y^2$
16. $b^4 - 16a^2$
17. $9x^2 - 16y^2$
18. $25z^2 - y^2$
19. $x^2y^2 - 4$
20. $a^2b^2 - 25$
21. $16 - x^2y^2$

4.2 Exercises

Factor.

22. $y^2 + 2y + 1$
23. $y^2 + 14y + 49$
24. $a^2 - 2a + 1$
25. $x^2 + 8x - 16$
26. $z^2 - 18z - 81$
27. $x^2 - 12x + 36$
28. $x^2 + 2xy + y^2$
29. $x^2 + 6xy + 9y^2$
30. $4a^2 + 4a + 1$
31. $25x^2 + 10x + 1$
32. $64a^2 - 16a + 1$
33. $9a^2 + 6a + 1$

Factor.

34. $16b^2 + 8b + 1$

35. $4a^2 - 20a + 25$

36. $4b^2 + 28b + 49$

37. $9a^2 - 42a + 49$

38. $25a^2 + 30ab + 9b^2$

39. $4a^2 - 12ab + 9b^2$

40. $49x^2 + 28xy + 4y^2$

41. $4y^2 - 36yz + 81z^2$

42. $64y^2 - 48yz + 9z^2$

4.3 Exercises

Factor.

43. $x(a + b) + 2(a + b)$

44. $a(x + y) + 4(x + y)$

45. $x(b + 2) - y(b + 2)$

46. $a(y - 4) - b(y - 4)$

47. $z(x - 3) - (x - 3)$

48. $a(y + 7) - (y + 7)$

49. $x(b - 2c) + y(b - 2c)$

50. $2x(x - 3) - (x - 3)$

51. $a(x - 2) + 5(2 - x)$

52. $a(x - 7) + b(7 - x)$

53. $b(y - 2) - 2a(y - 2)$

54. $x(a - 3) - 2y(a - 3)$

55. $b(y - 3) + 3(3 - y)$

56. $c(a - 2) - b(2 - a)$

57. $a(x - y) - 2(y - x)$

58. $3(a - b) - x(b - a)$

4.4 Exercises

Factor.

59. $5x^2 - 5$

60. $2x^2 - 18$

61. $x^3 + 4x^2 + 4x$

62. $y^3 - 10y^2 + 25y$

63. $x^4 + 2x^3 - 35x^2$

64. $a^4 - 11a^3 + 24a^2$

65. $5b^2 + 75b + 180$

66. $6y^2 - 48y + 72$

67. $3a^2 + 36a + 10$

68. $5a^2 - 30a + 4$

69. $2x^2y + 16xy - 66y$

70. $3a^2b + 21ab - 54b$

71. $x^3 - 6x^2 - 5x$

72. $b^3 - 8b^2 - 7b$

73. $3y^2 - 36$

74. $3y^2 - 147$

75. $20a^2 + 12a + 1$

76. $12a^2 - 36a + 27$

77. $x^2y^2 - 7xy^2 - 8y^2$

78. $a^2b^2 + 3a^2b - 88a^2$

79. $10a^2 - 5ab - 15b^2$

80. $16x^2 - 32xy + 12y^2$

81. $50 - 2x^2$

82. $72 - 2x^2$

83. $a^2b^2 - 10ab^2 + 25b^2$

84. $a^2b^2 + 6ab^2 + 9b^2$

85. $12a^3b - a^2b^2 - ab^3$

86. $2x^3y - 7x^2y^2 + 6xy^3$

87. $12a^3 - 12a^2 + 3a$

88. $18a^3 + 24a^2 + 8a$

89. $243 + 3a^2$

90. $75 + 27y^2$

91. $12a^3 - 46a^2 + 40a$

92. $24x^3 - 66x^2 + 15x$

93. $4a^3 + 20a^2 + 25a$

94. $2a^3 - 8a^2b + 8ab^2$

Factor.

95. $27a^2b - 18ab + 3b$

96. $a^2b^2 - 6ab^2 + 9b^2$

97. $48 - 12x - 6x^2$

98. $21x^2 - 11x^3 - 2x^4$

99. $x^4 - x^2y^2$

100. $b^4 - a^2b^2$

101. $18a^3 + 24a^2 + 8a$

102. $32xy^2 - 48xy + 18x$

103. $2b + ab - 6a^2b$

104. $20x - 11xy - 3xy^2$

105. $72xy^2 + 48xy + 8x$

106. $4x^2y + 8xy + 4y$

107. $15y^2 - 2xy^2 - x^2y^2$

108. $4x^4 - 38x^3 + 48x^2$

109. $3x^2 - 27y^2$

110. $x^4 - 25x^2$

111. $y^3 - 9y$

112. $a^4 - 16$

113. $15x^4y^2 - 13x^3y^3 - 20x^2y^4$

114. $45y^2 - 42y^3 - 24y^4$

115. $a(2x - 2) + b(2x - 2)$

116. $4a(x - 3) - 2b(x - 3)$

117. $x^2(x - 2) - (x - 2)$

118. $y^2(a - b) - (a - b)$

119. $a(x^2 - 4) + b(x^2 - 4)$

120. $x(a^2 - b^2) - y(a^2 - b^2)$

121. $4(x - 5) - x^2(x - 5)$

122. $y^2(a - b) - 9(a - b)$

123. $x^2(x - 2) + 4(2 - x)$

124. $a(2y^2 - 4) - b(4 - 2y^2)$

SECTION 5 Solving Equations

5.1 Objective To solve equations by factoring

Reference for Computer Tutor™

Recall that the Multiplication Property of Zero states that the product of a number and zero is zero.

If a is a real number, then $a \cdot 0 = 0 \cdot a = 0$.

Consider $x \cdot y = 0$. If this is a true equation, then either $x = 0$ or $y = 0$.

Principle of Zero Products

If the product of two factors is zero, then at least one of the factors must be zero.

If $a \cdot b = 0$, then $a = 0$ or $b = 0$.

The Principle of Zero Products is used in solving equations.

Solve: $(x - 2)(x - 3) = 0$

If $(x - 2)(x - 3) = 0$, then $(x - 2) = 0$ or $(x - 3) = 0$.

$$(x - 2)(x - 3) = 0$$

Rewrite each equation in the form *variable* = *constant*.

$$x - 2 = 0 \qquad x - 3 = 0$$
$$x = 2 \qquad x = 3$$

Write the solution.

The solutions are 2 and 3.

Check:

$$(x - 2)(x - 3) = 0 \qquad (x - 2)(x - 3) = 0$$
$$(2 - 2)(2 - 3) = 0 \qquad (3 - 2)(3 - 3) = 0$$
$$0(-1) = 0 \qquad (1)(0) = 0$$
$$0 = 0 \qquad 0 = 0$$

A true equation A true equation

An equation of the form $ax^2 + bx + c = 0$ is a **quadratic equation.** A quadratic equation is in **standard form** when the polynomial is in descending order and equal to zero.

$$3x^2 + 2x + 1 = 0$$
$$4x^2 - 3x + 2 = 0$$

Solve: $2x^2 + x = 6$

$$2x^2 + x = 6$$

Write the equation in standard form.

$$2x^2 + x - 6 = 0$$

Factor.

$$(2x - 3)(x + 2) = 0$$

Let each factor equal zero (the Principle of Zero Products).

$$2x - 3 = 0 \qquad x + 2 = 0$$

Rewrite each equation in the form *variable* = *constant*.

$$2x = 3 \qquad x = -2$$
$$x = \frac{3}{2}$$

Write the solution.

The solutions are $\frac{3}{2}$ and -2.

$\frac{3}{2}$ and -2 check as solutions.

Example 1

Solve: $x(x - 3) = 0$

Solution

$x(x - 3) = 0$

$x = 0 \qquad x - 3 = 0$

$x = 3$

The solutions are 0 and 3.

Example 2

Solve: $2x(x + 7) = 0$

Your solution

Example 3

Solve: $2x^2 - 50 = 0$

Solution

$2x^2 - 50 = 0$

$2(x^2 - 25) = 0$

$2(x + 5)(x - 5) = 0$

$x + 5 = 0 \qquad x - 5 = 0$

$x = -5 \qquad x = 5$

The solutions are -5 and 5.

Example 4

Solve: $4x^2 - 9 = 0$

Your solution

Example 5

Solve: $(x - 3)(x - 10) = -10$

Solution

$(x - 3)(x - 10) = -10$ Write in standard form.

$x^2 - 13x + 30 = -10$

$x^2 - 13x + 40 = 0$

$(x - 8)(x - 5) = 0$

$x - 8 = 0 \qquad x - 5 = 0$

$x = 8 \qquad x = 5$

The solutions are 8 and 5.

Example 6

Solve: $(x + 2)(x - 7) = 52$

Your solution

Solutions on p. A34

5.2 Objective To solve application problems

Use COMPUTER TUTOR™ DISK 6

Example 7

The sum of the squares of two consecutive positive even integers is equal to 100. Find the two integers.

Strategy

First positive even integer: n
Second positive even integer: $n + 2$

The sum of the square of the first positive even integer and the square of the second positive even integer is 100.

Solution

$n^2 + (n + 2)^2 = 100$

$n^2 + n^2 + 4n + 4 = 100$

$2n^2 + 4n + 4 = 100$

$2n^2 + 4n - 96 = 0$

$2(n^2 + 2n - 48) = 0$

$2(n - 6)(n + 8) = 0$

$n - 6 = 0 \qquad n + 8 = 0$
$n = 6 \qquad n = -8$

Since -8 is not a positive even integer, it is not a solution.

$n = 6$
$n + 2 = 6 + 2 = 8$

The two integers are 6 and 8.

Example 8

The sum of the squares of two consecutive positive integers is 61. Find the two integers.

Your strategy

Your solution

Solution on p. A35

Example 9

A stone is thrown into a well with an initial speed of 4 ft/s. The well is 420 ft deep. How many seconds later will the stone hit the bottom of the well? Use the equation $d = vt + 16t^2$, where d is the distance in feet, v is the initial speed, and t is the time in seconds.

Strategy

To find the time for the stone to drop to the bottom of the well, replace the variables d and v by their given values and solve for t.

Solution

$d = vt + 16t^2$

$420 = 4t + 16t^2$

$0 = -420 + 4t + 16t^2$

$16t^2 + 4t - 420 = 0$

$4(4t^2 + t - 105) = 0$

$4(4t + 21)(t - 5) = 0$

$4t + 21 = 0$ $\qquad$ $t - 5 = 0$

$4t = -21$ $\qquad$ $t = 5$

$t = -\frac{21}{4}$

Since the time cannot be a negative number, $-\frac{21}{4}$ is not a solution.

The time is 5 s.

Example 10

The length of a rectangle is 4 in. longer than twice the width. The area of the rectangle is 96 in.^{2}. Find the length and width of the rectangle.

Your strategy

Your solution

Solution on p. A35

5.1 Exercises

Solve.

1. $(y + 3)(y + 2) = 0$
2. $(y - 3)(y - 5) = 0$
3. $(z - 7)(z - 3) = 0$
4. $(z + 8)(z - 9) = 0$
5. $x(x - 5) = 0$
6. $x(x + 2) = 0$
7. $a(a - 9) = 0$
8. $a(a + 12) = 0$
9. $y(2y + 3) = 0$
10. $t(4t - 7) = 0$
11. $2a(3a - 2) = 0$
12. $4b(2b + 5) = 0$
13. $(b + 2)(b - 5) = 0$
14. $(b - 8)(b + 3) = 0$
15. $x^2 - 81 = 0$
16. $x^2 - 121 = 0$
17. $4x^2 - 49 = 0$
18. $16x^2 - 1 = 0$
19. $9x^2 - 1 = 0$
20. $16x^2 - 49 = 0$
21. $x^2 + 6x + 8 = 0$
22. $x^2 - 8x + 15 = 0$
23. $z^2 + 5z - 14 = 0$
24. $z^2 + z - 72 = 0$
25. $x^2 - 5x + 6 = 0$
26. $x^2 - 3x - 10 = 0$
27. $y^2 + 4y - 21 = 0$
28. $2y^2 - y - 1 = 0$
29. $2a^2 - 9a - 5 = 0$
30. $3a^2 + 14a + 8 = 0$
31. $6z^2 + 5z + 1 = 0$
32. $6y^2 - 19y + 15 = 0$
33. $x^2 - 3x = 0$
34. $a^2 - 5a = 0$
35. $x^2 - 7x = 0$
36. $2a^2 - 8a = 0$

Solve.

37. $a^2 + 5a = -4$

38. $a^2 - 5a = 24$

39. $y^2 - 5y = -6$

40. $y^2 - 7y = 8$

41. $2t^2 + 7t = 4$

42. $3t^2 + t = 10$

43. $3t^2 - 13t = -4$

44. $5t^2 - 16t = -12$

45. $x(x - 12) = -27$

46. $x(x - 11) = 12$

47. $y(y - 7) = 18$

48. $y(y + 8) = -15$

49. $p(p + 3) = -2$

50. $p(p - 1) = 20$

51. $y(y + 4) = 45$

52. $y(y - 8) = -15$

53. $x(x + 3) = 28$

54. $p(p - 14) = 15$

55. $(x + 8)(x - 3) = -30$

56. $(x + 4)(x - 1) = 14$

57. $(y + 3)(y + 10) = -10$

58. $(z - 5)(z + 4) = 52$

59. $(z - 8)(z + 4) = -35$

60. $(z - 6)(z + 1) = -10$

61. $(a + 3)(a + 4) = 72$

62. $(a - 4)(a + 7) = -18$

63. $(2x + 5)(x + 1) = -1$

64. $(z + 3)(z - 10) = -42$

65. $(y + 3)(2y + 3) = 5$

66. $(y + 5)(3y - 2) = -14$

5.2 Application Problems

Solve.

1. The square of a positive number is six more than five times the positive number. Find the number.

2. The square of a negative number is sixteen more than six times the negative number. Find the number.

3. The sum of two numbers is six. The sum of the squares of the two numbers is twenty. Find the two numbers.

4. The sum of two numbers is eight. The sum of the squares of the two numbers is thirty-four. Find the two numbers.

5. The sum of the squares of two consecutive positive integers is eighty-five. Find the two integers.

6. The sum of the squares of two consecutive positive even integers is one hundred. Find the two integers.

7. The sum of two numbers is ten. The product of the two numbers is twenty-one. Find the two numbers.

8. The sum of two numbers is twenty-three. The product of the two numbers is one hundred twenty. Find the two numbers.

9. The square of the sum of a number and three is one hundred forty-four. Find the number.

10. The square of the sum of a number and five is eighty-one. Find the number.

11. The product of two consecutive positive integers is two hundred ten. Find the integers.

12. The product of two consecutive odd positive integers is one hundred forty-three. Find the integers.

Solve.

13. The length of the base of a triangle is four times the height. The area of the triangle is 50 ft^2. Find the base and height of the triangle.

14. The height of a triangle is 3 m more than twice the length of the base. The area of the triangle is 76 m^2. Find the height of the triangle.

15. The length of a rectangle is three times the width. The area is 300 $in.^2$. Find the length and width of the rectangle.

16. The length of a rectangle is two more than twice the width. The area is 312 ft^2. Find the length and width of the rectangle.

17. The length of a rectangle is 5 in. more than twice the width. The area is 75 $in.^2$. Find the length and width of the rectangle.

18. The width of a rectangle is 5 ft less than the length. The area of the rectangle is 176 ft^2. Find the length and width of the rectangle.

19. The length of each side of a square is extended 2 in. The area of the resulting square is 144 $in.^2$. Find the length of a side of the original square.

20. The length of each side of a square is extended 5 in. The area of the resulting square is 64 $in.^2$. Find the length of a side of the original square.

21. An object is thrown downward, with an initial speed of 16 ft/s, from the top of a building 320 ft high. How many seconds later will the object hit the ground? Use the equation $d = vt + 16t^2$, where d is the distance in feet, v is the initial speed, and t is the time in seconds.

22. An object falls from an airplane that is flying at an altitude of 6400 ft. How many seconds later will the object hit the ground? Use the equation $16t^2 = d$, where d is the distance in feet and t is the time in seconds.

23. The radius of a circle is increased by 3 in., increasing the area by 100 $in.^2$. Find the radius of the original circle. Use 3.14 for π.

24. A circle has a radius of 10 in. Find the increase in area when the radius is increased by 2 in. Use 3.14 for π.

Calculators and Computers

Factoring

Remember, when factoring a polynomial, ask the following questions about the polynomial.

1. Is there a common factor? If so, factor out the common factor.
2. Is the polynomial the difference of two perfect squares? If so, factor.
3. Is the polynomial a perfect square trinomial? If so, factor.
4. Is the polynomial a trinomial which is the product of two binomials? If so, factor.
5. Is each factor irreducible over the integers? If not, factor.

Factoring polynomials is an important part of the study of algebra and is a learned skill requiring practice. To provide you with additional practice, the program FACTORING on the Student Disk will give you additional practice factoring a quadratic polynomial. You may choose to practice polynomials of the form

$$x^2 + bx + c \qquad \text{or}$$

$$ax^2 + bx + c$$

These choices, along with the option of quitting the program, are given on a menu screen. You may practice for as long as you like with any type of problem. At the end of each problem, you may select to return to the menu screen or to continue practicing.

The program will present you with a quadratic polynomial to factor. When you have tried to factor the polynomial using paper and pencil, press the RETURN key on the keyboard. The correct factorization will then be displayed.

Chapter Summary

KEY WORDS

The **greatest common factor** (GCF) of two or more integers is the greatest integer which is a factor of all the integers.

To **factor** a polynomial means to write the polynomial as a product of other polynomials.

To **factor** a trinomial of the form $ax^2 + bx + c$ means to express the trinomial as the product of two binomials.

A polynomial which does not factor using only integers is **irreducible over the integers.**

A product of a term and itself is a **perfect square.**

The **square root** of a perfect square is one of the two equal factors.

An equation of the form $ax^2 + bx + c = 0$ is a **quadratic equation.**

A quadratic equation is in **standard form** when the polynomial is in descending order and equal to zero. $ax^2 + bx + c = 0$ is in standard form.

ESSENTIAL RULES

Principle of Zero Products If the product of two factors is zero, then at least one of the factors must be zero.

If $a \cdot b = 0$, then $a = 0$ or $b = 0$.

The Basic Factoring Patterns for a trinomial are:

$x^2 + bx + c = (x + \square)(x + \square)$

$x^2 - bx + c = (x - \square)(x - \square)$

$x^2 - bx - c = (x - \square)(x + \square)$

$x^2 + bx - c = (x - \square)(x + \square)$

$ax^2 + bx + c = (\square x + \square)(\square x + \square)$

$ax^2 - bx + c = (\square x - \square)(\square x - \square)$

$ax^2 - bx - c = (\square x - \square)(\square x + \square)$

$ax^2 + bx - c = (\square x - \square)(\square x + \square)$

Review/Test

SECTION 1

1.1 Find the GCF of $12a^2b^3$ and $16ab^6$.

1.2 Factor $6x^3 - 8x^2 + 10x$.

SECTION 2

2.1a Factor $p^2 + 5p + 6$.

2.1b Factor $a^2 - 19a + 48$.

2.1c Factor $x^2 + 2x - 15$.

2.1d Factor $x^2 - 9x - 36$.

2.2a Factor $5x^2 - 45x - 15$.

2.2b Factor $2y^4 - 14y^3 - 16y^2$.

SECTION 3

3.1a Factor $2x^2 + 4x - 5$.

3.1b Factor $6x^2 + 19x + 8$.

3.2a Factor $8x^2 + 20x - 48$.

3.2b Factor $6x^2y^2 + 9xy^2 + 12y^2$.

Review/Test

SECTION **4**

4.1a Factor $b^2 - 16$.

4.1b Factor $4x^2 - 49y^2$.

4.2a Factor $p^2 + 12p + 36$.

4.2b Factor $4a^2 - 12ab + 9b^2$.

4.3a Factor $a(x - 2) + b(x - 2)$.

4.3b Factor $x(p + 1) - (p + 1)$.

4.4a Factor $3a^2 - 75$.

4.4b Factor $3x^2 + 12xy + 12y^2$.

SECTION **5**

5.1a Solve:
$(2a - 3)(a + 7) = 0$

5.1b Solve: $x(x - 8) = -15$

5.2 The length of a rectangle is 3 cm longer than twice the width. The area of the rectangle is 90 cm^2. Find the length and width of the rectangle.

Cumulative Review/Test

1. Subtract: $-2 - (-3) - 5 - (-11)$

2. Simplify $(3 - 7)^2 \div (-2) - 3 \cdot (-4)$.

3. Evaluate $-2a^2 \div 2b - c$ when $a = -4$, $b = 2$, and $c = -1$.

4. Simplify $-\frac{3}{4}(-20x^2)$.

5. Simplify $-2[4x - 2(3 - 2x) - 8x]$.

6. Solve: $-\frac{5}{7}x = -\frac{10}{21}$

7. Solve: $3x - 2 = 12 - 5x$

8. Solve: $-2 + 4[3x - 2(4 - x) - 3] = 4x + 2$

9. 120% of what number is 54?

10. Simplify $(-3a^3b^2)^2$.

11. Simplify $(x + 2)(x^2 - 5x + 4)$.

12. Simplify $(8x^2 + 4x - 3) \div (2x - 3)$.

13. Simplify $(x^{-4}y^3)^2$.

14. Find the GCF of $12x^3y^2$ and $42xy^6$.

15. Factor $15xy^2 - 20xy^4$.

16. Factor $x^2 - 5xy - 14y^2$.

Cumulative Review/Test

17. Factor $p^2 - 9p - 10$.

18. Factor $18a^3 + 57a^2 + 30a$.

19. Factor $36a^2 - 49b^2$.

20. Factor $4x^2 + 28xy + 49y^2$.

21. Factor $9x^2 + 15x - 14$.

22. Factor $18x^2 - 48xy + 32y^2$.

23. Factor $3y(x - 3) - 2(x - 3)$.

24. Solve: $3x^2 + 19x - 14 = 0$

25. A board 10 ft long is cut into two pieces. Four times the length of the shorter piece is 2 ft less than three times the length of the longer piece. Find the length of each piece.

26. A stereo which regularly sells for \$165 is on sale for \$99. Find the discount rate.

27. An investment of \$4000 is made at an annual simple interest rate of 8%. How much additional money must be invested at an annual simple interest rate of 11% so that the total interest earned is \$1035?

28. A family drove to a resort at an average speed of 42 mph and later returned over the same road at an average speed of 56 mph. Find the distance to the resort if the total driving time was 7 h.

29. Find three consecutive even integers such that five times the middle integer is twelve more than twice the sum of the first and third.

30. The length of the base of a triangle is three times the height. The area of the triangle is 24 in.^{2}. Find the length of the base of the triangle.

7

Algebraic Fractions

Objectives

- To simplify an algebraic fraction
- To multiply algebraic fractions
- To divide algebraic fractions
- To find the least common multiple (LCM) of two or more polynomials
- To express two fractions in terms of the LCM of their denominators
- To add or subtract algebraic fractions with like denominators
- To add or subtract algebraic fractions with unlike denominators
- To simplify a complex fraction
- To solve an equation containing fractions
- To solve a proportion
- To solve application problems
- To rewrite a literal equation in terms of one of the variables
- To solve work problems
- To solve uniform motion problems

Measurement of the Circumference of the Earth

Distances on the earth, the circumference of the earth, and the distance to the moon and stars are known to great precision. Eratosthenes, the fifth librarian of Alexandria (230 B.C.), laid the foundation of scientific geography with his determination of the circumference of the earth.

Eratosthenes was familiar with certain astronomical data that allowed him to calculate the circumference of the earth by using a proportion statement.

Eratosthenes knew that on a mid-summer day, the sun was directly overhead at Syrene, as shown in the diagram. At the same time, at Alexandria the sun was at a $7\frac{1}{2}$-degree angle from the zenith. The distance from Syrene to Alexandria was 5000 stadia (about 520 mi).

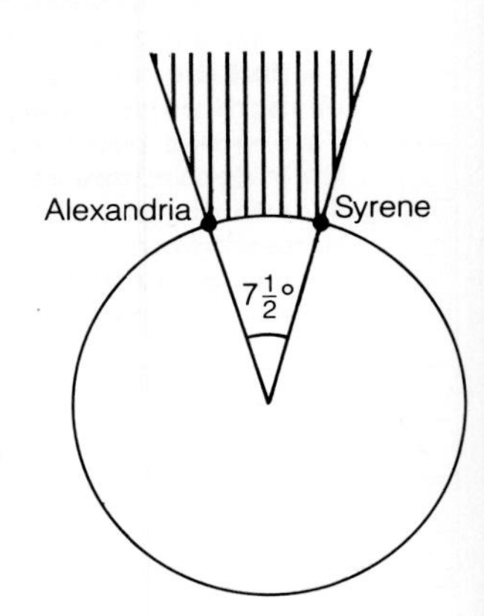

Knowing that the ratio of the $7\frac{1}{2}$-degree angle to one revolution (360°) is equal to the ratio of the arc length (520 mi) to the circumference, a proportion can be written and solved.

$$\frac{7\frac{1}{2}}{360°} = \frac{520}{C}$$

$$C = 24{,}960 \text{ miles}$$

This result, calculated over 2000 years ago, is very close to the accepted value of 24,800 miles.

SECTION 1 Multiplication and Division of Algebraic Fractions

1.1 Objective To simplify an algebraic fraction

Reference for Computer Tutor™

A fraction in which the numerator or denominator is a variable expression is called an **algebraic fraction.** Examples of algebraic fractions are shown at the right.

$$\frac{5}{z}, \quad \frac{x^2+1}{2x-1}, \quad \frac{y^2-3}{3xy+1}$$

Care must be exercised with algebraic fractions to assure that when the variables are replaced with numbers, the resulting denominator is not zero.

Consider the algebraic fraction at the right. The value of x cannot be 2, since the denominator would then be zero.

$$\frac{3x^2+1}{x^2-4}$$

$$\frac{3 \cdot 2^2+1}{2^2-4} = \frac{13}{0} \quad \text{Not a real number}$$

An algebraic fraction is in simplest form when the numerator and denominator have no common factors.

Simplify: $\dfrac{x^2-4}{x^2-2x-8}$

Factor the numerator and denominator.

$$\frac{x^2-4}{x^2-2x-8} = \frac{(x-2)(x+2)}{(x-4)(x+2)}$$

Cancel the common factors.

$$= \frac{(x-2)\overset{1}{\cancel{(x+2)}}}{(x-4)\underset{1}{\cancel{(x+2)}}}$$

Write the answer in simplest form.

$$= \frac{x-2}{x-4}$$

Simplify: $\dfrac{10+3x-x^2}{x^2-4x-5}$

Factor the numerator and the denominator.

$$\frac{10+3x-x^2}{x^2-4x-5} = \frac{(5-x)(2+x)}{(x-5)(x+1)}$$

Cancel the common factors.

Remember that $5 - x = -(x - 5)$.

Therefore, $\dfrac{5-x}{x-5} = \dfrac{-(x-5)}{x-5} = \dfrac{-1}{1} = -1$.

$$= \frac{\overset{-1}{\cancel{(5-x)}}(2+x)}{\underset{1}{\cancel{(x-5)}}(x+1)}$$

Write the answer in simplest form.

$$= -\frac{x+2}{x+1}$$

Example 1

Simplify: $\frac{4x^3y^4}{6x^4y}$

Solution

$$\frac{4x^3y^4}{6x^4y} = \frac{\overset{1}{\cancel{2}} \cdot 2x^3y^4}{\underset{1}{\cancel{2}} \cdot 3x^4y} = \frac{2y^3}{3x}$$ Use rules of exponents.

Example 2

Simplify: $\frac{6x^5y}{12x^2y^3}$

Your solution

Example 3

Simplify: $\frac{9 - x^2}{x^2 + x - 12}$

Solution

$$\frac{9 - x^2}{x^2 + x - 12} = \frac{\overset{-1}{\cancel{(3 - x)}}(3 + x)}{\underset{1}{\cancel{(x - 3)}}(x + 4)} = -\frac{x + 3}{x + 4}$$

Example 4

Simplify: $\frac{x^2 + 2x - 24}{16 - x^2}$

Your solution

Example 5

Simplify: $\frac{x^2 + 2x - 15}{x^2 - 7x + 12}$

Solution

$$\frac{x^2 + 2x - 15}{x^2 - 7x + 12} = \frac{(x + 5)\overset{1}{\cancel{(x - 3)}}}{\underset{1}{\cancel{(x - 3)}}(x - 4)} = \frac{x + 5}{x - 4}$$

Example 6

Simplify: $\frac{x^2 + 4x - 12}{x^2 - 3x + 2}$

Your solution

1.2 Objective To multiply algebraic fractions

Reference for Computer Tutor™

The product of two fractions is a fraction whose numerator is the product of the numerators of the two fractions and whose denominator is the product of the denominators of the two fractions.

$$\frac{a}{b} \cdot \frac{c}{d} = \frac{ac}{bd}$$

$$\frac{2}{3} \cdot \frac{4}{5} = \frac{8}{15}$$

$$\frac{3x}{y} \cdot \frac{2}{z} = \frac{6x}{yz}$$

$$\frac{x + 2}{x} \cdot \frac{3}{x - 2} = \frac{3x + 6}{x^2 - 2x}$$

Simplify: $\frac{x^2 + 3x}{x^2 - 3x - 4} \cdot \frac{x^2 - 5x + 4}{x^2 + 2x - 3}$

$$\frac{x^2 + 3x}{x^2 - 3x - 4} \cdot \frac{x^2 - 5x + 4}{x^2 + 2x - 3} =$$

Factor the numerator and denominator of each fraction.

$$\frac{x(x + 3)}{(x - 4)(x + 1)} \cdot \frac{(x - 4)(x - 1)}{(x + 3)(x - 1)} =$$

Multiply.
Cancel the common factors.

$$\frac{x\overset{1}{\cancel{(x + 3)}}\overset{1}{\cancel{(x - 4)}}\overset{1}{\cancel{(x - 1)}}}{\underset{1}{\cancel{(x - 4)}}(x + 1)\underset{1}{\cancel{(x + 3)}}\underset{1}{\cancel{(x - 1)}}} =$$

Write the answer in simplest form.

$$\frac{x}{x + 1}$$

Example 7

Simplify: $\frac{10x^2 - 15x}{12x - 8} \cdot \frac{3x - 2}{20x - 25}$

Solution

$$\frac{10x^2 - 15x}{12x - 8} \cdot \frac{3x - 2}{20x - 25} =$$

$$\frac{5x(2x - 3)}{4(3x - 2)} \cdot \frac{(3x - 2)}{5(4x - 5)} =$$

$$\frac{\overset{1}{\cancel{5}}x(2x - 3)\overset{1}{\cancel{(3x - 2)}}}{2 \cdot 2\underset{1}{\cancel{(3x - 2)}}\underset{1}{\cancel{5}}(4x - 5)} = \frac{x(2x - 3)}{4(4x - 5)}$$

Example 8

Simplify: $\frac{12x^2 + 3x}{10x - 15} \cdot \frac{8x - 12}{9x + 18}$

Your solution

Example 9

Simplify: $\frac{x^2 + x - 6}{x^2 + 7x + 12} \cdot \frac{x^2 + 3x - 4}{4 - x^2}$

Solution

$$\frac{x^2 + x - 6}{x^2 + 7x + 12} \cdot \frac{x^2 + 3x - 4}{4 - x^2} =$$

$$\frac{(x + 3)(x - 2)}{(x + 3)(x + 4)} \cdot \frac{(x + 4)(x - 1)}{(2 - x)(2 + x)} =$$

$$\frac{\overset{1}{\cancel{(x + 3)}}\overset{-1}{\cancel{(x - 2)}}\overset{1}{\cancel{(x + 4)}}(x - 1)}{\underset{1}{\cancel{(x + 3)}}\underset{1}{\cancel{(x + 4)}}\underset{1}{\cancel{(2 - x)}}(2 + x)} = -\frac{x - 1}{x + 2}$$

Example 10

Simplify: $\frac{x^2 + 2x - 15}{9 - x^2} \cdot \frac{x^2 - 3x - 18}{x^2 - 7x + 6}$

Your solution

Solutions on p. A37

1.3 Objective To divide algebraic fractions

The **reciprocal** of a fraction is a fraction with the numerator and denominator interchanged.

Fraction $\left\{ \begin{array}{ll} \frac{a}{b} & \frac{b}{a} \\ x^2 = \frac{x^2}{1} & \frac{1}{x^2} \\ \frac{x+2}{x} & \frac{x}{x+2} \end{array} \right\}$ Reciprocal

To divide two fractions, multiply by the reciprocal of the divisor.

$$\frac{a}{b} \div \frac{c}{d} = \frac{a}{b} \cdot \frac{d}{c} = \frac{ad}{bc}$$

$$\frac{4}{x} \div \frac{y}{5} = \frac{4}{x} \cdot \frac{5}{y} = \frac{20}{xy}$$

$$\frac{x+4}{x} \div \frac{x-2}{4} = \frac{x+4}{x} \cdot \frac{4}{x-2} = \frac{4(x+4)}{x(x-2)}$$

The basis for the division rule is shown at the right.

$$\frac{a}{b} \div \frac{c}{d} = \frac{\frac{a}{b}}{\frac{c}{d}} \cdot \frac{\frac{d}{c}}{\frac{d}{c}} = \frac{\frac{a}{b} \cdot \frac{d}{c}}{\frac{c}{d} \cdot \frac{d}{c}} = \frac{\frac{a}{b} \cdot \frac{d}{c}}{1} = \frac{a}{b} \cdot \frac{d}{c}$$

(with an arrow showing $\frac{a}{b} \div \frac{c}{d} = \frac{a}{b} \cdot \frac{d}{c}$)

Example 11

Simplify: $\frac{xy^2 - 3x^2y}{z^2} \div \frac{6x^2 - 2xy}{z^3}$

Solution

$$\frac{xy^2 - 3x^2y}{z^2} \div \frac{6x^2 - 2xy}{z^3} =$$

$$\frac{xy^2 - 3x^2y}{z^2} \cdot \frac{z^3}{6x^2 - 2xy} =$$

$$\frac{xy\overset{-1}{\cancel{(y - 3x)}} \cdot z^3}{z^2 \cdot 2x\underset{1}{\cancel{(3x - y)}}} = -\frac{yz}{2}$$

Example 12

Simplify: $\frac{a^2}{4bc^2 - 2b^2c} \div \frac{a}{6bc - 3b^2}$

Your solution

Example 13

Simplify: $\frac{2x^2 + 5x + 2}{2x^2 + 3x - 2} \div \frac{3x^2 + 13x + 4}{2x^2 + 7x - 4}$

Solution

$$\frac{2x^2 + 5x + 2}{2x^2 + 3x - 2} \div \frac{3x^2 + 13x + 4}{2x^2 + 7x - 4} =$$

$$\frac{2x^2 + 5x + 2}{2x^2 + 3x - 2} \cdot \frac{2x^2 + 7x - 4}{3x^2 + 13x + 4} =$$

$$\frac{(2x + 1)\overset{1}{\cancel{(x + 2)}}}{\underset{1}{\cancel{(2x - 1)}}\underset{1}{\cancel{(x + 2)}}} \cdot \frac{\overset{1}{\cancel{(2x - 1)}}\overset{1}{\cancel{(x + 4)}}}{(3x + 1)\underset{1}{\cancel{(x + 4)}}} = \frac{2x + 1}{3x + 1}$$

Example 14

Simplify: $\frac{3x^2 + 26x + 16}{3x^2 - 7x - 6} \div \frac{2x^2 + 9x - 5}{x^2 + 2x - 15}$

Your solution

Solutions on p. A37

SECTION 2 Expressing Fractions in Terms of the Least Common Multiple (LCM)

2.1 Objective To find the least common multiple (LCM) of two or more polynomials

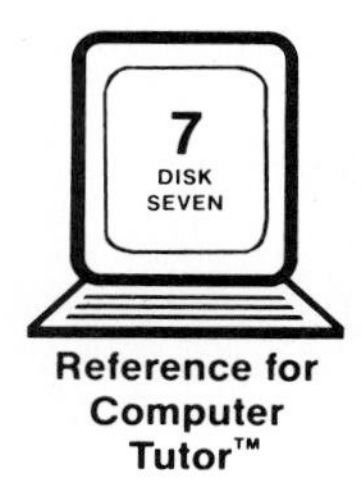

Reference for Computer Tutor™

The **least common multiple (LCM)** of two or more numbers is the smallest number that contains the prime factorization of each number.

The LCM of 12 and 18 is 36. 36 contains the prime factors of 12 and the prime factors of 18.

$12 = 2 \cdot 2 \cdot 3$
$18 = 2 \cdot 3 \cdot 3$

$$\text{LCM} = 36 = \overbrace{2 \cdot 2 \cdot 3}^{\text{Factors of 12}} \cdot 3 \quad \text{with } \underbrace{2 \cdot 3 \cdot 3}_{\text{Factors of 18}}$$

The least common multiple of two or more polynomials is the simplest polynomial that contains the factors of each polynomial.

To find the LCM of two or more polynomials, first factor each polynomial completely. The LCM is the product of each factor the greatest number of times it occurs in any one factorization.

Find the LCM of $4x^2 + 4x$ and $x^2 + 2x + 1$.

The LCM of the polynomials is the product of the LCM of the numerical coefficients and each variable factor the greatest number of times it occurs in any one factorization.

$4x^2 + 4x = 4x(x + 1) = 2 \cdot 2 \cdot x(x + 1)$
$x^2 + 2x + 1 = (x + 1)(x + 1)$

$$\text{LCM} = \overbrace{2 \cdot 2 \cdot x(x + 1)}^{\text{Factors of } 4x^2 + 4x}(x + 1) = 4x(x + 1)(x + 1)$$

(Factors of $x^2 + 2x + 1$: $(x + 1)(x + 1)$)

Example 1

Find the LCM of $4x^2y$ and $6xy^2$.

Solution

$4x^2y = 2 \cdot 2 \cdot x \cdot x \cdot y$ $6xy^2 = 2 \cdot 3 \cdot x \cdot y \cdot y$
$\text{LCM} = 2 \cdot 2 \cdot 3 \cdot x \cdot x \cdot y \cdot y = 12x^2y^2$

Example 2

Find the LCM of $8uv^2$ and $12uw$.

Your solution

Example 3

Find the LCM of $x^2 - x - 6$ and $9 - x^2$.

Solution

$x^2 - x - 6 = (x - 3)(x + 2)$
$9 - x^2 = -(x^2 - 9) = -(x + 3)(x - 3)$
$\text{LCM} = (x - 3)(x + 2)(x + 3)$

Example 4

Find the LCM of $m^2 - 6m + 9$ and $m^2 - 2m - 3$.

Your solution

...tions on p. A38

2.2 Objective To express two fractions in terms of the LCM of their denominators

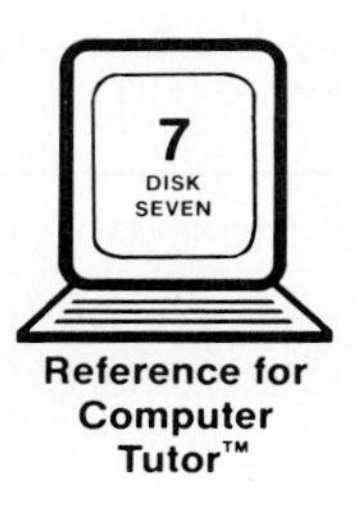

Reference for Computer Tutor™

When adding and subtracting fractions, it is frequently necessary to express two or more fractions in terms of a common denominator. This common denominator is the LCM of the denominators of the fractions.

Write the fractions $\frac{x+1}{4x^2}$ and $\frac{x-3}{6x^2-12x}$ in terms of the LCM of the denominators.

Find the LCM of the denominators.

The LCM is $12x^2(x-2)$.

For each fraction, multiply the numerator and denominator by the factors whose product with the denominator is the LCM.

$$\frac{x+1}{4x^2} = \frac{x+1}{4x^2} \cdot \frac{3(x-2)}{3(x-2)} = \frac{3x^2-3x-6}{12x^2(x-2)}$$

$$\frac{x-3}{6x^2-12x} = \frac{x-3}{6x(x-2)} \cdot \frac{2x}{2x} = \frac{2x^2-6x}{12x^2(x-2)}$$

LCM

Example 5

Write the fractions $\frac{x+2}{3x^2}$ and $\frac{x-1}{8xy}$ in terms of the LCM of the denominators.

Solution

The LCM is $24x^2y$.

$$\frac{x+2}{3x^2} = \frac{x+2}{3x^2} \cdot \frac{8y}{8y} = \frac{8xy+16y}{24x^2y}$$

$$\frac{x-1}{8xy} = \frac{x-1}{8xy} \cdot \frac{3x}{3x} = \frac{3x^2-3x}{24x^2y}$$

Example 6

Write the fractions $\frac{x-3}{4xy^2}$ and $\frac{2x+1}{9y^2z}$ in terms of the LCM of the denominators.

Your solution

Example 7

Write the fractions $\frac{2x-1}{2x-x^2}$ and $\frac{x}{x^2+x-6}$ in terms of the LCM of the denominators.

Solution

$$\frac{2x-1}{2x-x^2} = \frac{2x-1}{-(x^2-2x)} = -\frac{2x-1}{x^2-2x}$$

The LCM is $x(x-2)(x+3)$.

$$\frac{2x-1}{2x-x^2} = -\frac{2x-1}{x(x-2)} \cdot \frac{x+3}{x+3} = -\frac{2x^2+5x-3}{x(x-2)(x+3)}$$

$$\frac{x}{x^2+x-6} = \frac{x}{(x-2)(x+3)} \cdot \frac{x}{x} = \frac{x^2}{x(x-2)(x+3)}$$

Example 8

Write the fractions $\frac{x+4}{x^2-3x-10}$ and $\frac{2x}{25-x^2}$ in terms of the LCM of the denominators.

Your solution

2.1 Exercises

Find the LCM of the expressions.

1. $8x^3y$
   $12xy^2$

2. $6ab^2$
   $18ab^3$

3. $10x^4y^2$
   $15x^3y$

4. $12a^2b$
   $18ab^3$

5. $8x^2$
   $4x^2 + 8x$

6. $6y^2$
   $4y + 12$

7. $2x^2y$
   $3x^2 + 12x$

8. $4xy^2$
   $6xy^2 + 12y^2$

9. $9x(x + 2)$
   $12(x + 2)^2$

10. $8x^2(x - 1)^2$
    $10x^3(x - 1)$

11. $3x + 3$
    $2x^2 + 4x + 2$

12. $4x - 12$
    $2x^2 - 12x + 18$

13. $(x - 1)(x + 2)$
    $(x - 1)(x + 3)$

14. $(2x - 1)(x + 4)$
    $(2x + 1)(x + 4)$

15. $(2x + 3)^2$
    $(2x + 3)(x - 5)$

16. $(x - 7)(x + 2)$
    $(x - 7)^2$

17. $(x - 1)$
    $(x - 2)$
    $(x - 1)(x - 2)$

18. $(x + 4)(x - 3)$
    $x + 4$
    $x - 3$

19. $x^2 - x - 6$
    $x^2 + x - 12$

20. $x^2 + 3x - 10$
    $x^2 + 5x - 14$

21. $x^2 + 5x + 4$
    $x^2 - 3x - 28$

22. $x^2 - 10x + 21$
    $x^2 - 8x + 15$

23. $x^2 - 2x - 24$
    $x^2 - 36$

24. $x^2 + 7x + 10$
    $x^2 - 25$

25. $x^2 - 7x - 30$
    $x^2 - 5x - 24$

26. $2x^2 - 7x + 3$
    $2x^2 + x - 1$

27. $3x^2 - 11x + 6$
    $3x^2 + 4x - 4$

28. $2x^2 - 9x + 10$
    $2x^2 + x - 15$

29. $6 + x - x^2$
    $x + 2$
    $x - 3$

30. $15 + 2x - x^2$
    $x - 5$
    $x + 3$

31. $5 + 4x - x^2$
    $x - 5$
    $x + 1$

32. $x^2 + 3x - 18$
    $3 - x$
    $x + 6$

33. $x^2 - 5x + 6$
    $1 - x$
    $x - 6$

2.2 Exercises

Write each fraction in terms of the LCM of the denominators.

34. $\frac{4}{x}, \frac{3}{x^2}$

35. $\frac{5}{ab^2}, \frac{6}{ab}$

36. $\frac{x}{3y^2}, \frac{z}{4y}$

37. $\frac{5y}{6x^2}, \frac{7}{9xy}$

38. $\frac{y}{x(x-3)}, \frac{6}{x^2}$

39. $\frac{a}{y^2}, \frac{6}{y(y+5)}$

40. $\frac{9}{(x-1)^2}, \frac{6}{x(x-1)}$

41. $\frac{a^2}{y(y+7)}, \frac{a}{(y+7)^2}$

42. $\frac{3}{x-3}, \frac{5}{x(3-x)}$

43. $\frac{b}{y(y-4)}, \frac{b^2}{4-y}$

44. $\frac{3}{(x-5)^2}, \frac{2}{5-x}$

45. $\frac{3}{7-y}, \frac{2}{(y-7)^2}$

46. $\frac{3}{x^2+2x}, \frac{4}{x^2}$

47. $\frac{2}{y-3}, \frac{3}{y^3-3y^2}$

48. $\frac{x-2}{x+3}, \frac{x}{x-4}$

49. $\frac{x^2}{2x-1}, \frac{x+1}{x+4}$

50. $\frac{3}{x^2+x-2}, \frac{x}{x+2}$

51. $\frac{3x}{x-5}, \frac{4}{x^2-25}$

52. $\frac{5}{2x^2-9x+10}, \frac{x-1}{2x-5}$

53. $\frac{x-3}{3x^2+4x-4}, \frac{2}{x+2}$

54. $\frac{x}{x^2+x-6}, \frac{2x}{x^2-9}$

55. $\frac{x-1}{x^2+2x-15}, \frac{x}{x^2+6x+5}$

56. $\frac{x}{9-x^2}, \frac{x-1}{x^2-6x+9}$

57. $\frac{2x}{10+3x-x^2}, \frac{x+2}{x^2-8x+15}$

58. $\frac{3x}{x-5}, \frac{x}{x+4}, \frac{3}{20+x-x^2}$

59. $\frac{x+1}{x+5}, \frac{x+2}{x-7}, \frac{3}{35+2x-x^2}$

SECTION 3 Addition and Subtraction of Algebraic Fractions

3.1 Objective To add or subtract algebraic fractions with like denominators

Reference for Computer Tutor™

When adding algebraic fractions in which the denominators are the same, add the numerators. The denominator of the sum is the common denominator.

$$\frac{a}{b} + \frac{c}{b} = \frac{a+c}{b}$$

$$\frac{5x}{18} + \frac{7x}{18} = \frac{12x}{18} = \frac{2x}{3}$$

$$\frac{x}{x^2-1} + \frac{1}{x^2-1} = \frac{x+1}{x^2-1} = \frac{\overset{1}{\cancel{(x+1)}}}{(x-1)\underset{1}{\cancel{(x+1)}}} = \frac{1}{x-1}$$

Note that the sum is reduced to simplest form.

When subtracting algebraic fractions with like denominators, subtract the numerators. The denominator of the difference is the common denominator. Reduce the answer to simplest form.

$$\frac{2x}{x-2} - \frac{4}{x-2} = \frac{2x-4}{x-2} = \frac{2\overset{1}{\cancel{(x-2)}}}{\underset{1}{\cancel{x-2}}} = 2$$

$$\frac{3x-1}{x^2-5x+4} - \frac{2x+3}{x^2-5x+4} = \frac{(3x-1)-(2x+3)}{x^2-5x+4} = \frac{x-4}{x^2-5x+4} = \frac{\overset{1}{\cancel{(x-4)}}}{\underset{1}{\cancel{(x-4)}}(x-1)} = \frac{1}{x-1}$$

Example 1

Simplify: $\frac{7}{x^2} + \frac{9}{x^2}$

Solution

$$\frac{7}{x^2} + \frac{9}{x^2} = \frac{7+9}{x^2} = \frac{16}{x^2}$$

Example 2

Simplify: $\frac{3}{xy} + \frac{12}{xy}$

Your solution

Example 3

Simplify: $\frac{3x^2}{x^2-1} - \frac{x+4}{x^2-1}$

Solution

$$\frac{3x^2}{x^2-1} - \frac{x+4}{x^2-1} = \frac{3x^2-(x+4)}{x^2-1} = \frac{3x^2-x-4}{x^2-1} =$$

$$\frac{(3x-4)\overset{1}{\cancel{(x+1)}}}{(x-1)\underset{1}{\cancel{(x+1)}}} = \frac{3x-4}{x-1}$$

Example 4

Simplify: $\frac{2x^2}{x^2-x-12} - \frac{7x+4}{x^2-x-12}$

Your solution

Solutions on p. A38

Example 5

Simplify:

$$\frac{2x^2+5}{x^2+2x-3} - \frac{x^2-3x}{x^2+2x-3} + \frac{x-2}{x^2+2x-3}$$

Solution

$$\frac{2x^2+5}{x^2+2x-3} - \frac{x^2-3x}{x^2+2x-3} + \frac{x-2}{x^2+2x-3} =$$

$$\frac{(2x^2+5)-(x^2-3x)+(x-2)}{x^2+2x-3} =$$

$$\frac{2x^2+5-x^2+3x+x-2}{x^2+2x-3} =$$

$$\frac{x^2+4x+3}{x^2+2x-3} = \frac{\overset{1}{\cancel{(x+3)}}(x+1)}{\underset{1}{\cancel{(x+3)}}(x-1)} = \frac{x+1}{x-1}$$

Example 6

Simplify:

$$\frac{x^2-1}{x^2-8x+12} - \frac{2x+1}{x^2-8x+12} + \frac{x}{x^2-8x+12}$$

Your solution

Solution on p. A38

3.2 Objective To add or subtract algebraic fractions with unlike denominators

Reference for Computer Tutor™

Before two fractions with unlike denominators can be added or subtracted, each fraction must be expressed in terms of a common denominator. This common denominator is the LCM of the denominators of the fractions.

Simplify: $\frac{x-3}{x^2-2x} + \frac{6}{x^2-4}$

Find the LCM of the denominators. The LCM is $x(x-2)(x+2)$

Write each fraction in terms of the LCM. Multiply the factors in the numerator.

$$\frac{x-3}{x^2-2x} = \frac{x-3}{x(x-2)} \cdot \frac{x+2}{x+2} = \frac{x^2-x-6}{x(x-2)(x+2)}$$

$$\frac{6}{x^2-4} = \frac{6}{(x-2)(x+2)} \cdot \frac{x}{x} = \frac{6x}{x(x-2)(x+2)}$$

Add the fractions.

$$\frac{x-3}{x^2-2x} + \frac{6}{x^2-4} =$$

$$\frac{x^2-x-6}{x(x-2)(x+2)} + \frac{6x}{x(x-2)(x+2)} =$$

$$\frac{x^2+5x-6}{x(x-2)(x+2)} =$$

Factor the numerator to determine whether there are common factors in the numerator and denominator.

$$\frac{(x+6)(x-1)}{x(x-2)(x+2)}$$

Example 7

Simplify: $\frac{y}{x} - \frac{4y}{3x} + \frac{3y}{4x}$

Solution

The LCM of the denominators is $12x$.

$$\frac{y}{x} = \frac{y}{x} \cdot \frac{12}{12} = \frac{12y}{12x} \qquad \frac{4y}{3x} = \frac{4y}{3x} \cdot \frac{4}{4} = \frac{16y}{12x}$$

$$\frac{3y}{4x} = \frac{3y}{4x} \cdot \frac{3}{3} = \frac{9y}{12x}$$

$$\frac{y}{x} - \frac{4y}{3x} + \frac{3y}{4x} = \frac{12y}{12x} - \frac{16y}{12x} + \frac{9y}{12x} =$$

$$\frac{12y - 16y + 9y}{12x} = \frac{5y}{12x}$$

Example 8

Simplify: $\frac{z}{8y} - \frac{4z}{3y} + \frac{5z}{4y}$

Your solution

Example 9

Simplify: $\frac{2x}{x-3} - \frac{5}{3-x}$

Solution

The LCM of $x - 3$ and $3 - x$ is $x - 3$.
Remember: $3 - x = -(x - 3)$

$$\frac{2x}{x-3} = \frac{2x}{x-3} \cdot \frac{1}{1} = \frac{2x}{x-3}$$

$$\frac{5}{3-x} = \frac{5}{-(x-3)} \cdot \frac{-1}{-1} = \frac{-5}{x-3}$$

$$\frac{2x}{x-3} - \frac{5}{3-x} = \frac{2x}{x-3} - \frac{-5}{x-3} =$$

$$\frac{2x - (-5)}{x-3} = \frac{2x+5}{x-3}$$

Example 10

Simplify: $\frac{5x}{x-2} - \frac{3}{2-x}$

Your solution

Example 11

Simplify: $\frac{2x}{2x-3} - \frac{1}{x+1}$

Solution

The LCM is $(2x - 3)(x + 1)$.

$$\frac{2x}{2x-3} = \frac{2x}{2x-3} \cdot \frac{x+1}{x+1} = \frac{2x^2 + 2x}{(2x-3)(x+1)}$$

$$\frac{1}{x+1} = \frac{1}{x+1} \cdot \frac{2x-3}{2x-3} = \frac{2x-3}{(2x-3)(x+1)}$$

$$\frac{2x}{2x-3} - \frac{1}{x+1} = \frac{2x^2+2x}{(2x-3)(x+1)} - \frac{2x-3}{(2x-3)(x+1)} =$$

$$\frac{(2x^2+2x) - (2x-3)}{(2x-3)(x+1)} = \frac{2x^2+3}{(2x-3)(x+1)}$$

Example 12

Simplify: $\frac{4x}{3x-1} - \frac{9}{x+4}$

Your solution

Solutions on pp. A38–A39

Example 13

Simplify: $\frac{x+3}{x^2-2x-8}+\frac{3}{4-x}$

Solution

The LCM is $(x-4)(x+2)$.

$$\frac{x+3}{x^2-2x-8}=\frac{x+3}{(x-4)(x+2)}$$

$$\frac{3}{4-x}=\frac{3}{-(x-4)}\cdot\frac{-1\cdot(x+2)}{-1\cdot(x+2)}=\frac{-3(x+2)}{(x-4)(x+2)}$$

$$\frac{x+3}{x^2-2x-8}+\frac{3}{4-x}=\frac{x+3}{(x-4)(x+2)}+\frac{-3(x+2)}{(x-4)(x+2)}=$$

$$\frac{(x+3)+(-3)(x+2)}{(x-4)(x+2)}=\frac{x+3-3x-6}{(x-4)(x+2)}=$$

$$\frac{-2x-3}{(x-4)(x+2)}$$

Example 14

Simplify: $\frac{2x-1}{x^2-25}+\frac{2}{5-x}$

Your solution

Example 15

Simplify: $\frac{3x+2}{2x^2-x-1}-\frac{3}{2x+1}+\frac{4}{x-1}$

Solution

The LCM is $(2x+1)(x-1)$.

$$\frac{3x+2}{2x^2-x-1}=\frac{3x+2}{(2x+1)(x-1)}$$

$$\frac{3}{2x+1}=\frac{3}{2x+1}\cdot\frac{x-1}{x-1}=\frac{3x-3}{(2x+1)(x-1)}$$

$$\frac{4}{x-1}=\frac{4}{x-1}\cdot\frac{2x+1}{2x+1}=\frac{8x+4}{(2x+1)(x-1)}$$

$$\frac{3x+2}{2x^2-x-1}-\frac{3}{2x+1}+\frac{4}{x-1}=$$

$$\frac{3x+2}{(2x+1)(x-1)}-\frac{3x-3}{(2x+1)(x-1)}+\frac{8x+4}{(2x+1)(x-1)}=$$

$$\frac{(3x+2)-(3x-3)+(8x+4)}{(2x+1)(x-1)}=$$

$$\frac{3x+2-3x+3+8x+4}{(2x+1)(x-1)}=\frac{8x+9}{(2x+1)(x-1)}$$

Example 16

Simplify: $\frac{2x-3}{3x^2-x-2}+\frac{5}{3x+2}-\frac{1}{x-1}$

Your solution

Solutions on p. A39

3.1 Exercises

Simplify.

1. $\frac{3}{y^2} + \frac{8}{y^2}$

2. $\frac{6}{ab} - \frac{2}{ab}$

3. $\frac{3}{x + 4} - \frac{10}{x + 4}$

4. $\frac{x}{x + 6} - \frac{2}{x + 6}$

5. $\frac{3x}{2x + 3} + \frac{5x}{2x + 3}$

6. $\frac{6y}{4y + 1} - \frac{11y}{4y + 1}$

7. $\frac{2x + 1}{x - 3} + \frac{3x + 6}{x - 3}$

8. $\frac{4x + 3}{2x - 7} + \frac{3x - 8}{2x - 7}$

9. $\frac{5x - 1}{x + 9} - \frac{3x + 4}{x + 9}$

10. $\frac{6x - 5}{x - 10} - \frac{3x - 4}{x - 10}$

11. $\frac{x - 7}{2x + 7} - \frac{4x - 3}{2x + 7}$

12. $\frac{2n}{3n + 4} - \frac{5n - 3}{3n + 4}$

13. $\frac{x}{x^2 + 2x - 15} - \frac{3}{x^2 + 2x - 15}$

14. $\frac{3x}{x^2 + 3x - 10} - \frac{6}{x^2 + 3x - 10}$

15. $\frac{2x + 3}{x^2 - x - 30} - \frac{x - 2}{x^2 - x - 30}$

16. $\frac{3x - 1}{x^2 + 5x - 6} - \frac{2x - 7}{x^2 + 5x - 6}$

17. $\frac{4y + 7}{2y^2 + 7y - 4} - \frac{y - 5}{2y^2 + 7y - 4}$

18. $\frac{x + 1}{2x^2 - 5x - 12} + \frac{x + 2}{2x^2 - 5x - 12}$

19. $\frac{2x^2 + 3x}{x^2 - 9x + 20} + \frac{2x^2 - 3}{x^2 - 9x + 20} - \frac{4x^2 + 2x + 1}{x^2 - 9x + 20}$

20. $\frac{2x^2 + 3x}{x^2 - 2x - 63} - \frac{x^2 - 3x + 21}{x^2 - 2x - 63} - \frac{x - 7}{x^2 - 2x - 63}$

3.2 Exercises

Simplify.

21. $\frac{4}{x} + \frac{5}{y}$

22. $\frac{7}{a} + \frac{5}{b}$

23. $\frac{12}{x} - \frac{5}{2x}$

24. $\frac{5}{3a} - \frac{3}{4a}$

25. $\frac{1}{2x} - \frac{5}{4x} + \frac{7}{6x}$

26. $\frac{7}{4y} + \frac{11}{6y} - \frac{8}{3y}$

27. $\frac{5}{3x} - \frac{2}{x^2} + \frac{3}{2x}$

28. $\frac{6}{y^2} + \frac{3}{4y} - \frac{2}{5y}$

29. $\frac{2}{x} - \frac{3}{2y} + \frac{3}{5x} - \frac{1}{4y}$

30. $\frac{5}{2a} + \frac{7}{3b} - \frac{2}{b} - \frac{3}{4a}$

31. $\frac{2x+1}{3x} + \frac{x-1}{5x}$

32. $\frac{4x-3}{6x} + \frac{2x+3}{4x}$

33. $\frac{x-3}{6x} + \frac{x+4}{8x}$

34. $\frac{2x-3}{2x} + \frac{x+3}{3x}$

35. $\frac{2x+9}{9x} - \frac{x-5}{5x}$

36. $\frac{3y-2}{12y} - \frac{y-3}{18y}$

37. $\frac{x+4}{2x} - \frac{x-1}{x^2}$

38. $\frac{x-2}{3x^2} - \frac{x+4}{x}$

Simplify.

39. $\frac{x-10}{4x^2} + \frac{x+1}{2x}$

40. $\frac{x+5}{3x^2} + \frac{2x+1}{2x}$

41. $\frac{2x+1}{6x^2} - \frac{x-4}{4x}$

42. $\frac{x+3}{6x} - \frac{x-3}{8x^2}$

43. $\frac{x+2}{xy} - \frac{3x-2}{x^2y}$

44. $\frac{3x-1}{xy^2} - \frac{2x+3}{xy}$

45. $\frac{4x-3}{3x^2y} + \frac{2x+1}{4xy^2}$

46. $\frac{5x+7}{6xy^2} - \frac{4x-3}{8x^2y}$

47. $\frac{x-2}{8x^2} - \frac{x+7}{12xy}$

48. $\frac{3x-1}{6y^2} - \frac{x+5}{9xy}$

49. $\frac{4}{x-2} + \frac{5}{x+3}$

50. $\frac{2}{x-3} + \frac{5}{x-4}$

51. $\frac{6}{x-7} - \frac{4}{x+3}$

52. $\frac{3}{y+6} - \frac{4}{y-3}$

53. $\frac{2x}{x+1} + \frac{1}{x-3}$

54. $\frac{3x}{x-4} + \frac{2}{x+6}$

55. $\frac{4x}{2x-1} - \frac{5}{x-6}$

56. $\frac{6x}{x+5} - \frac{3}{2x+3}$

Simplify.

57. $\frac{2a}{a-7} + \frac{5}{7-a}$

58. $\frac{4x}{6-x} + \frac{5}{x-6}$

59. $\frac{x}{x^2-9} + \frac{3}{x-3}$

60. $\frac{y}{y^2-16} + \frac{1}{y-4}$

61. $\frac{2x}{x^2-x-6} - \frac{3}{x+2}$

62. $\frac{5x}{x^2+2x-8} - \frac{2}{x+4}$

63. $\frac{3x-1}{x^2-10x+25} - \frac{3}{x-5}$

64. $\frac{2a+3}{a^2-7a+12} - \frac{2}{a-3}$

65. $\frac{x+4}{x^2-x-42} + \frac{3}{7-x}$

66. $\frac{x+3}{x^2-3x-10} + \frac{2}{5-x}$

67. $\frac{1}{x+1} + \frac{x}{x-6} - \frac{5x-2}{x^2-5x-6}$

68. $\frac{x}{x-4} + \frac{5}{x+5} - \frac{11x-8}{x^2+x-20}$

69. $\frac{3x+1}{x-1} - \frac{x-1}{x-3} + \frac{x+1}{x^2-4x+3}$

70. $\frac{4x+1}{x-8} - \frac{3x+2}{x+4} - \frac{49x+4}{x^2-4x-32}$

71. $\frac{2x+9}{3-x} + \frac{x+5}{x+7} - \frac{2x^2+3x-3}{x^2+4x-21}$

72. $\frac{3x+5}{x+5} - \frac{x+1}{2-x} - \frac{4x^2-3x-1}{x^2+3x-10}$

SECTION 4 Complex Fractions

4.1 Objective To simplify a complex fraction

Reference for Computer Tutor™

A **complex fraction** is a fraction whose numerator or denominator contains one or more fractions. Examples of complex fractions are shown at the right.

$$\frac{3}{2-\frac{1}{2}}, \quad \frac{4+\frac{1}{x}}{3+\frac{2}{x}}, \quad \frac{\frac{1}{x-1}+x+3}{x-3+\frac{1}{x+4}}$$

Simplify: $\dfrac{1-\frac{4}{x^2}}{1+\frac{2}{x}}$

Find the LCM of the denominators of the fractions in the numerator and denominator.

The LCM of x^2 and x is x^2.

$$\frac{1-\frac{4}{x^2}}{1+\frac{2}{x}} =$$

Multiply the numerator and denominator of the complex fraction by the LCM. Then simplify.

$$\frac{1-\frac{4}{x^2}}{1+\frac{2}{x}} \cdot \frac{x^2}{x^2} = \frac{1\cdot x^2 - \frac{4}{x^2}\cdot x^2}{1\cdot x^2 + \frac{2}{x}\cdot x^2} =$$

$$\frac{x^2-4}{x^2+2x} = \frac{(x-2)\overset{1}{\cancel{(x+2)}}}{x\underset{1}{\cancel{(x+2)}}} =$$

$$\frac{x-2}{x}$$

Example 1

Simplify: $\dfrac{\frac{1}{x}+\frac{1}{2}}{\frac{1}{x^2}-\frac{1}{4}}$

Solution

The LCM of x, 2, x^2, and 4 is $4x^2$.

$$\frac{\frac{1}{x}+\frac{1}{2}}{\frac{1}{x^2}-\frac{1}{4}} = \frac{\frac{1}{x}+\frac{1}{2}}{\frac{1}{x^2}-\frac{1}{4}} \cdot \frac{4x^2}{4x^2} = \frac{\frac{1}{x}\cdot 4x^2 + \frac{1}{2}\cdot 4x^2}{\frac{1}{x^2}\cdot 4x^2 - \frac{1}{4}\cdot 4x^2} =$$

$$\frac{4x+2x^2}{4-x^2} = \frac{2x\overset{1}{\cancel{(2+x)}}}{(2-x)\underset{1}{\cancel{(2+x)}}} = \frac{2x}{2-x}$$

Example 2

Simplify: $\dfrac{\frac{1}{3}-\frac{1}{x}}{\frac{1}{9}-\frac{1}{x^2}}$

Your solution

Solution on p. A40

Example 3

Simplify: $\dfrac{1 - \frac{2}{x} - \frac{15}{x^2}}{1 - \frac{11}{x} + \frac{30}{x^2}}$

Solution

The LCM of x and x^2 is x^2.

$$\frac{1 - \frac{2}{x} - \frac{15}{x^2}}{1 - \frac{11}{x} + \frac{30}{x^2}} = \frac{1 - \frac{2}{x} - \frac{15}{x^2}}{1 - \frac{11}{x} + \frac{30}{x^2}} \cdot \frac{x^2}{x^2} =$$

$$\frac{1 \cdot x^2 - \frac{2}{x} \cdot x^2 - \frac{15}{x^2} \cdot x^2}{1 \cdot x^2 - \frac{11}{x} \cdot x^2 + \frac{30}{x^2} \cdot x^2} = \frac{x^2 - 2x - 15}{x^2 - 11x + 30} =$$

$$\frac{\overset{1}{\cancel{(x - 5)}}(x + 3)}{\underset{1}{\cancel{(x - 5)}}(x - 6)} = \frac{x + 3}{x - 6}$$

Example 4

Simplify: $\dfrac{1 + \frac{4}{x} + \frac{3}{x^2}}{1 + \frac{10}{x} + \frac{21}{x^2}}$

Your solution

Example 5

Simplify: $\dfrac{x - 8 + \frac{20}{x + 4}}{x - 10 + \frac{24}{x + 4}}$

Solution

The LCM is $x + 4$.

$$\frac{x - 8 + \frac{20}{x + 4}}{x - 10 + \frac{24}{x + 4}} = \frac{x - 8 + \frac{20}{x + 4}}{x - 10 + \frac{24}{x + 4}} \cdot \frac{x + 4}{x + 4} =$$

$$\frac{(x - 8)(x + 4) + \frac{20}{x + 4} \cdot (x + 4)}{(x - 10)(x + 4) + \frac{24}{x + 4} \cdot (x + 4)} =$$

$$\frac{x^2 - 4x - 32 + 20}{x^2 - 6x - 40 + 24} = \frac{x^2 - 4x - 12}{x^2 - 6x - 16} =$$

$$\frac{(x - 6)\overset{1}{\cancel{(x + 2)}}}{(x - 8)\underset{1}{\cancel{(x + 2)}}} = \frac{x - 6}{x - 8}$$

Example 6

Simplify: $\dfrac{x + 3 - \frac{20}{x - 5}}{x + 8 + \frac{30}{x - 5}}$

Your solution

4.1 Exercises

Simplify.

1. $\dfrac{1 + \dfrac{3}{x}}{1 - \dfrac{9}{x^2}}$

2. $\dfrac{1 + \dfrac{4}{x}}{1 - \dfrac{16}{x^2}}$

3. $\dfrac{2 - \dfrac{8}{x + 4}}{3 - \dfrac{12}{x + 4}}$

4. $\dfrac{5 - \dfrac{25}{x + 5}}{1 - \dfrac{3}{x + 5}}$

5. $\dfrac{1 + \dfrac{5}{y - 2}}{1 - \dfrac{2}{y - 2}}$

6. $\dfrac{2 - \dfrac{11}{2x - 1}}{3 - \dfrac{17}{2x - 1}}$

7. $\dfrac{4 - \dfrac{2}{x + 7}}{5 + \dfrac{1}{x + 7}}$

8. $\dfrac{5 + \dfrac{3}{x - 8}}{2 - \dfrac{1}{x - 8}}$

9. $\dfrac{1 - \dfrac{1}{x} - \dfrac{6}{x^2}}{1 - \dfrac{9}{x^2}}$

10. $\dfrac{1 + \dfrac{4}{x} + \dfrac{4}{x^2}}{1 - \dfrac{2}{x} - \dfrac{8}{x^2}}$

11. $\dfrac{1 - \dfrac{5}{x} - \dfrac{6}{x^2}}{1 + \dfrac{6}{x} + \dfrac{5}{x^2}}$

12. $\dfrac{1 - \dfrac{7}{a} + \dfrac{12}{a^2}}{1 + \dfrac{1}{a} - \dfrac{20}{a^2}}$

13. $\dfrac{1 - \dfrac{6}{x} + \dfrac{8}{x^2}}{\dfrac{4}{x^2} + \dfrac{3}{x} - 1}$

14. $\dfrac{1 + \dfrac{3}{x} - \dfrac{18}{x^2}}{\dfrac{21}{x^2} - \dfrac{4}{x} - 1}$

15. $\dfrac{x - \dfrac{4}{x + 3}}{1 + \dfrac{1}{x + 3}}$

16. $\dfrac{y + \dfrac{1}{y - 2}}{1 + \dfrac{1}{y - 2}}$

Simplify.

17. $$\frac{1 - \dfrac{x}{2x + 1}}{x - \dfrac{1}{2x + 1}}$$

18. $$\frac{1 - \dfrac{2x - 2}{3x - 1}}{x - \dfrac{4}{3x - 1}}$$

19. $$\frac{x - 5 + \dfrac{14}{x + 4}}{x + 3 - \dfrac{2}{x + 4}}$$

20. $$\frac{a + 4 + \dfrac{5}{a - 2}}{a + 6 + \dfrac{15}{a - 2}}$$

21. $$\frac{x + 3 - \dfrac{10}{x - 6}}{x + 2 - \dfrac{20}{x - 6}}$$

22. $$\frac{x - 7 + \dfrac{5}{x - 1}}{x - 3 + \dfrac{1}{x - 1}}$$

23. $$\frac{y - 6 + \dfrac{22}{2y + 3}}{y - 5 + \dfrac{11}{2y + 3}}$$

24. $$\frac{x + 2 - \dfrac{12}{2x - 1}}{x + 1 - \dfrac{9}{2x - 1}}$$

25. $$\frac{x - \dfrac{2}{2x - 3}}{2x - 1 - \dfrac{8}{2x - 3}}$$

26. $$\frac{x + 3 - \dfrac{18}{2x + 1}}{x - \dfrac{6}{2x + 1}}$$

27. $$\frac{\dfrac{1}{x} - \dfrac{2}{x - 1}}{\dfrac{3}{x} + \dfrac{1}{x - 1}}$$

28. $$\frac{\dfrac{3}{n + 1} + \dfrac{1}{n}}{\dfrac{2}{n + 1} + \dfrac{3}{n}}$$

29. $$\frac{\dfrac{3}{2x - 1} - \dfrac{1}{x}}{\dfrac{4}{x} + \dfrac{2}{2x - 1}}$$

30. $$\frac{\dfrac{4}{3x + 1} + \dfrac{3}{x}}{\dfrac{6}{x} - \dfrac{2}{3x + 1}}$$

31. $$\frac{\dfrac{3}{b - 4} - \dfrac{2}{b + 1}}{\dfrac{5}{b + 1} - \dfrac{1}{b - 4}}$$

32. $$\frac{\dfrac{5}{x - 5} - \dfrac{3}{x - 1}}{\dfrac{6}{x - 1} + \dfrac{2}{x - 5}}$$

SECTION 5 Solving Equations Containing Fractions

5.1 Objective To solve an equation containing fractions

Reference for Computer Tutor™

To solve an equation containing fractions, **clear denominators** by multiplying each side of the equation by the LCM of the denominators. Then solve for the variable.

Solve: $\frac{3x-1}{4} + \frac{2}{3} = \frac{7}{6}$

The LCM is 12.

$$\frac{3x-1}{4} + \frac{2}{3} = \frac{7}{6}$$

Multiply each side of the equation by the LCM of the denominators.

$$12\left(\frac{3x-1}{4} + \frac{2}{3}\right) = 12 \cdot \frac{7}{6}$$

Simplify using the Distributive Property and the Properties of Fractions.

$$12\left(\frac{3x-1}{4}\right) + 12 \cdot \frac{2}{3} = 12 \cdot \frac{7}{6}$$

$$\frac{\overset{3}{\cancel{12}}}{1}\left(\frac{3x-1}{\underset{1}{\cancel{4}}}\right) + \frac{\overset{4}{\cancel{12}}}{1} \cdot \frac{2}{\underset{1}{\cancel{3}}} = \frac{\overset{2}{\cancel{12}}}{1} \cdot \frac{7}{\underset{1}{\cancel{6}}}$$

Solve for x.

$$9x - 3 + 8 = 14$$
$$9x + 5 = 14$$
$$9x = 9$$
$$x = 1$$

1 checks as a solution.
The solution is 1.

Occasionally, a value of the variable that appears to be a solution will make one of the denominators zero. In this case, the equation has no solution for that value of the variable.

Solve: $\frac{2x}{x-2} = 1 + \frac{4}{x-2}$

The LCM is $x - 2$.

$$\frac{2x}{x-2} = 1 + \frac{4}{x-2}$$

Multiply each side of the equation by the LCM of the denominators.

$$(x-2)\frac{2x}{x-2} = (x-2)\left(1 + \frac{4}{x-2}\right)$$

Simplify using the Distributive Property and Properties of Fractions.

$$(x-2)\left(\frac{2x}{x-2}\right) = (x-2) \cdot 1 + (x-2) \cdot \frac{4}{x-2}$$

$$\frac{\overset{1}{\cancel{x-2}}}{1} \cdot \frac{2x}{\underset{1}{\cancel{x-2}}} = (x-2) + \frac{\overset{1}{\cancel{x-2}}}{1} \cdot \frac{4}{\underset{1}{\cancel{x-2}}}$$

Solve for x.

$$2x = x - 2 + 4$$
$$2x = x + 2$$
$$x = 2$$

When x is replaced by 2, the denominators of $\frac{2x}{x-2}$ and $\frac{4}{x-2}$ are zero.

Therefore, the equation has no solution

Example 1

Solve: $\frac{x}{x+4} = \frac{2}{x}$

Solution

$$\frac{x}{x+4} = \frac{2}{x} \quad \text{The LCM is } x(x+4).$$

$$\frac{x(x+4)}{1} \cdot \frac{x}{x+4} = \frac{x(x+4)}{1} \cdot \frac{2}{x}$$

$$\frac{x\cancel{(x+4)}}{1} \cdot \frac{x}{\cancel{x+4}} = \frac{\cancel{x}(x+4)}{1} \cdot \frac{2}{\cancel{x}}$$

$$x^2 = (x+4)2$$

$$x^2 = 2x + 8 \quad \text{Quadratic Equation}$$

$$x^2 - 2x - 8 = 0$$

$$(x-4)(x+2) = 0$$

$$x - 4 = 0 \qquad x + 2 = 0$$

$$x = 4 \qquad x = -2$$

Both 4 and -2 check as solutions.

The solutions are 4 and -2.

Example 2

Solve: $\frac{x}{x+6} = \frac{3}{x}$

Your solution

Example 3

Solve: $\frac{3x}{x-4} = 5 + \frac{12}{x-4}$

Solution

$$\frac{3x}{x-4} = 5 + \frac{12}{x-4} \quad \text{The LCM is } x - 4.$$

$$\frac{(x-4)}{1} \cdot \frac{3x}{x-4} = \frac{(x-4)}{1}\left(5 + \frac{12}{x-4}\right)$$

$$\frac{\cancel{(x-4)}}{1} \cdot \frac{3x}{\cancel{x-4}} = \frac{(x-4)}{1} \cdot 5 + \frac{\cancel{(x-4)}}{1} \cdot \frac{12}{\cancel{x-4}}$$

$$3x = (x-4)5 + 12$$

$$3x = 5x - 20 + 12$$

$$3x = 5x - 8$$

$$-2x = -8$$

$$x = 4$$

4 does not check as a solution.

The equation has no solution.

Example 4

Solve: $\frac{5x}{x+2} = 3 - \frac{10}{x+2}$

Your solution

Solutions on p. A40

5.1 Exercises

Solve.

1. $\frac{2x}{3} - \frac{5}{2} = -\frac{1}{2}$

2. $\frac{x}{3} - \frac{1}{4} = \frac{1}{12}$

3. $\frac{x}{3} - \frac{1}{4} = \frac{x}{4} - \frac{1}{6}$

4. $\frac{2y}{9} - \frac{1}{6} = \frac{y}{9} + \frac{1}{6}$

5. $\frac{2x - 5}{8} + \frac{1}{4} = \frac{x}{8} + \frac{3}{4}$

6. $\frac{3x + 4}{12} - \frac{1}{3} = \frac{5x + 2}{12} - \frac{1}{2}$

7. $\frac{6}{2a + 1} = 2$

8. $\frac{12}{3x - 2} = 3$

9. $\frac{9}{2x - 5} = -2$

10. $\frac{6}{4 - 3x} = 3$

11. $2 + \frac{5}{x} = 7$

12. $3 + \frac{8}{n} = 5$

13. $1 - \frac{9}{x} = 4$

14. $3 - \frac{12}{x} = 7$

15. $\frac{2}{y} + 5 = 9$

16. $\frac{6}{x} + 3 = 11$

17. $\frac{3}{x - 2} = \frac{4}{x}$

18. $\frac{5}{x + 3} = \frac{3}{x - 1}$

Solve.

19. $\frac{2}{3x - 1} = \frac{3}{4x + 1}$

20. $\frac{5}{3x - 4} = \frac{-3}{1 - 2x}$

21. $\frac{-3}{2x + 5} = \frac{2}{x - 1}$

22. $\frac{4}{5y - 1} = \frac{2}{2y - 1}$

23. $\frac{4x}{x - 4} + 5 = \frac{5x}{x - 4}$

24. $\frac{2x}{x + 2} - 5 = \frac{7x}{x + 2}$

25. $2 + \frac{3}{a - 3} = \frac{a}{a - 3}$

26. $\frac{x}{x + 4} = 3 - \frac{4}{x + 4}$

27. $\frac{x}{x - 1} = \frac{8}{x + 2}$

28. $\frac{x}{x + 12} = \frac{1}{x + 5}$

29. $\frac{2x}{x + 4} = \frac{3}{x - 1}$

30. $\frac{5}{3n - 8} = \frac{n}{n + 2}$

31. $x + \frac{6}{x - 2} = \frac{3x}{x - 2}$

32. $x - \frac{6}{x - 3} = \frac{2x}{x - 3}$

33. $\frac{8}{y} = \frac{2}{y - 2} + 1$

34. $\frac{8}{r} + \frac{3}{r - 1} = 3$

35. $\frac{x}{x + 2} + \frac{2}{x - 2} = \frac{x + 6}{x^2 - 4}$

36. $\frac{x}{x + 4} = \frac{11}{x^2 - 16} + 2$

SECTION 6 Ratio and Proportion

6.1 Objective To solve a proportion

Reference for Computer Tutor™

Quantities such as 4 meters, 15 seconds, and 8 gallons are number quantities written with units. In these examples the units are meters, seconds, and gallons.

A **ratio** is the quotient of two quantities which have the same unit.

The length of a living room is 16 ft and the width is 12 ft. The ratio of the length to the width is written:

$\frac{16\text{ ft}}{12\text{ ft}} = \frac{16}{12} = \frac{4}{3}$

A ratio is in simplest form when the two numbers do not have a common factor. Note that the units are not written.

A **rate** is the quotient of two quantities which have different units.

There are 2 lb of salt in 8 gal of water. The salt-to-water rate is:

$\frac{2\text{ lb}}{8\text{ gal}} = \frac{1\text{ lb}}{4\text{ gal}}$

A rate is in simplest form when the two numbers do not have a common factor. The units are written as part of the rate.

A **proportion** is an equation which states the equality of two ratios or rates.
Examples of proportions are shown at the right.

$$\frac{30\text{ mi}}{4\text{ h}} = \frac{15\text{ mi}}{2\text{ h}}$$
$$\frac{4}{6} = \frac{8}{12}$$
$$\frac{3}{4} = \frac{x}{8}$$

Solve the proportion $\frac{4}{x} = \frac{2}{3}$.

$$\frac{4}{x} = \frac{2}{3}$$

Multiply each side of the proportion by the LCM of the denominators.

$$3x\left(\frac{4}{x}\right) = 3x\left(\frac{2}{3}\right)$$

Solve the equation.

$$12 = 2x$$
$$6 = x$$

The solution is 6.

Example 1

Solve the proportion $\frac{8}{x+3} = \frac{4}{x}$.

Solution

$$\frac{8}{x+3} = \frac{4}{x}$$
$$x(x+3)\frac{8}{x+3} = x(x+3)\frac{4}{x}$$
$$8x = 4(x+3)$$
$$8x = 4x + 12$$
$$4x = 12$$
$$x = 3$$

The solution is 3.

Example 2

Solve the proportion $\frac{2}{x+3} = \frac{6}{5x+5}$.

Your solution

Solution on p. A41

6.2 Objective To solve application problems

Use COMPUTER TUTOR™ DISK 7

Example 3

The monthly loan payment for a car is \$28.35 for each \$1000 borrowed. At this rate, find the monthly payment for a \$6000 car loan.

Strategy

To find the monthly payment, write and solve a proportion using P to represent the monthly car payment.

Solution

$$\frac{\$28.35}{\$1000} = \frac{P}{\$6000}$$

$$6000\left(\frac{28.35}{1000}\right) = 6000\left(\frac{1}{6000}P\right)$$

$$170.10 = P$$

The monthly payment is \$170.10.

Example 4

Sixteen ceramic tiles are required to tile a 9-square-foot area. At this rate, how many square feet can be tiled using 256 ceramic tiles?

Your strategy

Your solution

Example 5

An investment of \$500 earns \$60 each year. At the same rate, how much additional money must be invested to earn \$90 each year?

Strategy

To find the additional amount of money which must be invested, write and solve a proportion using x to represent the additional money. Then $500 + x$ is the total amount invested.

Solution

$$\frac{\$60}{\$500} = \frac{\$90}{\$500 + x}$$

$$\frac{3}{25} = \frac{90}{500 + x}$$

$$25(500 + x)\left(\frac{3}{25}\right) = 25(500 + x)\left(\frac{90}{500 + x}\right)$$

$$(500 + x)3 = 25(90)$$

$$1500 + 3x = 2250$$

$$3x = 750$$

$$x = 250$$

An additional \$250 must be invested.

Example 6

Three ounces of a medication are required for a 150-pound adult. At the same rate, how many additional ounces of medication are required for a 200-pound adult?

Your strategy

Your solution

Solutions on p. A41

6.1 Exercises

Solve.

1. $\frac{x}{12} = \frac{3}{4}$
2. $\frac{6}{x} = \frac{2}{3}$
3. $\frac{4}{9} = \frac{x}{27}$
4. $\frac{16}{9} = \frac{64}{x}$
5. $\frac{x+3}{12} = \frac{5}{6}$
6. $\frac{3}{5} = \frac{x-4}{10}$
7. $\frac{18}{x+4} = \frac{9}{5}$
8. $\frac{2}{11} = \frac{20}{x-3}$
9. $\frac{2}{x} = \frac{4}{x+1}$
10. $\frac{16}{x-2} = \frac{8}{x}$
11. $\frac{x+3}{4} = \frac{x}{8}$
12. $\frac{x-6}{3} = \frac{x}{5}$
13. $\frac{2}{x-1} = \frac{6}{2x+1}$
14. $\frac{9}{x+2} = \frac{3}{x-2}$
15. $\frac{2x}{7} = \frac{x-2}{14}$

6.2 Application Problems

Solve.

1. A salt water solution is made by dissolving 2 lb of salt in 5 gal of water. At this rate, how many pounds of salt are required for 25 gal of water?

2. A building contractor estimates that three overhead lights are needed for every 250 ft^2 of floor space. Using this estimate, how many light fixtures are necessary for a 10,000-square-foot office building?

3. A pre-election survey showed that 2 out of every 5 voters would vote in an election. At this rate, how many people would be expected to vote in a city of 25,000?

4. A quality control inspector found 2 defective electric blenders in a shipment of 100 blenders. At this rate, how many blenders would be defective in a shipment of 5000?

Solve.

5. A landscape designer estimates that 6 pieces of lumber are necessary for each 25 ft^2 of patio wood decking. Using this estimate, how many square feet of decking can be made from 36 pieces of lumber?

6. A carpet manufacturer uses 2 lb of wool for every 3 lb of nylon in a certain grade of carpet. At this rate, how many pounds of nylon are required for 250 lb of wool?

7. The license fee for a car which cost $5500 was $66. At the same rate, what is the license fee for a car which costs $7500?

8. The real estate tax for a home which costs $75,000 is $750. At this rate, what is the value of a home for which the real estate tax is $562.50?

9. In a wildlife preserve, 16 deer are captured, tagged, and then released. Later, 30 deer are captured. Two of the 30 are found to have tags. Estimate the number of deer in the preserve.

10. In a lake, 50 fish are caught, tagged, and then released. Later 70 fish are caught. Five of the 70 are found to have tags. Estimate the number of fish in the lake.

11. A stock investment of 100 shares paid a dividend of $124. At this rate, how many additional shares are required to earn a dividend of $186?

12. A chef estimates that 50 lb of vegetables will serve 130 people. Using this estimate, how many additional pounds will be necessary to serve 156 people?

13. A caterer estimates that 3 gal of fruit punch will serve 40 people. How much additional punch is necessary to serve 60 people?

14. A farmer estimates that 5600 bushels of wheat can be harvested from 160 acres of land. Using this estimate, how many additional acres are needed to harvest 8120 bushels of wheat?

SECTION 7 Literal Equations

7.1 Objective — To rewrite a literal equation in terms of one of the variables

Reference for Computer Tutor™

A **literal equation** is an equation which contains more than one variable. Examples of literal equations are shown at the right.

$$2x + 3y = 6$$

$$4w - 2x + z = 0$$

Formulas are used to express a relationship among physical quantities. A **formula** is a literal equation which states rules about measurements. Examples of formulas are shown at the right.

$$\frac{1}{R_1} + \frac{1}{R_2} = \frac{1}{R} \quad \text{(Physics)}$$

$$s = a + (n - 1)d \quad \text{(Mathematics)}$$

$$A = P + Prt \quad \text{(Business)}$$

The Addition and Multiplication Properties can be used to solve a literal equation for one of the variables. The goal is to rewrite the equation so that the letter being solved for is alone on one side of the equation and all the other numbers and variables are on the other side.

Solve $A = P(1 + i)$ for i.

The goal is to rewrite the equation so that i is on one side of the equation and all other variables are on the other side.

Use the Distributive Property to remove parentheses.

$$A = P(1 + i)$$
$$A = P + Pi$$

Add the opposite of P to each side of the equation.

$$A + (-P) = P + (-P) + Pi$$
$$A - P = Pi$$

Multiply each side of the equation by the reciprocal of P.

$$\frac{1}{P}(A - P) = \frac{1}{P}(Pi)$$
$$\frac{A - P}{P} = i$$

Example 1

Solve $3x - 4y = 12$ for y.

Solution

$$3x - 4y = 12$$
$$3x + (-3x) - 4y = -3x + 12$$
$$-4y = -3x + 12$$
$$\left(-\frac{1}{4}\right)(-4y) = \left(-\frac{1}{4}\right)(-3x + 12)$$
$$y = \frac{3}{4}x - 3$$

Example 2

Solve $5x - 2y = 10$ for y.

Your solution

Solution on p. A42

Example 3

Solve $I = \frac{E}{R + r}$ for R.

Solution

$$I = \frac{E}{R + r}$$
$$(R + r)I = (R + r)\frac{E}{R + r}$$
$$RI + rI = E$$
$$RI + rI + (-rI) = E + (-rI)$$
$$RI = E - rI$$
$$\frac{1}{I}(RI) = \frac{1}{I}(E - rI)$$
$$R = \frac{E - rI}{I}$$

Example 4

Solve $s = \frac{A + L}{2}$ for L.

Your solution

Example 5

Solve $L = a(1 + ct)$ for c.

Solution

$$L = a(1 + ct)$$
$$L = a + act$$
$$L + (-a) = a + (-a) + act$$
$$L - a = act$$
$$\frac{1}{at}(L - a) = \frac{1}{at}(act)$$
$$\frac{L - a}{at} = c$$

Example 6

Solve $S = a + (n - 1)d$ for n.

Your solution

Example 7

Solve $S = C - rC$ for C.

Solution

$$S = C - rC$$
$$S = (1 - r)C \qquad \text{Factoring}$$
$$\frac{1}{1 - r} \cdot S = \frac{1}{1 - r}(1 - r)C$$
$$\frac{S}{1 - r} = C$$

Example 8

Solve $S = C + rC$ for C.

Your solution

7.1 Exercises

Solve for y.

1. $3x + y = 10$
2. $2x + y = 5$
3. $4x - y = 3$
4. $5x - y = 7$
5. $3x + 2y = 6$
6. $2x + 3y = 9$
7. $2x - 5y = 10$
8. $5x - 2y = 4$
9. $2x + 7y = 14$
10. $6x - 5y = 10$
11. $x + 3y = 6$
12. $x + 2y = 8$
13. $x - 4y = 12$
14. $x - 3y = 9$
15. $7x - 2y - 14 = 0$
16. $2x - 9y - 18 = 0$
17. $3x - y + 7 = 0$
18. $2x - y + 5 = 0$

Solve for x.

19. $x + 3y = 6$
20. $x + 6y = 10$
21. $3x - y = 3$
22. $2x - y = 6$
23. $2x + 5y = 10$
24. $4x + 3y = 12$
25. $x - 2y + 1 = 0$
26. $x - 4y - 3 = 0$
27. $5x + 4y + 20 = 0$
28. $3x + 5y + 15 = 0$
29. $3x - 2y - 15 = 0$
30. $5x - 8y + 10 = 0$

Solve the formula for the given variable.

31. $A = \frac{1}{2}bh$; h (Geometry)

32. $P = a + b + c$; b (Geometry)

33. $d = rt$; t (Physics)

34. $E = IR$; R (Physics)

35. $PV = nRT$; T (Chemistry)

36. $A = bh$; h (Geometry)

37. $P = 2l + 2w$; l (Geometry)

38. $F = \frac{9}{5}C + 32$; C (Temperature Conversion)

39. $A = \frac{1}{2}h(b_1 + b_2)$; b_1 (Geometry)

40. $C = \frac{5}{9}(F - 32)$; F (Temperature Conversion)

41. $V = \frac{1}{3}Ah$; h (Geometry)

42. $P = R - C$; C (Business)

43. $R = \frac{C - S}{t}$; S (Business)

44. $P = \frac{R - C}{n}$; R (Business)

45. $A = P + Prt$; P (Business)

46. $T = fm - gm$; m (Engineering)

47. $A = Sw + w$; w (Physics)

48. $a = S - Sr$; S (Mathematics)

SECTION 8 Application Problems

8.1 Objective To solve work problems

Reference for Computer Tutor™

If a painter can paint a room in 4 h, then in 1 h the painter can paint $\frac{1}{4}$ of the room. The painter's rate of work is $\frac{1}{4}$ of the room each hour. The **rate of work** is that part of a task which is completed in one unit of time.

A pipe can fill a tank in 30 min. This pipe can fill $\frac{1}{30}$ of the tank in 1 min. The rate of work is $\frac{1}{30}$ of the tank each minute. If a second pipe can fill the tank in x min, the rate of work for the second pipe is $\frac{1}{x}$ of the tank each minute.

In solving a work problem, the goal is to determine the time it takes to complete a task. The basic equation that is used to solve work problems is:

Rate of work × time worked = part of task completed

For example, if a faucet can fill a sink in 6 min, then in 5 min the faucet will fill $\frac{1}{6} \times 5 = \frac{5}{6}$ of the sink. In 5 min the faucet completes $\frac{5}{6}$ of the task.

A painter can paint a wall in 20 min. The painter's apprentice can paint the same wall in 30 min. How long will it take to paint the wall when they work together?

STRATEGY FOR SOLVING A WORK PROBLEM

▷ For each person or machine, write a numerical or variable expression for the rate of work, the time worked, and the part of the task completed. The results can be recorded in a table.

Unknown time to paint the wall working together: t

	Rate of work	·	Time worked	=	Part of task completed
Painter	$\frac{1}{20}$	·	t	=	$\frac{t}{20}$
Apprentice	$\frac{1}{30}$	·	t	=	$\frac{t}{30}$

▷ Determine how the parts of the task completed are related. Use the fact that the sum of the parts of the task completed must equal 1, the complete task.

The sum of the part of the task completed by the painter and the part of the task completed by the apprentice is 1.

$$\frac{t}{20} + \frac{t}{30} = 1$$
$$60\left(\frac{t}{20} + \frac{t}{30}\right) = 60 \cdot 1$$
$$3t + 2t = 60$$
$$5t = 60$$
$$t = 12$$

Working together, they will paint the wall in 12 min.

Example 1

A small water pipe takes three times longer to fill a tank than does a large water pipe. With both pipes open it takes 4 h to fill the tank. Find the time it would take the small pipe working alone to fill the tank.

Strategy

▷ Time for large pipe to fill the tank: t
Time for small pipe to fill the tank: $3t$

	Rate	Time	Part
Small pipe	$\frac{1}{3t}$	4	$\frac{4}{3t}$
Large pipe	$\frac{1}{t}$	4	$\frac{4}{t}$

▷ The sum of the parts of the task completed by each pipe must equal one.

Solution

$$\frac{4}{3t} + \frac{4}{t} = 1$$
$$3t\left(\frac{4}{3t} + \frac{4}{t}\right) = 3t \cdot 1$$
$$4 + 12 = 3t$$
$$16 = 3t$$
$$\frac{16}{3} = t$$

$$3t = 3\left(\frac{16}{3}\right) = 16$$

The small pipe working alone takes 16 h to fill the tank.

Example 2

Two computer printers that work at the same rate are working together to print the payroll checks for a large corporation. After working together for 2 h, one of the printers quits. The second requires 3 more hours to complete the payroll checks. Find the time it would take one printer working alone to print the payroll.

Your strategy

Your solution

Solution on p. A43

8.2 Objective To solve uniform motion problems

Reference for Computer Tutor™

A car that travels constantly in a straight line at 30 mph is in uniform motion. **Uniform motion** means that the speed of an object does not change.

The basic equation used to solve uniform motion problems is:

Distance = rate × time

An alternate form of this equation can be written by solving the equation for time.

$$\frac{\textbf{Distance}}{\textbf{Rate}} = \textbf{time}$$

This form of the equation is useful when the total time of travel for two objects or the time of travel between two points is known.

The speed of a boat in still water is 20 mph. The boat traveled 75 mi down a river in the same amount of time as it traveled 45 mi up the river. Find the rate of the river's current.

STRATEGY FOR SOLVING A UNIFORM MOTION PROBLEM

▷ For each object, write a numerical or variable expression for the distance, rate, and time. The results can be recorded in a table.

The unknown rate of the river's current: r

	Distance	÷	Rate	=	Time
Down river	75	÷	$20 + r$	=	$\frac{75}{20 + r}$
Up river	45	÷	$20 - r$	=	$\frac{45}{20 - r}$

▷ Determine how the times traveled by each object are related. For example, it may be known that the times are equal or the total time may be known.

The time down the river is equal to the time up the river.

$$\frac{75}{20 + r} = \frac{45}{20 - r}$$

$$(20 + r)(20 - r)\frac{75}{20 + r} = (20 + r)(20 - r)\frac{45}{20 - r}$$

$$(20 - r)75 = (20 + r)45$$

$$1500 - 75r = 900 + 45r$$

$$-120r = -600$$

$$r = 5$$

The rate of the river's current is 5 mph.

Example 3

A cyclist rode the first 20 mi of a trip at a constant rate. For the next 16 mi, the cyclist reduced the speed by 2 mph. The total time for the 36 mi was 4 h. Find the rate of the cyclist for each leg of the trip.

Strategy

▷ Rate for the first 20 mi: r

	Distance	Rate	Time
1st 20 mi	20	r	$\frac{20}{r}$
Next 16 mi	16	$r-2$	$\frac{16}{r-2}$

▷ The total time for the trip was 4 h.

Solution

$$\frac{20}{r} + \frac{16}{r-2} = 4$$

$$r(r-2)\left[\frac{20}{r} + \frac{16}{r-2}\right] = r(r-2) \cdot 4$$

$$(r-2)20 + 16r = 4r^2 - 8r$$

$$20r - 40 + 16r = 4r^2 - 8r$$

$$36r - 40 = 4r^2 - 8r \quad \text{Quadratic Equation}$$

$$0 = 4r^2 - 44r + 40$$

$$0 = 4(r^2 - 11r + 10)$$

$$0 = 4(r-10)(r-1)$$

$$r - 10 = 0 \qquad r - 1 = 0$$

$$r = 10 \qquad r = 1$$

The solution $r = 1$ mph is not possible, since the rate on the last 16 mi would then be -1 mph.

10 mph was the rate for the first 20 mi.
8 mph was the rate for the next 16 mi.

Example 4

The total time for a sailboat to sail back and forth across a lake 6 km wide was 2 h. The rate sailing back was three times the rate sailing across. Find the rate across the lake.

Your strategy

Your solution

Solution on p. A43

8.1 Application Problems

Solve.

1. A park fountain has two sprinklers which are used to fill a fountain. One sprinkler can fill the fountain in 3 h, while the second sprinkler can fill the fountain in 6 h. How long will it take to fill the fountain with both sprinklers operating?

2. A new printing press can complete the weekly edition of a news magazine in 10 h. An older printing press requires 15 h to do the same task. How long would it take to print the weekly edition with both presses operating?

3. One member of a gardening crew can mow a lawn in 20 min, while the second member of the crew requires 30 min to mow the same lawn. How long would it take to mow the lawn when they work together?

4. Two farmers are plowing a field. One farmer, using an old tractor and working alone, requires 12 h to plow the field. A second farmer, using a modern tractor, can plow the same field in 4 h. How long would it take to plow the field with both tractors working together?

5. A business report for a company can be printed in 55 min using one computer. A second computer can print the report in 66 min. How long will it take to print the report with both computer's operating?

6. A small air conditioner will cool a room 5° in 75 min. A larger air conditioner will cool the room 5° in 50 min. How long would it take to cool the room 5° with both air conditioners operating?

7. A new machine can fill soda bottles three times faster than an old machine. With both machines working together, they can complete the task in 9 h. How long would it take the new machine working alone to complete the task?

8. An experienced painter can paint a fence twice as fast as an inexperienced painter. Working together, the painters require 4 h to paint the fence. How long would it take the experienced painter working alone to paint the fence?

9. A plumber can install a garbage disposal in 45 min. With the plumber's assistant helping, the task would take 30 min. How long would it take the assistant working alone to complete the task?

10. A mason can construct a retaining wall in 10 h. With the mason's apprentice assisting, the task would take 6 h. How long would it take the apprentice working alone to construct the wall?

Solve.

11. With cold and hot water running, a bathtub can be filled in 6 min. The hot water faucet working alone requires 15 min to fill the tub. How long would it take the cold water faucet working alone to fill the tub?

12. Two solar heating panels will raise the temperature of water 1° in 60 min. One panel, working alone, requires 90 min to raise the temperature of the water 1°. How long would it take the second panel working alone to heat the water?

13. An electrician requires 2 h to install an electrical motor, while an apprentice requires 6 h to install the motor. The electrician worked for one hour and then stopped. How long would it take the apprentice to complete the installation?

14. One member of a telephone repair crew can wire new telephone circuits in 4 h, while it would take 6 h for the second member of the crew to do the same job. After working alone for 2 h, the first crew member quits and the second member completes the task. How long does it take the second member to complete the task?

15. One welder requires 2 h to make the welds on a steel frame, while a second welder requires 4 h to do the same job. The first welder worked for one hour and then quit. How long would it take the second welder to complete the welds?

16. One cement mason can lay a cement foundation in 8 h, while it takes a second 12 h to do the same task. After working alone for 4 h, the first mason quits. How long would it take the second mason to complete the task?

17. A large and small inlet pipe work together to fill a pool. The larger pipe requires 8 h working alone to fill the pool. After both pipes have been operating for 2 h, the larger pipe is turned off. The small pipe requires 9 more hours to fill the pool. How long would it take the smaller pipe working alone to fill the pool?

18. Two bricklayers who work at the same rate are building a brick wall. After working together for 7 h, one bricklayer quit. The second bricklayer required 14 more hours to finish the wall. How long would it take one of the bricklayers working alone to build the wall?

19. A large and small refrigerator compressor are working together to cool a food storage locker. The large compressor requires 6 h working alone to cool the locker. After both compressors are on for one hour, the large compressor is shut off. The small compressor continues operating for 9 more hours to cool the locker. How long would it take the small compressor working alone to cool the storage locker?

20. Two carpenters who work at the same rate are framing the outside walls of a house. After working together for 10 h, one of the carpenters quits. The second carpenter requires 20 more hours to finish the house frame. Find the time it would take one of the carpenters working alone to frame the house.

8.2 Application Problems

Solve.

21. An express train travels 300 mi in the same amount of time that a freight train travels 180 mi. The rate of the express train is 20 mph faster than the freight train. Find the rate of each train.

22. A postal clerk on vacation took an 8-mile cruise in a sailboat in the same amount of time as a 20-mile cruise on a power boat. The rate of the power boat is 12 mph faster than the rate of the sailboat. Find the rate of each boat.

23. A twin-engine plane can fly 660 mi in the same amount of time as it takes a single-engine plane to fly 330 mi. The rate of the twin-engine plane is 100 mph faster than the single-engine plane. Find the rate of the twin-engine plane.

24. The rate of a motorcycle is 36 mph faster than the rate of a bicycle. The motorcycle travels 192 mi in the same amount of time as the bicycle travels 48 mi. Find the rate of the motorcycle.

25. A sales accountant traveled 1800 mi by jet and 300 mi on a prop plane. The rate of the jet is four times the rate of the prop plane. The entire trip took a total of 5 h. Find the rate of each.

26. A motorist drove 90 mi before running out of gas and then walking 5 mi to a gas station. The rate of the motorist in the car was nine times the rate walking. The time spent walking and driving was 3 h. Find the rate at which the motorist walks.

27. A computer representative traveled 135 mi by train and then an additional 855 mi by plane. The rate of the plane was three times the rate of the train and the total time for the trip was 6 h. Find the rate of the plane.

28. A marketing manager traveled 1080 mi on a corporate jet and then an additional 180 mi by helicopter. The rate of the jet is four times the rate of the helicopter. The entire trip took 5 h. Find the rate of the jet.

29. A freight train and a passenger train leave a town at 10 A.M., and head for a town 300 mi away. The rate of the passenger train is twice the rate of the freight train. The passenger train arrives 5 h ahead of the freight train. Find the rate of each train.

30. A single-engine plane and a corporate jet leave an airport at 1 P.M., and head for another airport 660 mi away. The rate of the corporate jet is three times the rate of the single-engine plane. The single-engine plane arrives 4 h after the corporate jet. Find the rate of each plane.

Solve.

31. A light plane can fly at a rate of 100 mi in calm air. Traveling with the wind, the plane flew 360 mi in the same amount of time as it flew 240 mi against the wind. Find the rate of the wind.

32. A tour boat used for river excursions can travel 6 mph in calm water. The amount of time it takes to travel 12 mi against the river's current is the same as the amount of time it takes to travel 24 mi with the current. Find the rate of the current.

33. A twin-engine plane can travel 200 mph in calm air. Flying with the wind, the plane can fly 660 mi in the same amount of time it takes to fly 540 mi against the wind. Find the rate of the wind.

34. A rowing team can row 20 mph in calm water. Rowing with the current of a river, the team can row 25 mi in the same amount of time as they can row 15 mi against the current. Find the rate of the current.

35. A balloonist flew 10 mi on the first leg of a trip before changing direction and flying an additional 8 mi. Because of the wind, the change in direction caused the rate of the balloon to decrease by 1 mph. The total flying time was 4 h. Find the rate of the balloon on the first leg of the trip.

36. A hiker walked 9 mi at a constant rate and then reduced this rate by 1 mph. Another 4 mi was walked at this reduced rate. The time required to walk the 4 mi was one hour less than the time required to walk the 9 mi. Find the rate at which the hiker walked the first 9 mi.

37. A canoeist can paddle at a rate of 7 mph in calm water. Traveling with the current, the canoe traveled 40 mi in the same amount of time as it traveled 16 mi against the current. Find the rate of the current.

38. A jet can fly 600 mph in calm air. Traveling with the wind, the plane can fly 2100 mi in the same amount of time as it flies 1500 mi against the wind. Find the rate of the wind.

39. A small plane flew 200 mi at a constant rate, and then reduced the speed by 10 mph. An additional 90 mi was flown at the reduced speed. The total flying time was 3 h. Find the rate of the plane for the first 200 mi.

40. A jogger ran 12 mi at a constant rate and then reduced the jogging rate by 1 mph. An additional 5 mi was run at the reduced rate. The total time jogging was 3 h. Find the rate of the jogger for the first 12 mi.

Calculators and Computers

Simplifying an Algebraic Fraction

The first three sections of Chapter 7 present operations with algebraic fractions. Many students find performing operations on algebraic fractions difficult, especially adding and subtracting those with unlike denominators. The difficulty often lies in the fact that, in any one problem, there are a number of steps which must be performed, and it is essential that the student understand the function of each step of the process. It is helpful to keep in mind that addition, subtraction, multiplication, and division performed on algebraic fractions involves the same steps as these operations performed on numerical fractions. It is the fractions themselves that make the procedure look more difficult, and the greater amount of time which it takes to perform them that makes them seem so much more complicated. Therefore, you are encouraged to practice as much as possible.

The program SIMPLIFY AN ALGEBRAIC EXPRESSION on the Student Disk will allow you to practice reducing an algebraic fraction to simplest form. The program allows you to choose one of three levels of difficulty. Level one contains the easiest problems and level three are the most difficult.

After you have chosen a level of difficulty, the program will display a problem. Using paper and pencil, simplify the expression. Then press the RETURN key. The correct solution will be displayed. After you have completed a problem, you may continue with problems of the same difficulty, return to the menu and change the level, or quit the program.

Simplifying these expressions is an important part of your algebra training. Continued practice of this skill will pay rewards in your future math classes.

Chapter Summary

KEY WORDS

An **algebraic fraction** is a fraction in which the numerator or denominator is a variable expression.

An algebraic fraction is in **simplest form** when the numerator and denominator have no common factors.

The **reciprocal** of a fraction is a fraction with the numerator and denominator interchanged.

The **least common multiple** (LCM) of two or more numbers is the smallest number that contains the prime factorization of each number.

A **complex fraction** is a fraction whose numerator or denominator contains one or more fractions.

A **ratio** is the quotient of two quantities which have the same unit.

A **rate** is the quotient of two quantities which have different units.

A **proportion** is an equation which states the equality of two ratios or rates.

A **literal equation** is an equation which contains more than one variable.

A **formula** is a literal equation which states rules about measurements.

ESSENTIAL RULES

To multiply fractions: $\frac{a}{b} \cdot \frac{c}{d} = \frac{ac}{bd}$

To divide fractions: $\frac{a}{b} \div \frac{c}{d} = \frac{a}{b} \cdot \frac{d}{c}$

To add fractions: $\frac{a}{c} + \frac{b}{c} = \frac{a + b}{c}$

To subtract fractions: $\frac{a}{c} - \frac{b}{c} = \frac{a - b}{c}$

Equation for Work Problems: **Rate of work × time worked = part of task completed**

Uniform Motion Equation: **Distance = rate × time**

Review/Test

SECTION 1

1.1a Simplify: $\dfrac{16x^5y}{24x^2y^4}$

1.1b Simplify: $\dfrac{x^2 + 4x - 5}{1 - x^2}$

1.2 Simplify:
$\dfrac{x^3y^4}{x^2 - 4x + 4} \cdot \dfrac{x^2 - x - 2}{x^6y^4}$

1.3 Simplify:
$\dfrac{x^2 + 3x + 2}{x^2 + 5x + 4} \div \dfrac{x^2 - x - 6}{x^2 + 2x - 15}$

SECTION 2

2.1 Find the LCM of $6x - 3$ and $2x^2 + x - 1$.

2.2 Write each fraction in terms of the LCM of the denominators.
$\dfrac{3}{x^2 - 2x}, \quad \dfrac{x}{x^2 - 4}$

SECTION 3

3.1 Simplify:
$\dfrac{2x}{x^2 + 3x - 10} - \dfrac{4}{x^2 + 3x - 10}$

3.2 Simplify: $\dfrac{2}{2x - 1} - \dfrac{3}{3x + 1}$

Review/Test

SECTION 4

4.1 Simplify: $\dfrac{1 + \frac{1}{x} - \frac{12}{x^2}}{1 + \frac{2}{x} - \frac{8}{x^2}}$

SECTION 5

5.1a Solve: $\frac{6}{x} - 2 = 1$

5.1b Solve: $\frac{2x}{x+1} - 3 = \frac{-2}{x+1}$

SECTION 6

6.1 Solve the proportion.
$\frac{3}{x+4} = \frac{5}{x+6}$

6.2 A salt water solution is formed by mixing 4 lb of salt with 10 gal of water. At this rate, how many additional pounds of salt are required for 15 gal of water?

SECTION 7

7.1a Solve $3x - 8y = 16$ for y.

7.1b Solve $d = s + rt$ for t.

SECTION 8

8.1 A pool can be filled with one pipe in 6 h, while a second pipe requires 12 h to fill the pool. How long would it take to fill the pool with both pipes turned on?

8.2 A small plane can fly at 110 mph in calm air. Flying with the wind the plane can fly 260 mi in the same amount of time as it can fly 180 mi against the wind. Find the rate of the wind.

Cumulative Review/Test

1. Simplify
$\left(\frac{2}{3}\right)^2 \div \left(\frac{3}{2} - \frac{2}{3}\right) + \frac{1}{2}$.

2. Evaluate $-a^2 + (a - b)^2$ when $a = -2$ and $b = 3$.

3. Simplify
$-2x - (-3y) + 7x - 5y$.

4. Simplify
$2[3x - 7(x - 3) - 8]$.

5. Solve: $4 - \frac{2}{3}x = 7$

6. Solve:
$3[x - 2(x - 3)] = 2(3 - 2x)$

7. Find $16\frac{2}{3}\%$ of 60.

8. Simplify $(a^2b^5)(ab^2)$.

9. Simplify $(a - 3b)(a + 4b)$.

10. Simplify $\frac{15b^4 - 5b^2 + 10b}{5b}$.

11. Simplify $(x^3 - 8) \div (x - 2)$.

12. Factor $12x^2 - x - 1$.

13. Factor $y^2 - 7y + 6$.

14. Factor $2a^3 + 7a^2 - 15a$.

15. Factor $4b^2 - 100$.

16. Solve:
$(x + 3)(2x - 5) = 0$

Cumulative Review/Test

17. Simplify: $\frac{12x^4y^2}{18xy^7}$

18. Simplify: $\frac{x^2 - 7x + 10}{25 - x^2}$

19. Simplify:
$\frac{x^2 - x - 56}{x^2 + 8x + 7} \div \frac{x^2 - 13x + 40}{x^2 - 4x - 5}$

20. Simplify: $\frac{2}{2x - 1} - \frac{1}{x + 1}$

21. Simplify: $\frac{1 - \frac{2}{x} - \frac{15}{x^2}}{1 - \frac{25}{x^2}}$

22. Solve: $\frac{3x}{x - 3} - 2 = \frac{10}{x - 3}$

23. Solve the proportion.

$\frac{2}{x - 2} = \frac{12}{x + 3}$

24. Solve $f = v + at$ for t.

25. Translate "the difference between five times a number and thirteen is the opposite of eight" into an equation and solve.

26. A silversmith mixes 60 g of an alloy which is 40% silver with 120 g of another silver alloy. The resulting alloy is 60% silver. Find the percent of silver in the 120 g alloy.

27. The length of the base of a triangle is 2 in. less than twice the height. The area of the triangle is 30 in.^{2}. Find the base and height of the triangle.

28. A life insurance policy costs \$16 for every \$1000 of coverage. At this rate, how much additional money would a policy of \$5000 cost?

29. One water pipe can fill a tank in 9 min while a second pipe requires 18 min to fill the tank. How long would it take both pipes working together to fill the tank?

30. The rower of a boat can row at a rate of 5 mph in calm water. Rowing with the current, the boat travels 14 mi in the same amount of time as it travels 6 mi against the current. Find the rate of the current.

8

Graphs and Linear Equations

Objectives

- To graph points on a rectangular coordinate system
- To determine a solution of a linear equation in two variables
- To solve application problems
- To graph an equation of the form $y = mx + b$
- To graph an equation of the form $Ax + By = C$
- To find the x- and y-intercepts of a straight line
- To find the slope of a straight line
- To graph a line using the slope and y-intercept
- To find the equation of a line using the equation $y = mx + b$
- To find the equation of a line using the point-slope formula

Magic Squares

A magic square is a square array of distinct integers so arranged that the numbers along any row, column, or main diagonal have the same sum. An example of a magic square is shown at the right.

8	3	4
1	5	9
6	7	2

The oldest known example of a magic square comes from China. Estimates are that this magic square is over 4000 years old. It is shown at the right.

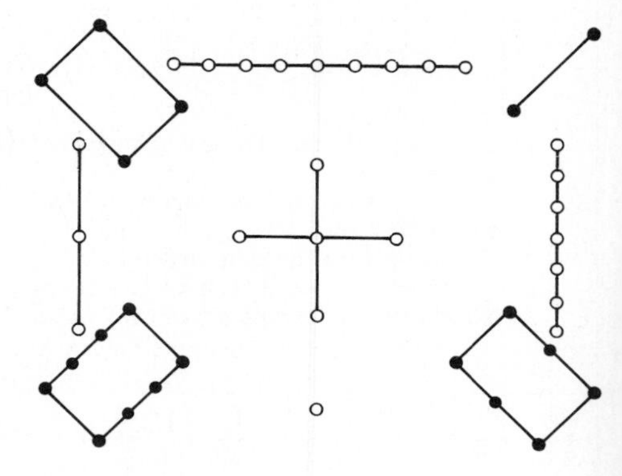

There is a simple way to produce a magic square with an odd number of cells. Start by writing a 1 in the top middle cell. The rule then is to proceed diagonally upward to the right with the successive integers.

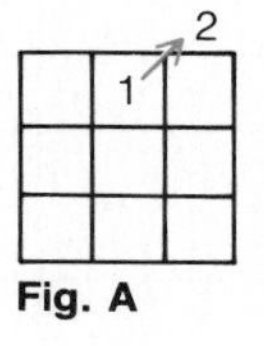

Fig. A

	1	
		2

Fig. B

When the rule takes you outside the square, write the number by either shifting across the square from right to left or down the square from top to bottom, as the case may be. For example, in Fig. B the second number (2) is outside the square above a column. As the '2' is above a column, it should be shifted down to the bottom cell in that column. In Fig. C, the '3' is outside the square to the right of a column and should therefore be shifted all the way to the left.

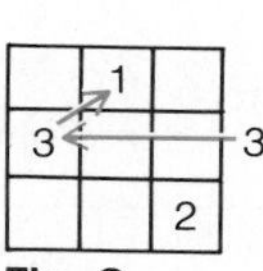

Fig. C

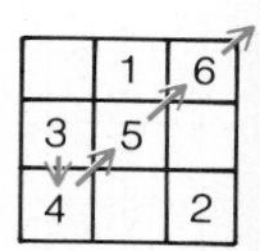

Fig. D

	1	6
3	5	7
4		2

Fig. E

8	1	6
3	5	7
4		2

Fig. F

If the rule takes you to a square which is already filled (as shown in Fig. D), then write the number in the cell directly below the last number written. Continue until the entire square is filled.

8	1	6
3	5	7
4	9	2

Fig. G

8	1	6
3	5	7
4	9	2

Fig. H

It is possible to begin a magic square with any integer and proceed by using the above rule and consecutive integers.

For an odd magic square beginning with one, the sum of a row, column, or diagonal is $\frac{n(n^2 + 1)}{2}$, where n is the number of rows.

SECTION 1 The Rectangular Coordinate System

1.1 Objective To graph points on a rectangular coordinate system

Reference for Computer Tutor™

A **rectangular coordinate system** is formed by two number lines, one horizontal and one vertical, that intersect at the zero point of each line. The point of intersection is called the **origin.** The two lines are called the **coordinate axes,** or simply **axes.**

The axes determine a plane and divide the plane into four regions, called **quadrants.** The quadrants are numbered counterclockwise from I to IV.

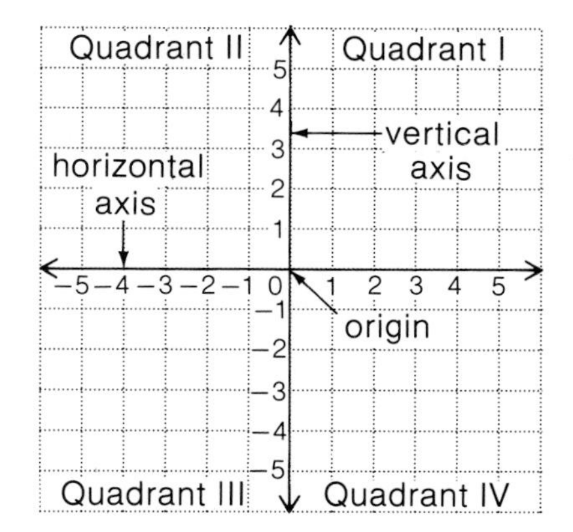

Each point in the plane can be identified by a pair of numbers called an **ordered pair.** The first number of the pair measures a horizontal distance and is called the **abscissa.** The second number of the pair measures a vertical distance and is called the **ordinate.** The **coordinates** of a point are the numbers in the ordered pair associated with the point.

horizontal distance → vertical distance
ordered pair → (3 , 4)
abscissa ← ordinate

The **graph of an ordered pair** is a point in the plane. The graphs of the points (2,3) and (3,2) are shown at the right. Notice that they are different points. The order in which the numbers in an ordered pair appear *is* important.

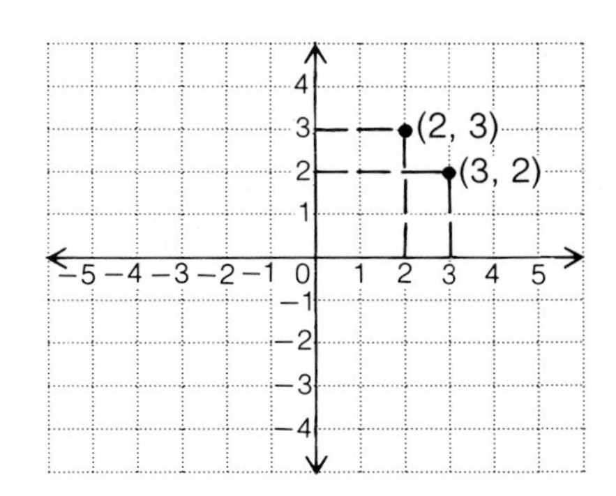

Example 1 Graph the ordered pairs (−2,−3), (3,−2), (1,3), and (4,1).

Solution

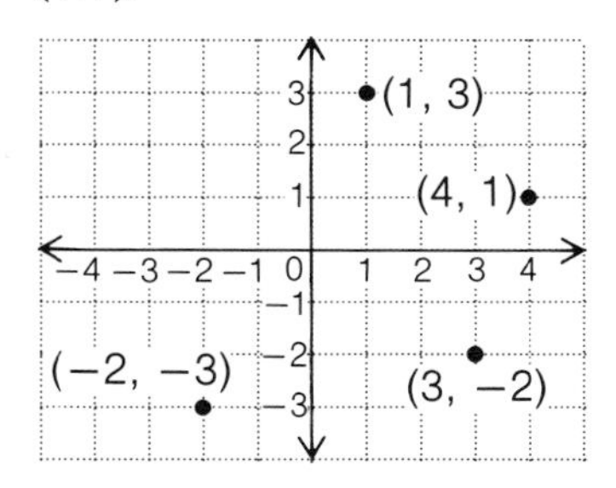

Example 2 Graph the ordered pairs (−1,3), (1,4), (−4,0), and (−2,−1).

Your solution

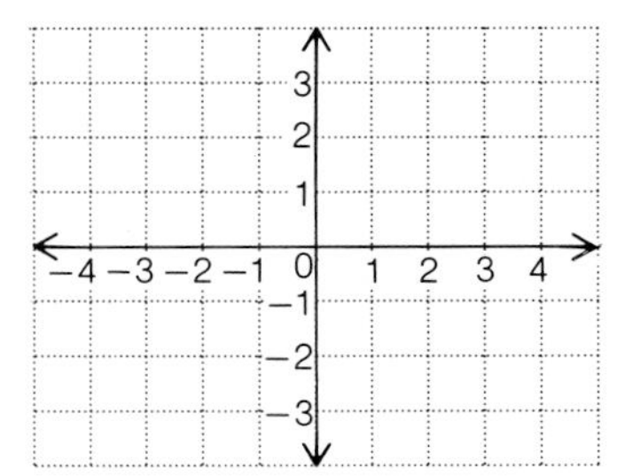

Solution on p. A45

Example 3 Find the coordinates of each of the points.

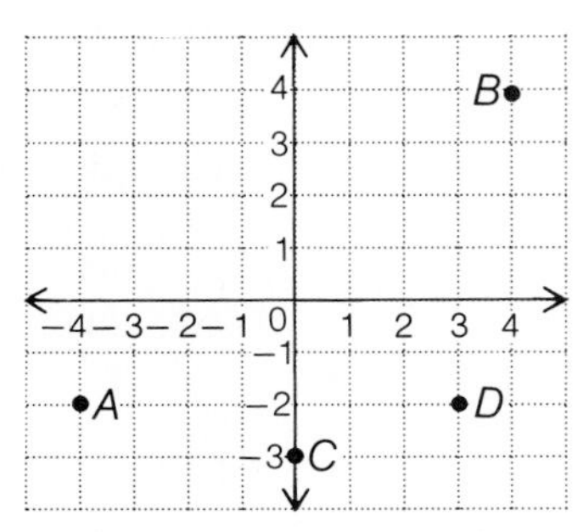

Solution
$A(-4,-2)$
$B(4,4)$
$C(0,-3)$
$D(3,-2)$

Example 4 Find the coordinates of each of the points.

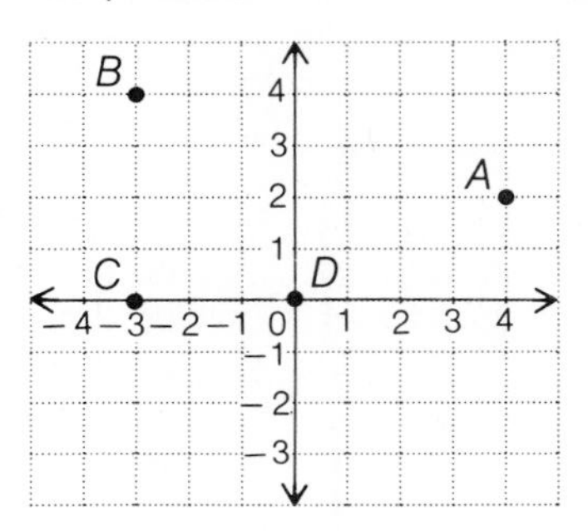

Your solution

Example 5
a) Name the abscissas of points A and C.
b) Name the ordinates of points B and D.

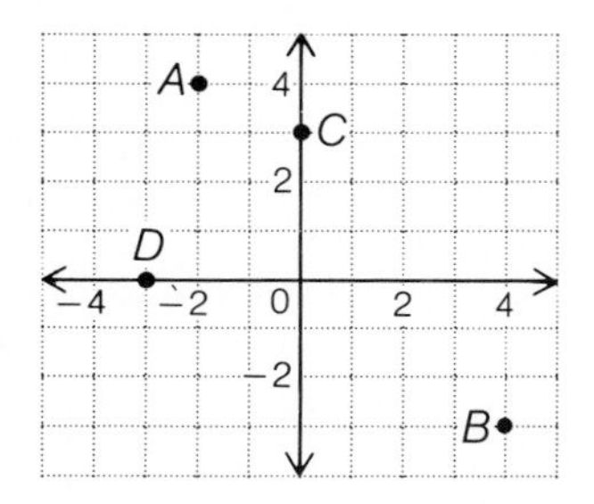

Solution
a) Abscissa of point A: -2
Abscissa of point C: 0
b) Ordinate of point B: -3
Ordinate of point D: 0

Example 6
a) Name the abscissas of points A and C.
b) Name the ordinates of points B and D.

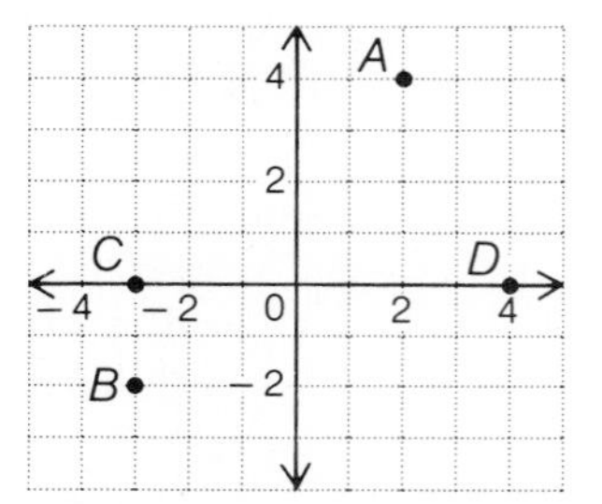

Your solution

1.2 Objective To determine a solution of a linear equation in two variables

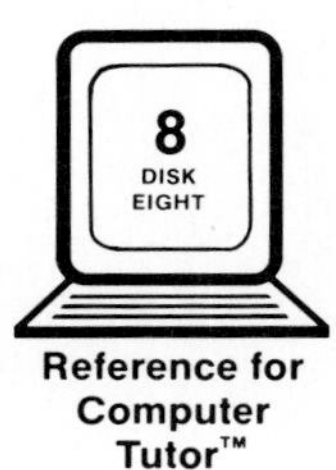

Reference for Computer Tutor™

An equation of the form $y = mx + b$, where m and b are constants, is a **linear equation in two variables.** Examples of linear equations are shown at the right. Note that the exponent of each variable is always one.

$y = 3x + 4$	$(m = 3, \quad b = 4)$
$y = 2x - 3$	$(m = 2, \quad b = -3)$
$y = -\frac{2}{3}x + 1$	$(m = -\frac{2}{3}, b = 1)$
$y = -2x$	$(m = -2, \quad b = 0)$
$y = x + 2$	$(m = 1, \quad b = 2)$

A **solution of an equation in two variables** is an ordered pair of numbers (x,y) which makes the equation a true statement.

Is $(1,-2)$ a solution of $y = 3x - 5$?

Replace x with 1, the abscissa.
Replace y with -2, the ordinate.

$$\begin{array}{c|c} \multicolumn{2}{c}{y = 3x - 5} \\ \hline -2 & 3(1) - 5 \\ & 3 - 5 \end{array}$$

$$-2 = -2$$

Compare the results. If the results are equal, the given ordered pair is a solution. If the results are not equal, the given ordered pair is not a solution.

Yes, $(1,-2)$ is a solution of the equation $y = 3x - 5$.

Besides the ordered pair $(1,-2)$, there are many other ordered pair solutions of the equation $y = 3x - 5$. For example, the method used above can be used to show that $(2,1)$, $(-1,-8)$, $\left(\frac{2}{3},-3\right)$ and $(0,-5)$ are also solutions.

In general, a linear equation in two variables has an infinite number of solutions. By choosing any value for x and substituting that value into the linear equation, a corresponding value of y can be found.

Find the ordered pair solution of $y = 2x - 5$ corresponding to $x = 1$.

Substitute 1 for x.
Solve for y.

$$\begin{aligned} y &= 2x - 5 \\ &= 2 \cdot 1 - 5 \\ &= 2 - 5 \\ &= -3 \end{aligned}$$

The ordered pair solution is $(1,-3)$.

Example 7 Is $(-3,2)$ a solution of $y = 2x + 2$?

Solution

$$\begin{array}{c|c} \multicolumn{2}{c}{y = 2x + 2} \\ \hline 2 & 2(-3) + 2 \\ & -6 + 2 \\ & -4 \end{array}$$

$$2 \neq -4$$

No, $(-3,2)$ is not a solution of $y = 2x + 2$.

Example 8 Is $(2,-4)$ a solution of $y = -\frac{1}{2}x - 3$?

Your solution

Example 9 Find the ordered pair solution of $y = \frac{2}{3}x - 1$ corresponding to $x = 3$.

Solution

$$\begin{aligned} y &= \tfrac{2}{3}x - 1 \\ &= \tfrac{2}{3}(3) - 1 \\ &= 2 - 1 \\ &= 1 \end{aligned}$$

The ordered pair solution is $(3,1)$.

Example 10 Find the ordered pair solution of $y = -\frac{1}{4}x + 1$ corresponding to $x = 4$.

Your solution

Solutions on p. A45

1.3 Objective To solve application problems

Reference for Computer Tutor™

A rectangular coordinate system is frequently used in business, science, and mathematics to show a relationship between two variables. One variable is represented along the horizontal axis and the other variable is represented along the vertical axis.

For example, a physics student measured the speed of impact of a solid ball dropped from various heights above the ground. The results were recorded as the following ordered pairs: (1,8), (4,16), (9,24). The abscissa of an ordered pair is the height in feet and the ordinate of an ordered pair is the speed in feet per second. For example, the ordered pair (4,16) corresponds to dropping the ball from a height of 4 ft and recording the speed as 16 ft/s. The results are graphed at the right.

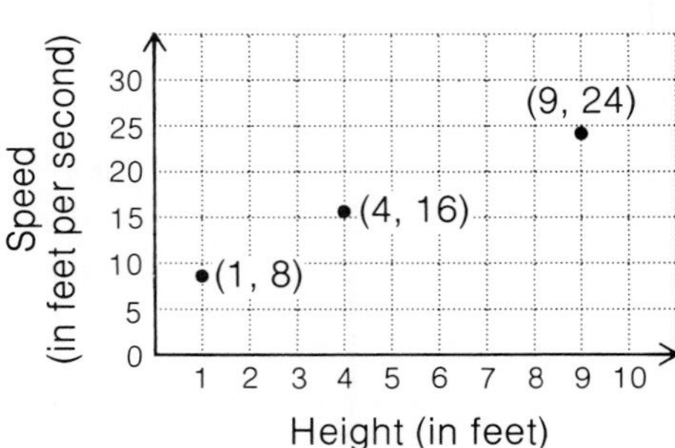

Example 11 A chemist measured the temperature of a chemical reaction at different times and recorded the results as the following ordered pairs: (5,10), (10,15), and (20,25). The abscissa of an ordered pair is time measured in minutes, and the ordinate of an ordered pair is temperature measured in degrees. Graph the ordered pairs on the rectangular coordinate system.

Strategy Graph the ordered pairs (5,10), (10,15), and (20,25).

Solution

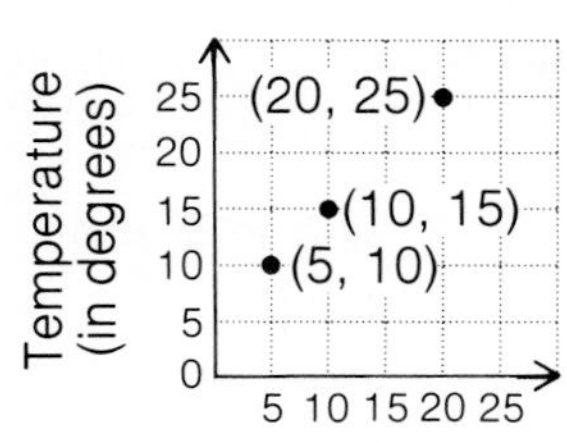

Example 12 A biochemist recorded the number of bacteria in a culture at different times as the following ordered pairs: (1,8), (3,10), and (4,11). The abscissa of an ordered pair is the time in hours, and the ordinate of an ordered pair is the number of bacteria. Graph the ordered pairs on the rectangular coordinate system.

Your strategy

Your solution

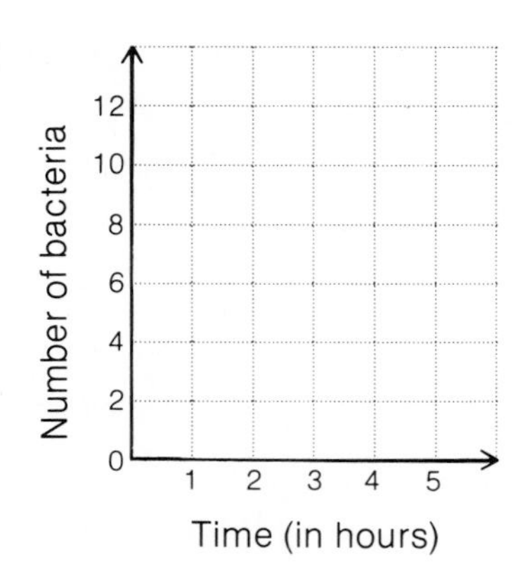

Solution on p. A45

1.1 Exercises

1. Graph the ordered pairs $(-2,1)$, $(3,-5)$, $(-2,4)$, and $(0,3)$.

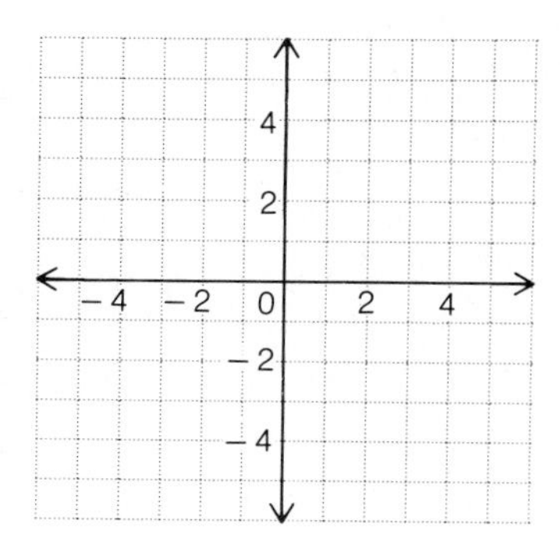

2. Graph the ordered pairs $(5,-1)$, $(-3,-3)$, $(-1,0)$, and $(1,-1)$.

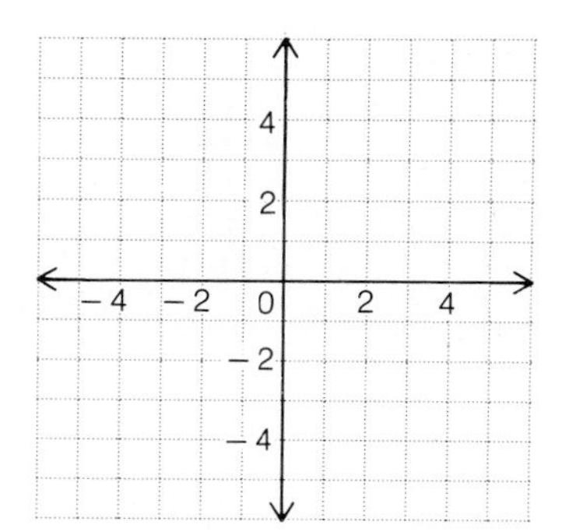

3. Graph the ordered pairs $(0,0)$, $(0,-5)$, $(-3,0)$, and $(0,2)$.

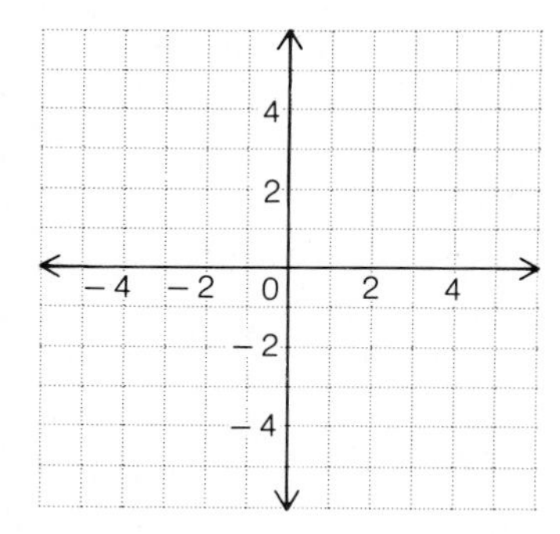

4. Graph the ordered pairs $(-4,5)$, $(-3,1)$, $(3,-4)$, and $(5,0)$.

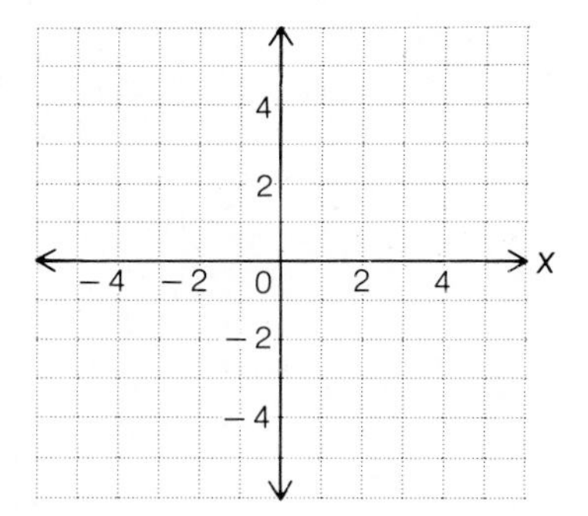

5. Graph the ordered pairs $(-1,4)$, $(-2,-3)$, $(0,2)$, and $(4,0)$.

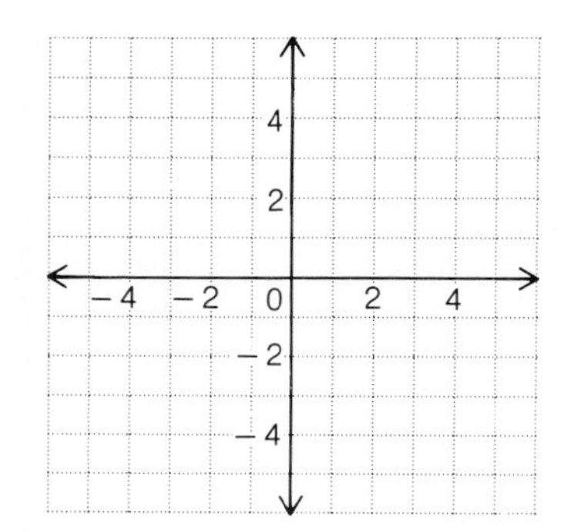

6. Graph the ordered pairs $(5,2)$, $(-4,-1)$, $(0,0)$, and $(0,3)$.

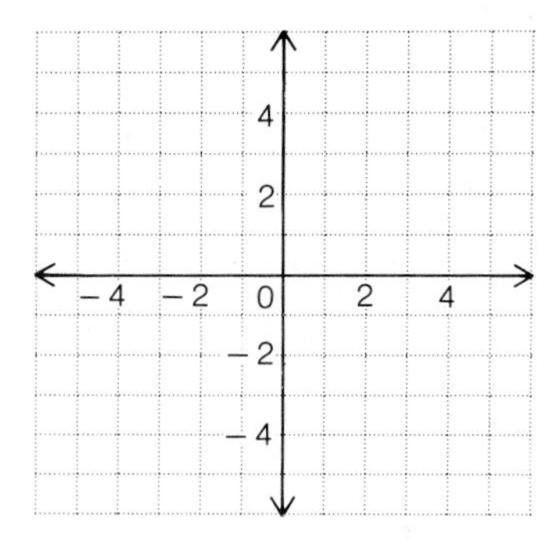

7. Find the coordinates of each of the points.

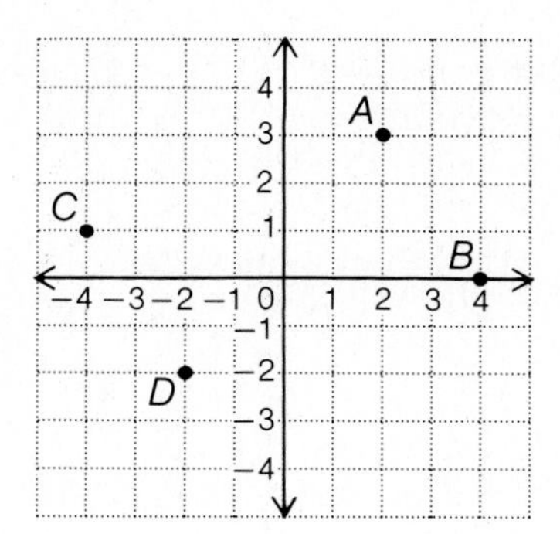

8. Find the coordinates of each of the points.

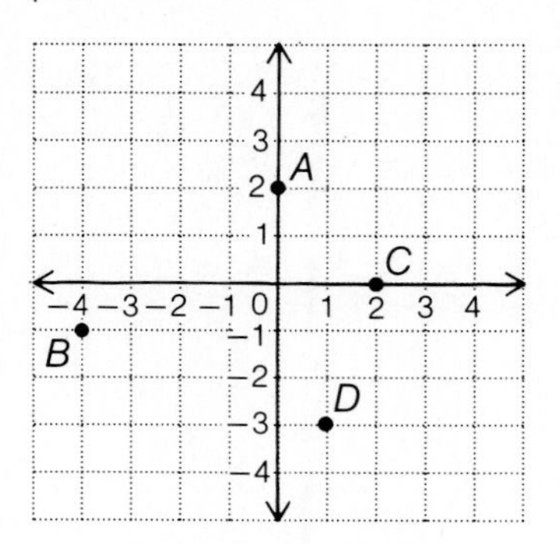

9. Find the coordinates of each of the points.

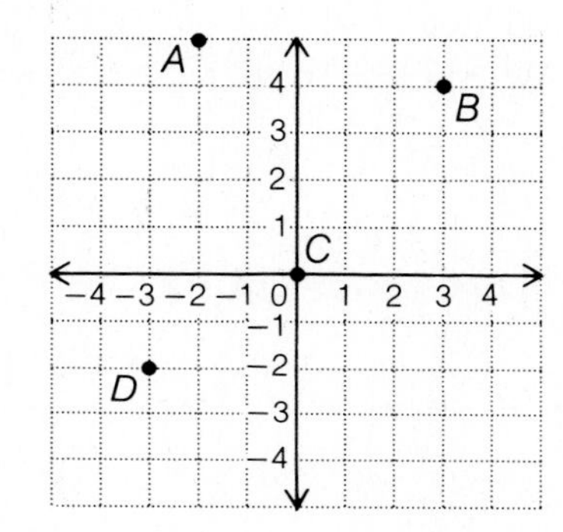

10. Find the coordinates of each of the points.

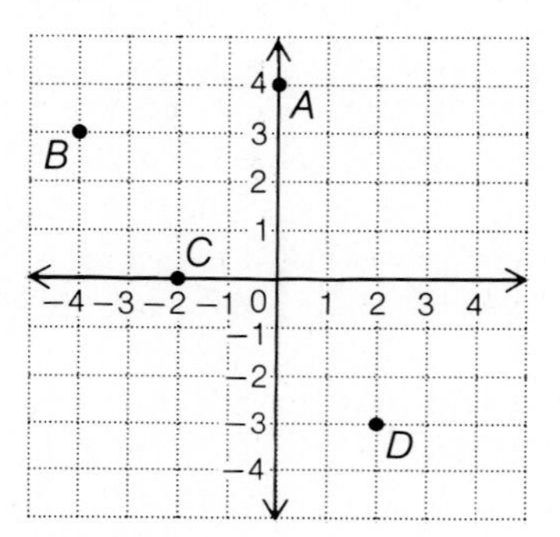

11. a) Name the abscissas of points *A* and *C*.
b) Name the ordinates of points *B* and *D*.

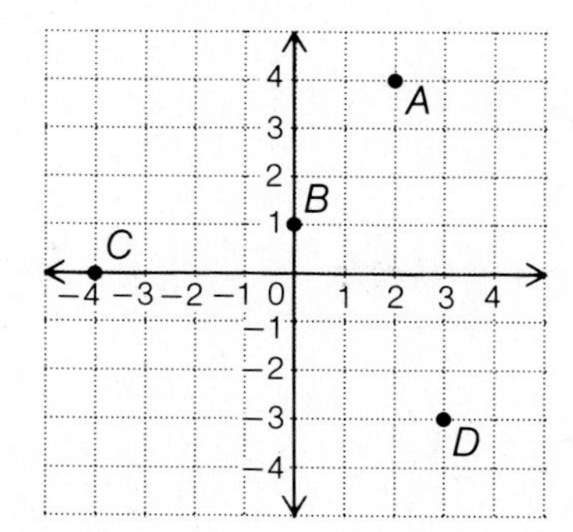

12. a) Name the abscissas of points *A* and *C*.
b) Name the ordinates of points *B* and *D*.

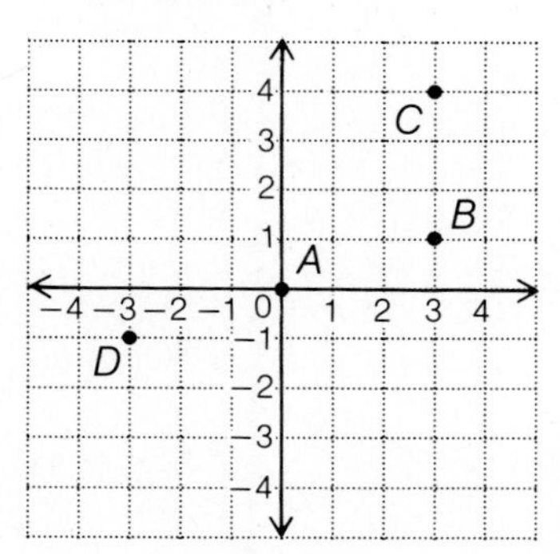

1.2 Exercises

13. Is (3,4) a solution of $y = -x + 7$?

14. Is $(2,-3)$ a solution of $y = x + 5$?

15. Is $(-1,2)$ a solution of $y = \frac{1}{2}x - 1$?

16. Is $(1,-3)$ a solution of $y = -2x - 1$?

17. Is (4,1) a solution of $y = \frac{1}{4}x + 1$?

18. Is $(-5,3)$ a solution of $y = -\frac{2}{5}x + 1$?

19. Is (0,4) a solution of $y = \frac{3}{4}x + 4$?

20. Is $(-2,0)$ a solution of $y = -\frac{1}{2}x - 1$?

21. Is (0,0) a solution of $y = 3x + 2$?

22. Is (0,0) a solution of $y = -\frac{3}{4}x$?

23. Find the ordered pair solution of $y = 3x - 2$ corresponding to $x = 3$.

24. Find the ordered pair solution of $y = 4x + 1$ corresponding to $x = -1$.

25. Find the ordered pair solution of $y = \frac{2}{3}x - 1$ corresponding to $x = 6$.

26. Find the ordered pair solution of $y = \frac{3}{4}x - 2$ corresponding to $x = 4$.

27. Find the ordered pair solution of $y = -3x + 1$ corresponding to $x = 0$.

28. Find the ordered pair solution of $y = \frac{2}{5}x - 5$ corresponding to $x = 0$.

29. Find the ordered pair solution of $y = \frac{2}{5}x + 2$ corresponding to $x = -5$.

30. Find the ordered pair solution of $y = -\frac{1}{6}x - 2$ corresponding to $x = 12$.

1.3 Application Problems

Solve.

1. A physics student, measuring the amount of frictional force on a smooth surface, recorded the following ordered pairs: (1,2), (2,3), (3,5), (3,6). The first component of an ordered pair is the distance in feet the object slid across the surface, and the second component is the force in pounds used to move the object. Graph the ordered pairs.

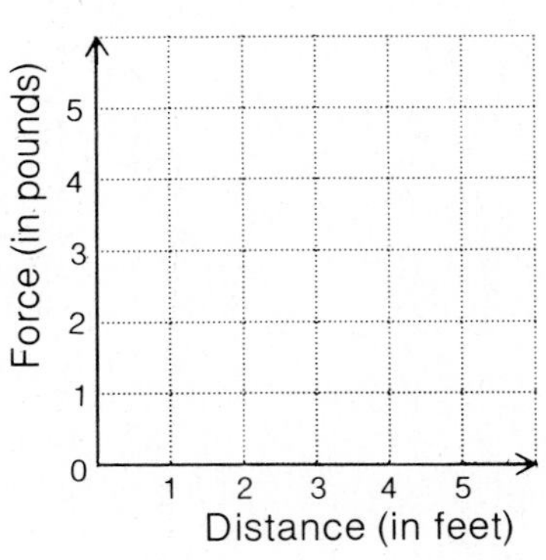

2. For ten consecutive winter days, a meteorologist measured the temperature in a town. The results were recorded as the following ordered pairs: (−3,3), (−1,2), (0,4), (1,1). The first component of an ordered pair is the temperature in degrees and the second component is the number of days that temperature was recorded during the ten-day period. Graph the ordered pairs.

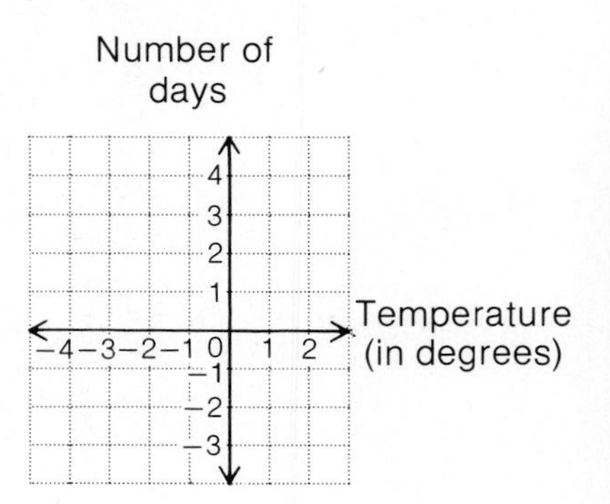

3. The magnification of an object at various distances from a lens was recorded as the following ordered pairs: (3,5), (5,10), (7,15), (9,20). The first component of an ordered pair is the magnification and the second component is the distance from the object in centimeters. Graph the ordered pairs.

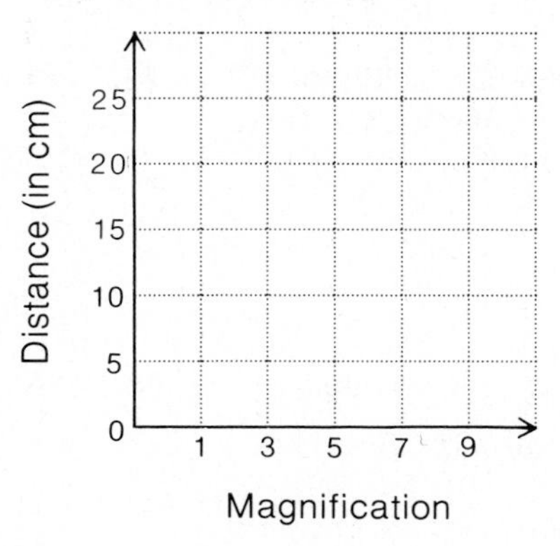

4. The speed of a ball as it rolls down a ramp is recorded as the following ordered pairs: (0,0), (1,2), (2,4), (3,6). The first component of an ordered pair is the time in seconds and the second component is the speed in feet per second. Graph the ordered pairs.

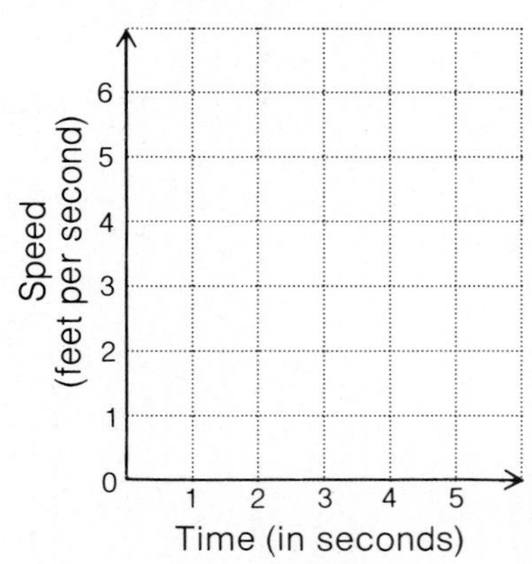

SECTION 2 Graphs of Straight Lines

2.1 Objective To graph an equation of the form $y = mx + b$

Reference for Computer Tutor™

The **graph of an equation in two variables** is a drawing of the ordered pair solutions of the equation. For a linear equation in two variables, the graph is a straight line.

To graph a linear equation, find ordered pair solutions of the equation. Do this by choosing any value of x and finding the corresponding value of y. Repeat this procedure, choosing different values for x, until you have found the number of solutions desired.

Since the graph of a linear equation in two variables is a straight line, and a straight line is determined by two points, it is necessary to find only two solutions. However, it is recommended that at least three points be used to insure accuracy.

Graph $y = 2x + 1$.

Choose any values of x and then find the corresponding values of y. The numbers 0, 2, and -1 were chosen arbitrarily for x. It is convenient to record these solutions in a table.

x	$y = 2x + 1$	y
0	$2 \cdot 0 + 1$	1
2	$2 \cdot 2 + 1$	5
-1	$2(-1) + 1$	-1

The horizontal axis is the x-axis. The vertical axis is the y-axis. Graph the ordered pair solutions (0,1), (2,5), and $(-1,-1)$. Draw a line through the ordered pair solutions.

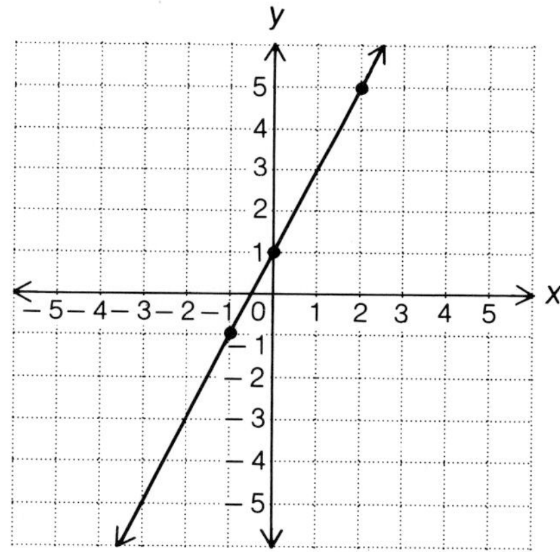

Remember that a graph is a drawing of the ordered pair solutions of the equation. Therefore, every point on the graph is a solution of the equation and every solution of the equation is a point on the graph.

Note that $(-2,-3)$ and (1,3) are points on the graph and that these points are solutions of the equation $y = 2x + 1$.

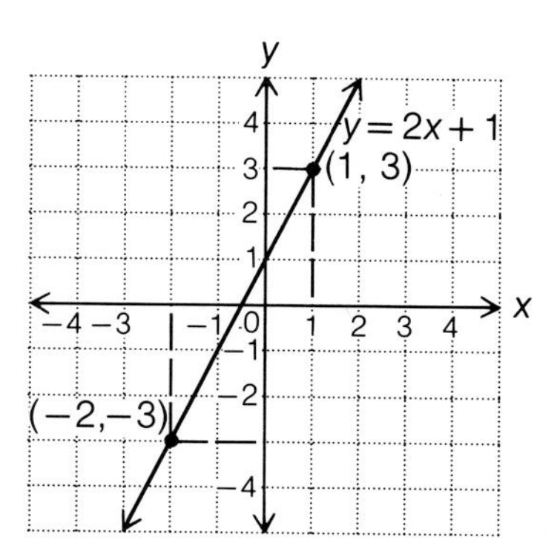

Graph $y = \frac{1}{3}x - 1$.

Step 1 Find at least three solutions.
When m is a fraction, choose values of x that will simplify the evaluation.
Display the ordered pairs in a table.

x	y
0	−1
3	0
−3	−2

Step 2 Graph the ordered pairs on a rectangular coordinate system and draw a straight line through the points.

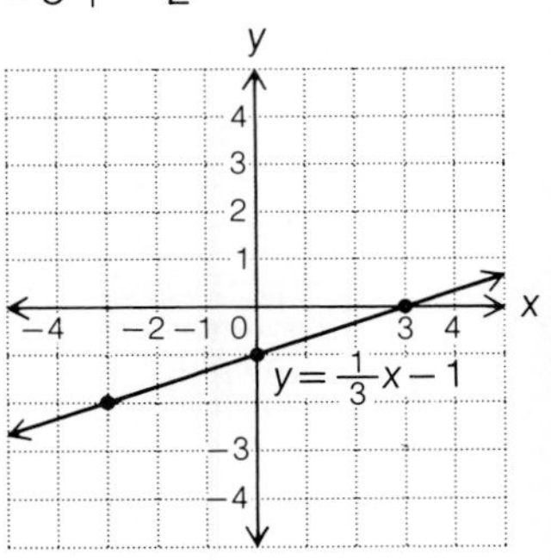

Example 1 Graph $y = 3x - 2$.

Solution

x	y
0	−2
1	1
−1	−5
2	4

Example 2 Graph $y = 3x + 1$.

Your solution

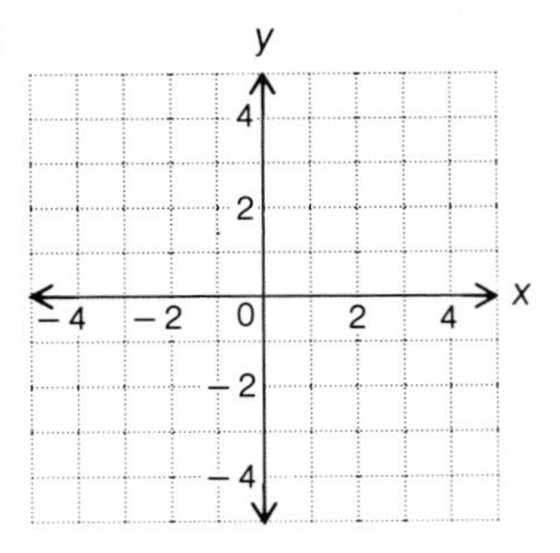

Example 3 Graph $y = 2x$.

Solution

x	y
0	0
2	4
−2	−4
1	2

Example 4 Graph $y = -2x$.

Your solution

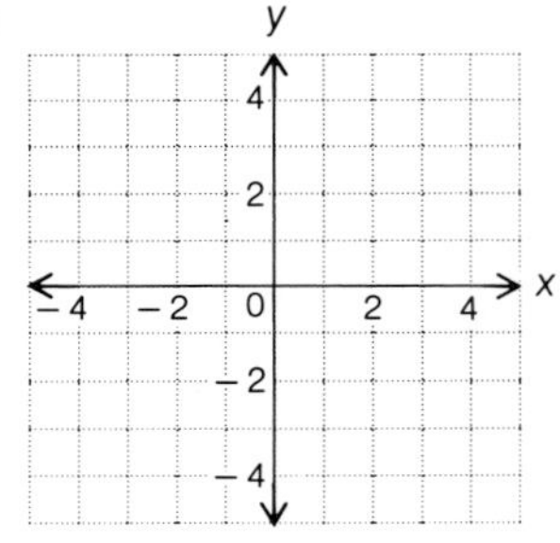

Example 5 Graph $y = \frac{1}{2}x - 1$.

Solution

x	y
0	−1
2	0
−2	−2
4	1

Example 6 Graph $y = \frac{1}{3}x - 3$.

Your solution

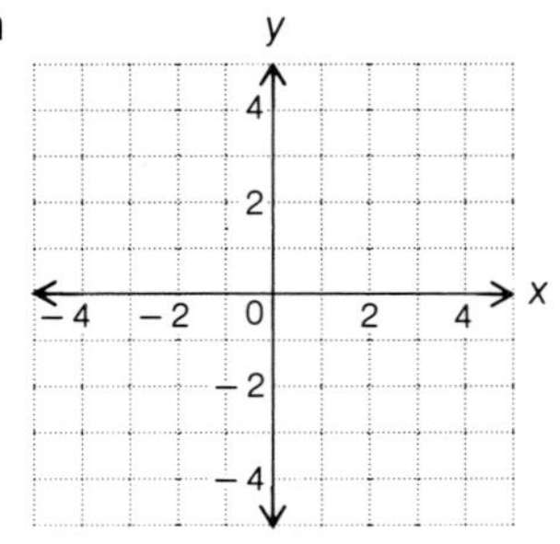

2.2 Objective To graph an equation of the form $Ax + By = C$

Reference for Computer Tutor™

An equation in the form $Ax + By = C$, where A, B, and C are constants, is also a linear equation. Examples of these equations are shown at the right.

$2x + 3y = 6 \quad (A = 2, B = 3, \quad C = 6)$
$x - 2y = -4 \quad (A = 1, B = -2, C = -4)$
$2x + y = 0 \quad (A = 2, B = 1, \quad C = 0)$
$4x - 5y = 2 \quad (A = 4, B = -5, C = 2)$

To graph an equation of the form $Ax + By = C$, first solve the equation for y. Then follow the same procedure used for graphing an equation of the form $y = mx + b$.

Graph $3x + 4y = 12$.

Step 1 Solve the equation for y.

$$3x + 4y = 12$$
$$4y = -3x + 12$$
$$y = -\frac{3}{4}x + 3$$

Step 2 Find at least three solutions. Display the ordered pairs in a table.

x	y
0	3
4	0
−4	6

Step 3 Graph the ordered pairs on a rectangular coordinate system and draw a straight line through the points.

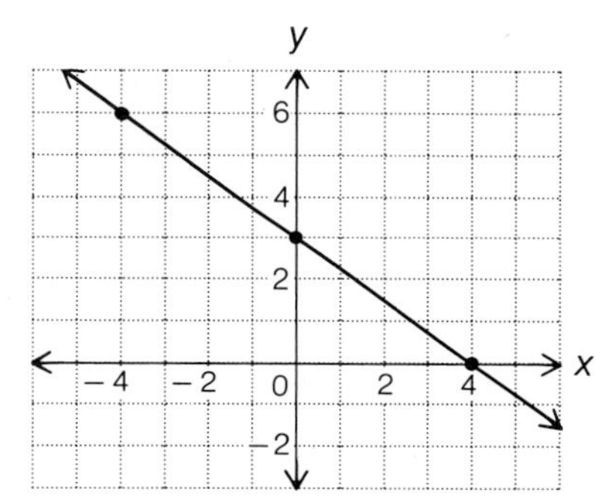

The graph of an equation in which one of the variables is missing is either a horizontal or a vertical line.

The equation $y = 2$ could be written $0 \cdot x + y = 2$. No matter what value of x is chosen, y is always 2. Some solutions to the equation are (3,2), (−1,2), (0,2), and (−4,2). The graph is shown at the right.

The **graph of $y = b$** is a horizontal line passing through point $(0,b)$.

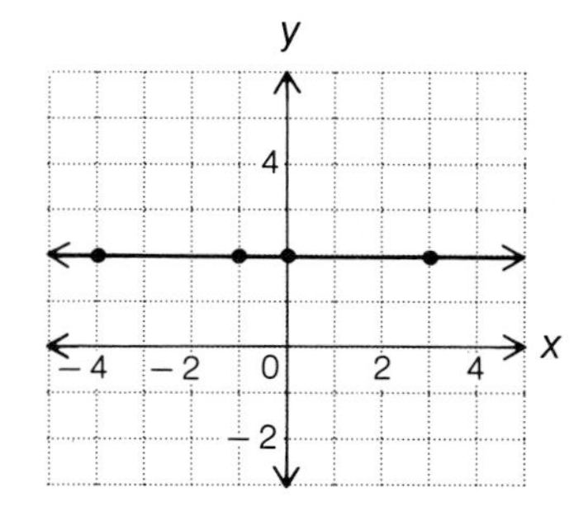

The equation $x = -2$ could be written $x + 0 \cdot y = -2$. No matter what value of y is chosen, x is always −2. Some solutions of the equation are (−2,3), (−2,−2), (−2,0), and (−2,2). The graph is shown at the right.

The **graph of $x = a$** is a vertical line passing through point $(a,0)$.

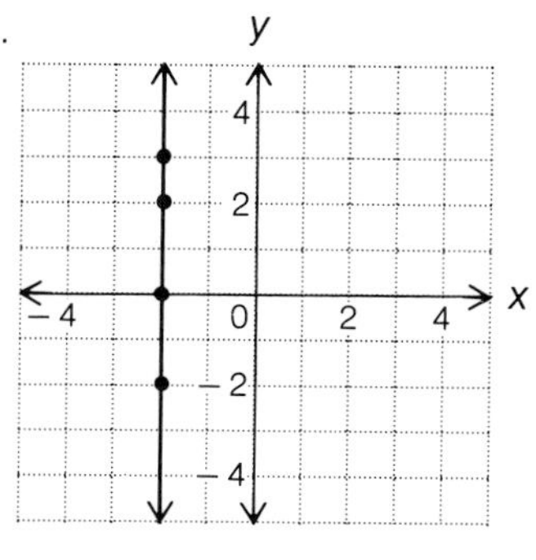

Example 7 Graph $2x - 5y = 10$.

Solution

$$2x - 5y = 10$$
$$-5y = -2x + 10$$
$$y = \frac{2}{5}x - 2$$

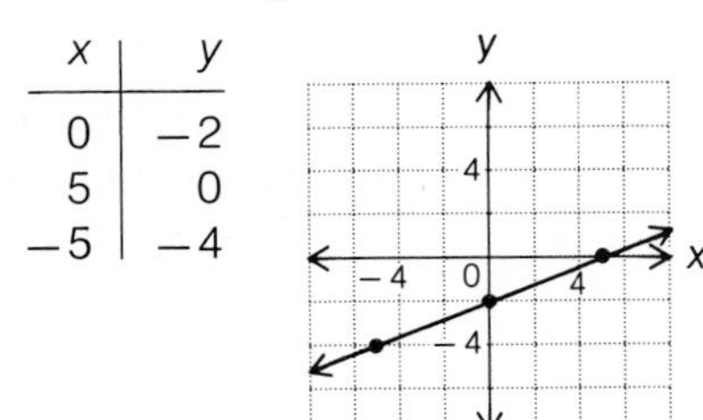

x	y
0	−2
5	0
−5	−4

Example 8 Graph $5x - 2y = 10$.

Your solution

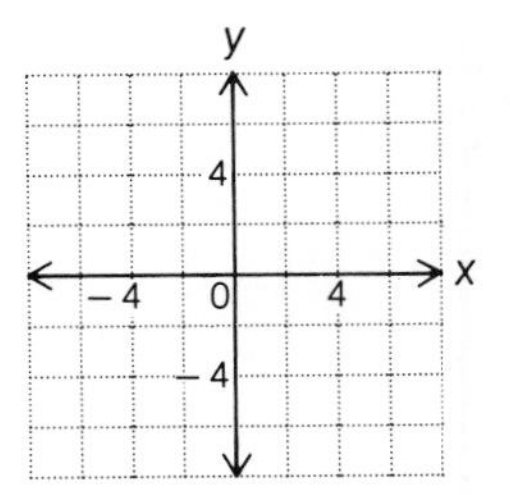

Example 9 Graph $x + 2y = 6$.

Solution

$$x + 2y = 6$$
$$2y = -x + 6$$
$$y = -\frac{1}{2}x + 3$$

x	y
0	3
2	2
−2	4
4	1

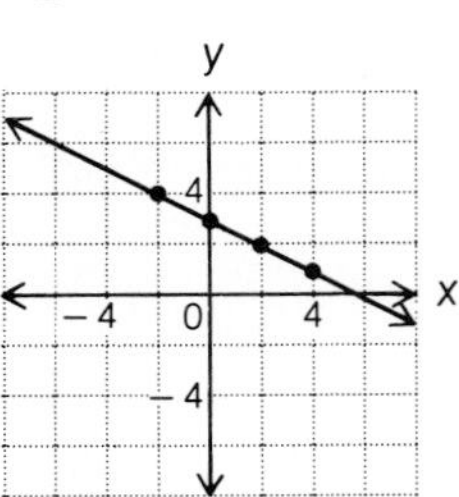

Example 10 Graph $x - 3y = 9$.

Your solution

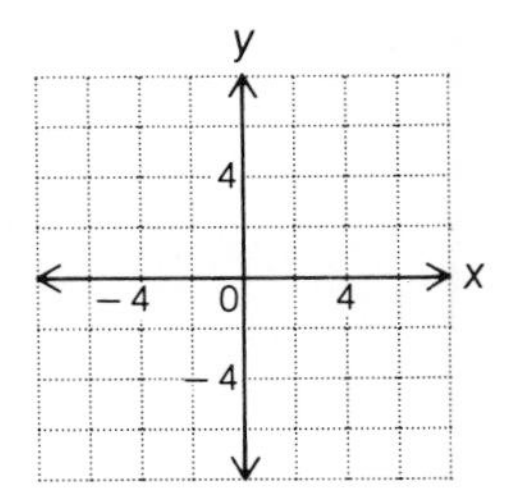

Example 11 Graph $y = -2$.

Solution The graph of an equation of the form $y = b$ is a horizontal line passing through point $(0,b)$.

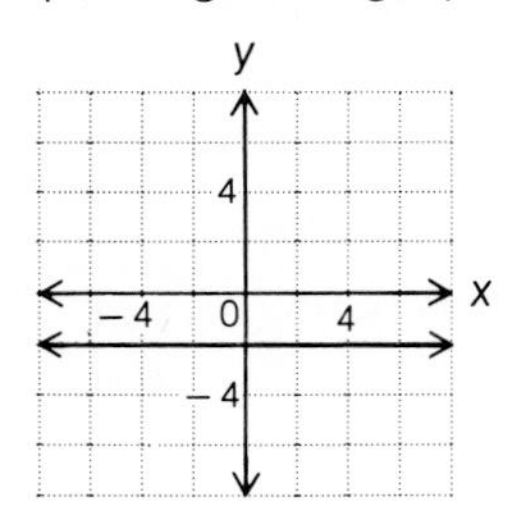

Example 12 Graph $y = 3$.

Your solution

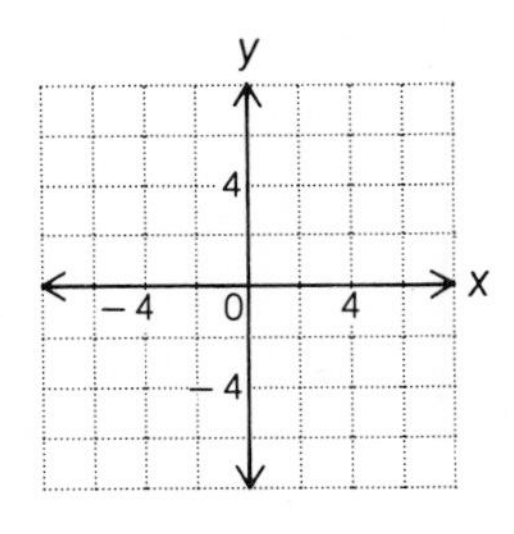

Example 13 Graph $x = 3$.

Solution The graph of an equation of the form $x = a$ is a vertical line passing through point $(a,0)$.

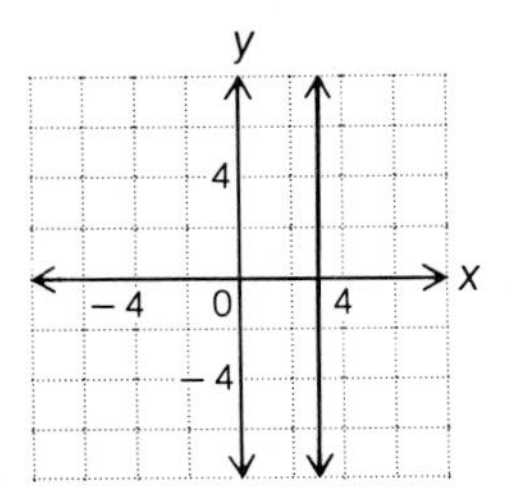

Example 14 Graph $x = -4$.

Your solution

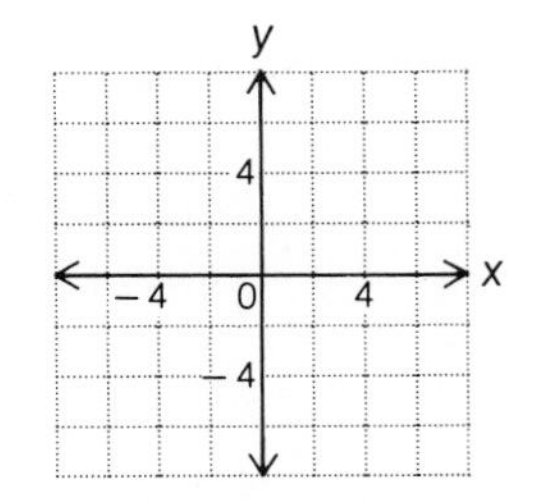

2.1 Exercises

Graph.

1. $y = 2x - 3$

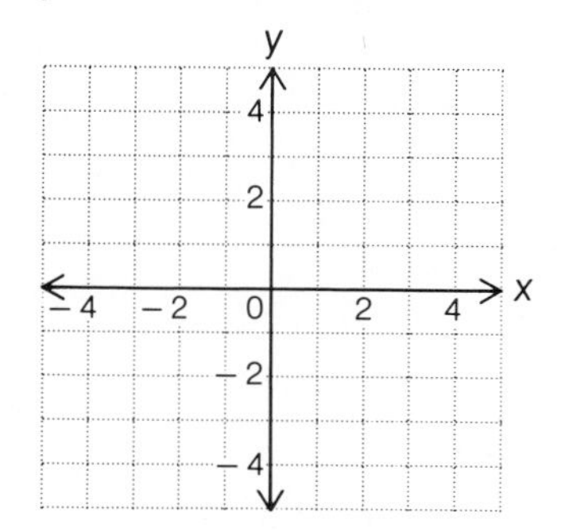

2. $y = -2x + 2$

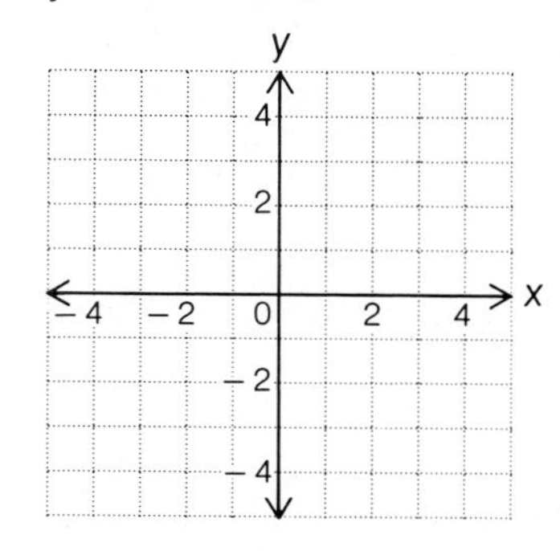

3. $y = \frac{1}{3}x$

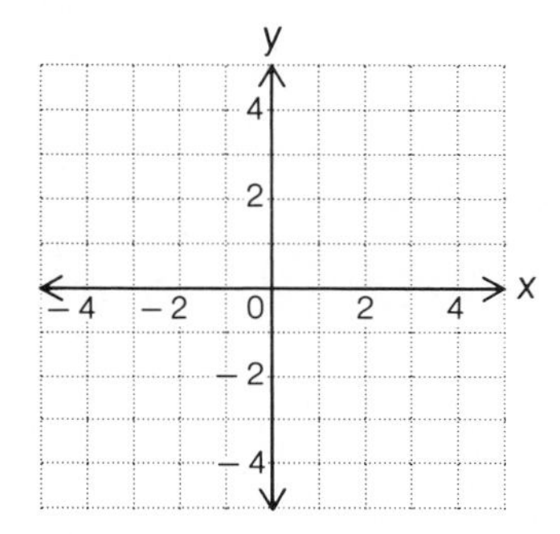

4. $y = -3x$

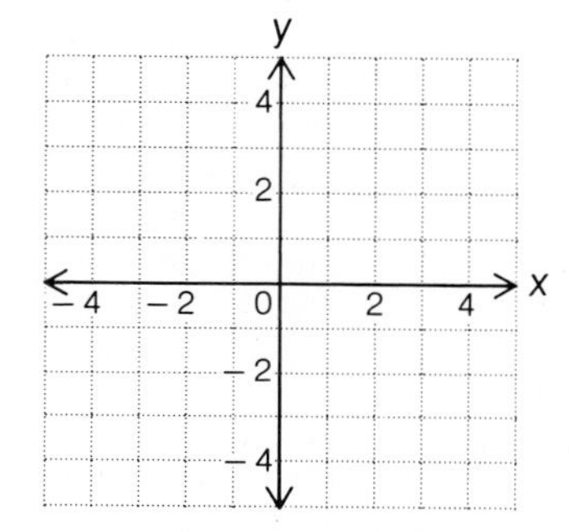

5. $y = \frac{2}{3}x - 1$

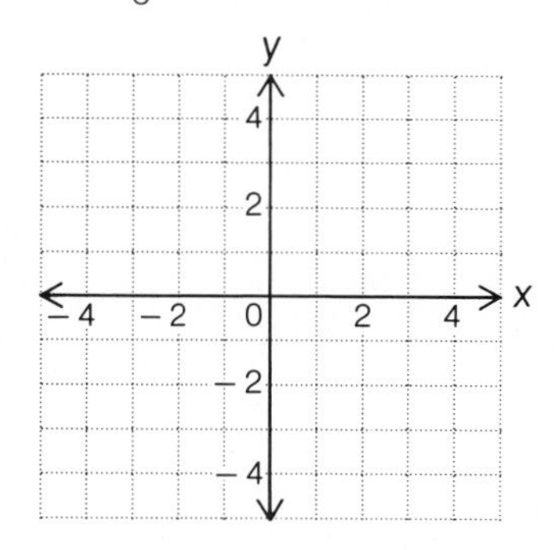

6. $y = \frac{3}{4}x + 2$

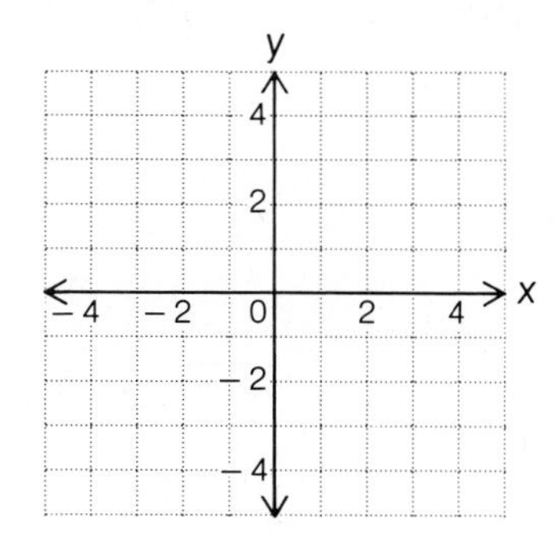

7. $y = -\frac{1}{4}x + 2$

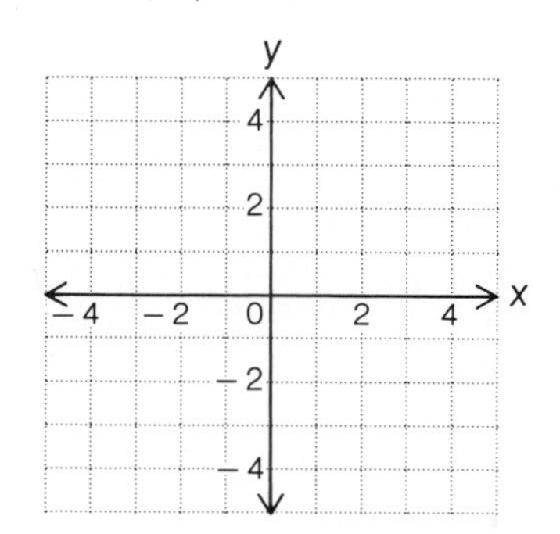

8. $y = -\frac{1}{3}x + 1$

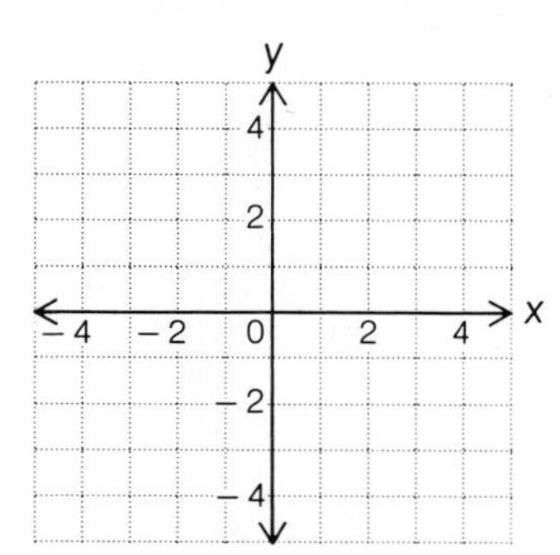

Graph.

9. $y = -\frac{2}{5}x + 1$

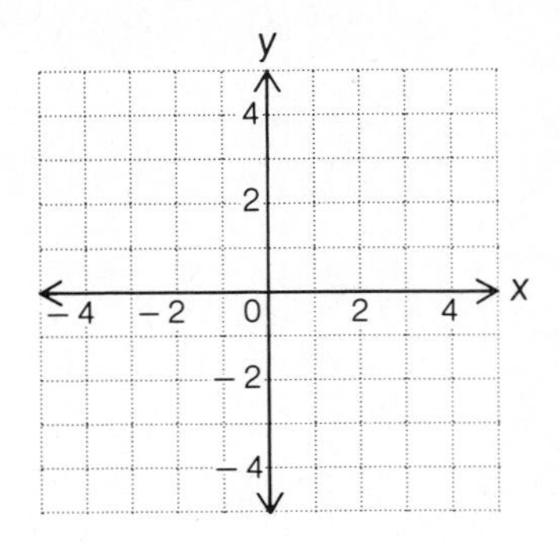

10. $y = -\frac{1}{2}x + 3$

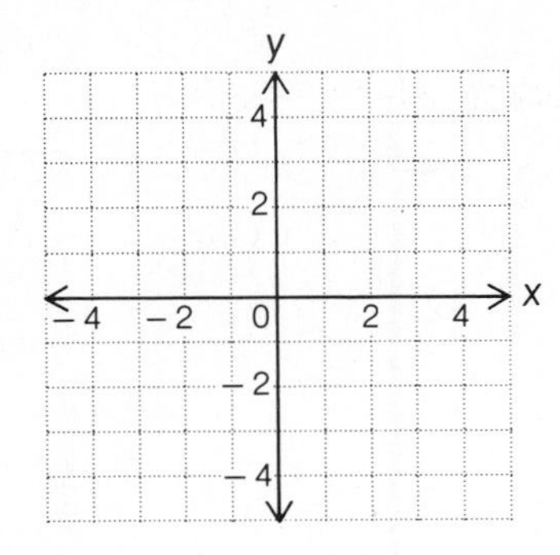

11. $y = 2x - 4$

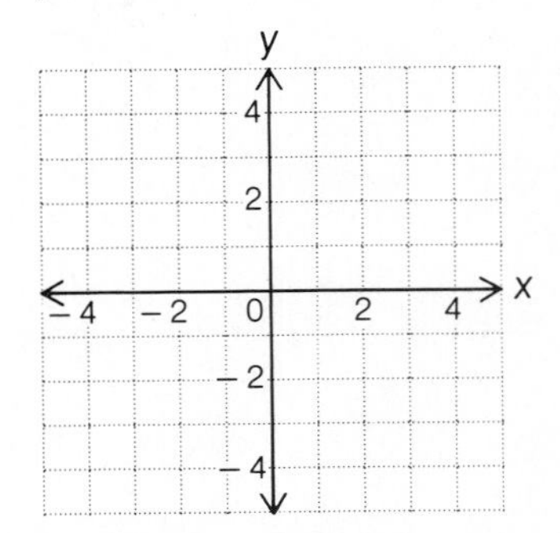

12. $y = 3x - 4$

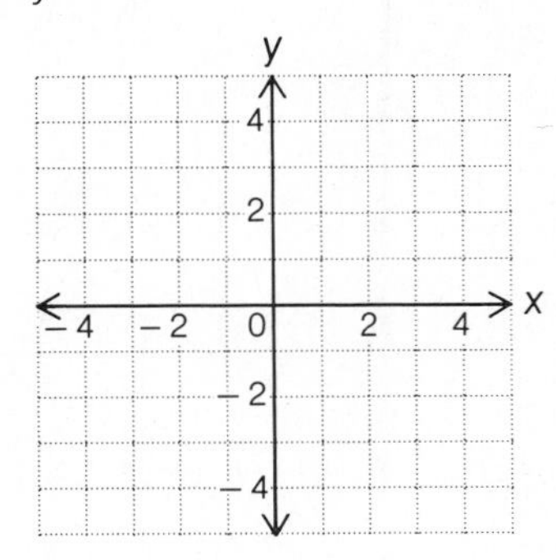

13. $y = -x + 2$

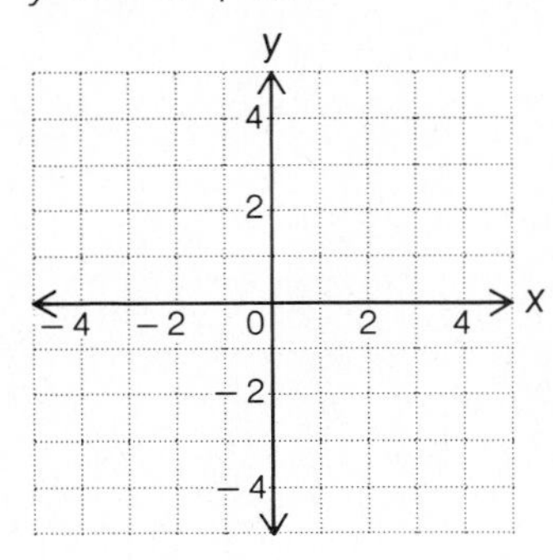

14. $y = -x - 1$

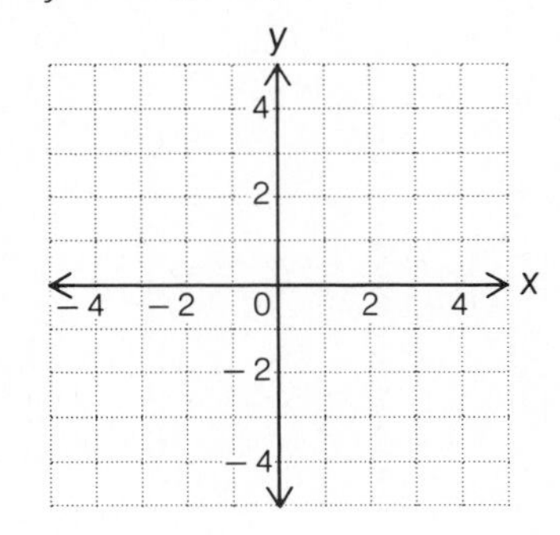

15. $y = -\frac{2}{3}x + 1$

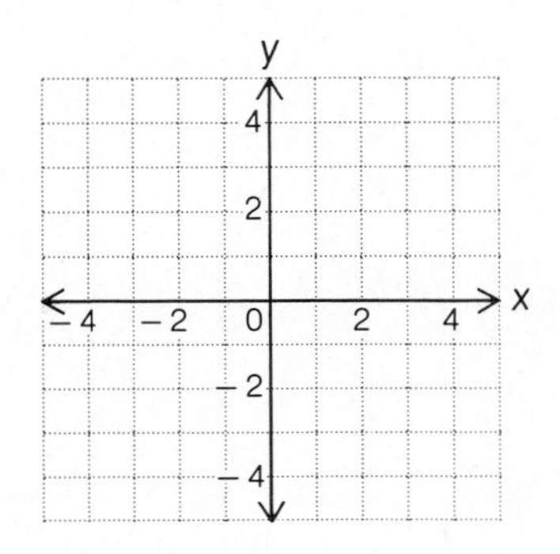

16. $y = 5x - 4$

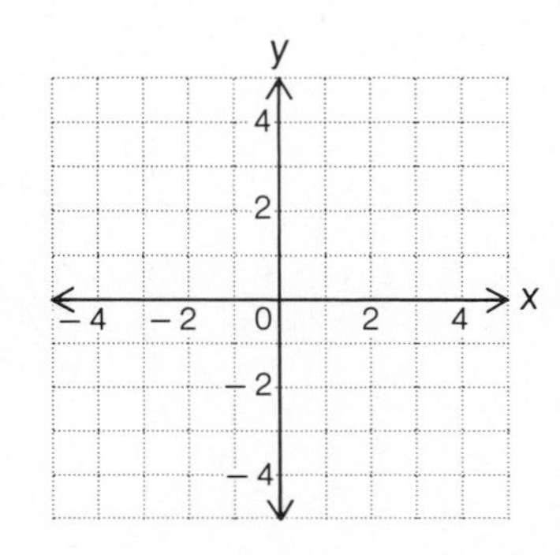

2.2 Exercises

Graph.

17. $3x + y = 3$

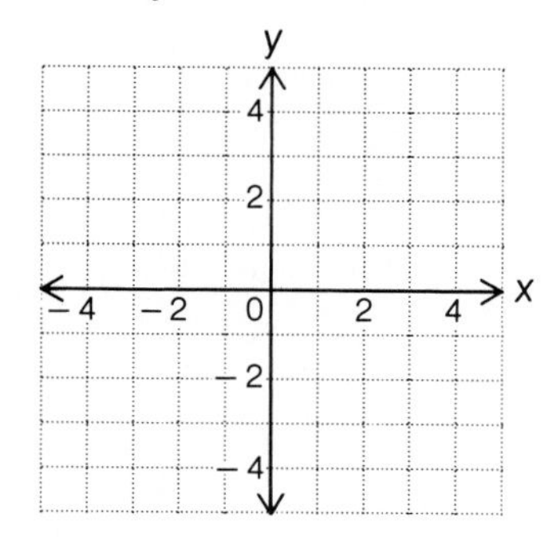

18. $2x + y = 4$

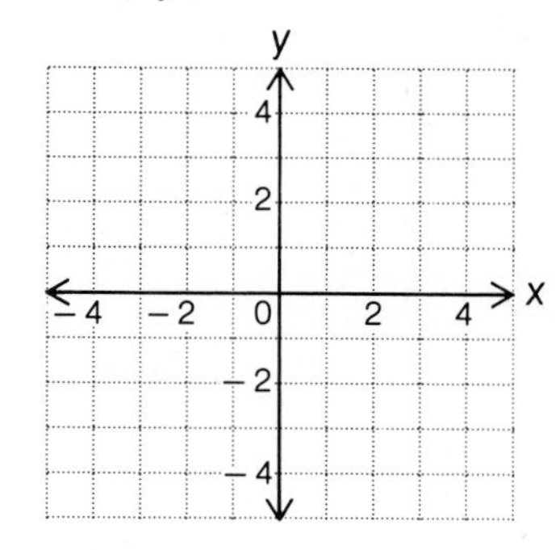

19. $2x + 3y = 6$

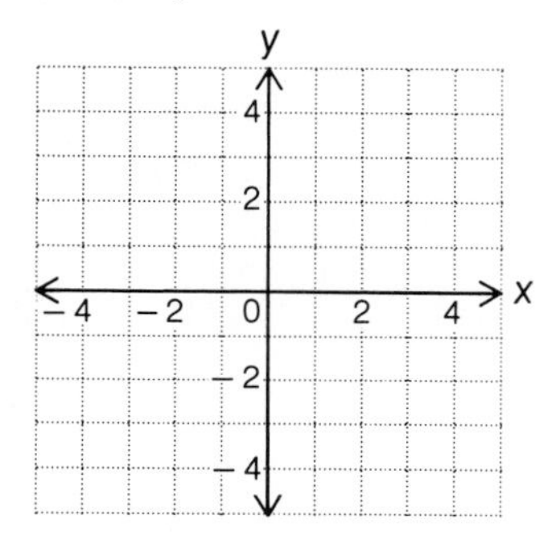

20. $3x + 2y = 4$

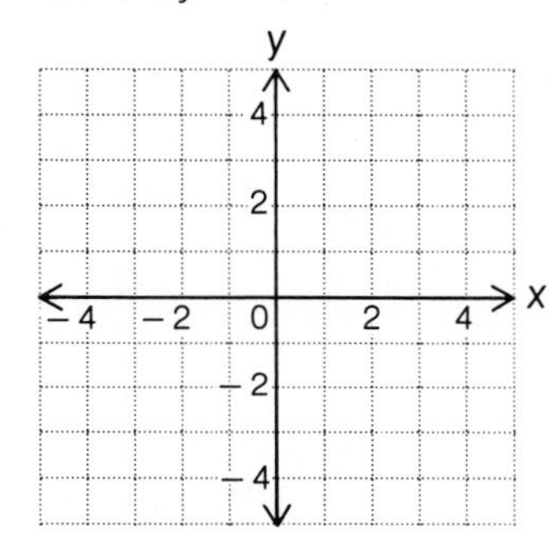

21. $x - 2y = 4$

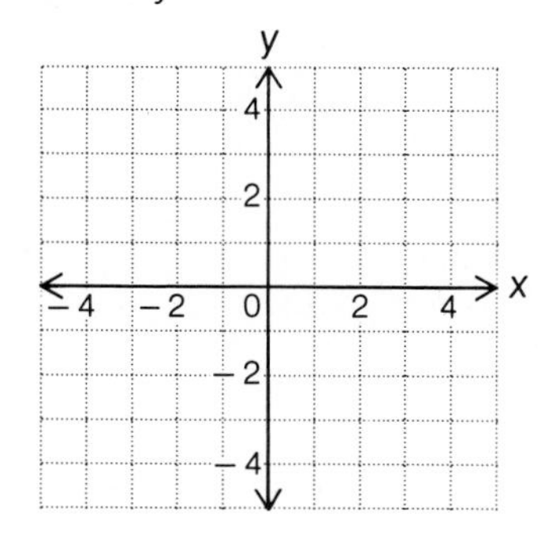

22. $x - 3y = 6$

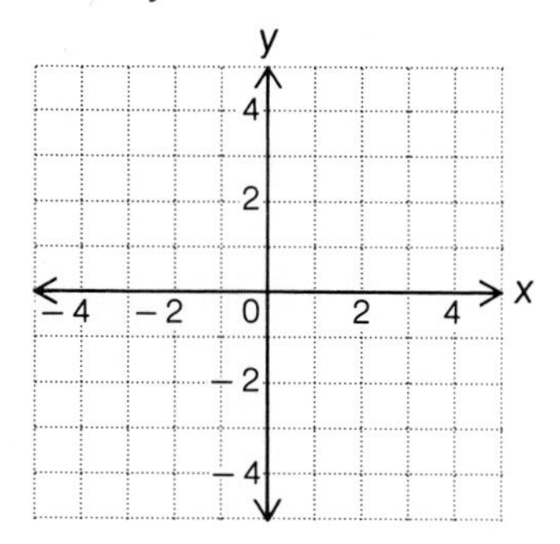

23. $y = 4$

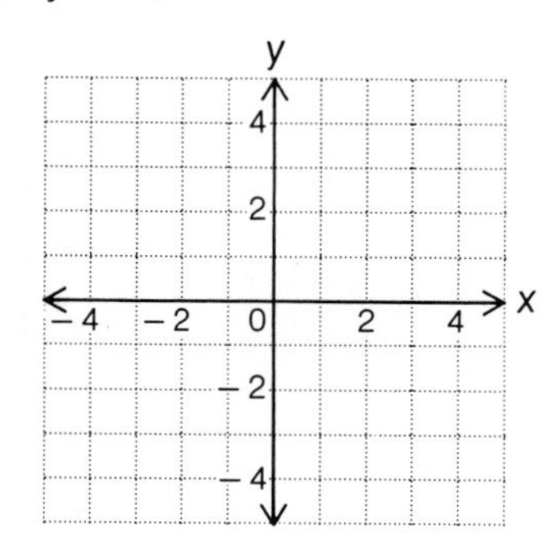

24. $x = -2$

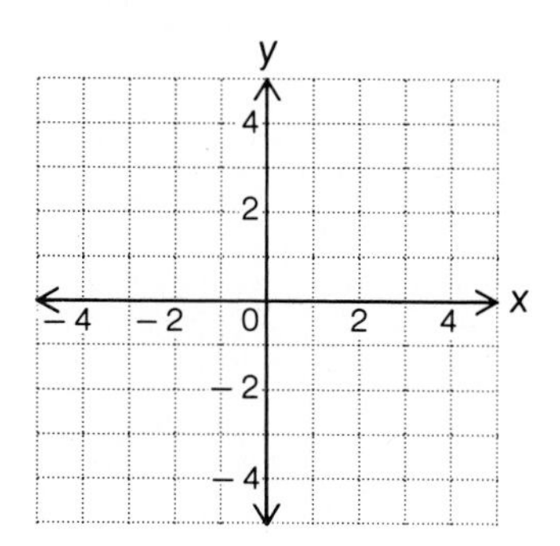

Graph.

25. $2x - 3y = 6$

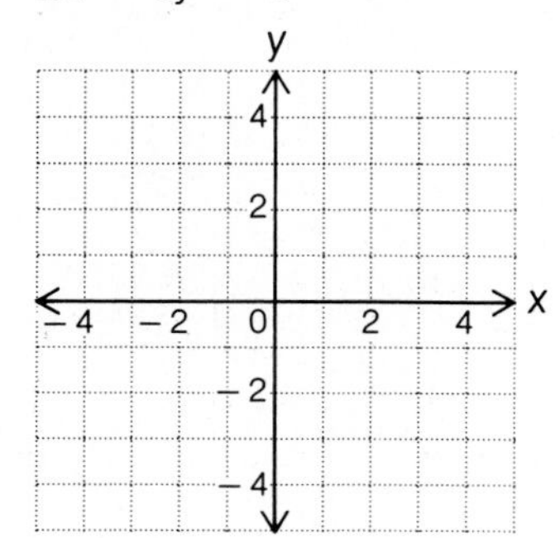

26. $3x - 2y = 8$

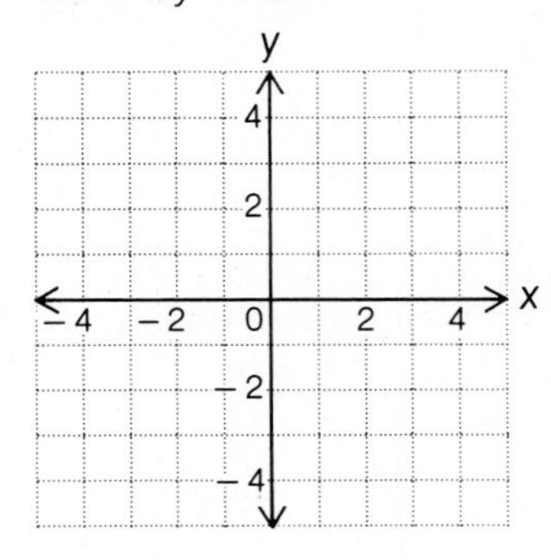

27. $2x + 5y = 10$

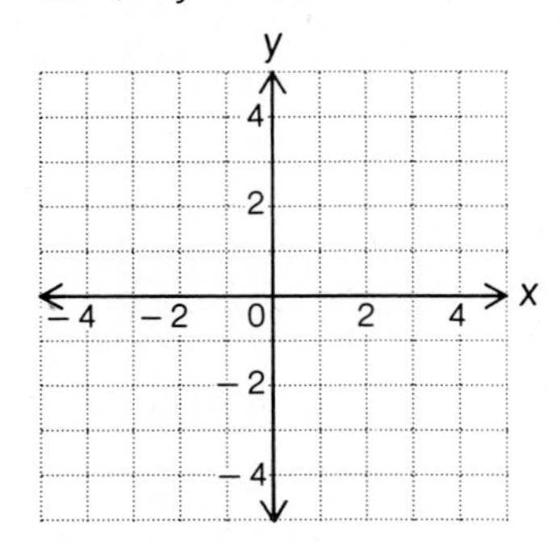

28. $3x + 4y = 12$

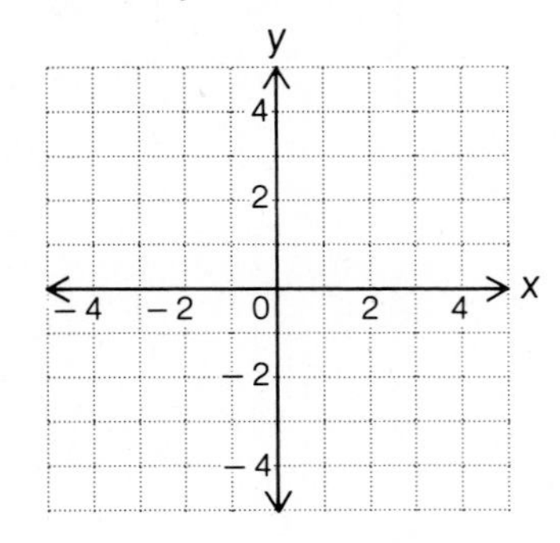

29. $x = 3$

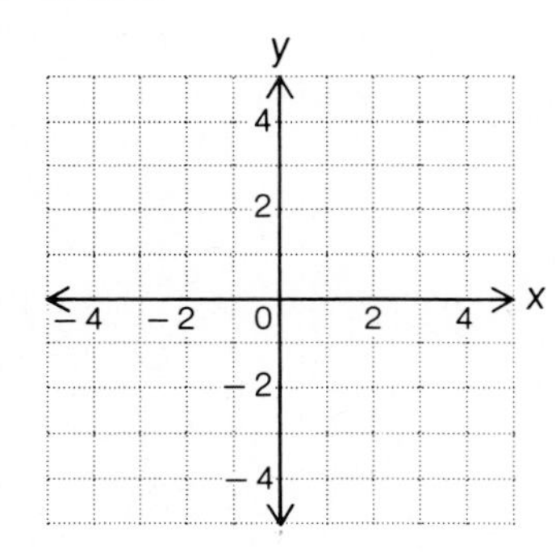

30. $y = -4$

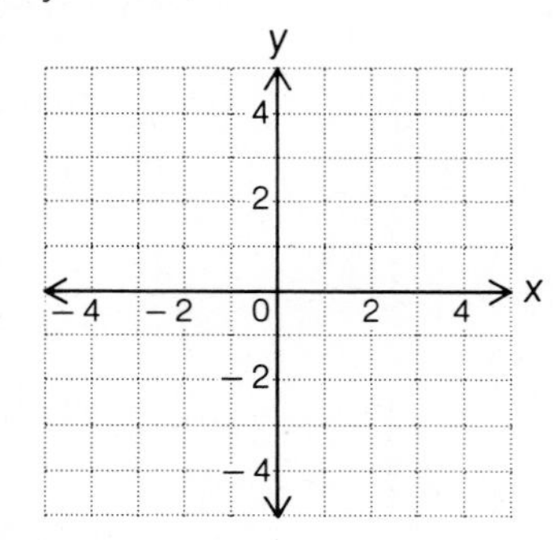

31. $x - 3y = 6$

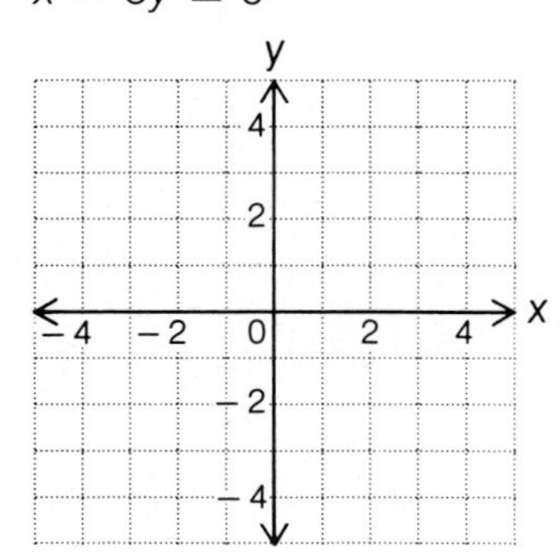

32. $4x - 3y = 12$

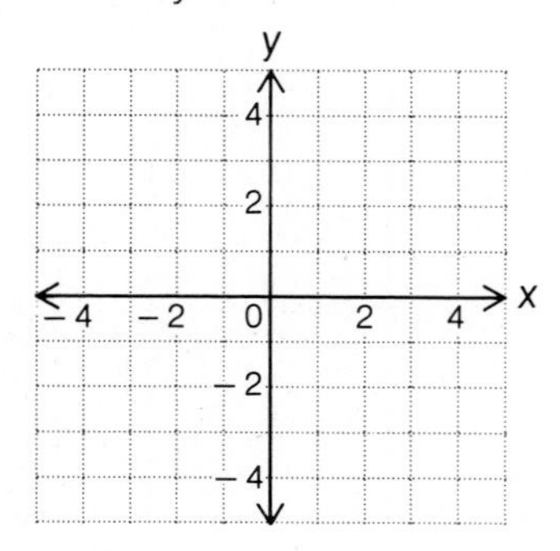

SECTION 3 Intercepts and Slopes of Straight Lines

3.1 Objective To find the x- and y-intercepts of a straight line

Reference for Computer Tutor™

The graph of the equation $2x + 3y = 6$ is shown at the right. The graph crosses the x-axis at the point (3,0). This point is called the ***x*-intercept.** The graph also crosses the y-axis at the point (0,2). This point is called the ***y*-intercept.**

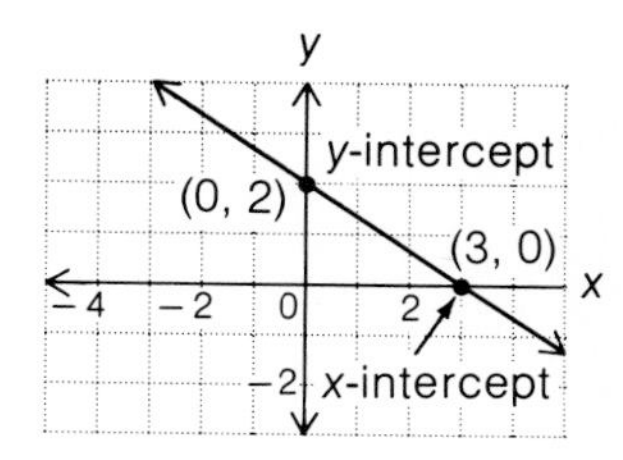

Find the x-intercept and the y-intercept of the graph of the equation $2x - 3y = 12$.

To find the x-intercept, let $y = 0$. (Any point on the x-axis has y-coordinate 0.)

$$2x - 3y = 12$$
$$2x - 3(0) = 12$$
$$2x = 12$$
$$x = 6$$

The x-intercept is (6,0).

To find the y-intercept, let $x = 0$. (Any point on the y-axis has x-coordinate 0.)

$$2x - 3y = 12$$
$$2(0) - 3y = 12$$
$$-3y = 12$$
$$y = -4$$

The y-intercept is (0,−4).

Find the y-intercept of $y = 3x + 4$.

Let $x = 0$.

$$y = 3x + 4$$
$$= 3(0) + 4$$
$$= 4$$

The y-intercept is (0,4).

For any equation of the form $y = mx + b$, the y-intercept is $(0,b)$.

A linear equation can be graphed by finding the x- and y-intercepts and then drawing a line through the two points.

Example 1 Find the x- and y-intercepts for $x - 2y = 4$. Graph the line.

Solution x-intercept:

$$x - 2y = 4$$
$$x - 2(0) = 4$$
$$x = 4$$

(4,0)

y-intercept:

$$x - 2y = 4$$
$$0 - 2y = 4$$
$$-2y = 4$$
$$y = -2$$

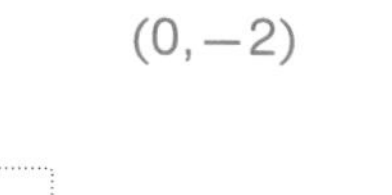

(0,−2)

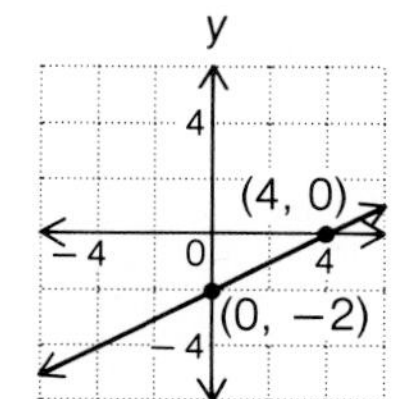

Example 2 Find the x- and y-intercepts for $4x - y = 4$. Graph the line.

Your solution

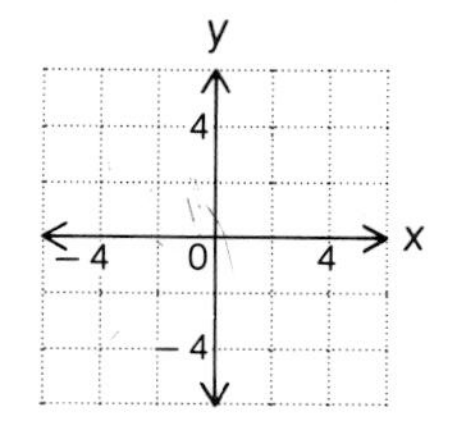

Solution on p. A47

Example 3 Find the x- and y-intercepts for $y = 2x - 4$. Graph the line.

Solution

x-intercept:

$$y = 2x - 4$$
$$0 = 2x - 4$$
$$-2x = -4$$
$$x = 2$$

(2,0)

y-intercept:

$(0,b)$

$b = -4$

$(0,-4)$

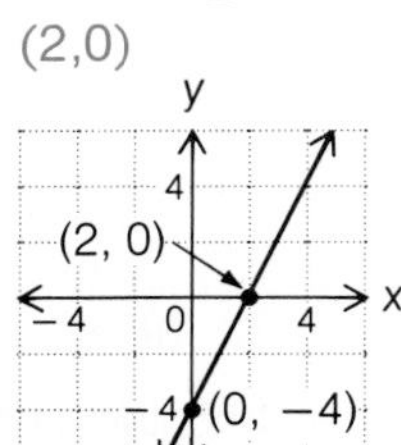

Example 4 Find the x- and y-intercepts for $y = 3x - 6$. Graph the line.

Your solution

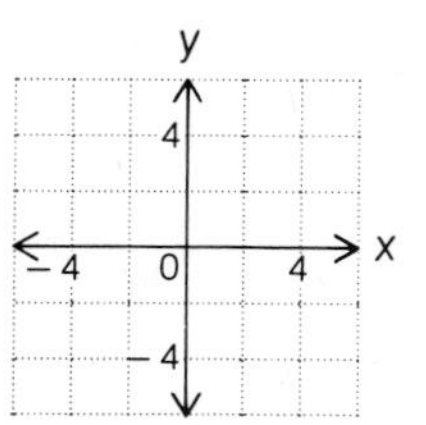

Solution on p. A47

3.2 Objective To find the slope of a straight line

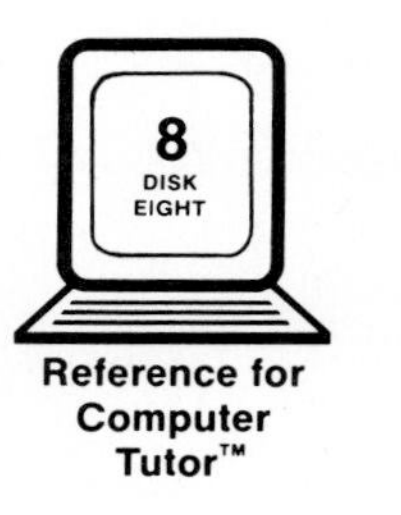

Reference for Computer Tutor™

The graphs of $y = \frac{2}{3}x + 1$ and $y = 2x + 1$ are shown at the right. Each graph crosses the y-axis at the point (0,1), but the graphs have different slants. The **slope** of a line is a measure of the slant of a line. The symbol for slope is m.

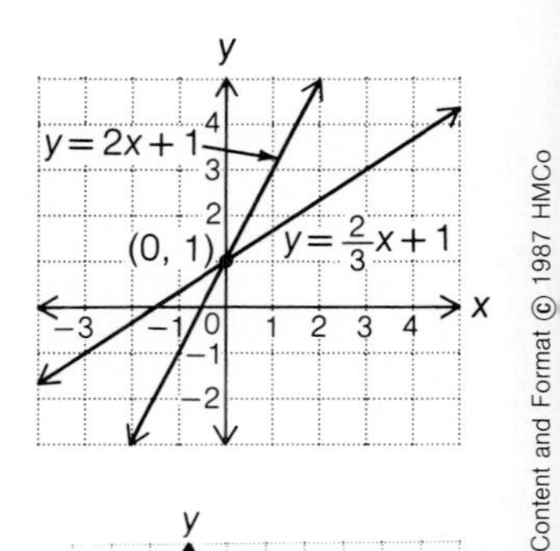

The slope of a line containing two points is the ratio of the change in the y values of the two points to the change in the x values. The line containing the points $(-2,-3)$ and $(6,1)$ is graphed at the right. The change in the y values is the difference between the two ordinates.

Change in $y = 1 - (-3) = 4$

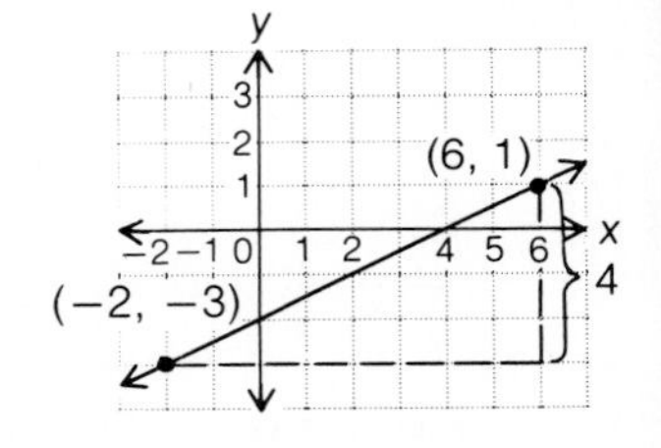

The change in the x values is the difference between the two abscissas.

Change in $x = 6 - (-2) = 8$

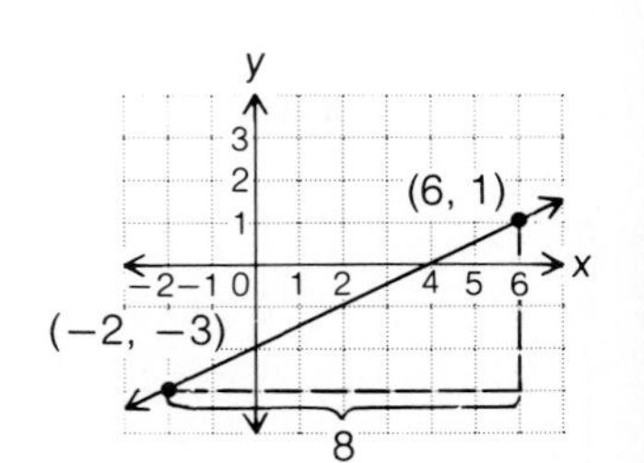

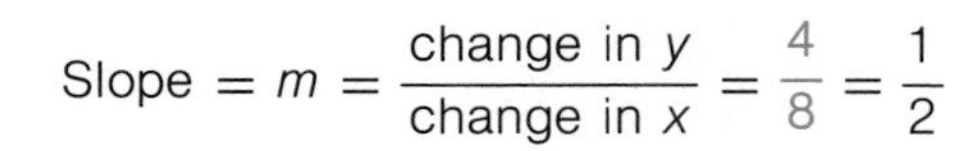

$$\text{Slope} = m = \frac{\text{change in } y}{\text{change in } x} = \frac{4}{8} = \frac{1}{2}$$

A formula for the slope of a line containing two points, P_1 and P_2, whose coordinates are (x_1,y_1) and (x_2,y_2), is given by:

$$\textbf{Slope} = m = \frac{y_2 - y_1}{x_2 - x_1}$$

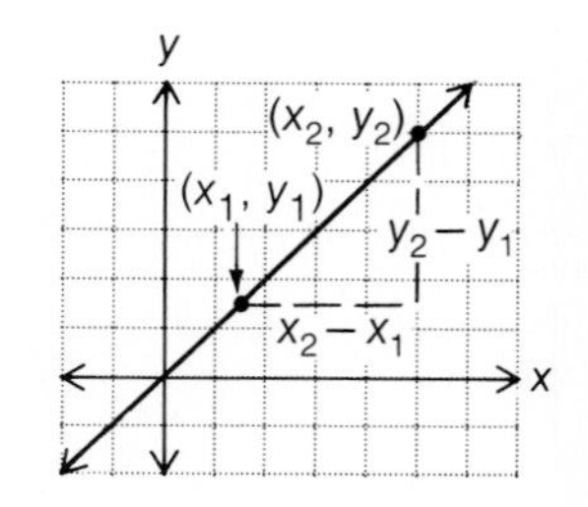

Find the slope of the line containing the points (−1,1) and (2,3).

Let P_1 = (−1,1) and P_2 = (2,3).
(It does not matter which point is named P_1 or P_2; the slope will be the same.)

$$m = \frac{y_2 - y_1}{x_2 - x_1} = \frac{3 - 1}{2 - (-1)} = \frac{2}{3}$$

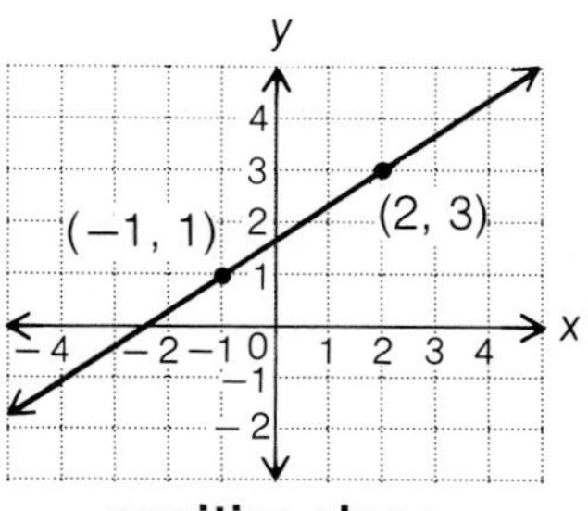

positive slope

A line which slants upward to the right always has a **positive slope.**

Find the slope of the line containing the points (−3,4) and (2,−2).

Let P_1 = (−3,4) and P_2 = (2,−2).

$$m = \frac{y_2 - y_1}{x_2 - x_1} = \frac{-2 - 4}{2 - (-3)} = \frac{-6}{5} = -\frac{6}{5}$$

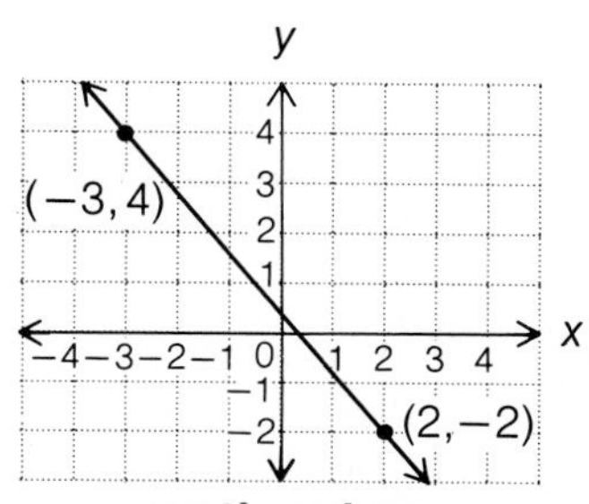

negative slope

A line which slants downward to the right always has a **negative slope.**

Find the slope of the line containing the points (−1,3) and (4,3).

Let P_1 = (−1,3) and P_2 = (4,3).

$$m = \frac{y_2 - y_1}{x_2 - x_1} = \frac{3 - 3}{4 - (-1)} = \frac{0}{5} = 0$$

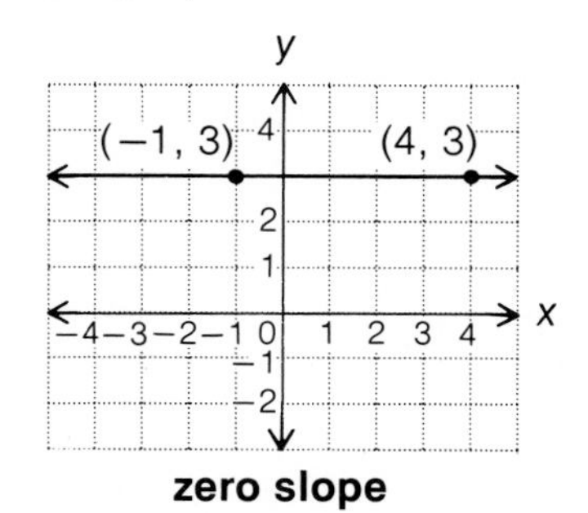

zero slope

A horizontal line has **zero slope.**

Find the slope of a line containing the points (2,−2) and (2,4).

Let P_1 = (2,−2) and P_2 = (2,4).

$$m = \frac{y_2 - y_1}{x_2 - x_1} = \frac{4 - (-2)}{2 - 2} = \frac{6}{0}$$ Not a real number

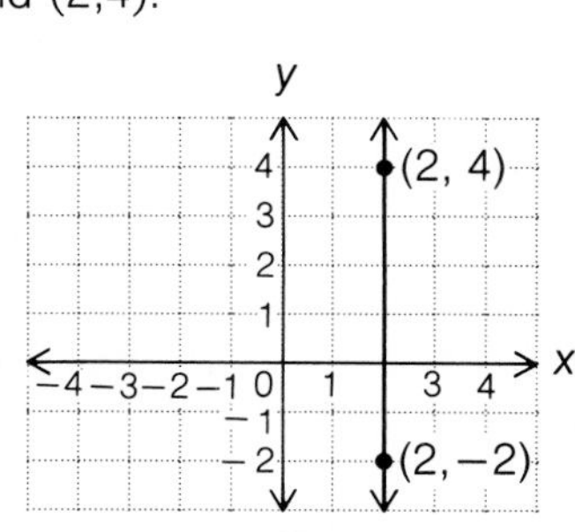

no slope

A vertical line has **no slope.**

Example 5 Find the slope of the line containing the points $(-2,-1)$ and $(3,4)$.

Solution Let $P_1 = (-2,-1)$ and $P_2 = (3,4)$.

$$m = \frac{y_2 - y_1}{x_2 - x_1} = \frac{4 - (-1)}{3 - (-2)} = \frac{5}{5} = 1$$

The slope is 1.

Example 6 Find the slope of the line containing the points $(-1,2)$ and $(1,3)$.

Your solution

Example 7 Find the slope of the line containing the points $(-3,1)$ and $(2,-2)$.

Solution Let $P_1 = (-3,1)$ and $P_2 = (2,-2)$.

$$m = \frac{y_2 - y_1}{x_2 - x_1} = \frac{-2 - 1}{2 - (-3)} = \frac{-3}{5}$$

The slope is $-\frac{3}{5}$.

Example 8 Find the slope of the line containing the points $(1,2)$ and $(4,-5)$.

Your solution

Example 9 Find the slope of the line containing the points $(-1,4)$ and $(-1,0)$.

Solution Let $P_1 = (-1,4)$ and $P_2 = (-1,0)$.

$$m = \frac{y_2 - y_1}{x_2 - x_1} = \frac{0 - 4}{-1 - (-1)} = \frac{-4}{0}$$

The line has no slope.

Example 10 Find the slope of the line containing the points $(2,3)$ and $(2,7)$.

Your solution

Example 11 Find the slope of the line containing the points $(-1,2)$ and $(4,2)$.

Solution Let $P_1 = (-1,2)$ and $P_2 = (4,2)$.

$$m = \frac{y_2 - y_1}{x_2 - x_1} = \frac{2 - 2}{4 - (-1)} = \frac{0}{5} = 0$$

The line has zero slope.

Example 12 Find the slope of the line containing the points $(1,-3)$ and $(-5,-3)$.

Your solution

Solutions on pp. A47–A48

3.3 Objective To graph a line using the slope and y-intercept

The graph of the equation $y = \frac{2}{3}x + 1$ is shown at the right. The points $(-3,-1)$ and $(3,3)$ are on the graph. The slope of the line is:

$$m = \frac{3-(-1)}{3-(-3)} = \frac{4}{6} = \frac{2}{3}$$

Note that the slope of the line has the same value as the coefficient of x.

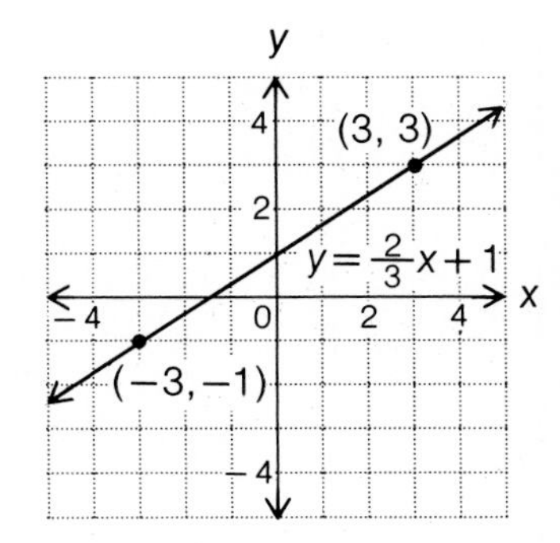

For any equation of the form $y = mx + b$, the slope of the line is m, the coefficient of x. The y-intercept is $(0,b)$. Thus, an equation of the form $y = mx + b$ is called the **slope-intercept form of a straight line.**

Find the slope and the y-intercept of the line $y = -\frac{3}{4}x + 1$.

$$y = \boxed{m}\,x \boxed{+\,b}$$

$$y = \boxed{-\tfrac{3}{4}}\,x \boxed{+\,1}$$

Slope $= m = -\frac{3}{4}$ — y-intercept $= (0, b) = (0, 1)$

The slope is $-\frac{3}{4}$. The y-intercept is $(0,1)$.

When the equation of a straight line is in the form $y = mx + b$, the graph can be drawn using the slope and y-intercept. First locate the y-intercept. Use the slope to find a second point on the line. Then draw a line through the two points.

Graph $y = 2x - 3$.

y-intercept $= (0,b) = (0,-3)$

$$m = 2 = \frac{2}{1} = \frac{\text{change in } y}{\text{change in } x}$$

Beginning at the y-intercept, move right 1 unit (change in x) and then up 2 units (change in y).

$(1,-1)$ is a second point on the graph.

Draw a line through the two points $(0,-3)$ and $(1,-1)$.

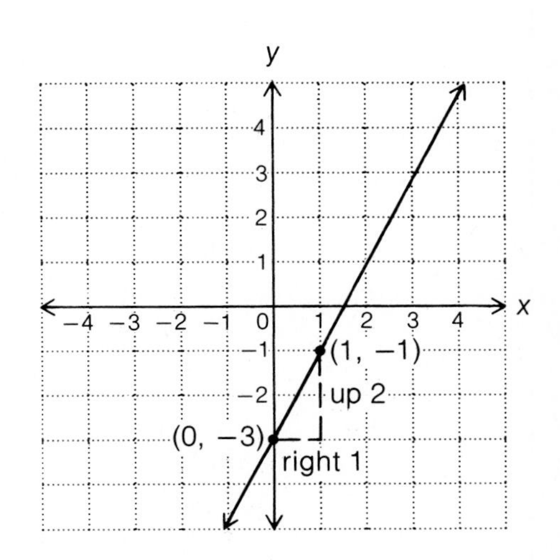

Example 13 Graph $y = -\frac{2}{3}x + 1$ by using the slope and y-intercept.

Solution y-intercept $= (0,b) = (0,1)$

$m = -\frac{2}{3} = \frac{-2}{3}$ (Move right 3 units, then down 2 units.)

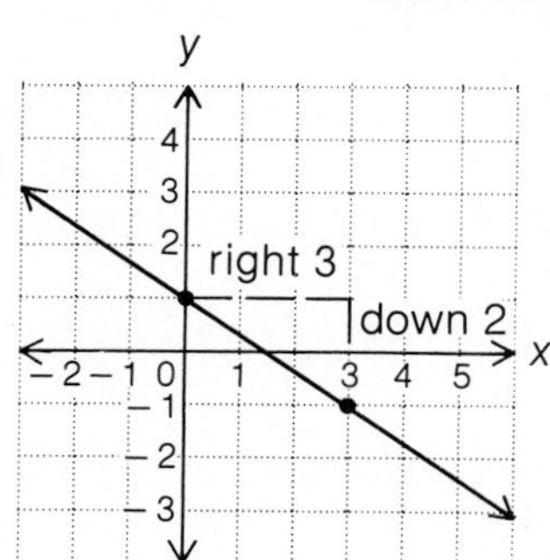

Example 14 Graph $y = -\frac{1}{4}x - 1$ by using the slope and y-intercept.

Your solution

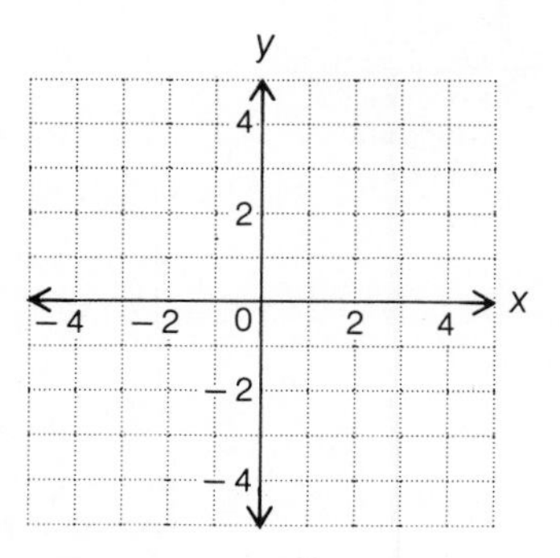

Example 15 Graph $y = -\frac{3}{4}x$ by using the slope and y-intercept.

Solution y-intercept $= (0,b) = (0,0)$

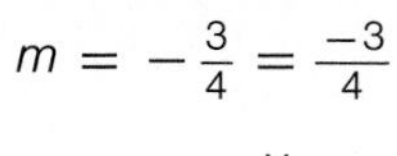

$m = -\frac{3}{4} = \frac{-3}{4}$

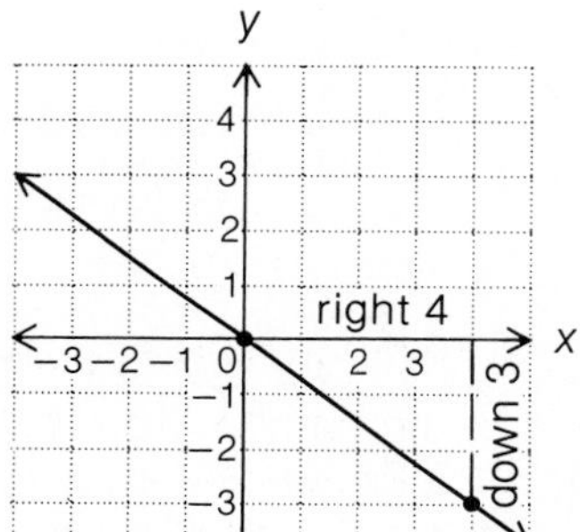

Example 16 Graph $y = -\frac{3}{5}x$ by using the slope and y-intercept.

Your solution

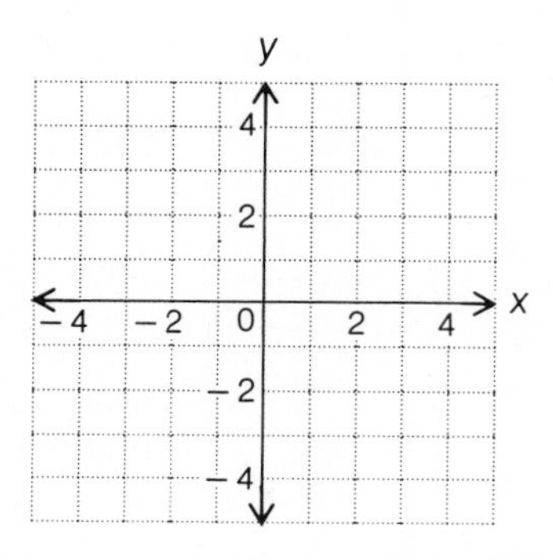

Example 17 Graph $2x - 3y = 6$ by using the slope and y-intercept.

Solution Solve the equation for y.

$$2x - 3y = 6$$
$$-3y = -2x + 6$$
$$y = \frac{2}{3}x - 2$$

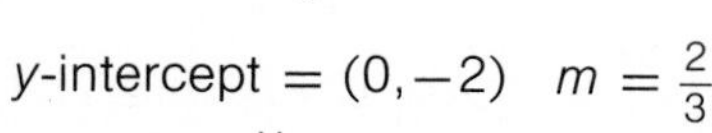

y-intercept $= (0,-2)$ $m = \frac{2}{3}$

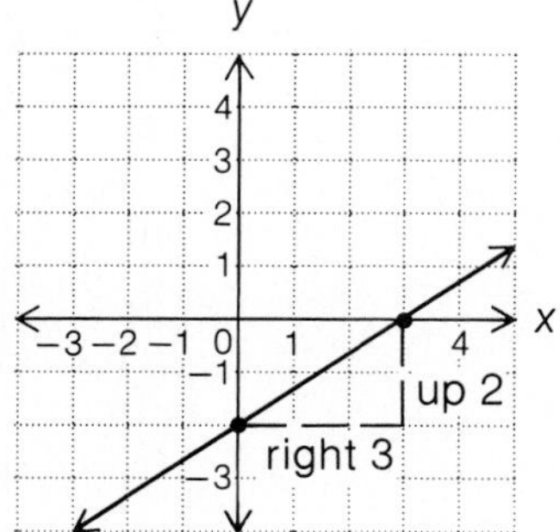

Example 18 Graph $x - 2y = 4$ by using the slope and y-intercept.

Your solution

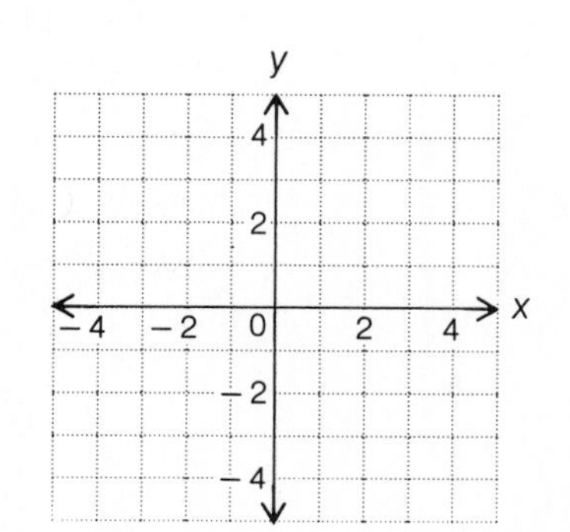

Solutions on p. A48

3.1 Exercises

Find the x- and y-intercepts.

1. $x - y = 3$

2. $3x + 4y = 12$

3. $y = 3x - 6$

4. $y = 2x + 10$

5. $x - 5y = 10$

6. $3x + 2y = 12$

7. $y = 3x + 12$

8. $y = 5x + 10$

9. $2x - 3y = 0$

10. $3x + 4y = 0$

11. $y = -\frac{1}{2}x + 3$

12. $y = \frac{2}{3}x - 4$

Find the x- and y-intercepts and graph.

13. $5x + 2y = 10$

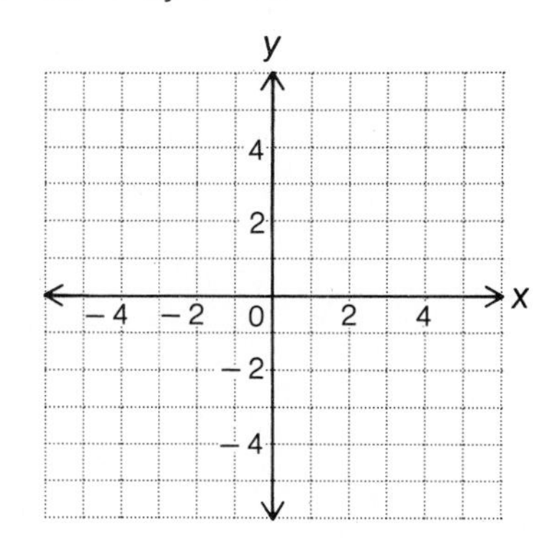

14. $x - 3y = 6$

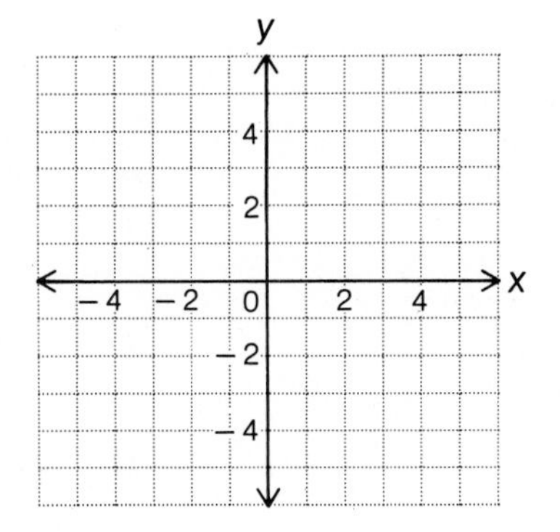

15. $y = \frac{3}{4}x - 3$

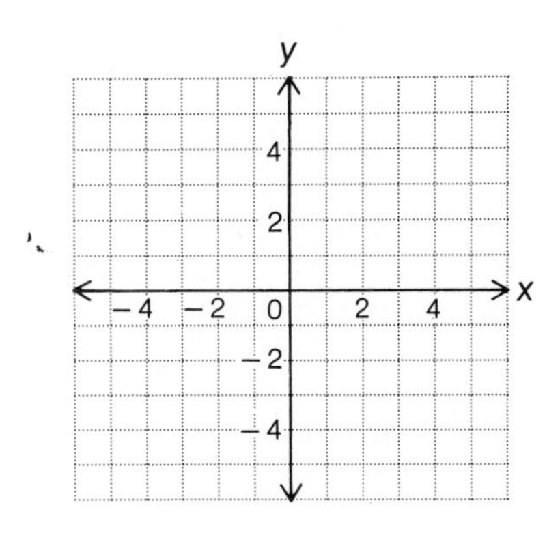

16. $y = \frac{2}{5}x - 2$

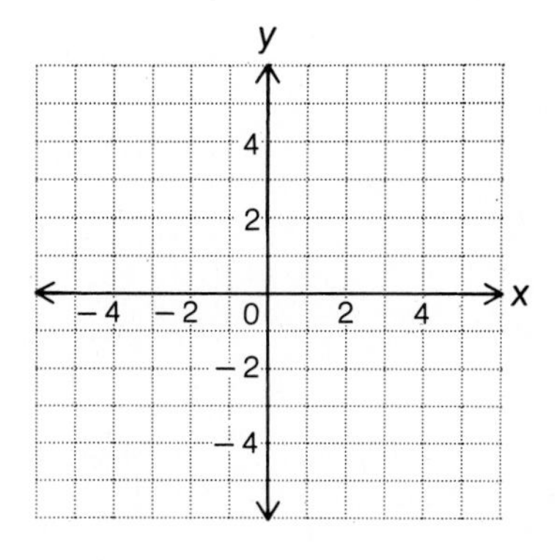

3.2 Exercises

Find the slope of the line containing the points.

17. $P_1(4,2), P_2(3,4)$

18. $P_1(2,1), P_2(3,4)$

19. $P_1(-1,3), P_2(2,4)$

20. $P_1(-2,1), P_2(2,2)$

21. $P_1(2,4), P_2(4,-1)$

22. $P_1(1,3), P_2(5,-3)$

23. $P_1(-2,3), P_2(2,1)$

24. $P_1(5,-2), P_2(1,0)$

25. $P_1(8,-3), P_2(4,1)$

26. $P_1(0,3), P_2(2,-1)$

27. $P_1(3,-4), P_2(3,5)$

28. $P_1(-1,2), P_2(-1,3)$

29. $P_1(4,-2), P_2(3,-2)$

30. $P_1(5,1), P_2(-2,1)$

31. $P_1(0,-1), P_2(3,-2)$

32. $P_1(3,0), P_2(2,-1)$

33. $P_1(-2,3), P_2(1,3)$

34. $P_1(4,-1), P_2(-3,-1$

35. $P_1(-2,4), P_2(-1,-1)$

36. $P_1(6,-4), P_2(4,-2)$

37. $P_1(-2,-3), P_2(-2,1$

38. $P_1(5,1), P_2(5,-2)$

39. $P_1(-1,5), P_2(5,1)$

40. $P_1(-1,5), P_2(7,1)$

3.3 Exercises

Graph by using the slope and y-intercept.

41. $y = 3x + 1$

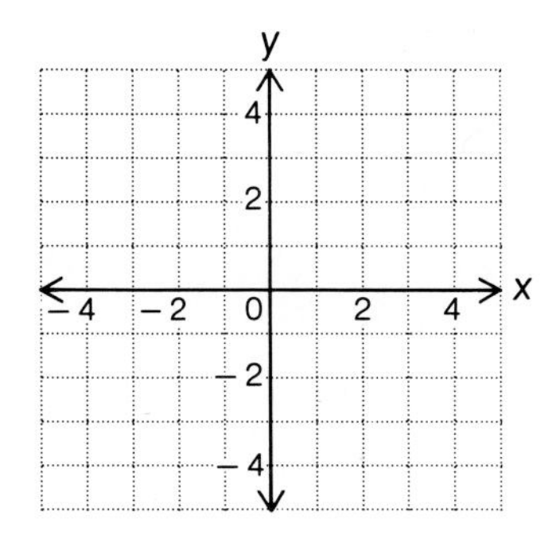

42. $y = -2x - 1$

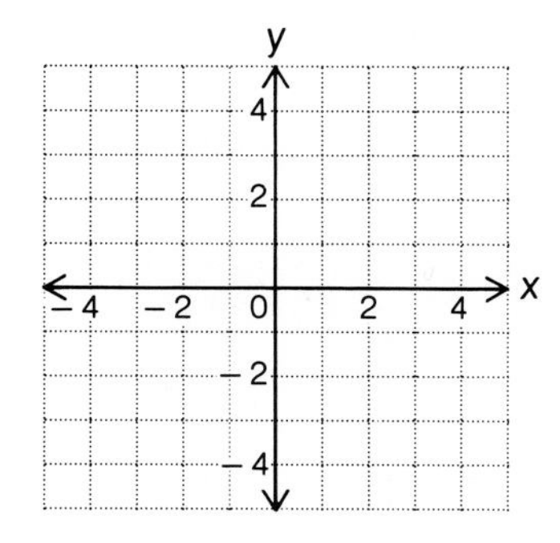

43. $y = \frac{2}{5}x - 2$

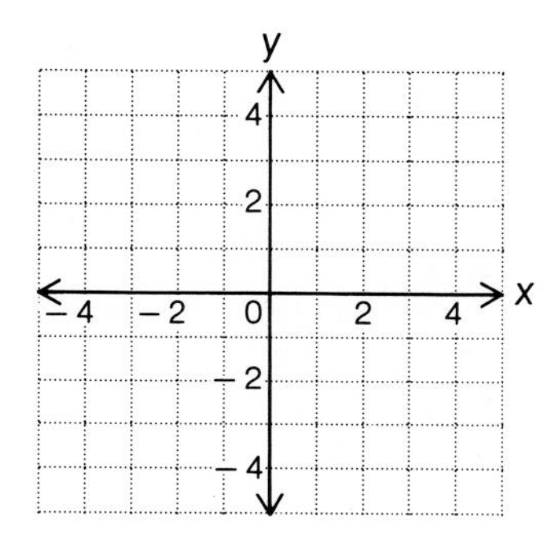

44. $y = \frac{3}{4}x + 1$

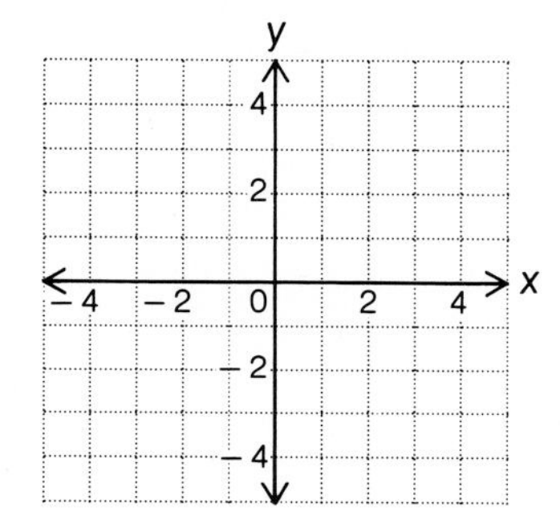

45. $2x + y = 3$

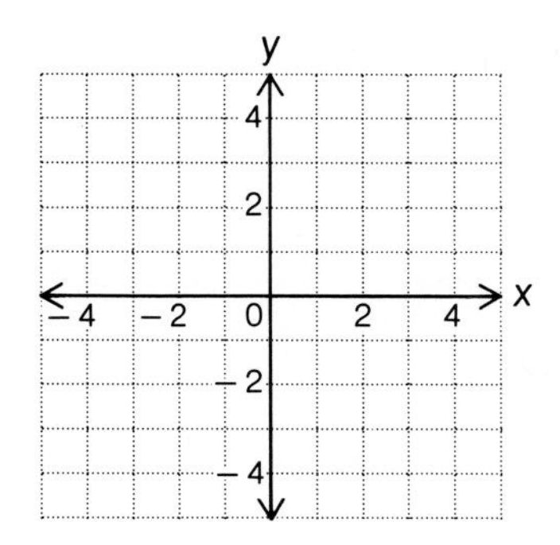

46. $3x - y = 1$

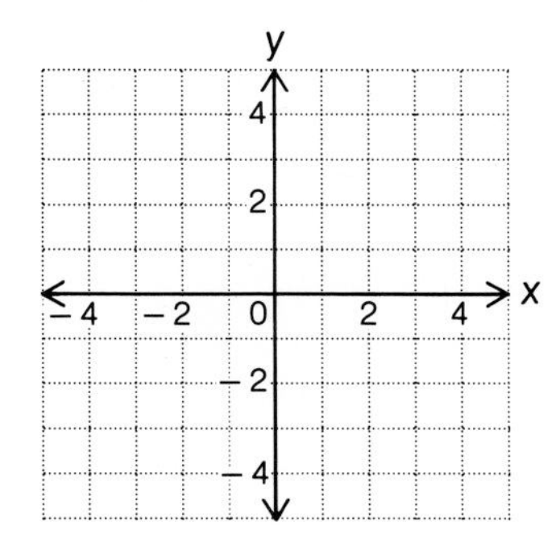

47. $x - 2y = 4$

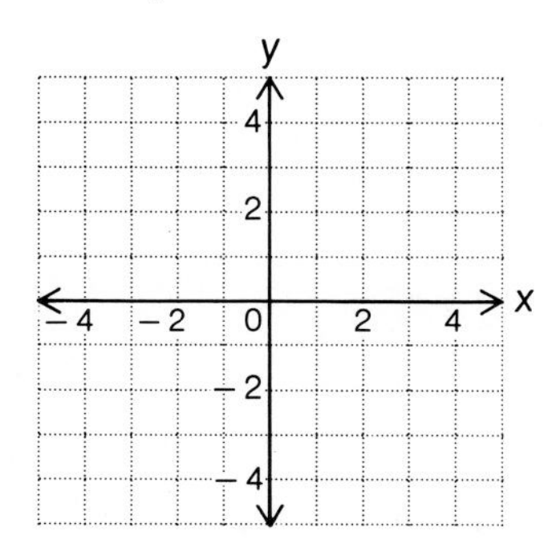

48. $x + 3y = 6$

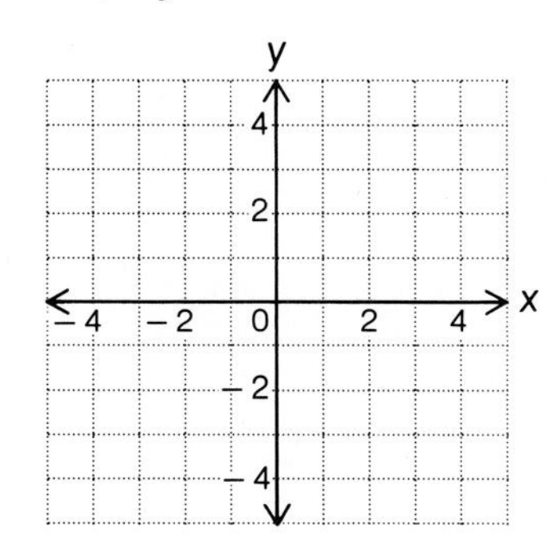

Graph by using the slope and y-intercept.

49. $y = \frac{2}{3}x$

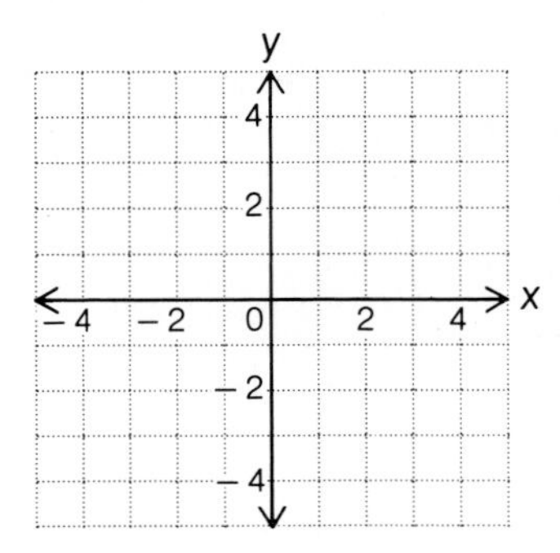

50. $y = \frac{1}{2}x$

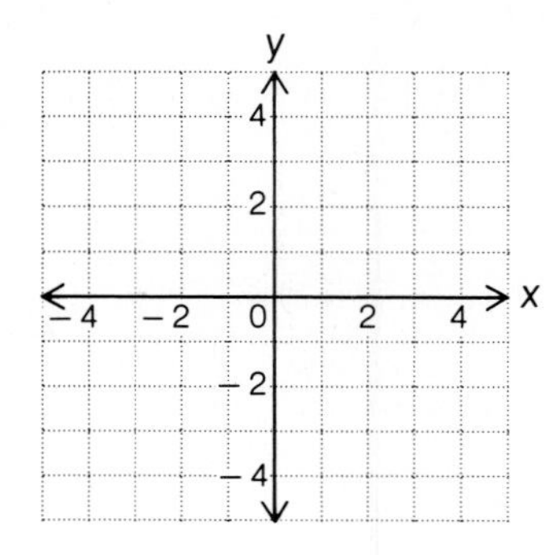

51. $y = -x + 1$

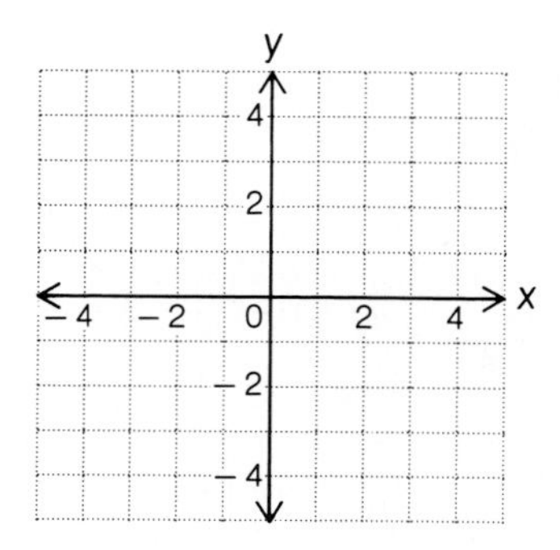

52. $y = -x - 3$

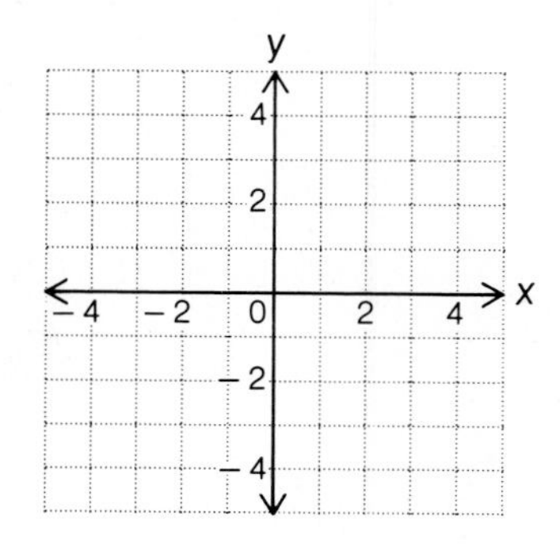

53. $3x - 4y = 12$

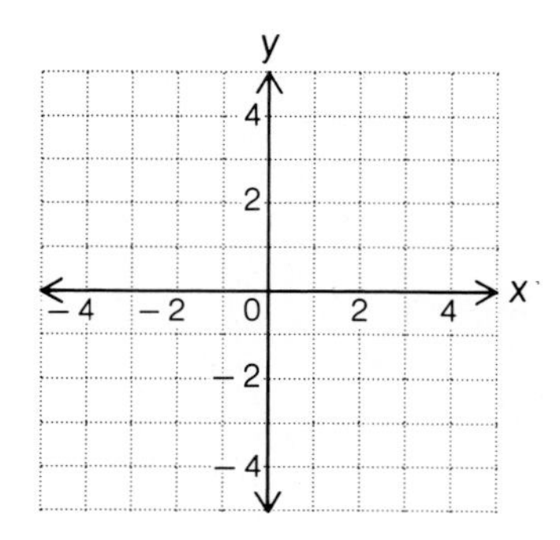

54. $5x - 2y = 10$

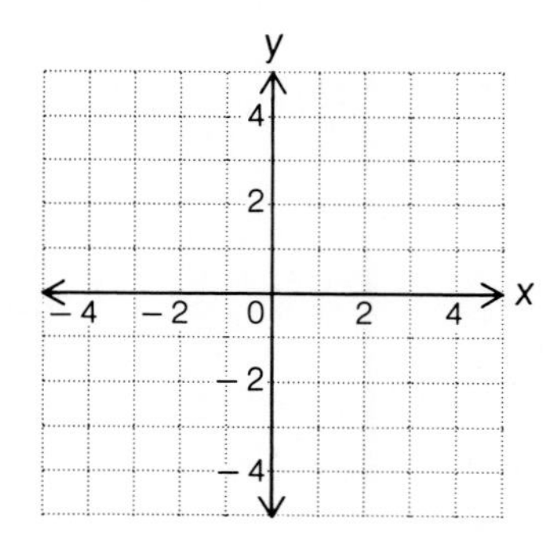

55. $y = -4x + 2$

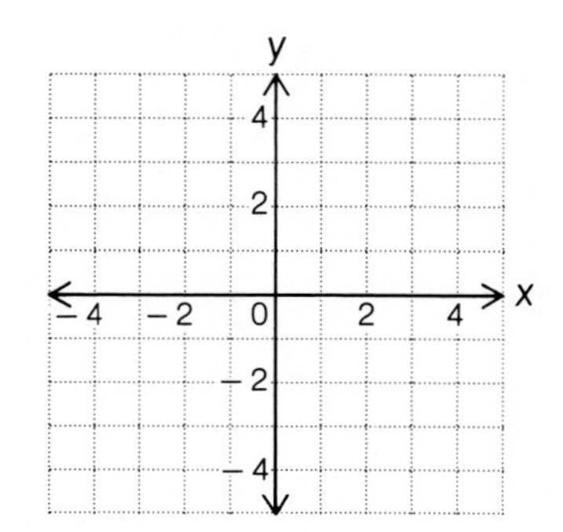

56. $4x - 5y = 20$

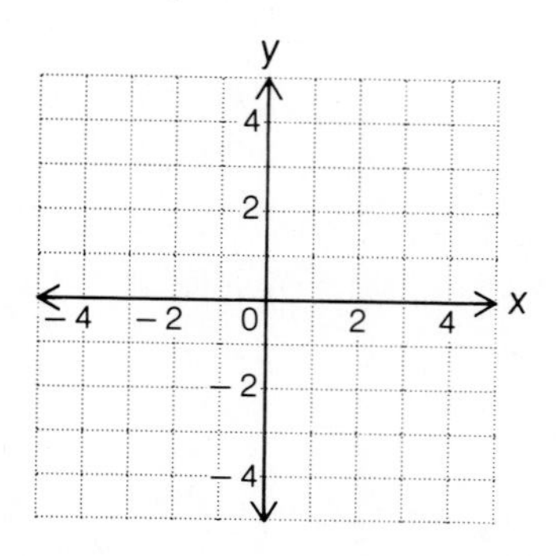

SECTION 4 Equations of Straight Lines

4.1 Objective To find the equation of a line using the equation $y = mx + b$

Reference for Computer Tutor™

When the slope of a line and a point on the line are known, the equation of the line can be written using the slope-intercept form, $y = mx + b$.

Find the equation of the line which has slope 3 and y-intercept (0,2).

$y = mx + b$

The given slope, 3, is m. Replace m with 3. — $y = 3x + b$

The given point, (0,2), is the y-intercept. Replace b with 2. — $y = 3x + 2$

Find the equation of the line which has slope $\frac{1}{2}$ and contains point $(-2,4)$.

$y = mx + b$

The given slope, $\frac{1}{2}$, is m. — $y = \frac{1}{2}x + b$

Replace m with $\frac{1}{2}$.

The given point, $(-2,4)$ is a solution of the equation of the line. Replace x and y in the equation with the coordinates of the point. — $4 = \frac{1}{2}(-2) + b$

Solve for b, the y-intercept. — $4 = -1 + b$

$5 = b$

Write the equation of the line by replacing m and b in the equation by their values. — $y = mx + b$

$y = \frac{1}{2}x + 5$

Example 1 Find the equation of the line which contains the point $(3,-3)$ and has slope $\frac{2}{3}$.

Solution

$$y = \frac{2}{3}x + b$$
$$-3 = \frac{2}{3}(3) + b$$
$$-3 = 2 + b$$
$$-5 = b$$
$$y = \frac{2}{3}x - 5$$

Example 2 Find the equation of the line which contains the point $(4,-2)$ and has slope $\frac{3}{2}$.

Your solution

Solution on p. A49

4.2 Objective To find the equation of a line using the point-slope formula

Reference for Computer Tutor™

An alternate method for finding the equation of a line, given the slope and a point on the line, involves use of the point-slope formula. The point-slope formula is derived from the formula for slope.

Let (x_1,y_1) be the given point on the line and (x,y) be any other point on the line.

Formula for slope $\quad \frac{y - y_1}{x - x_1} = m$

Multiply both sides of the equation by $(x - x_1)$. $\quad \frac{y - y_1}{x - x_1}(x - x_1) = m(x - x_1)$

Simplify. $\quad y - y_1 = m(x - x_1)$

The **point-slope formula** is $\boldsymbol{y - y_1 = m(x - x_1)}$.

In this equation, m is the slope and (x_1,y_1) is the given point.

Find the equation of the line which passes through point $(-3,2)$ and has slope $\frac{2}{3}$.

$(x_1,y_1) = (-3,2) \qquad m = \frac{2}{3}$

$$y - y_1 = m(x - x_1)$$
$$y - 2 = \frac{2}{3}[x - (-3)]$$
$$y - 2 = \frac{2}{3}(x + 3)$$
$$y - 2 = \frac{2}{3}x + 2$$
$$y = \frac{2}{3}x + 4$$

The equation of the line is $y = \frac{2}{3}x + 4$.

Example 3 Use the point-slope formula to find the equation of the line which passes through point $(-2,-1)$ and has slope $\frac{3}{2}$.

Solution $m = \frac{3}{2} \qquad (x_1,y_1) = (-2,-1)$

$$y - y_1 = m(x - x_1)$$
$$y - (-1) = \frac{3}{2}[x - (-2)]$$
$$y + 1 = \frac{3}{2}(x + 2)$$
$$y + 1 = \frac{3}{2}x + 3$$
$$y = \frac{3}{2}x + 2$$

The equation of the line is $y = \frac{3}{2}x + 2$.

Example 4 Use the point-slope formula to find the equation of the line which passes through point $(4,-2)$ and has slope $\frac{3}{4}$.

Your solution

Solution on p. A49

4.1 Exercises

Use the slope-intercept form.

1. Find the equation of the line which contains the point (0,2) and has slope 2.

2. Find the equation of the line which contains the point (0,−1) and has slope −2.

3. Find the equation of the line which contains the point (−1,2) and has slope −3.

4. Find the equation of the line which contains the point (2,−3) and has slope 3.

5. Find the equation of the line which contains the point (3,1) and has slope $\frac{1}{3}$.

6. Find the equation of the line which contains the point (−2,3) and has slope $\frac{1}{2}$.

7. Find the equation of the line which contains the point (4,−2) and has slope $\frac{3}{4}$.

8. Find the equation of the line which contains the point (2,3) and has slope $-\frac{1}{2}$.

9. Find the equation of the line which contains the point (5,−3) and has slope $-\frac{3}{5}$.

10. Find the equation of the line which contains the point (5,−1) and has slope $\frac{1}{5}$.

11. Find the equation of the line which contains the point (2,3) and has slope $\frac{1}{4}$.

12. Find the equation of the line which contains the point (−1,2) and has slope $-\frac{1}{2}$.

4.2 Exercises

Use the point-slope formula.

13. Find the equation of the line which passes through point $(1, -1)$ and has slope 2.

14. Find the equation of the line which passes through the point $(2,3)$ and has slope -1.

15. Find the equation of the line which passes through point $(-2,1)$ and has slope -2.

16. Find the equation of the line which passes through point $(-1, -3)$ and has slope -3.

17. Find the equation of the line which passes through point $(0,0)$ and has slope $\frac{2}{3}$.

18. Find the equation of the line which passes through point $(0,0)$ and has slope $-\frac{1}{5}$.

19. Find the equation of the line which passes through point $(2,3)$ and has slope $\frac{1}{2}$.

20. Find the equation of the line which passes through point $(3, -1)$ and has slope $\frac{2}{3}$.

21. Find the equation of the line which passes through point $(-4,1)$ and has slope $-\frac{3}{4}$.

22. Find the equation of the line which passes through point $(-5,0)$ and has slope $-\frac{1}{5}$.

23. Find the equation of the line which passes through point $(-2,1)$ and has slope $\frac{3}{4}$.

24. Find the equation of the line which passes through point $(3, -2)$ and has slope $\frac{1}{6}$.

Calculators and Computers

Graphs of Straight Lines

The program GRAPHS OF STRAIGHT LINES on the Student Disk graphs lines of the form

$$y = mx + b$$

where m is the slope of the line and b is the y-intercept.

This program will allow you to examine the effect on the graph when changes are made in the slope m while the y-intercept remains constant. You can also examine the effect changes in the y-intercept will have on the graph.

After each line is drawn, you will have the option of erasing the line, drawing another line, or quitting. Here is an example to get you started:

Graph $y = x + 1 \quad (m = 1, \quad b = 1);$
$y = 2x + 1 \quad (m = 2, \quad b = 1);$
$y = -x + 1 \quad (m = -1, b = 1);$
$y = -2x + 1 \quad (m = -2, b = 1).$

For each of these graphs, the y-intercept is 1 and the graph passes through the point (0,1).

Now erase the graphs and try these.

Graph $y = 2x - 1 \quad (m = 2, b = -1);$
$y = 2x + 2 \quad (m = 2, b = 2);$
$y = 2x - 4 \quad (m = 2, b = -4).$

The three graphs are parallel since the slope of each line is 2. The y-intercept changes.

Try this program with your own values. Remember that when entering a fractional slope or y-intercept, enter it as a decimal. For example, to see the graph of $y = \frac{3}{4}x + 2$, enter the slope $\frac{3}{4}$ as 0.75.

Chapter Summary

KEY WORDS

A **rectangular coordinate system** is formed by two number lines, one horizontal and one vertical, that intersect at the zero point of each line.

The number lines which make up a rectangular coordinate system are called the **coordinate axes** or simply **axes.**

The **origin** is the point of intersection of the two coordinate axes.

A rectangular coordinate system divides the plane into four regions, called **quadrants.**

An **ordered pair** (a,b) is used to locate a point in a plane.

The first number in an ordered pair is called the **abscissa.**

The second number in an ordered pair is called the **ordinate.**

The **coordinates** of a point are the numbers in the ordered pair associated with the point.

An equation of the form $y = mx + b$, where m and b are constants, is a **linear equation in two variables.**

The point at which a graph crosses the x-axis is called the **x-intercept.**

The point at which a graph crosses the y-axis is called the **y-intercept.**

The **slope** of a line is the measure of the slant of a line. The symbol for slope is ***m.***

A line which slants upward to the right has a **positive slope.**

A line which slants downward to the right has a **negative slope.**

A horizontal line has **zero slope.**

A vertical line has **no slope.**

ESSENTIAL RULES

Slope of a straight line: $\text{slope} = m = \dfrac{y_2 - y_1}{x_2 - x_1}$

Slope-intercept form of a straight line: $y = mx + b$

Point-slope form of a straight line: $y - y_1 = m(x - x_1)$

Review/Test

SECTION 1

1.1 Graph the ordered pairs (−3,1) and (0,2).

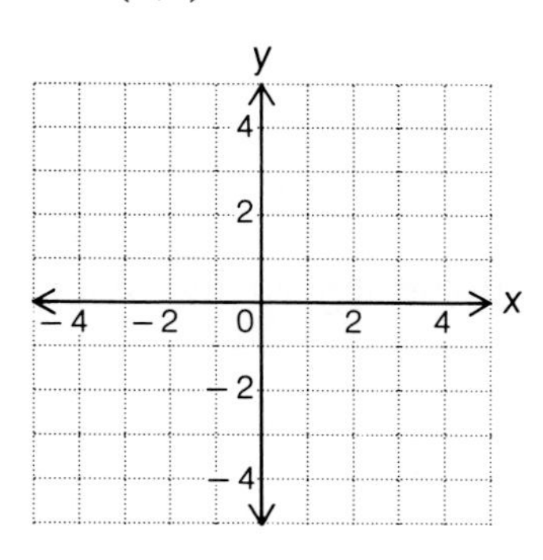

1.2 Find the ordered pair solution of $y = -\frac{2}{3}x + 2$ corresponding to $x = 3$.

1.3 The costs for an amount of energy were recorded as the following ordered pairs: (1,5), (2,10), (4,20), and (5,25), where the abscissa is the number of kilowatt-hours used and the ordinate is the cost in cents. Graph the ordered pairs.

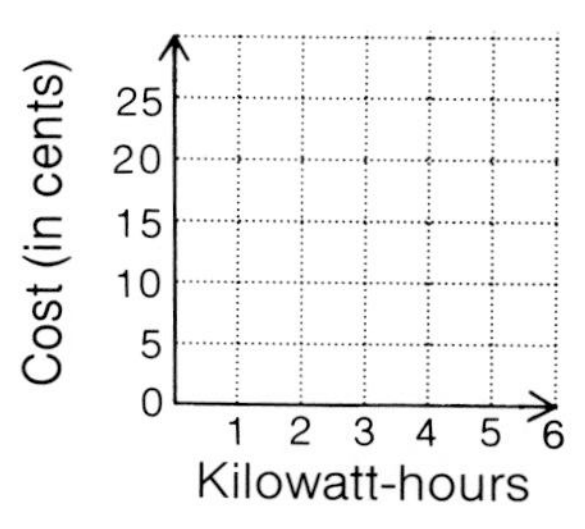

SECTION 2

2.1a Graph $y = 3x + 1$.

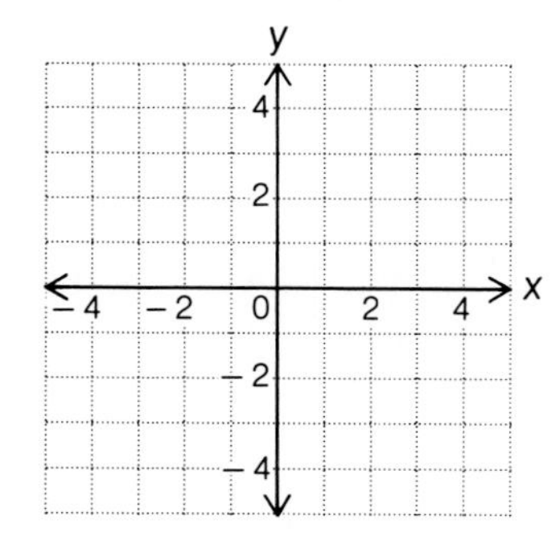

2.1b Graph $y = -\frac{3}{4}x + 3$.

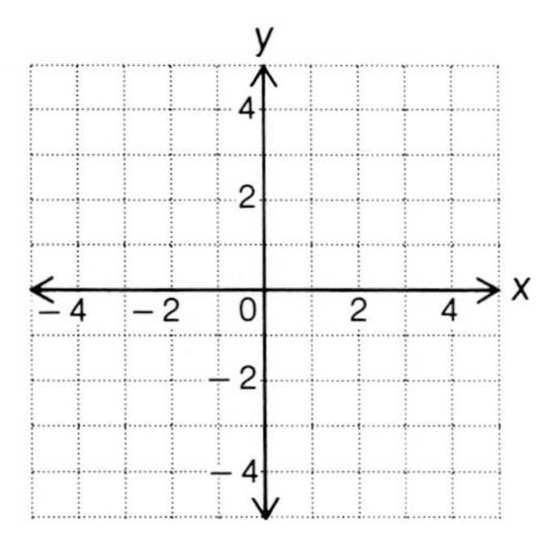

2.2a Graph $3x - 2y = 6$.

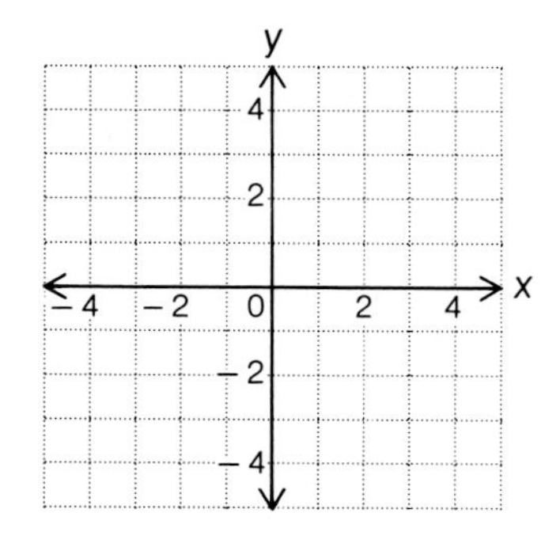

2.2b Graph $x + 3 = 0$.

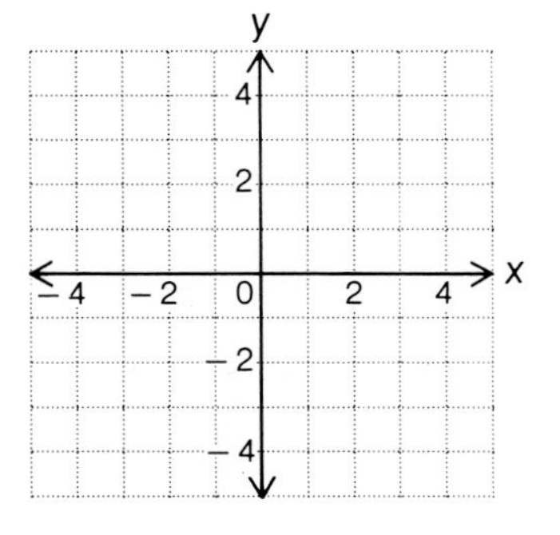

Review/Test

SECTION 3

3.1a Find the x- and y-intercepts for $6x - 4y = 12$.

3.1b Find the x- and y-intercepts for $y = \frac{1}{2}x + 1$.

3.2a Find the slope of the line containing the points $(2, -3)$ and $(4, 1)$.

3.2b Find the slope of the line containing the points $(3, -4)$ and $(1, -4)$.

3.3a Graph the line which has slope $-\frac{2}{3}$ and y-intercept $(0, 4)$.

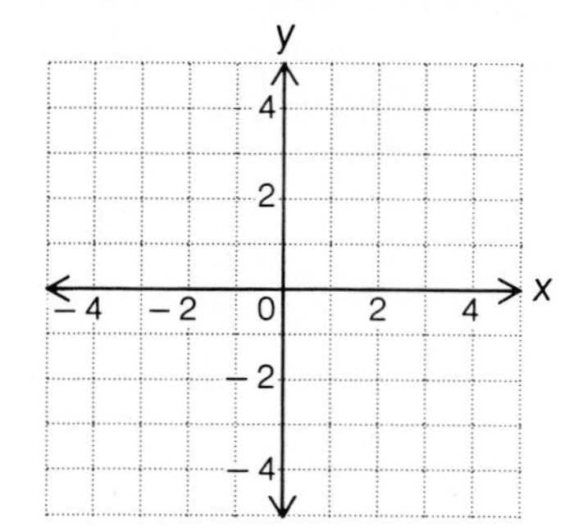

3.3b Graph the line which has slope 2 and y-intercept -2.

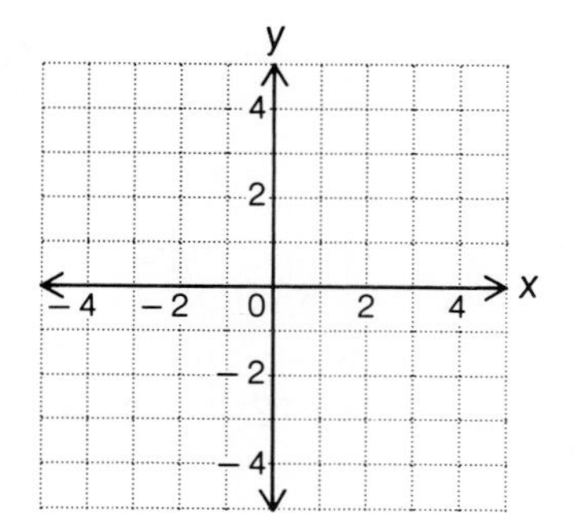

SECTION 4

4.1a Find the equation of the line which contains the point $(0, -1)$ and has slope 3.

4.1b Find the equation of the line which contains the point $(-3, 1)$ and has slope $\frac{2}{3}$.

4.2a Find the equation of the line which contains the point $(2, 3)$ and has slope $\frac{1}{2}$.

4.2b Find the equation of the line which contains the point $(-1, 2)$ and has slope $-\frac{2}{3}$.

Cumulative Review/Test

1. Simplify
$12 - 18 \div 3 \cdot (-2)^2$.

2. Evaluate $\frac{a - b}{a^2 - c}$ when
$a = -2$, $b = 3$, and $c = -4$.

3. Simplify
$4(2 - 3x) - 5(x - 4)$.

4. Solve: $2x - \frac{2}{3} = \frac{7}{3}$

5. Solve:
$3x - 2[x - 3(2 - 3x)] = x - 7$

6. Write $6\frac{2}{3}\%$ as a fraction.

7. Simplify
$(-2x^2y)^3(2xy^2)^2$.

8. Simplify $\frac{-15x^7}{5x^5}$.

9. Divide:
$(x^2 - 4x - 21) \div (x - 7)$

10. Factor $5x^2 + 15x + 10$.

11. Factor $x(a + 2) + y(a + 2)$.

12. Solve: $x(x - 2) = 8$

13. Simplify:
$\frac{x^5y^3}{x^2 - x - 6} \cdot \frac{x^2 - 9}{x^2y^4}$

14. Simplify:
$\frac{3x}{x^2 + 5x - 24} - \frac{9}{x^2 + 5x - 24}$

15. Solve: $3 - \frac{1}{x} = \frac{5}{x}$

16. Solve $4x - 5y = 15$ for y.

Cumulative Review/Test

17. Find the ordered pair solution of $y = 2x - 1$ corresponding to $x = -2$.

18. Find the slope of the line containing the points (2,3) and (−2,3).

19. Find the equation of the line which contains the point (2, −1) and has slope $\frac{1}{2}$.

20. Find the equation of the line which contains the point (0,2) and has slope −3.

21. Find the equation of the line which contains the point (−1,0) and has slope 2.

22. Find the equation of the line which contains the point (6,1) and has slope $\frac{2}{3}$.

23. A suit which regularly sells for $89 is on sale for 30% off the regular price. Find the sale price.

24. A gold coin is 60 years older than a silver coin. Fifteen years ago the gold coin was twice as old as the silver coin was then. Find the present ages of the two coins.

25. The real estate tax for a home which costs $50,000 is $625. At this rate, what is the value of a home for which the real estate tax is $1375?

26. An electrician requires 6 h to wire a garage. An apprentice can do the same job in 10 h. How long would it take to wire the garage when both the electrician and the apprentice are working?

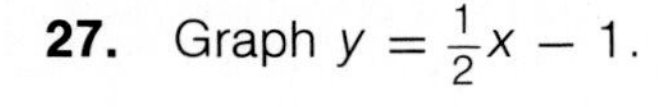

27. Graph $y = \frac{1}{2}x - 1$.

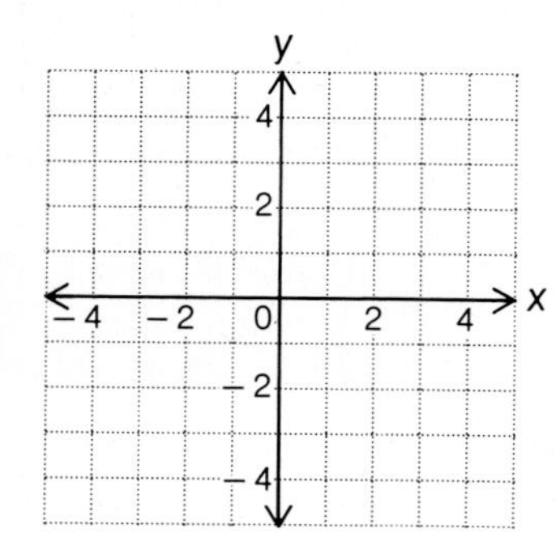

28. Graph the line which has slope $\frac{1}{2}$ and y-intercept 2.

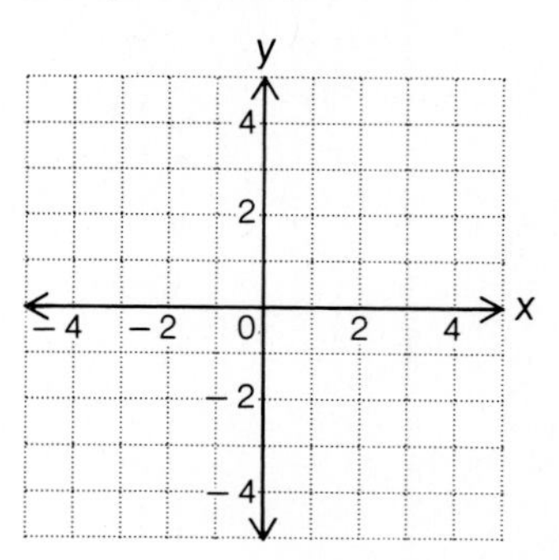

9

Systems of Linear Equations

Objectives

- To determine if a given ordered pair is a solution of a system of linear equations
- To solve a system of linear equations by graphing
- To solve a system of linear equations by the substitution method
- To solve a system of linear equations by the addition method
- To solve rate-of-wind or current problems
- To solve application problems using two variables

Input-Output Analysis

In 1973, the Nobel Prize in Economics was awarded for applications of mathematics to economics. The technique was to examine various sectors of an economy (steel industry, oil, farms, autos, and many others) and determine how each sector interacted with the others. Over 500 sectors of the economy were studied.

The interaction of each sector with the other was written as a series of equations. This series of equations is called a *system of equations*.

Using a computer, economists searched for a solution to the system of equations which would determine the output levels various sectors would have to meet to satisfy the requests from other sectors. The method is called Input-Output Analysis.

This chapter begins the study of systems of equations.

SECTION 1 Solving Systems of Linear Equations by Graphing

1.1 Objective To determine if a given ordered pair is a solution of a system of linear equations

Reference for Computer Tutor™

Equations considered together are called a **system of equations.** A system of equations is shown at the right.

$$2x + y = 3$$
$$x + y = 1$$

A **solution of a system of equations** is an ordered pair which is a solution of each equation of the system.

Is $(2, -1)$ a solution of the system
$2x + y = 3$
$x + y = 1$?

$2x + y = 3$	
$2(2) + (-1)$	3
$4 + (-1)$	3
$3 = 3$	

$x + y = 1$	
$2 + (-1)$	1
$1 = 1$	

Yes, since $(2, -1)$ is a solution of each equation, it is the solution of the system of equations.

Example 1 Is $(1, -3)$ a solution of the system
$3x + 2y = -3$
$x - 3y = 6$?

Solution

$3x + 2y = -3$	
$3 \cdot 1 + 2(-3)$	-3
$3 + (-6)$	-3
$-3 = -3$	

$x - 3y = 6$	
$1 - 3(-3)$	6
$1 - (-9)$	6
$10 \neq 6$	

No, $(1, -3)$ is not a solution of the system of equations.

Example 2 Is $(-1, -2)$ a solution of the system
$2x - 5y = 8$
$-x + 3y = -5$?

Your solution

Solution on p. A51

1.2 Objective To solve a system of linear equations by graphing

Reference for Computer Tutor™

The solution of a system of linear equations can be found by graphing the two lines on the same coordinate system. The point of intersection of the lines is the ordered pair which is a solution of each equation of the system. It is the solution of the system of equations.

Solve by graphing: $2x + 3y = 6$
$2x + y = -2$

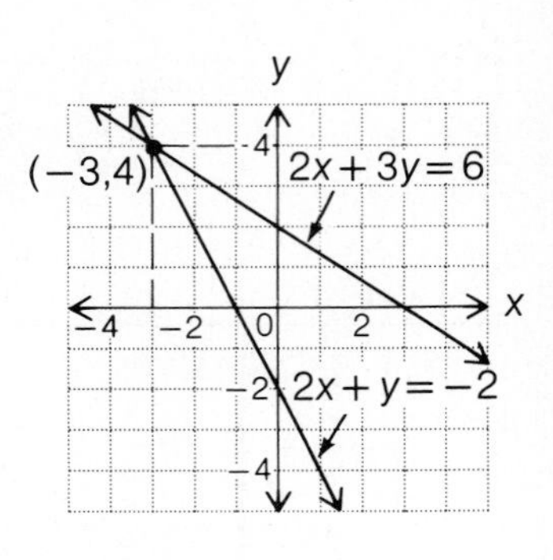

Graph each line.

Find the point of intersection.

$(-3,4)$ is a solution of each equation.

The solution is $(-3,4)$.

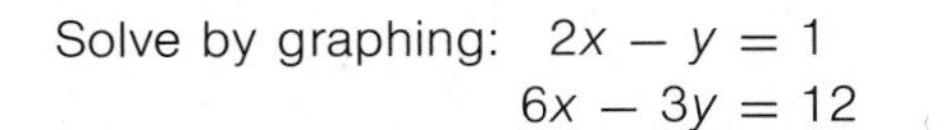

Solve by graphing: $2x - y = 1$
$6x - 3y = 12$

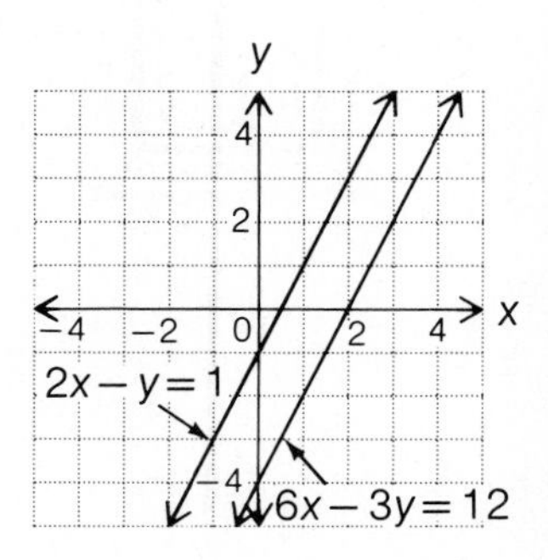

Graph each line.

The lines are parallel and therefore do not intersect. The system of equations has no solution.

Example 3 Solve by graphing:
$x - 2y = 2$
$x + y = 5$

Solution

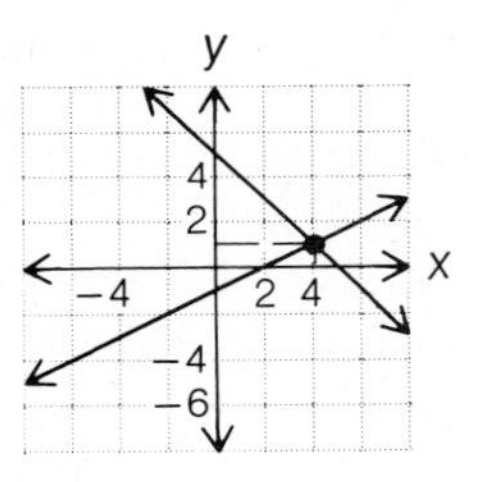

The solution is $(4,1)$.

Example 4 Solve by graphing:
$x + 3y = 3$
$-x + y = 5$

Your solution

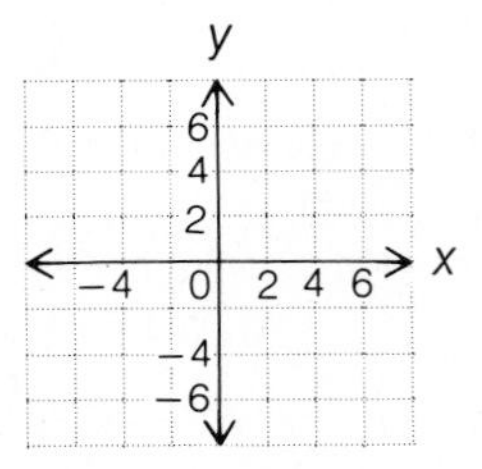

Example 5 Solve by graphing:
$4x - 2y = 6$
$y = 2x - 3$

Solution

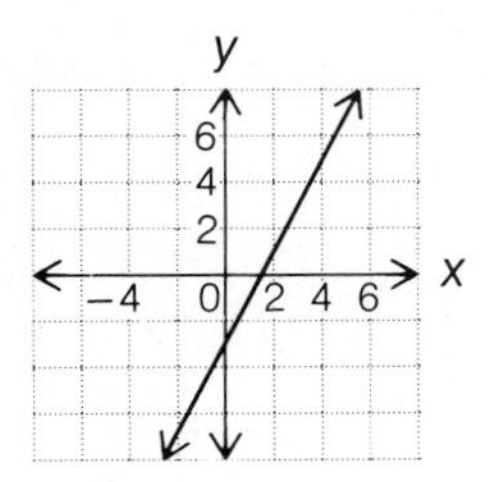

The two equations represent the same line. Any ordered pair which is a solution of one equation is also a solution of the other equation.

Example 6 Solve by graphing:
$y = 3x - 1$
$6x - 2y = -6$

Your solution

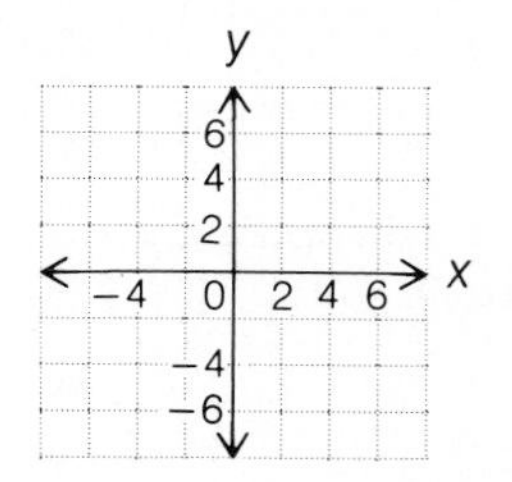

Solutions on p. A51

1.1 Exercises

1. Is (2,3) a solution of the system
$$3x + 4y = 18$$
$$2x - y = 1?$$

2. Is (2,−1) a solution of the system
$$x - 2y = 4$$
$$2x + y = 3?$$

3. Is (1,−2) a solution of the system
$$3x - y = 5$$
$$2x + 5y = -8?$$

4. Is (−1,−1) a solution of the system
$$x - 4y = 3$$
$$3x + y = 2?$$

5. Is (4,3) a solution of the system
$$5x - 2y = 14$$
$$x + y = 8?$$

6. Is (2,5) a solution of the system
$$3x + 2y = 16$$
$$2x - 3y = 4?$$

7. Is (−1,3) a solution of the system
$$4x - y = -5$$
$$2x + 5y = 13?$$

8. Is (4,−1) a solution of the system
$$x - 4y = 9$$
$$2x - 3y = 11?$$

9. Is (0,0) a solution of the system
$$4x + 3y = 0$$
$$2x - y = 1?$$

10. Is (2,0) a solution of the system
$$3x - y = 6$$
$$x + 3y = 2?$$

11. Is (2,−3) a solution of the system
$$y = 2x - 7$$
$$3x - y = 9?$$

12. Is (−1,−2) a solution of the system
$$3x - 4y = 5$$
$$y = x - 1?$$

13. Is (5,2) a solution of the system
$$y = 2x - 8$$
$$y = 3x - 13?$$

14. Is (−4,3) a solution of the system
$$y = 2x + 11$$
$$y = 5x - 19?$$

15. Is (−2,−3) a solution of the system
$$3x - 4y = 6$$
$$2x - 7y = 17?$$

16. Is (0,0) a solution of the system
$$y = 2x$$
$$3x + 5y = 0?$$

17. Is (0,−3) a solution of the system
$$4x - 3y = 9$$
$$2x + 5y = 15?$$

18. Is (4,0) a solution of the system
$$2x + 3y = 8$$
$$x - 5y = 4?$$

1.2 Exercises

Solve by graphing.

19. $x - y = 3$
$x + y = 5$

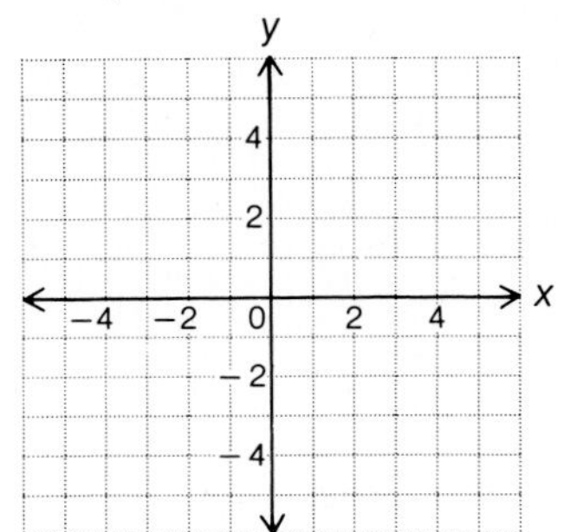

20. $2x - y = 4$
$x + y = 5$

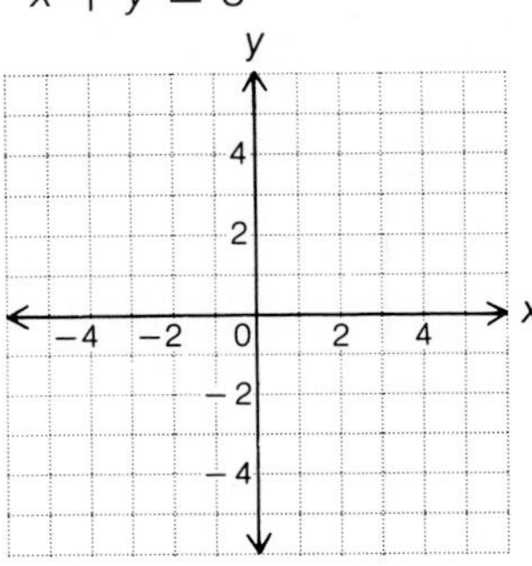

21. $x + 2y = 6$
$x - y = 3$

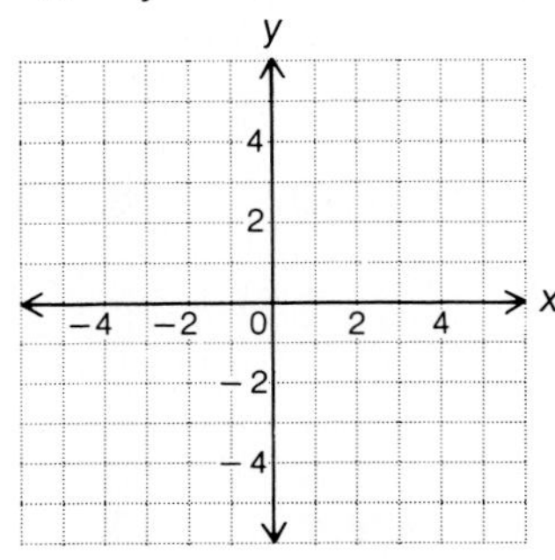

22. $3x - y = 3$
$2x + y = 2$

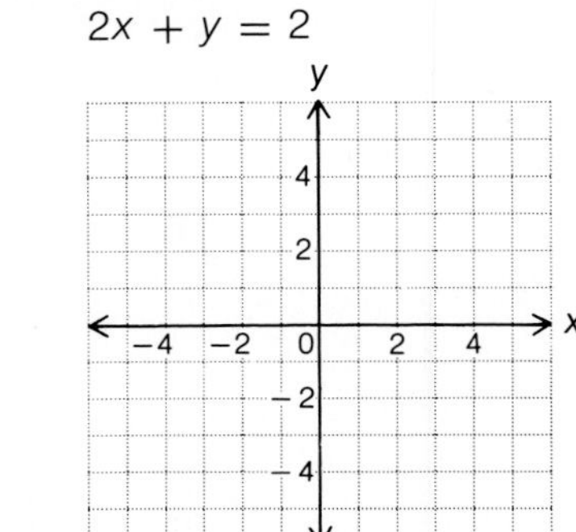

23. $3x - 2y = 6$
$y = 3$

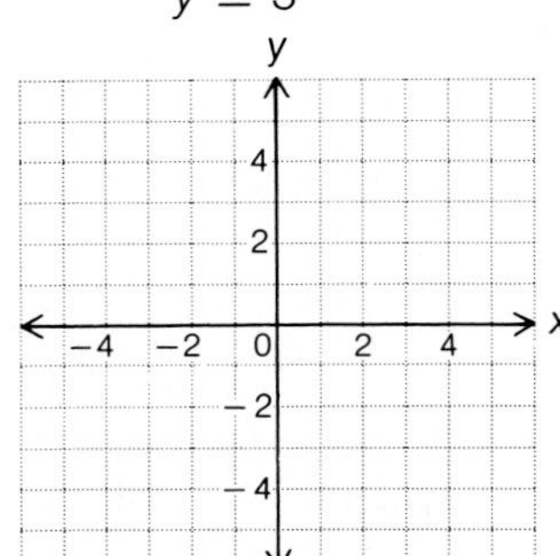

24. $x = 2$
$3x + 2y = 4$

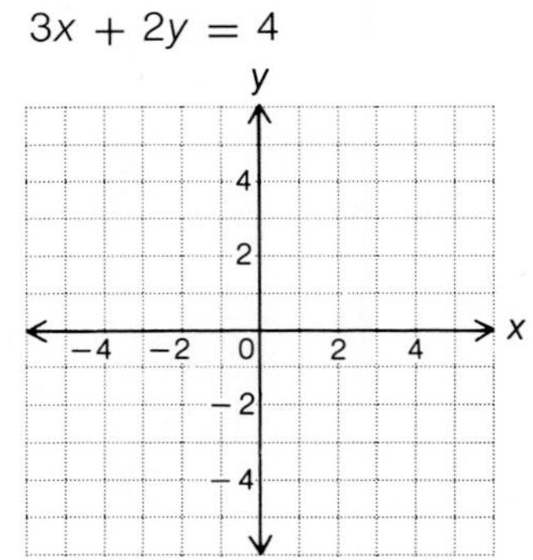

25. $x = 3$
$y = -2$

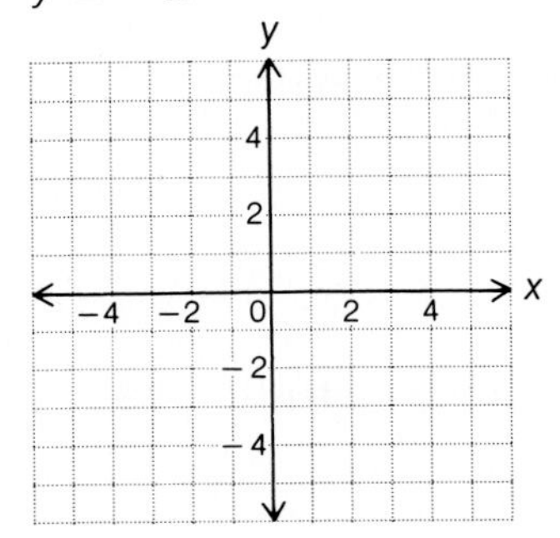

26. $x + 1 = 0$
$y - 3 = 0$

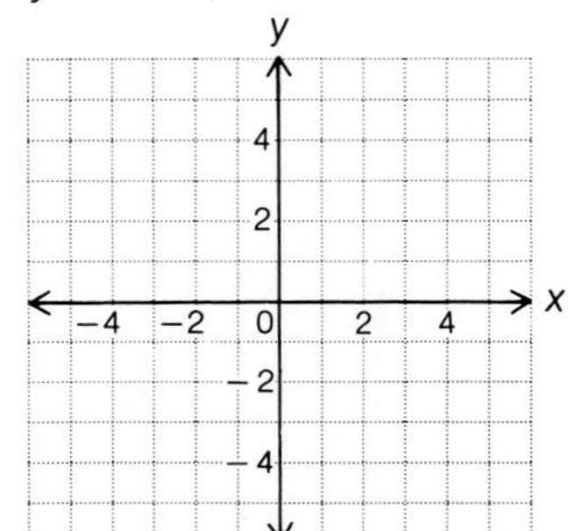

Solve by graphing.

27. $y = 2x - 6$
$x + y = 0$

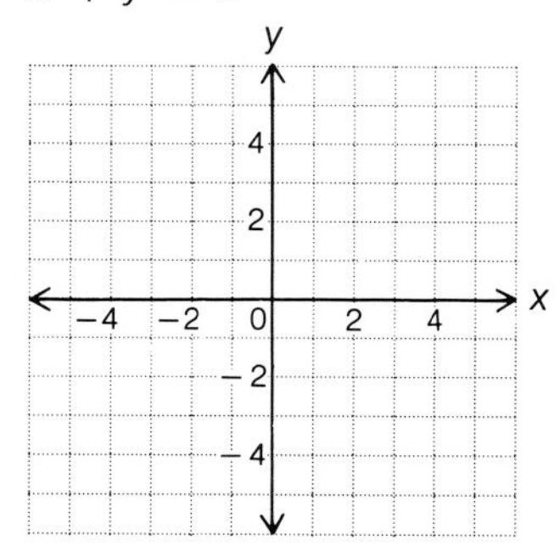

28. $5x - 2y = 11$
$y = 2x - 5$

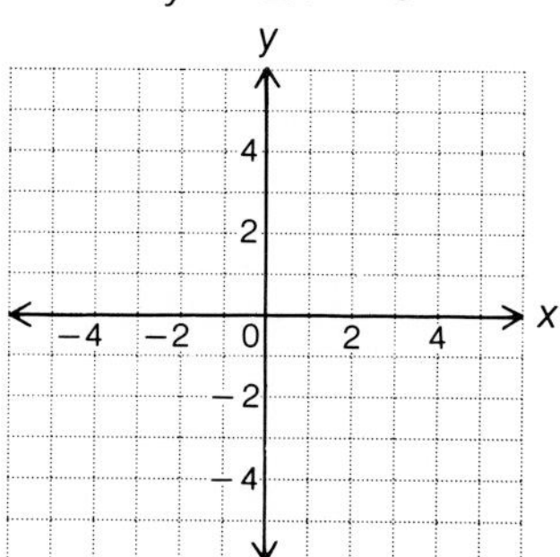

29. $2x + y = -2$
$6x + 3y = 6$

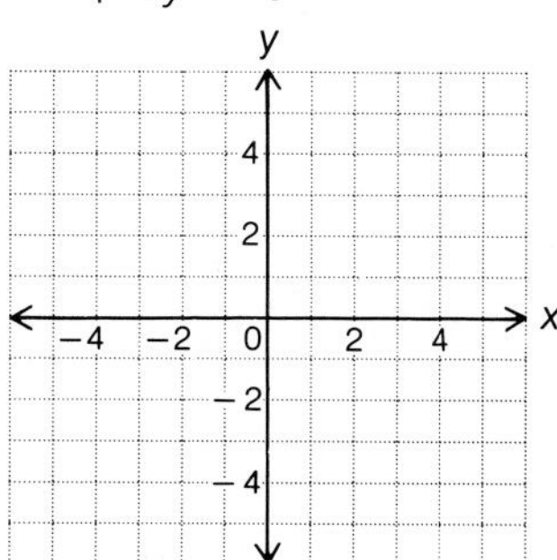

30. $x + y = 5$
$3x + 3y = 6$

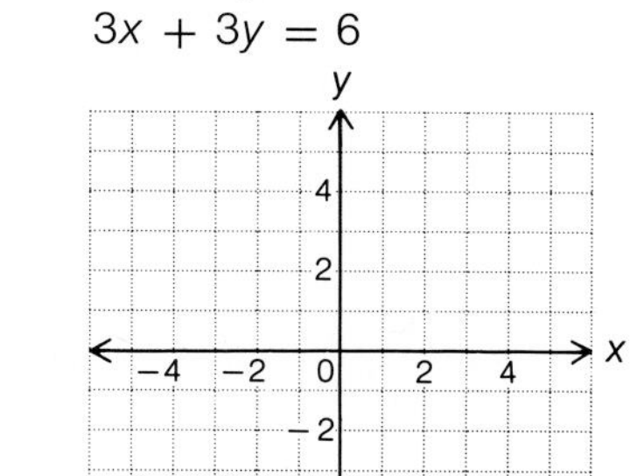

31. $4x - 2y = 4$
$y = 2x - 2$

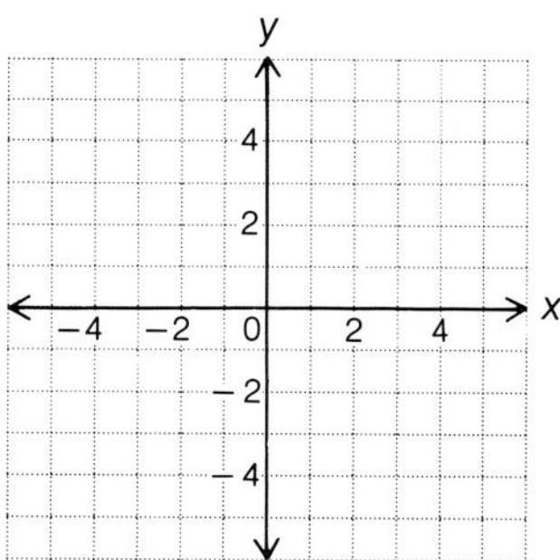

32. $2x + 6y = 6$
$y = -\frac{1}{3}x + 1$

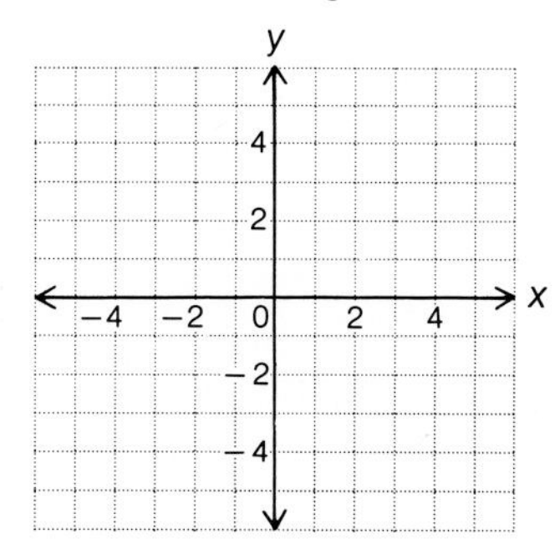

33. $x - y = 5$
$2x - y = 6$

34. $5x - 2y = 10$
$3x + 2y = 6$

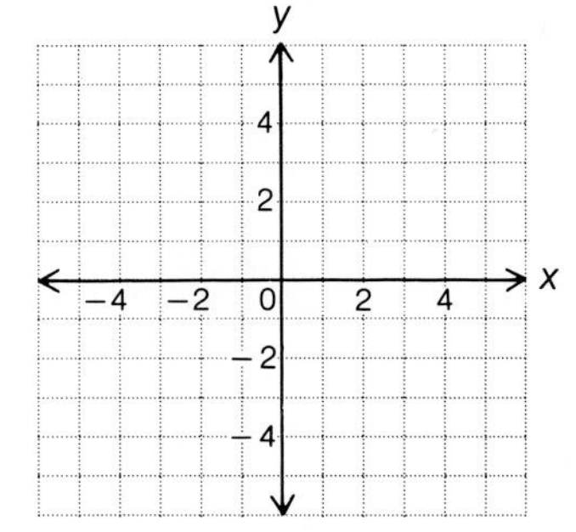

Solve by graphing.

35. $3x + 4y = 0$
$2x - 5y = 0$

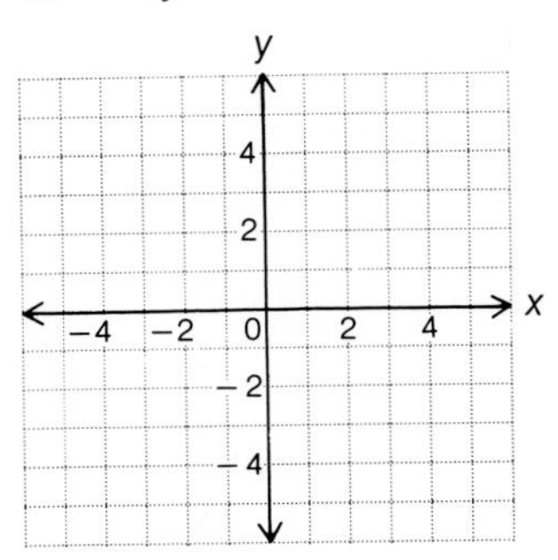

36. $2x - 3y = 0$
$y = -\frac{1}{3}x$

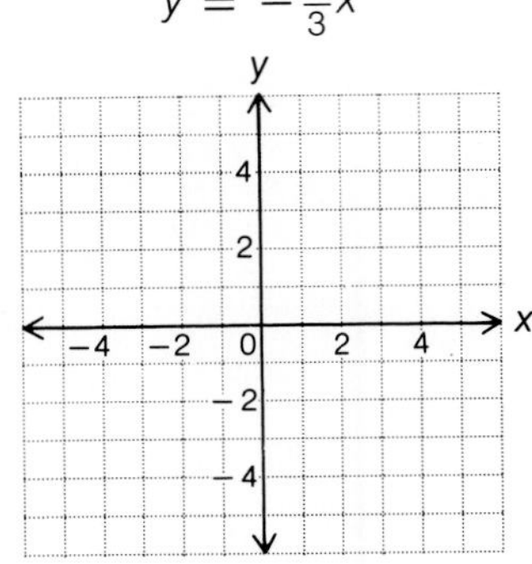

37. $x - 3y = 3$
$2x - 6y = 12$

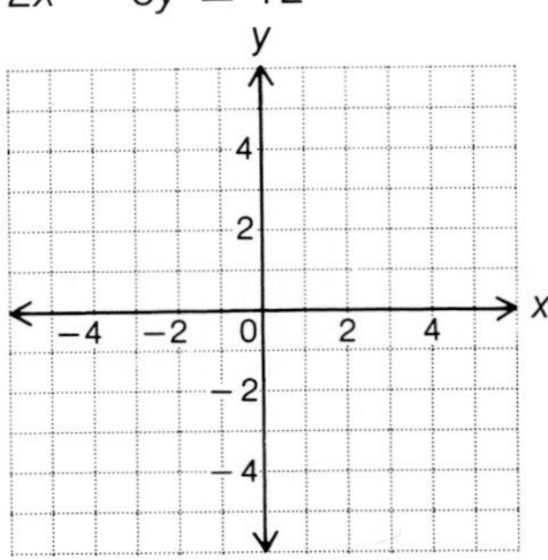

38. $4x + 6y = 12$
$6x + 9y = 18$

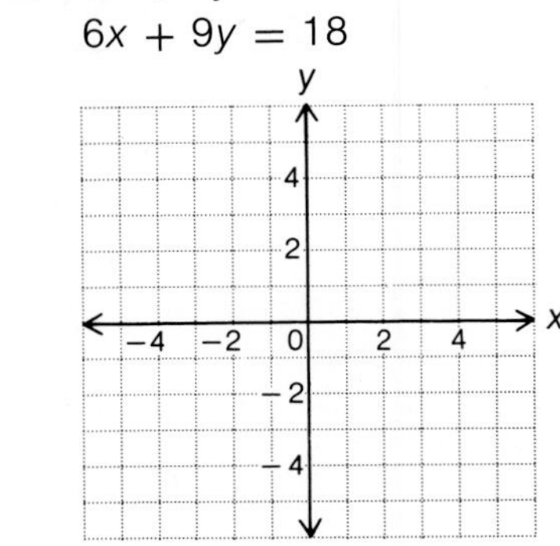

39. $3x + 2y = -4$
$x = 2y + 4$

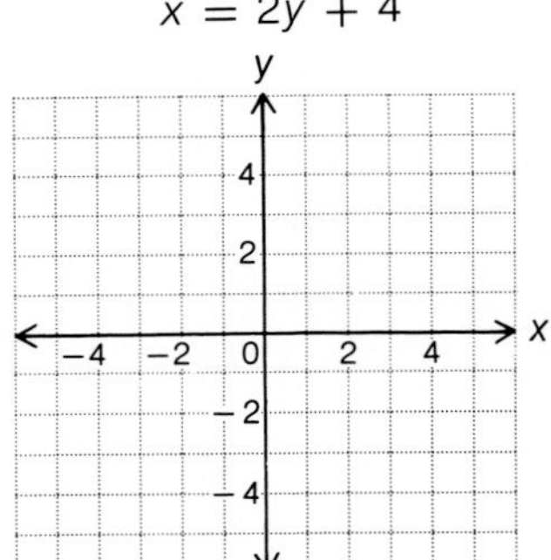

40. $5x + 2y = -14$
$3x - 4y = 2$

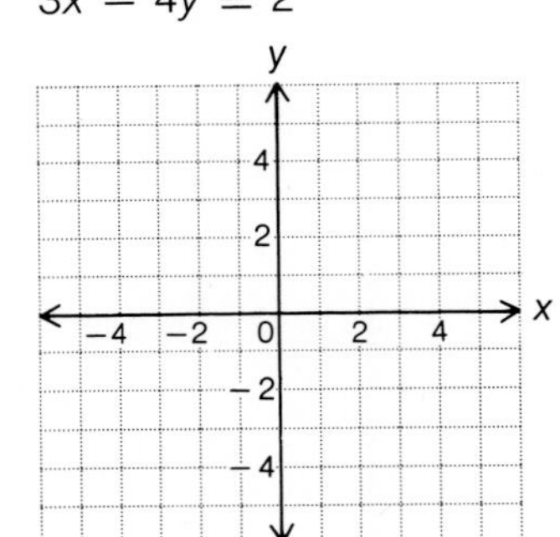

41. $4x - y = 5$
$3x - 2y = 5$

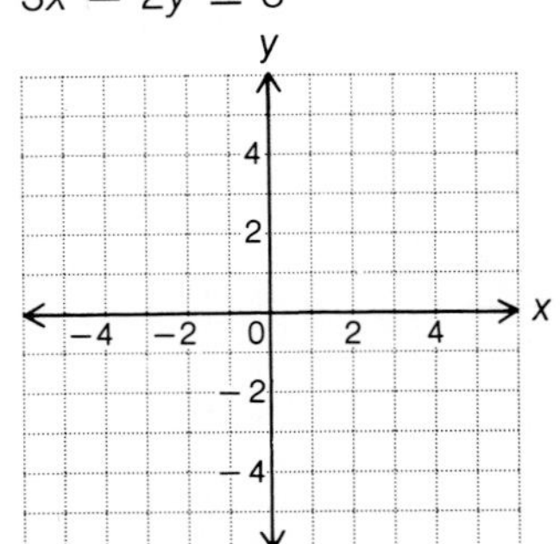

42. $2x - 3y = 9$
$4x + 3y = -9$

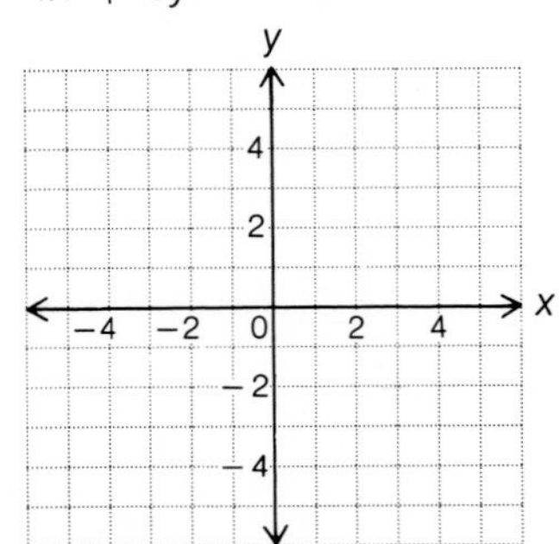

SECTION 2 Solving Systems of Linear Equations by the Substitution Method

2.1 Objective To solve a system of linear equations by the substitution method

Reference for Computer Tutor™

A graphical solution of a system of equations may give only an approximate solution of the system. For example, the point $\left(\frac{1}{4}, \frac{1}{2}\right)$ would be difficult to read from the graph. An algebraic method, called the **substitution method,** can be used to find an exact solution of a system.

In the system of equations at the right, equation (2) states that $y = 3x - 9$. Substitute $3x - 9$ for y in equation (1).

$$\begin{aligned} (1) \quad 2x + 5y &= -11 \\ (2) \quad y &= 3x - 9 \end{aligned}$$

$$2x + 5(3x - 9) = -11$$

Solve for x.

$$\begin{aligned} 2x + 15x - 45 &= -11 \\ 17x - 45 &= -11 \\ 17x &= 34 \\ x &= 2 \end{aligned}$$

Substitute the value of x into equation (2) and solve for y.

$$\begin{aligned} (2) \quad y &= 3x - 9 \\ y &= 3 \cdot 2 - 9 \\ y &= 6 - 9 \\ y &= -3 \end{aligned}$$

The solution is $(2, -3)$.

Solve: $5x + y = 4$
$2x - 3y = 5$

$$\begin{aligned} (1) \quad 5x + y &= 4 \\ (2) \quad 2x - 3y &= 5 \end{aligned}$$

Solve equation (1) for y. Equation (1) is chosen because it is the easier equation to solve for one variable in terms of the other.

$$\begin{aligned} 5x + y &= 4 \\ y &= -5x + 4 \end{aligned}$$

Substitute $-5x + 4$ for y in equation (2).

$$2x - 3(-5x + 4) = 5$$

Solve for x.

$$\begin{aligned} 2x + 15x - 12 &= 5 \\ 17x - 12 &= 5 \\ 17x &= 17 \\ x &= 1 \end{aligned}$$

Substitute the value of x in equation (1) and solve for y.

$$\begin{aligned} 5x + y &= 4 \\ 5(1) + y &= 4 \\ 5 + y &= 4 \\ y &= -1 \end{aligned}$$

The solution is $(1, -1)$.

Example 1 Solve by substitution:

$$3x + 4y = -2$$
$$-x + 2y = 4$$

Solution Solve equation (2) for x.

$$-x + 2y = 4$$
$$-x = -2y + 4$$
$$x = 2y - 4$$

Substitute in equation (1).

$$3(2y - 4) + 4y = -2$$
$$6y - 12 + 4y = -2$$
$$10y - 12 = -2$$
$$10y = 10$$
$$y = 1$$

Substitute in equation (2).

$$-x + 2y = 4$$
$$-x + 2(1) = 4$$
$$-x + 2 = 4$$
$$-x = 2$$
$$x = -2$$

The solution is $(-2,1)$.

Example 2 Solve by substitution:

$$7x - y = 4$$
$$3x + 2y = 9$$

Your solution

Example 3 Solve by substitution:

$$4x + 2y = 5$$
$$y = -2x + 1$$

Solution

$$4x + 2y = 5$$
$$4x + 2(-2x + 1) = 5$$
$$4x - 4x + 2 = 5$$
$$2 = 5$$

This is not a true equation. The lines are parallel and therefore do not intersect. The system does not have a solution.

Example 4 Solve by substitution:

$$3x - y = 4$$
$$y = 3x + 2$$

Your solution

Example 5 Solve by substitution:

$$6x - 2y = 4$$
$$y = 3x - 2$$

Solution

$$6x - 2y = 4$$
$$6x - 2(3x - 2) = 4$$
$$6x - 6x + 4 = 4$$
$$4 = 4$$

This is a true equation. The two equations represent the same line. Any ordered pair that is a solution of one equation is also a solution of the other equation.

Example 6 Solve by substitution:

$$y = -2x + 1$$
$$6x + 3y = 3$$

Your solution

2.1 Exercises

Solve by substitution.

1. $2x + 3y = 7$
   $x = 2$

2. $y = 3$
   $3x - 2y = 6$

3. $y = x - 3$
   $x + y = 5$

4. $y = x + 2$
   $x + y = 6$

5. $x = y - 2$
   $x + 3y = 2$

6. $x = y + 1$
   $x + 2y = 7$

7. $2x + 3y = 9$
   $y = x - 2$

8. $3x + 2y = 11$
   $y = x + 3$

9. $3x - y = 2$
   $y = 2x - 1$

10. $2x - y = -5$
    $y = x + 4$

11. $x = 2y - 3$
    $2x - 3y = -5$

12. $x = 3y - 1$
    $3x + 4y = 10$

13. $y = 4 - 3x$
    $3x + y = 5$

14. $y = 2 - 3x$
    $6x + 2y = 7$

15. $x = 3y + 3$
    $2x - 6y = 12$

16. $x = 2 - y$
    $3x + 3y = 6$

17. $3x + 5y = -6$
    $x = 5y + 3$

18. $y = 2x + 3$
    $4x - 3y = 1$

19. $4x - 3y = -1$
    $y = 2x - 3$

20. $3x - 7y = 28$
    $x = 3 - 4y$

21. $7x + y = 14$
    $2x - 5y = -33$

Solve by substitution.

22. $3x + y = 4$
$4x - 3y = 1$

23. $x - 4y = 9$
$2x - 3y = 11$

24. $3x - y = 6$
$x + 3y = 2$

25. $4x - y = -5$
$2x + 5y = 13$

26. $3x - y = 5$
$2x + 5y = -8$

27. $3x + 4y = 18$
$2x - y = 1$

28. $4x + 3y = 0$
$2x - y = 0$

29. $5x + 2y = 0$
$x - 3y = 0$

30. $6x - 3y = 6$
$2x - y = 2$

31. $3x + y = 4$
$9x + 3y = 12$

32. $x - 5y = 6$
$2x - 7y = 9$

33. $x + 7y = -5$
$2x - 3y = 5$

34. $y = 2x + 11$
$y = 5x - 19$

35. $y = 2x - 8$
$y = 3x - 13$

36. $y = -4x + 2$
$y = -3x - 1$

37. $x = 3y + 7$
$x = 2y - 1$

38. $x = 4y - 2$
$x = 6y + 8$

39. $x = 3 - 2y$
$x = 5y - 10$

40. $y = 2x - 7$
$y = 4x + 5$

41. $3x - y = 11$
$2x + 5y = -4$

42. $-x + 6y = 8$
$2x + 5y = 1$

SECTION 3 Solving Systems of Linear Equations by the Addition Method

3.1 Objective To solve a system of linear equations by the addition method

Reference for Computer Tutor™

Another algebraic method for solving a system of equations is called the **addition method.** It is based on the Addition Property of Equations.

Note, for the system of equations at the right, the effect of adding equation (2) to equation (1). Since $2y$ and $-2y$ are opposites, adding the equations results in an equation with only one variable.

$$\begin{aligned} (1)\quad 3x + 2y &= 4 \\ (2)\quad 4x - 2y &= 10 \\ 7x + 0y &= 14 \\ 7x &= 14 \end{aligned}$$

The solution of the resulting equation is the first component of the ordered pair solution of the system.

$$\begin{aligned} 7x &= 14 \\ x &= 2 \end{aligned}$$

The second component is found by substituting the value of x into equation (1) or (2) and then solving for y. Equation (1) is used here.

$$\begin{aligned} (1)\quad 3x + 2y &= 4 \\ 3\cdot 2 + 2y &= 4 \\ 6 + 2y &= 4 \\ 2y &= -2 \\ y &= -1 \end{aligned}$$

The solution is $(2, -1)$.

Sometimes adding the two equations does not eliminate one of the variables. In this case, use the Multiplication Property of Equations to rewrite one or both of the equations, so that when the equations are added, one of the variables is eliminated.

To do this, first choose which variable to eliminate. The coefficients of that variable must be opposites. Multiply each equation by a constant which will produce coefficients which are opposites.

Solve: $3x + 2y = 7$
$5x - 4y = 19$

$$\begin{aligned} (1)\quad 3x + 2y &= 7 \\ (2)\quad 5x - 4y &= 19 \end{aligned}$$

Eliminate y. Multiply equation (1) by 2.

$$\begin{aligned} 2(3x + 2y) &= 2\cdot 7 \\ 5x - 4y &= 19 \end{aligned}$$

Now the coefficients of the y terms are opposites.

$$\begin{aligned} 6x + 4y &= 14 \\ 5x - 4y &= 19 \end{aligned}$$

Add the equations.
Solve for x.

$$\begin{aligned} 11x + 0y &= 33 \\ 11x &= 33 \\ x &= 3 \end{aligned}$$

Substitute the value of x into one of the equations and solve for y. Equation (2) is used here.

$$\begin{aligned} (2)\quad 5x - 4y &= 19 \\ 5\cdot 3 - 4y &= 19 \\ 15 - 4y &= 19 \\ -4y &= 4 \\ y &= -1 \end{aligned}$$

The solution is $(3, -1)$.

Solve: $5x + 6y = 3$
$2x - 5y = 16$

(1) $5x + 6y = 3$
(2) $2x - 5y = 16$

Eliminate x. Multiply equation (1) by 2 and equation (2) by -5. Note how the constants are selected.

$2(5x + 6y) = 2 \cdot 3$
$-5(2x - 5y) = -5 \cdot 16$

The negative is used so that the coefficients will be opposites.

Now the coefficients of the x terms are opposites.

$10x + 12y = 6$
$-10x + 25y = -80$

Add the equations.
Solve for y.

$0x + 37y = -74$
$37y = -74$
$y = -2$

Substitute the value of y into one of the equations and solve for x. Equation (1) is used here.

(1) $5x + 6y = 3$
$5x + 6(-2) = 3$
$5x - 12 = 3$
$5x = 15$
$x = 3$

The solution is $(3, -2)$.

Solve: $5x = 2y + 19$
$3x + 4y = 1$

(1) $5x = 2y + 19$
(2) $3x + 4y = 1$

Write equation (1) in the form $Ax + By = C$.

$5x - 2y = 19$
$3x + 4y = 1$

Eliminate y. Multiply equation (1) by 2.

$2(5x - 2y) = 2 \cdot 19$
$3x + 4y = 1$

Now the coefficients of the y terms are opposites.

$10x - 4y = 38$
$3x + 4y = 1$

Add the equations.
Solve for x.

$13x + 0y = 39$
$13x = 39$
$x = 3$

Substitute the value of x into one of the equations and solve for y. Equation (1) is used here.

$5x = 2y + 19$
$5 \cdot 3 = 2y + 19$
$15 = 2y + 19$
$-4 = 2y$
$-2 = y$

The solution is $(3, -2)$.

Solve: $2x + y = 2$
$4x + 2y = 5$

(1) $2x + y = 2$
(2) $4x + 2y = 5$

Eliminate y. Multiply equation (1) by -2.

$$-2(2x + y) = -2 \cdot 2$$
$$4x + 2y = 5$$

$$-4x - 2y = -4$$
$$4x + 2y = 5$$

Add the equations.

$$0x + 0y = 1$$
$$0 = 1$$

This is not a true equation. The lines are parallel and therefore do not intersect. The system does not have a solution.

Example 1 Solve by the addition method:
$$2x + 4y = 7$$
$$5x - 3y = -2$$

Solution Eliminate x.

$$5(2x + 4y) = 5 \cdot 7$$
$$-2(5x - 3y) = -2 \cdot (-2)$$

$$10x + 20y = 35$$
$$-10x + 6y = 4$$

Add the equations.

$$26y = 39$$
$$y = \frac{39}{26} = \frac{3}{2}$$

Replace y in equation (1).

$$2x + 4\left(\frac{3}{2}\right) = 7$$
$$2x + 6 = 7$$
$$2x = 1$$
$$x = \frac{1}{2}$$

The solution is $\left(\frac{1}{2}, \frac{3}{2}\right)$.

Example 2 Solve by the addition method:
$$x - 2y = 1$$
$$2x + 4y = 0$$

Your solution

Solution on p. A53

Example 3 Solve by the addition method:

$$6x + 9y = 15$$
$$4x + 6y = 10$$

Solution Eliminate x.

$$4(6x + 9y) = 4 \cdot 15$$
$$-6(4x + 6y) = -6 \cdot 10$$

$$24x + 36y = 60$$
$$-24x - 36y = -60$$

Add the equations.

$$0x + 0y = 0$$
$$0 = 0$$

This is a true equation. The two equations represent the same line. Any ordered pair that is a solution of one equation is also a solution of the other equation.

Example 4 Solve by the addition method:

$$2x - 3y = 4$$
$$-4x + 6y = -8$$

Your solution

Example 5 Solve by the addition method:

$$2x = y + 8$$
$$3x + 2y = 5$$

Solution Write equation (1) in the form $Ax + By = C$.

$$2x = y + 8$$
$$2x - y = 8$$

Eliminate y.

$$2(2x - y) = 2 \cdot 8$$
$$3x + 2y = 5$$

$$4x - 2y = 16$$
$$3x + 2y = 5$$

Add the equations.

$$7x = 21$$
$$x = 3$$

Replace x in equation (1).

$$2 \cdot 3 = y + 8$$
$$6 = y + 8$$
$$-2 = y$$

The solution is $(3, -2)$.

Example 6 Solve by the addition method:

$$4x + 5y = 11$$
$$3y = x + 10$$

Your solution

3.1 Exercises

Solve by the addition method.

1. $x + y = 4$
   $x - y = 6$

2. $2x + y = 3$
   $x - y = 3$

3. $x + y = 4$
   $2x + y = 5$

4. $x - 3y = 2$
   $x + 2y = -3$

5. $2x - y = 1$
   $x + 3y = 4$

6. $x - 2y = 4$
   $3x + 4y = 2$

7. $4x - 5y = 22$
   $x + 2y = -1$

8. $3x - y = 11$
   $2x + 5y = 13$

9. $2x - y = 1$
   $4x - 2y = 2$

10. $x + 3y = 2$
    $3x + 9y = 6$

11. $4x + 3y = 15$
    $2x - 5y = 1$

12. $3x - 7y = 13$
    $6x + 5y = 7$

13. $2x - 3y = 1$
    $4x - 6y = 2$

14. $2x + 4y = 6$
    $3x + 6y = 9$

15. $5x - 2y = -1$
    $x + 3y = -5$

16. $4x - 3y = 1$
    $8x + 5y = 13$

17. $5x + 7y = 10$
    $3x - 14y = 6$

18. $7x + 10y = 13$
    $4x + 5y = 6$

19. $3x - 2y = 0$
    $6x + 5y = 0$

20. $5x + 2y = 0$
    $3x + 5y = 0$

21. $2x - 3y = 16$
    $3x + 4y = 7$

Solve by the addition method.

22. $3x + 4y = 10$
$4x + 3y = 11$

23. $5x + 3y = 7$
$2x + 5y = 1$

24. $-2x + 7y = 9$
$3x + 2y = -1$

25. $7x - 2y = 13$
$5x + 3y = 27$

26. $12x + 5y = 23$
$2x - 7y = 39$

27. $8x - 3y = 11$
$6x - 5y = 11$

28. $4x - 8y = 36$
$3x - 6y = 27$

29. $5x + 15y = 20$
$2x + 6y = 8$

30. $y = 2x - 3$
$3x + 4y = -1$

31. $3x = 2y + 7$
$5x - 2y = 13$

32. $2y = 4 - 9x$
$9x - y = 25$

33. $2x + 9y = 16$
$5x = 1 - 3y$

34. $3x - 4 = y + 18$
$4x + 5y = -21$

35. $2x + 3y = 7 - 2x$
$7x + 2y = 9$

36. $5x - 3y = 3y + 4$
$4x + 3y = 11$

37. $3x + y = 1$
$5x + y = 2$

38. $2x - y = 1$
$2x - 5y = -1$

39. $4x + 3y = 3$
$x + 3y = 1$

40. $2x - 5y = 4$
$x + 5y = 1$

41. $3x - 4y = 1$
$4x + 3y = 1$

42. $2x - 7y = -17$
$3x + 5y = 17$

SECTION 4 Application Problems in Two Variables

4.1 Objective To solve rate-of-wind or current problems

Reference for Computer Tutor™

Motion problems which involve an object moving with or against a wind or current normally require two variables to solve.

Flying with the wind, a small plane can fly 600 mi in 3 h. Against the wind, the plane can fly the same distance in 4 h. Find the rate of the plane in calm air and the rate of the wind.

STRATEGY FOR SOLVING RATE-OF-WIND OR CURRENT PROBLEMS

▷ Choose one variable to represent the rate of the object in calm conditions and a second variable to represent the rate of the wind or current. Using these variables, express the rate of the object with and against the wind or current. Use the equation $d = rt$ to write expressions for the distance traveled by the object. The results can be recorded in a table.

Rate of plane in calm air: p
Rate of wind: w

	Rate	·	Time	=	Distance
With the wind	$p + w$	·	3	=	$3(p + w)$
Against the wind	$p - w$	·	4	=	$4(p - w)$

▷ Determine how the expressions for distance are related.

The distance traveled with the wind is 600 mi. $3(p + w) = 600$
The distance traveled against the wind is 600 mi. $4(p - w) = 600$

Solve the system of equations.

$3(p + w) = 600$ $\quad \frac{1}{3} \cdot 3(p + w) = \frac{1}{3} \cdot 600$ $\quad p + w = 200$

$4(p - w) = 600$ $\quad \frac{1}{4} \cdot 4(p - w) = \frac{1}{4} \cdot 600$ $\quad p - w = 150$

$2p = 350$
$p = 175$

$p + w = 200$
$175 + w = 200$
$w = 25$

The rate of the plane in calm air is 175 mph.
The rate of the wind is 25 mph.

Example 1

A 450-mile trip from one city to another takes 3 h when a plane is flying with the wind. The return trip, against the wind, takes 5 h. Find the rate of the plane in still air and the rate of the wind.

Strategy

▷ Rate of the plane in still air: p
Rate of the wind: w

	Rate	Time	Distance
With wind	$p + w$	3	$3(p + w)$
Against wind	$p - w$	5	$5(p - w)$

▷ The distance traveled with the wind is 450 mi.
The distance traveled against the wind is 450 mi.

Solution

$$3(p + w) = 450 \qquad \frac{1}{3} \cdot 3(p + w) = \frac{1}{3} \cdot 450$$

$$5(p - w) = 450 \qquad \frac{1}{5} \cdot 5(p - w) = \frac{1}{5} \cdot 450$$

$$p + w = 150$$
$$p - w = 90$$
$$2p = 240$$
$$p = 120$$

$$p + w = 150$$
$$120 + w = 150$$
$$w = 30$$

The rate of the plane in still air is 120 mph.
The rate of the wind is 30 mph.

Example 2

A canoeist paddling with the current can travel 15 mi in 3 h. Against the current it takes 5 h to travel the same distance. Find the rate of the current and the rate of the canoeist in calm water.

Your strategy

Your solution

Solution on p. A54

4.2 Objective To solve application problems using two variables

Reference for Computer Tutor™

The application problems in this section are varieties of those problems solved earlier in the text. Each of the strategies for the problems in this section will result in a system of equations.

A jeweler purchased 5 oz of a gold alloy and 20 oz of a silver alloy for a total cost of \$540. The next day, at the same prices per ounce, the jeweler purchased 4 oz of the gold alloy and 25 oz of the silver alloy for a total cost of \$450. Find the cost per ounce of the gold and silver alloys.

STRATEGY FOR SOLVING AN APPLICATION PROBLEM IN TWO VARIABLES

▷ Choose one variable to represent one of the unknown quantities and a second variable to represent the other unknown quantity. Write numerical or variable expressions for all the remaining quantities. These results can be recorded in two tables, one for each of the conditions.

Cost per ounce of gold: g
Cost per ounce of silver: s

First Day

	Amount	·	Unit Cost	=	Value
Gold	5	·	g	=	$5g$
Silver	20	·	s	=	$20s$

Second Day

	Amount	·	Unit Cost	=	Value
Gold	4	·	g	=	$4g$
Silver	25	·	s	=	$25s$

▷ Determine a system of equations. The strategies presented in Chapter 4 can be used to determine the relationships between the expressions in the tables. Each table will give one equation of the system.

The total value of the purchase on the first day was \$540. $5g + 20s = 540$
The total value of the purchase on the second day was \$450. $4g + 25s = 450$

Solve the system of equations.

$$5g + 20s = 540 \qquad 4(5g + 20s) = 4 \cdot 540 \qquad 20g + 80s = 2160$$
$$4g + 25s = 450 \qquad -5(4g + 25s) = -5 \cdot 450 \qquad -20g - 125s = -2250$$

$$-45s = -90$$
$$s = 2$$

$$5g + 20s = 540$$
$$5g + 20(2) = 540$$
$$5g + 40 = 540$$
$$5g = 500$$
$$g = 100$$

The cost per ounce of the gold alloy was \$100.
The cost per ounce of the silver alloy was \$2.

Example 3

In five years, an oil painting will be twice as old as a water color painting will be then. Five years ago, the oil painting was three times as old as the water color was then. Find the present age of each painting.

Strategy

▷ Present age of oil painting: x
Present age of water color: y

	Present	Future
Oil	x	$x + 5$
Water Color	y	$y + 5$

	Present	Past
Oil	x	$x - 5$
Water Color	y	$y - 5$

▷ In five years, twice the age of the water color will be the age of the oil painting. Five years ago, three times the age of the water color was the age of the oil painting.

Solution

$$2(y + 5) = x + 5 \qquad 2y + 10 = x + 5$$
$$3(y - 5) = x - 5 \qquad 3y - 15 = x - 5$$

$$2y + 5 = x$$
$$3y - 15 = x - 5$$

$$3y - 15 = (2y + 5) - 5$$
$$3y - 15 = 2y$$
$$y - 15 = 0$$
$$y = 15$$

$$2y + 5 = x$$
$$2(15) + 5 = x$$
$$30 + 5 = x$$
$$35 = x$$

The present age of the oil painting is 35 years.
The present age of the water color is 15 years.

Example 4

Two coin banks contain only dimes and quarters. In the first bank, the total value of the coins is \$4.80. In the second bank, there are twice as many quarters as in the first bank and one half the number of dimes. The total value of the coins in the second bank is \$8.40. Find the number of dimes and the number of quarters in the first bank.

Your strategy

Your solution

Solution on p. A54

4.1 Application Problems

Solve.

1. Paddling with the current, a canoeist can go 24 mi in 3 h. Against the current it takes 4 h to go the same distance. Find the rate of the canoeist in calm water and the rate of the current.

2. A pilot flying with the wind flew the 750 mi between two cities in 3 h. The return trip against the wind took 5 h. Find the rate of the plane in calm air and the rate of the wind.

3. A motorboat traveling with the current can go 100 mi in 4 h. Against the current it takes 5 h to go the same distance. Find the rate of the motorboat in still water and the rate of the current.

4. A plane flying with a tailwind flew 360 mi in 2 h. Against the wind, it took 3 h to fly the same distance. Find the rate of the plane in calm air and the rate of the wind.

5. A rowing team rowing with the current traveled 16 mi in 2 h. Against the current, the team rowed 8 mi in 2 h. Find the rate of the rowing team in calm water and the rate of the current.

6. A small plane flew 300 mi with the wind in 2 h. Against the wind, it took 3 h to travel the same distance. Find the rate of the plane in calm air and the rate of the wind.

7. A small plane flew 260 mi in 2 h with the wind. Flying against the wind, the plane flew 180 mi in 2 h. Find the rate of the plane in calm air and the rate of the wind.

8. A motorboat traveling with the current went 30 mi in 3 h. Traveling against the current the boat went 12 mi in 3 h. Find the rate of the boat in calm water and the rate of the current.

9. A crew can row 60 km downstream in 3 h. Rowing upstream, against the current, the crew traveled 24 km in 3 h. Find the rowing rate of the crew in calm water and the rate of the current.

10. A plane flew 2000 km in 5 h traveling with the wind. Against the wind, the plane could fly only 1500 km in the same amount of time. Find the rate of the plane in calm air and the rate of the wind.

4.2 Application Problems

Solve.

11. A business manager had two reports photocopied. The first report, which cost \$3 to photocopy, included 50 black-and-white pages and 10 color pages. The total cost for photocopying the 75 black-and-white and the 20 color pages in the second report was \$5. Find the cost per copy for a black-and-white page and for a color page.

12. A computer store received two shipments of calculators. The value of the first shipment, which contained 10 scientific and 15 business calculators, was \$425. The value of the second shipment, which contained 8 scientific and 20 business calculators, was \$460. Find the cost of a scientific and the cost of a business calculator.

13. A metallurgist made two purchases. The first purchase, which cost \$110, included 20 kg of a tin alloy and 25 kg of an aluminum alloy. The second purchase, which cost \$60, included 10 kg of the tin alloy and 15 kg of the aluminum alloy. Find the cost per kilogram of the tin and the aluminum alloys.

14. For \$28, a customer purchased 2 lb of kona-blend coffee and 3 lb of a mocha-blend coffee. A second customer purchased 4 lb of the kona coffee and 2 lb of the mocha coffee for a total of \$32. Find the cost per pound of the kona coffee and the mocha coffee.

15. Two coin banks contain only nickels and quarters. The total value of the coins in the first bank is \$3.30. In the second bank there are two fewer quarters than in the first bank and twice as many nickels. The total value of the coins in the second bank is \$3.10. Find the number of nickels and the number of quarters in the first bank.

16. Two coin banks contain only nickels and dimes. The total value of the coins in the first bank is \$4. In the second bank there are 10 more nickels than in the first bank and one half as many dimes. The total value of the coins in the second bank is \$3.50. Find the number of nickels and the number of dimes in the first bank.

17. The total value of the dimes and quarters in a coin bank is \$3.70. If the quarters were dimes and the dimes were quarters, the total value of the coins would be \$4. Find the number of dimes and the number of quarters in the bank.

18. The total value of the nickels and dimes in a coin bank is \$5. If the nickels were dimes and the dimes were nickels, the total value of the coins would be \$4. Find the number of nickels and the number of dimes in the bank.

19. One year ago, an adult was five times the age a child was then. One year from now the adult will be four times the age the child will be then. Find the present ages of the adult and the child.

20. If twice the age of a stamp is added to three times the age of a coin, the result is 100. The difference between five times the age of the stamp and twice the age of the coin is three. Find the age of each.

Calculators and Computers

Systems of Equations

In Chapter 9, three methods of determining the solutions to systems of linear equations are presented: solving by graphing, solving by the substitution method, and solving by the addition method. As stated in the text, solving a system by graphing is not the most efficient method of finding a solution. However, doing so should give you a good visual understanding of the concept of solutions of systems of linear equations.

The substitution method is used most often when one variable is given in terms of the other; for example, in the equation $y = 2x + 5$, y is given in terms of x. Therefore, the value $2x + 5$ is easily substituted for y in another equation.

The addition method is used when neither of the equations is easily solved for one of the variables. Students sometimes find this method difficult at first because of the number of steps involved in finding a solution. Remember that the goal here is the same as it was when finding the solution to one equation: to rewrite the equation in the form variable = constant.

By using the addition method, the system of equations $\begin{array}{l} ax + by = c \\ dx + ey = f \end{array}$ can be solved.

The solution is $x = \dfrac{ce - bf}{ae - bd}$ and $y = \dfrac{af - cd}{ae - bd}$, $ae - bd \neq 0$.

Using this solution, a system of equations can be solved by using a calculator. It is helpful to observe that the denominators for each expression are identical. The calculation for the denominator is done first and then stored in the calculator's memory. If the value of the denominator is zero, then the system is dependent and this calculator method cannot be used.

Solve: $\begin{array}{l} 2x - 5y = 9 \\ 4x + 3y = 2 \end{array}$

Make a list of the values of a, b, c, d, e, and f.

$a = 2$ $\quad b = -5$ $\quad c = 9$

$d = 4$ $\quad e = 3$ $\quad f = 2$

Calculate the denominator $D = ae - bd$. $\quad D = 2 \cdot 3 - (-5) \cdot 4 = 6 + 20 = 26$

Store the result in memory. $\quad$ Press M+.

Find x. Replace the letters by the given values. $x = \frac{ce - bf}{D} = \frac{9 \cdot 3 - (-5) \cdot 2}{26}$

Calculate x. 9 [×] 3 [−] [(] 5 [+/−] [×] 2 [)] [÷] [MR] [=]

The result in the display should be 1.423077.

Find y. Replace the letters by the given values. $y = \frac{af - cd}{D} = \frac{2 \cdot 2 - 9 \cdot 4}{26}$

Calculate y. 2 [×] 2 [−] [(] 9 [×] 4 [)] [÷] [MR] [=]

The result in the display should be −1.230769.

The solution of the system is (1.423077, −1.230769).

The keys [M+] (store in memory) and [MR] (recall from memory) were used for this illustration. Some calculators use the keys [STO] (store in memory) and [RCL] (recall from memory). If your calculator uses these keys, then use these keys in place of the keys shown in the illustration.

Chapter Summary

KEY WORDS

Equations considered together are called a **system of equations.**

A **solution of a system of equations** in two variables is an ordered pair which is a solution of each equation of the system.

ESSENTIAL RULES

A system of equations can be solved by **the graphing method, the substitution method, or the addition method.**

Review/Test

SECTION 1

1.1a Is $(-2,3)$ a solution of the system
$$2x + 5y = 11$$
$$x + 3y = 7?$$

1.1b Is $(1,-3)$ a solution of the system
$$3x - 2y = 9$$
$$4x + y = 1?$$

1.2 Solve by graphing:
$$3x + 2y = 6$$
$$5x + 2y = 2$$

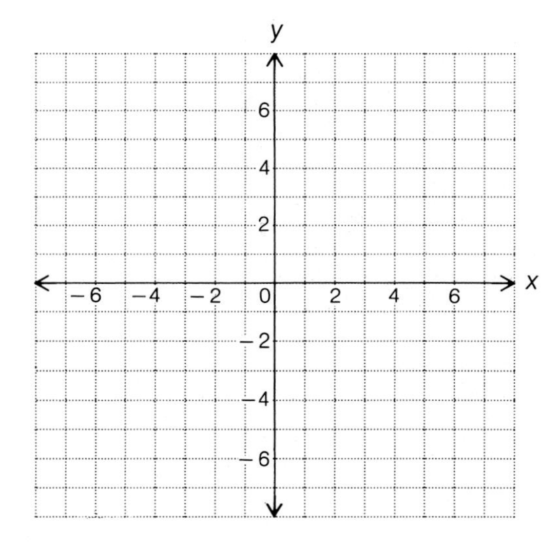

Review/Test

SECTION 2

2.1a Solve by substitution.

$$4x - y = 11$$
$$y = 2x - 5$$

2.1b Solve by substitution.

$$x = 2y + 3$$
$$3x - 2y = 5$$

2.1c Solve by substitution.

$$3x + 5y = 1$$
$$2x - y = 5$$

2.1d Solve by substitution.

$$3x - 5y = 13$$
$$x + 3y = 1$$

SECTION 3

3.1a Solve by the addition method.

$$4x + 3y = 11$$
$$5x - 3y = 7$$

3.1b Solve by the addition method.

$$2x - 5y = 6$$
$$4x + 3y = -1$$

3.1c Solve by the addition method.

$$7x + 3y = 11$$
$$2x - 5y = 9$$

3.1d Solve by the addition method.

$$5x + 6y = -7$$
$$3x + 4y = -5$$

SECTION 4

4.1 With the wind, a plane flies 240 mi in 2 h. Against the wind, the plane requires 3 h to fly the same distance. Find the rate of the plane in calm air and the rate of the wind.

4.2 For the first performance of a play in a community theater, 50 reserved-seat tickets and 80 general-admission tickets were sold. The total receipts were $980. For the second performance, 60 reserved-seat tickets and 90 general-admission tickets were sold. The total receipts were $1140. Find the price of a reserved-seat ticket and the price of a general-admission ticket.

Cumulative Review/Test

1. Evaluate $\frac{a^2 - b^2}{2a}$ when $a = 4$ and $b = -2$.

2. Solve: $-\frac{3}{4}x = \frac{9}{8}$

3. Solve: $4 - 3(2 - 3x) = 7x - 9$

4. Simplify: $(2a^2 - 3a + 1)(2 - 3a)$

5. Simplify: $\frac{(-2x^2y)^4}{-8x^3y^2}$

6. Simplify: $(4b^2 - 8b + 4) \div (2b - 3)$

7. Simplify $\frac{8x^{-2}y^5}{-2xy^4}$.

8. Factor $4x^2y^4 - 64y^2$.

9. Solve: $(x - 5)(x + 2) = -6$

10. Simplify: $\frac{x^2 - 6x + 8}{2x^3 + 6x^2} \div \frac{2x - 8}{4x^3 + 12x^2}$

11. Simplify: $\frac{x - 1}{x + 2} + \frac{2x + 1}{x^2 + x - 2}$

12. Simplify: $\frac{x + 4 - \frac{7}{x - 2}}{x + 8 + \frac{21}{x - 2}}$

13. Solve: $\frac{x}{2x - 3} + 2 = \frac{-7}{2x - 3}$

14. Solve $A = P + Prt$ for r.

15. Find the x- and y-intercept for $2x - 3y = 12$.

16. Find the slope of the line containing the points $(2, -3)$ and $(-3, 4)$.

Cumulative Review/Test

17. Find the equation of the line which contains the point $(-2,3)$ and has slope $-\frac{3}{2}$.

18. Is $(2,0)$ a solution of the system

$$5x - 3y = 10$$
$$4x + 7y = 8?$$

19. Solve by substitution.

$$3x - 5y = -23$$
$$x + 2y = -4$$

20. Solve by the addition method.

$$5x - 3y = 29$$
$$4x + 7y = -5$$

21. A total of \$8750 is invested into two simple interest accounts. On one account the annual simple interest rate is 9.6%, while on the second account the annual simple interest rate is 7.2%. How much should be invested in each account so that the total interest earned by each account is the same?

22. A passenger train leaves a train depot one-half hour after a freight train leaves the same depot. The freight train is traveling 8 mph slower than the passenger train. Find the rate of each train if the passenger train overtakes the freight train in 3 h.

23. The length of each side of a square is extended 4 in. The area of the resulting square is 144 in.^{2}. Find the length of a side of the original square.

24. A plane can travel 160 mph in calm air. Flying with the wind, the plane can fly 570 mi in the same amount of time as it takes to fly 390 mi against the wind. Find the rate of the wind.

25. Graph $2x - 3y = 6$.

26. Solve by graphing:

$$3x + 2y = 6$$
$$3x - 2y = 6$$

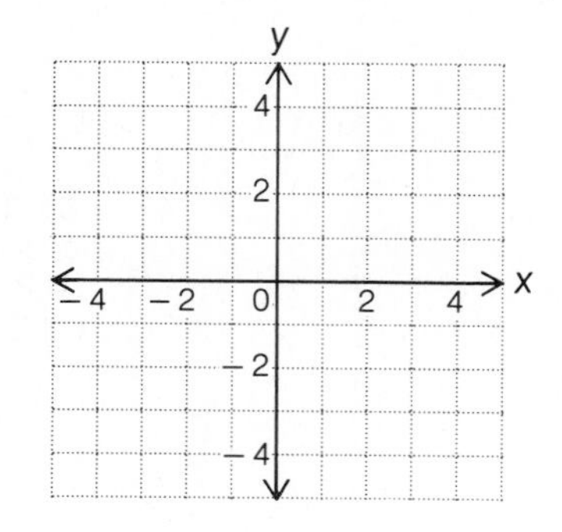

27. With the current, a motorboat can travel 48 mi in 3 h. Against the current, the boat requires 4 h to travel the same distance. Find the rate of the boat in calm water.

28. Two coin banks contain only dimes and nickels. In the first bank, the total value of the coins is \$5.50. In the second bank, there are one half as many dimes as in the first bank and 10 less nickels. The total value of the coins in the second bank is \$3. Find the number of dimes in the first bank.

10

Inequalities

Objectives

- To write a set using the roster method
- To write a set using set builder notation
- To graph the solution set of an inequality on the number line
- To solve an inequality using the Addition Property of Inequalities
- To solve an inequality using the Multiplication Property of Inequalities
- To solve application problems
- To solve general inequalities
- To solve application problems
- To graph an inequality in two variables

Calculations of Pi

There are many early references of estimated values for pi. Among one of the earliest is from the Rhind Papyrus, which was found in Egypt in the 1800's. Scientists have estimated that these tablets were written around 1600 B.C. The Rhind Papyrus contains the estimate 3.1604 for pi.

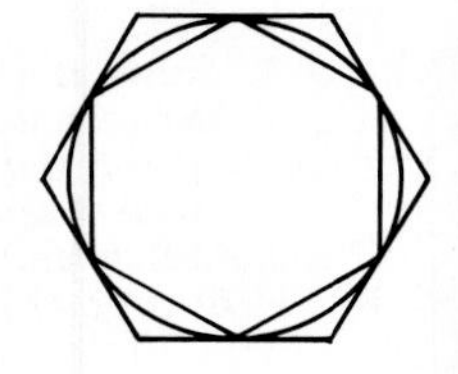

One of the most famous calculations of pi came around 240 B.C. and was calculated by Archimedes. The calculation was based on finding the perimeter of inscribed and circumscribed six-sided polygons (or hexagons). Once the perimeter for a hexagon figure was calculated, known formulas could be used to calculate the perimeter of polygons with twice the number of sides. Continuing in this way, Archimedes calculated the perimeters for the polygons with 12, 24, 48, and 96 sides.

His calculations resulted in a value of pi between $3\frac{10}{71}$ and $3\frac{1}{7}$. You might recognize $3\frac{1}{7}$ or $\frac{22}{7}$ as an approximation for pi still used today.

After Archimedes' work, calculations to improve the accuracy of pi were continued. One French mathematician, using Archimedes' method, estimated pi by using a polygon of 393,216 sides. A mathematician from the Netherlands estimated pi by using a polygon with over one million sides.

Around the 1650's, new mathematical methods were developed to estimate the value of pi. These methods started yielding estimates of pi accurate to over 70 places. By the 1850's an estimate for pi was accurate to 200 places.

Today, using more refined mathematical methods and a computer, estimates of the value of pi now exceed one million places.

In 1914, a Scientific American contained the following short note:

> "See, I have a rhyme assisting my feeble brain,
> its tasks oftimes resisting."

Can you see what this note has to do with the estimates for the value of pi?

(Each word length represents a digit in the approximation 3.141592653579.)

SECTION 1 Sets

1.1 Objective To write a set using the roster method

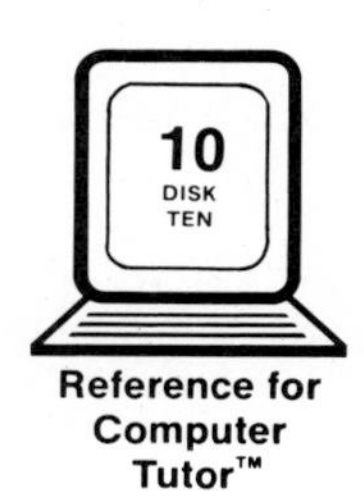

Reference for Computer Tutor™

A **set** is a collection of objects. The objects in a set are called the **elements** of the set.

The **roster method** of writing a set encloses a list of the elements in braces.

The set of the last three letters of the alphabet is written {x,y,z}.

The set of the positive integers less than 5 is written {1,2,3,4}.

Use the roster method to write the set of integers between 0 and 10.

A set is designated by a capital letter. Note that 0 and 10 are not elements of the set.

$A = \{1,2,3,4,5,6,7,8,9\}$

Use the roster method to write the set of natural numbers.

The three dots mean that the pattern of numbers continues without end.

$A = \{1,2,3,4,...\}$

The symbol "$\in$" means "is an element of."

$2 \in B$ is read "2 is an element of set B."

Given $A = \{3,5,9\}$, then $3 \in A$, $5 \in A$, and $9 \in A$.

The **empty set, or null set,** is the set which contains no elements. The symbol $\emptyset$ or { } is used to represent the empty set.

The set of people who have run a two-minute mile is the empty set.

The **union** of two sets, written $A \cup B$, is the set that contains the elements of A and the elements of B.

Find $A \cup B$, given $A = \{1,2,3,4\}$ and $B = \{3,4,5,6\}$.

The union of A and B contains all the elements of A and all the elements of B. The elements which are in both A and B are listed only once.

$A \cup B = \{1,2,3,4,5,6\}$

The **intersection** of two sets, written $A \cap B$, is the set that contains the elements which are common to both A and B.

Find $A \cap B$, given $A = \{1,2,3,4\}$ and $B = \{3,4,5,6\}$.

The intersection of A and B contains the elements common to A and B.

$A \cap B = \{3,4\}$

Example 1 Use the roster method to write the set of the odd positive integers less than 12.

Solution $A = \{1,3,5,7,9,11\}$

Example 2 Use the roster method to write the set of the odd negative integers greater than -10.

Your solution

Example 3 Use the roster method to write the set of the even positive integers.

Solution $A = \{2,4,6,...\}$

Example 4 Use the roster method to write the set of the odd positive integers.

Your solution

Example 5 Find $D \cup E$, given $D = \{6,8,10,12\}$ and $E = \{-8,-6,10,12\}$.

Solution $D \cup E = \{-8,-6,6,8,10,12\}$

Example 6 Find $A \cup B$, given $A = \{-2,-1,0,1,2\}$ and $B = \{0,1,2,3,4\}$.

Your solution

Example 7 Find $A \cap B$, given $A = \{5,6,9,11\}$ and $B = \{5,9,13,15\}$.

Solution $A \cap B = \{5,9\}$

Example 8 Find $C \cap D$, given $C = \{10,12,14,16\}$ and $D = \{10,16,20,26\}$.

Your solution

Example 9 Find $A \cap B$, given $A = \{1,2,3,4\}$ and $B = \{8,9,10,11\}$.

Solution $A \cap B = \varnothing$

Example 10 Find $A \cap B$, given $A = \{-5,-4,-3,-2\}$ and $B = \{2,3,4,5\}$.

Your solution

Solutions on p. A56

1.2 Objective To write a set using set builder notation

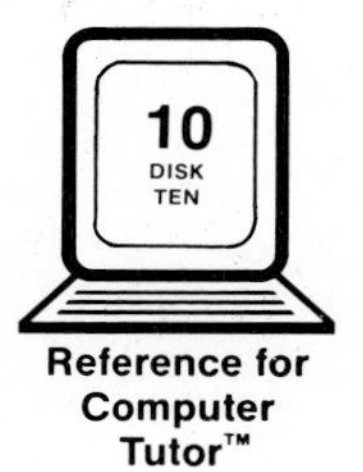

Reference for Computer Tutor™

Another method of representing sets is called **set builder notation.** Using set builder notation, the set of all positive integers less than 10 would be written:

$\{x \mid x < 10, x \text{ is a positive integer}\}$, read "the set of all x such that x is less than 10 and x is a positive integer."

Use set builder notation to write the set of real numbers greater than 4.

"$x \in$ real numbers" is read "x is an element of the real numbers."

$\{x \mid x > 4, x \in \text{real numbers}\}$

Example 11 Use set builder notation to write the set of negative integers greater than -100.

Solution $\{x \mid x > -100, x \text{ is a negative integer}\}$

Example 12 Use set builder notation to write the set of positive even integers less than 59.

Your solution

Example 13 Use set builder notation to write the set of real numbers less than 60.

Solution $\{x \mid x < 60, x \in \text{real numbers}\}$

Example 14 Use set builder notation to write the set of real numbers greater than -3.

Your solution

Solutions on p. A56

1.3 Objective To graph the solution set of an inequality on the number line

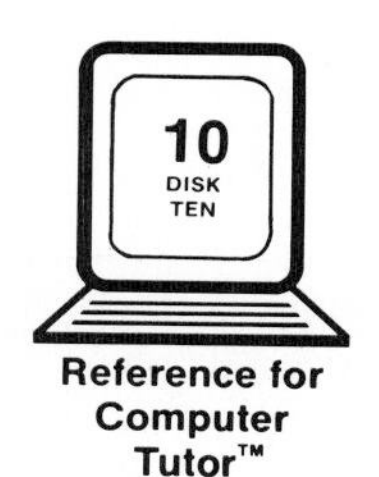

Reference for Computer Tutor™

An expression which contains the symbol $>$, $<$, $\geq$ (is greater than or equal to), or $\leq$ (is less than or equal to) is called an **inequality.** An inequality expresses the relative order of two mathematical expressions. The expressions can be either numerical or variable expressions.

$4 > 2$
$3x \leq 7$
$x^2 - 2x > y + 4$
Inequalities

The **solution set of an inequality** is a set of real numbers and can be graphed on the number line.

Graph the solution set of $x > 1$.

The solution set is the real numbers greater than 1. The circle on the graph indicates that 1 is not included in the solution set.

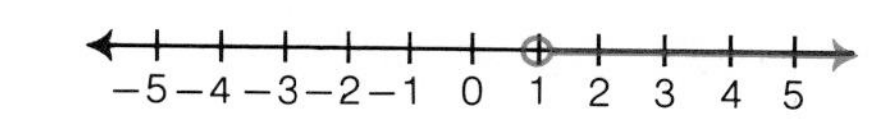

Graph the solution set of $x \geq 1$.

The dot at 1 indicates that 1 is included in the solution set.

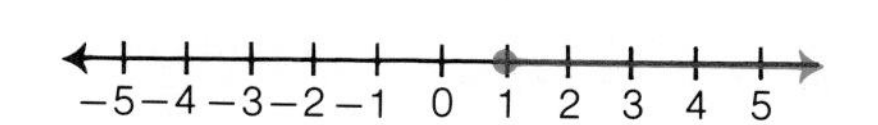

Graph the solution set of $x < -1$.

The numbers less than -1 are to the left of -1 on the number line.

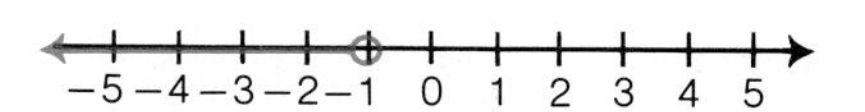

The union of two sets is the set that contains all the elements of each set.

Graph the solution set of $(x > 4) \cup (x < 1)$.

The solution set is the numbers greater than 4 and the numbers less than 1.

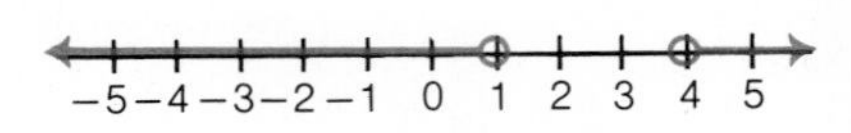

The intersection of two sets is the set that contains the elements common to both sets.

Graph the solution set of $(x > -1) \cap (x < 2)$.

The solution set is the numbers between -1 and 2.

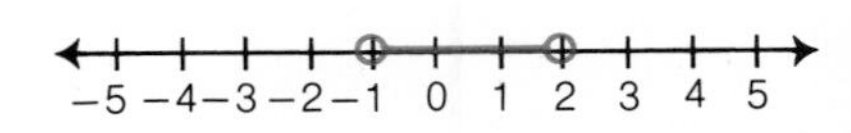

Example 15 Graph the solution set of $x < 5$.

Solution The solution set is the numbers less than 5.

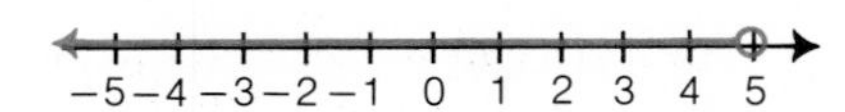

Example 16 Graph the solution set of $x > -2$.

Your solution

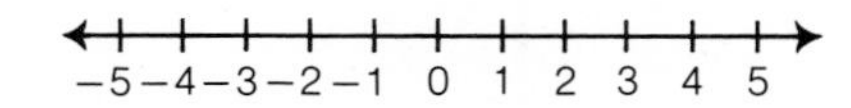

Example 17 Graph the solution set of $(x > -2) \cap (x < 1)$.

Solution The solution set is the numbers between -2 and 1.

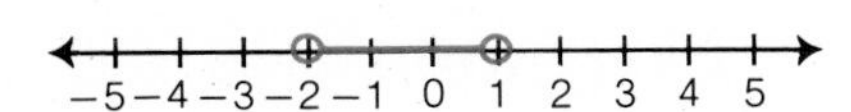

Example 18 Graph the solution set of $(x > -1) \cup (x < -3)$.

Your solution

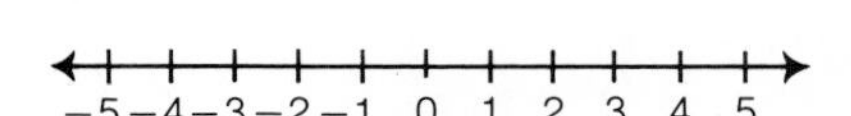

Example 19 Graph the solution set of $(x \leq 5) \cup (x \geq -3)$.

Solution The solution set is the real numbers.

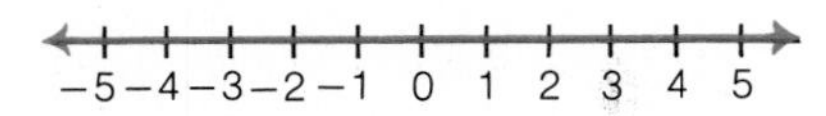

Example 20 Graph the solution set of $(x < 5) \cup (x \geq -2)$.

Your solution

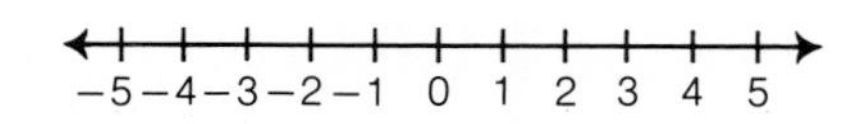

Example 21 Graph the solution set of $(x > 3) \cup (x < 1)$.

Solution The solution set is the numbers greater than 3 and the numbers less than 1.

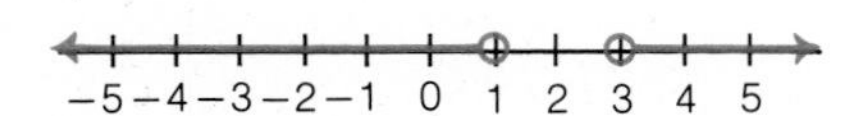

Example 22 Graph the solution set of $(x \leq 4) \cap (x \geq -4)$.

Your solution

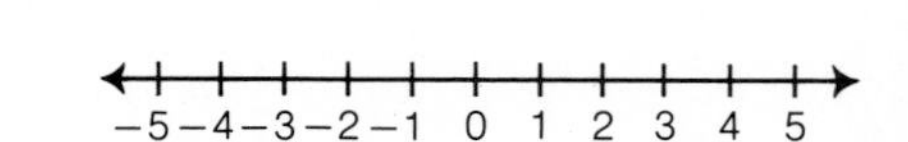

Solutions on p. A56

1.1 Exercises

Use the roster method to write the set.

1. the integers between 15 and 22
2. the integers between -10 and -4
3. the odd integers between 8 and 18
4. the even integers between -11 and -1
5. the letters of the alphabet between a and d
6. the letters of the alphabet between p and v
7. all perfect squares less than 50
8. positive integers less than 20 that are divisible by 4

Find $A \cup B$.

9. $A = \{3,4,5\}$ $B = \{4,5,6\}$
10. $A = \{-3,-2,-1\}$ $B = \{-2,-1,0\}$
11. $A = \{-10,-9,-8\}$ $B = \{8,9,10\}$
12. $A = \{a,b,c\}$ $B = \{x,y,z\}$
13. $A = \{a,b,d,e\}$ $B = \{c,d,e,f\}$
14. $A = \{m,n,p,q\}$ $B = \{m,n,o\}$
15. $A = \{1,3,7,9\}$ $B = \{7,9,11,13\}$
16. $A = \{-3,-2,-1\}$ $B = \{-1,1,2\}$

Find $A \cap B$.

17. $A = \{3,4,5\}$ $B = \{4,5,6\}$
18. $A = \{-4,-3,-2\}$ $B = \{-6,-5,-4\}$
19. $A = \{-4,-3,-2\}$ $B = \{2,3,4\}$
20. $A = \{1,2,3,4\}$ $B = \{1,2,3,4\}$
21. $A = \{a,b,c,d,e\}$ $B = \{c,d,e,f,g\}$
22. $A = \{m,n,o,p\}$ $B = \{k,l,m,n,\}$
23. $A = \{1,7,9,11\}$ $B = \{7,11,17\}$
24. $A = \{3,6,9,12\}$ $B = \{6,12,18\}$

1.2 Exercises

Use set builder notation to write the set.

25. negative integers greater than -5

26. positive integers less than 5

27. integers greater than 30

28. integers less than -70

29. even integers greater than 5

30. odd integers less than -2

31. the real numbers greater than 8

32. the real numbers less than 57

33. the real numbers greater than -5

34. the real numbers less than -63

1.3 Exercises

Graph the solution set.

35. $x > 2$

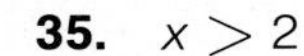

−5 −4 −3 −2 −1 0 1 2 3 4 5

36. $x \geq -1$

−5 −4 −3 −2 −1 0 1 2 3 4 5

37. $x \leq 0$

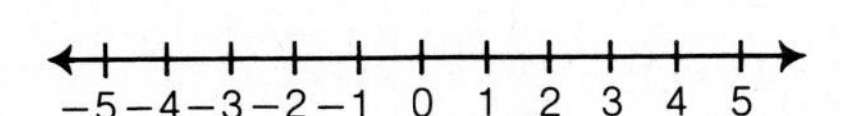

38. $x < 4$

−5 −4 −3 −2 −1 0 1 2 3 4 5

39. $(x > -2) \cup (x < -4)$

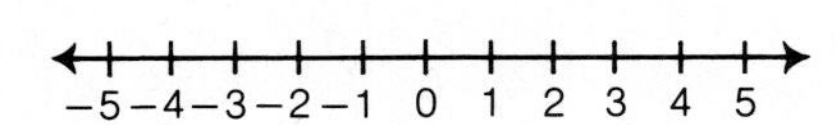

40. $(x > 4) \cup (x < -2)$

−5 −4 −3 −2 −1 0 1 2 3 4 5

41. $(x > -2) \cap (x < 4)$

42. $(x > -3) \cap (x < 3)$

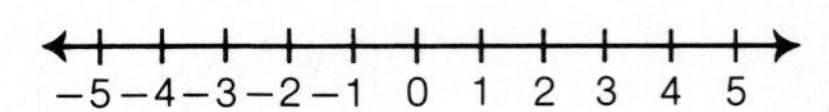

43. $(x \geq -2) \cup (x < 4)$

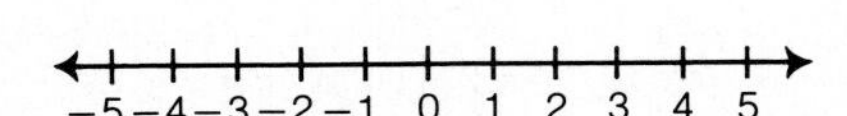

44. $(x > 0) \cup (x \leq 4)$

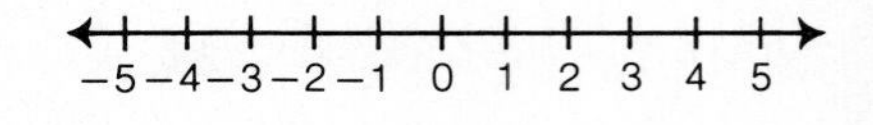

SECTION 2 The Addition and Multiplication Properties of Inequalities

2.1 Objective To solve an inequality using the Addition Property of Inequalities

Reference for Computer Tutor™

The **solution set of an inequality** is a set of numbers, each element of which, when substituted for the variable, results in a true inequality.

The inequality at the right is true if the variable is replaced by 7, 9.3, or $\frac{15}{2}$.

$$x + 3 > 8$$

$$\left.\begin{aligned} 7 + 3 &> 8 \\ 9.3 + 3 &> 8 \\ \tfrac{15}{2} + 3 &> 8 \end{aligned}\right\} \text{True inequalities}$$

The inequality is false if the variable is replaced by 4, 1.5, or $-\frac{1}{2}$.

$$\left.\begin{aligned} 4 + 3 &> 8 \\ 1.5 + 3 &> 8 \\ -\tfrac{1}{2} + 3 &> 8 \end{aligned}\right\} \text{False inequalities}$$

There are many values of the variable x that will make the inequality $x + 3 > 8$ true. The solution set of $x + 3 > 8$ is any number greater than 5.

The graph of the solution set of $x + 3 > 8$

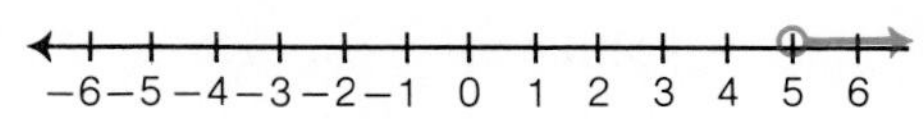

In solving an inequality, the goal is to rewrite the given inequality in the form *variable* > *constant* or *variable* < *constant*. The Addition Property of Inequalities is used to rewrite an inequality in this form.

Addition Property of Inequalities

The same number can be added to each side of an inequality without changing the solution set of the inequality.

If $a > b$, then $a + c > b + c$.

If $a < b$, then $a + c < b + c$.

The Addition Property of Inequalities also holds true for an inequality containing the symbol $\geq$ or $\leq$.

The Addition Property of Inequalities is used when, in order to rewrite an inequality in the form *variable* > *constant* or *variable* < *constant,* a term must be removed from one side of the inequality. Add the opposite of the term to each side of the inequality.

To rewrite the inequality at the right, add the opposite of the constant term 4 to each side of the inequality. Then simplify.

$$x + 4 < 5$$

$$x + 4 + (-4) < 5 + (-4)$$

$$x < 1$$

The graph of the solution set of $x + 4 < 5$

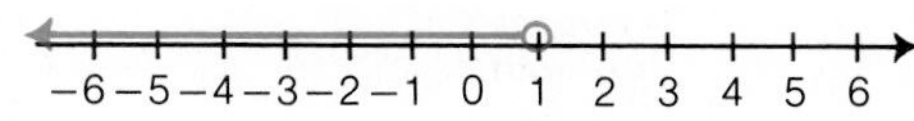

Solve: $5x - 6 \le 4x - 4$

Step 1 Add the opposite of the variable term $4x$ to each side of the inequality. Simplify.

$$5x - 6 \le 4x - 4$$
$$5x + (-4x) - 6 \le 4x + (-4x) - 4$$
$$x - 6 \le -4$$

Step 2 Add the opposite of the constant term -6 to each side of the inequality. Simplify.

$$x - 6 + 6 \le -4 + 6$$
$$x \le 2$$

Example 1 Solve and graph the solution set of $x + 5 > 3$.

Solution

$$x + 5 > 3$$
$$x + 5 + (-5) > 3 + (-5)$$
$$x > -2$$

−5 −4 −3 −2 −1 0 1 2 3 4 5

Example 2 Solve and graph the solution set of $x + 2 < -2$.

Your solution

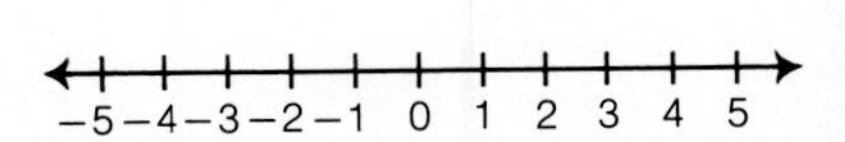

Example 3 Solve: $7x - 14 \le 6x - 16$

Solution

$$7x - 14 \le 6x - 16$$
$$7x + (-6x) - 14 \le 6x + (-6x) - 16$$
$$x - 14 \le -16$$
$$x - 14 + 14 \le -16 + 14$$
$$x \le -2$$

Example 4 Solve: $5x + 3 > 4x + 5$

Your solution

2.2 Objective To solve an inequality using the Multiplication Property of Inequalities

Reference for Computer Tutor™

In solving an inequality, the goal is to rewrite the given inequality in the form *variable* $>$ *constant* or *variable* $<$ *constant*. The Multiplication Property of Inequalities is used when, in order to rewrite an inequality in this form, a coefficient must be removed from one side of the inequality.

Multiplication Property of Inequalities

Rule 1

Each side of an inequality can be multiplied by the same positive number without changing the solution set of the inequality.

If $a > b$ and $c > 0$, then $ac > bc$.

$$5 > 4$$
$$5(2) > 4(2)$$
$$10 > 8 \quad \text{A true inequality}$$

If $a < b$ and $c > 0$, then $ac < bc$.

$$6 < 9$$
$$6(3) < 9(3)$$
$$18 < 27 \quad \text{A true inequality}$$

Rule 2

If each side of an inequality is multiplied by the same negative number and the inequality symbol is reversed, then the solution set of the inequality is not changed.

If $a > b$ and $c < 0$, then $ac < bc$.

$$5 > 4$$
$$5(-2) < 4(-2)$$
$$-10 < -8 \quad \text{A true inequality}$$

If $a < b$ and $c < 0$, then $ac > bc$.

$$6 < 9$$
$$6(-3) > 9(-3)$$
$$-18 > -27 \quad \text{A true inequality}$$

The Multiplication Property of Inequalities also holds true for an inequality containing the symbol $\geq$ or $\leq$.

To rewrite the inequality at the right, multiply each side of the inequality by the reciprocal of the coefficient -2.

$$-2x \leq 8$$

Since $-\frac{1}{2}$ is a negative number, the inequality symbol must be reversed.

$$-\frac{1}{2}(-2x) \geq -\frac{1}{2}(8)$$

Simplify.

$$x \geq -4$$

The graph of the solution set of $-2x \leq 8$

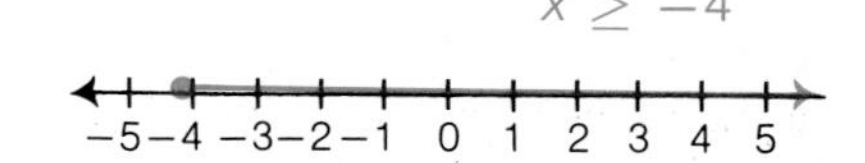

Solve: $-5x \leq 8$

Multiply each side of the inequality by the reciprocal of the coefficient -5.

$$-5x \leq 8$$

Since $-\frac{1}{5}$ is a negative number, the inequality symbol must be reversed.

$$-\frac{1}{5}(-5x) \geq -\frac{1}{5}(8)$$

Simplify.

$$x \geq -\frac{8}{5}$$

Example 5 Solve and graph the solution set of $7x < -14$.

Solution

$$7x < -14$$
$$\frac{1}{7}(7x) < \frac{1}{7}(-14)$$
$$x < -2$$

−5 −4 −3 −2 −1 0 1 2 3 4 5

Example 6 Solve and graph the solution set of $3x < 9$.

Your solution

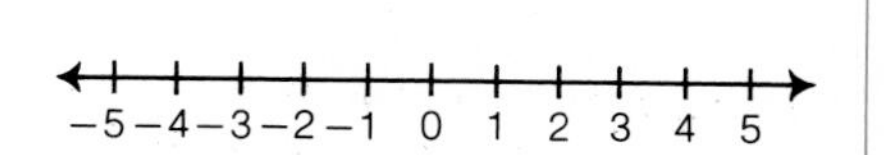

Solution on p. A56

Example 7 Solve: $-\frac{5}{8}x \le \frac{5}{12}$

Solution

$$-\frac{5}{8}x \le \frac{5}{12}$$

$$-\frac{8}{5}\left(-\frac{5}{8}x\right) \ge -\frac{8}{5}\left(\frac{5}{12}\right)$$

$$x \ge -\frac{2}{3}$$

Example 8 Solve: $-\frac{3}{4}x \ge 18$

Your solution

2.3 Objective To solve application problems

Use COMPUTER TUTOR™ DISK 10

Example 9 A student must have at least 450 points out of 500 points on five tests to receive an A in a course. One student's results on the first four tests were 94, 87, 77, and 95. What scores on the last test will enable the student to receive an A in the course?

Strategy To find the scores, write and solve an inequality using N to represent the score on the last test.

Solution

Total number of points on the 5 tests	is greater than or equal to	450

$$94 + 87 + 77 + 95 + N \ge 450$$

$$353 + N \ge 450$$

$$353 + (-353) + N \ge 450 + (-353)$$

$$N \ge 97$$

The student's score on the last test must be equal to or greater than 97.

Example 10 An appliance dealer will make a profit on the sale of a television set if the cost of the new set is less than 70% of the selling price. What minimum selling price will enable the dealer to make a profit on a television set that costs the dealer $314?

Your strategy

Your solution

2.1 Exercises

Solve and graph the solution set.

1. $x + 1 < 3$

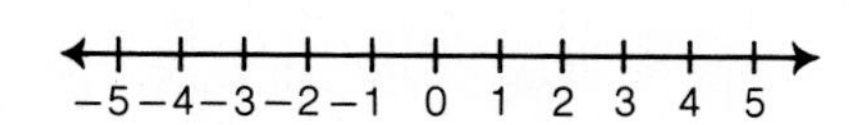

2. $y + 2 < 2$

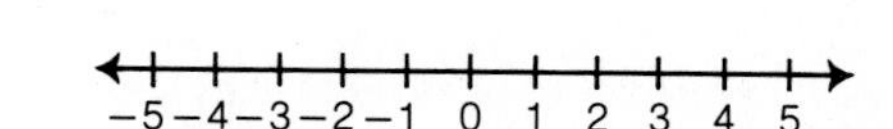

3. $x - 5 > -2$

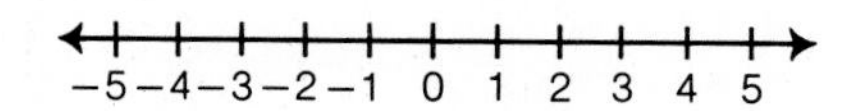

4. $x - 3 > -2$

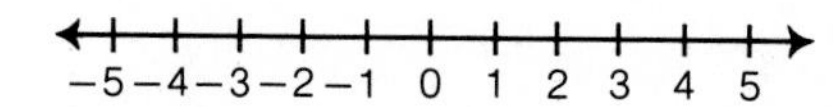

5. $n + 4 \geq 7$

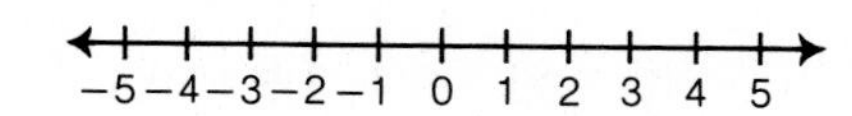

6. $x + 5 \geq 3$

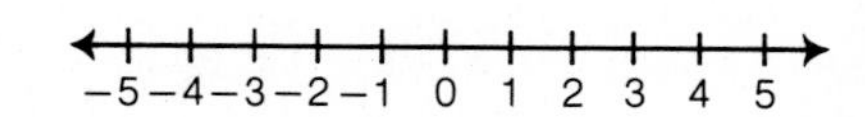

7. $x - 6 \leq -10$

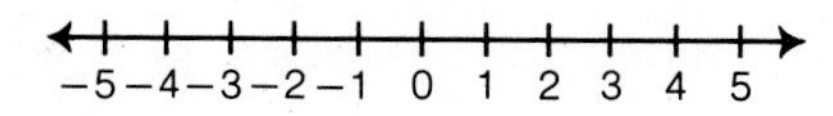

8. $y - 8 \leq -11$

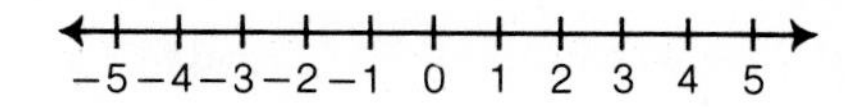

9. $5 + x \geq 4$

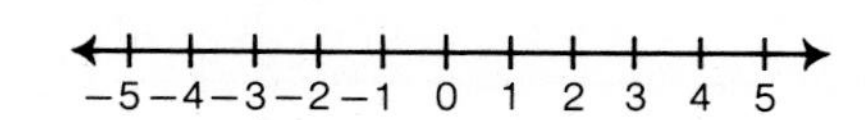

10. $-2 + n \geq 0$

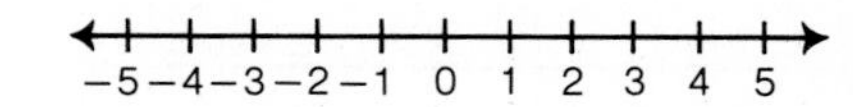

Solve.

11. $y - 3 \geq -12$

12. $x + 8 \geq -14$

13. $3x - 5 < 2x + 7$

14. $5x + 4 < 4x - 10$

15. $8x - 7 \geq 7x - 2$

16. $3n - 9 \geq 2n - 8$

17. $2x + 4 < x - 7$

18. $9x + 7 < 8x - 7$

19. $4x - 8 \leq 2 + 3x$

20. $5b - 9 < 3 + 4b$

21. $6x + 4 \geq 5x - 2$

22. $7x - 3 \geq 6x - 2$

23. $2x - 12 > x - 10$

24. $3x + 9 > 2x + 7$

25. $d + \frac{1}{2} < \frac{1}{3}$

26. $x - \frac{3}{8} < \frac{5}{6}$

27. $x + \frac{5}{8} \geq -\frac{2}{3}$

28. $y + \frac{5}{12} \geq -\frac{3}{4}$

Solve.

29. $x - \frac{3}{8} < \frac{1}{4}$

30. $y + \frac{5}{9} \leq \frac{5}{6}$

31. $2x - \frac{1}{2} < x + \frac{3}{4}$

32. $6x - \frac{1}{3} \leq 5x - \frac{1}{2}$

33. $3x + \frac{5}{8} > 2x + \frac{5}{6}$

34. $4b - \frac{7}{12} \geq 3b - \frac{9}{16}$

35. $3.8x < 2.8x - 3.8$

36. $1.2x < 0.2x - 7.3$

37. $x + 5.8 \leq 4.6$

38. $n - 3.82 \leq 3.95$

39. $x - 3.5 < 2.1$

40. $x - 0.23 \leq 0.47$

41. $1.33x - 1.62 > 0.33x - 3.1$

42. $2.49x + 1.35 \geq 1.49x - 3.45$

2.2 Exercises

Solve and graph the solution set.

43. $3x < 12$

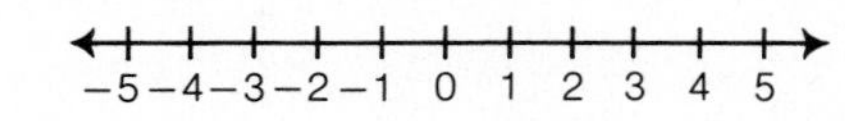

44. $8x \leq -24$

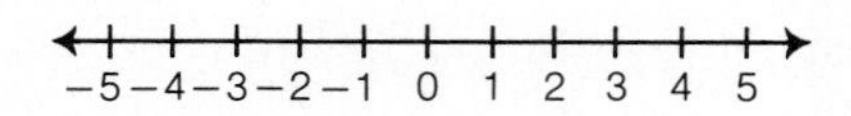

45. $5y \geq 15$

−5 −4 −3 −2 −1 0 1 2 3 4 5

46. $24x > -48$

−5 −4 −3 −2 −1 0 1 2 3 4 5

47. $16x \leq 16$

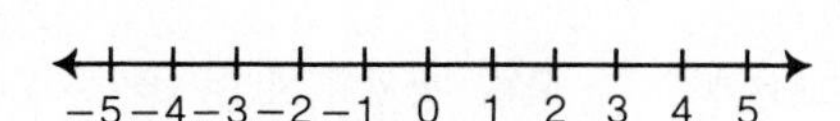

48. $3x > 0$

−5 −4 −3 −2 −1 0 1 2 3 4 5

49. $-8x > 8$

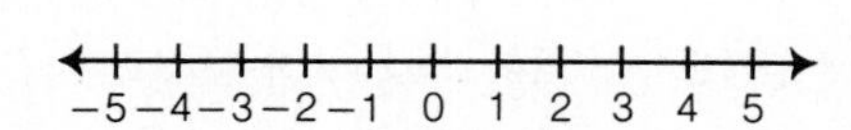

50. $-2n \leq -8$

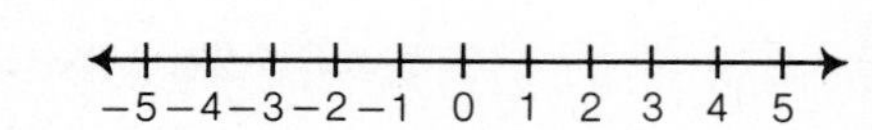

51. $-6b > 24$

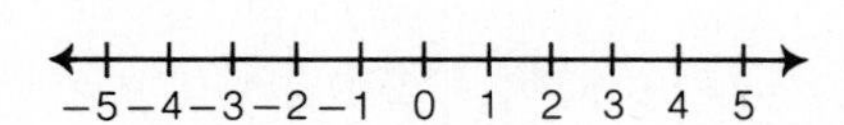

52. $-4x < 8$

Solve.

53. $-5y \geq 20$

54. $3x < 5$

55. $7x > 2$

56. $6x \leq -1$

57. $2x \leq -5$

58. $\frac{5}{6}n < 15$

59. $\frac{3}{4}x < 12$

60. $\frac{2}{3}y \geq 4$

61. $\frac{5}{8}x \geq 10$

62. $-\frac{2}{3}x \leq 4$

63. $-\frac{3}{7}x \leq 6$

64. $-\frac{2}{11}b \geq -6$

65. $-\frac{4}{7}x \geq -12$

66. $\frac{2}{3}n < \frac{1}{2}$

67. $\frac{3}{5}x < \frac{7}{10}$

68. $-\frac{2}{3}x \geq \frac{4}{7}$

69. $-\frac{3}{8}x \geq \frac{9}{14}$

70. $-\frac{3}{5}x < -\frac{6}{7}$

71. $-\frac{4}{5}x < -\frac{8}{15}$

72. $-\frac{3}{4}y \geq -\frac{5}{8}$

73. $-\frac{8}{9}x \geq -\frac{16}{27}$

74. $1.5x \leq 6.30$

75. $2.3x \leq 5.29$

76. $-3.5d > 7.35$

77. $-0.24x > 0.768$

78. $4.25m > -34$

79. $-3.9x \geq -19.5$

80. $0.035x < -0.0735$

81. $0.07x < -0.378$

82. $-11.7x \leq 4.68$

83. $0.685y \geq -2.15775$

84. $1.38n > -0.9936$

85. $-5.24x < 43.0728$

86. $-0.0663b < 0.19227$

87. $13.58x \leq 95.06$

88. $18.92x < 264.88$

2.3 Application Problems

Solve.

1. Eight less than a number is greater than ten. Find the smallest integer that will satisfy the inequality.

2. Four less than the product of six and a number is greater than three plus the product of five and the number. Find the smallest integer that will satisfy the inequality.

3. Three eighths of a number is less than or equal to nine sixteenths. Find the largest number that will satisfy the inequality.

4. Two thirds of a number is greater than or equal to seven twelfths. Find the smallest number that will satisfy the inequality.

5. A student must have at least 320 points out of 400 points on four tests to receive a B in a history course. One student's results on the first three tests were 82, 71, and 94. What scores on the last test will enable the student to receive at least a B in the course?

6. A student received a 63, a 73, a 94, and an 80 on four tests in a psychology course. What scores on the last test will enable the student to receive a minimum of 400 points?

7. A marketing representative receives payments of $1560, $1980, and $1270 in commissions during a three-month period. What is the minimum commission the marketing representative must earn during the fourth month in order to earn a minimum of $6000 during the four-month period?

8. A plumber's annual income for the last four years was $37,520, $29,860, $45,832, and $26,440. What is the minimum annual income the plumber must earn this year in order to have earned $150,000 during the five-year period?

9. An appliance dealer will make a profit on the sale of a refrigerator if the cost of the new refrigerator is less than 88% of the selling price. What minimum selling price will enable the dealer to make a profit on a refrigerator that costs the dealer $540?

10. A new-car dealer will make a profit on the sale of a car if the cost of the new car is less than $\frac{13}{16}$ of the selling price. What minimum selling price will enable the dealer to make a profit on a car that costs the dealer $7000?

SECTION 3 General Inequalities

3.1 Objective To solve general inequalities

Reference for Computer Tutor™

In solving an inequality, frequently the application of both the Addition and Multiplication Properties of Inequalities is required.

Solve: $3x - 2 < 5x + 4$

Step 1 Add the opposite of the variable term $5x$ to each side of the inequality. Simplify.

$$3x - 2 < 5x + 4$$
$$3x + (-5x) - 2 < 5x + (-5x) + 4$$
$$-2x - 2 < 4$$

Step 2 Add the opposite of the constant term -2 to each side of the inequality. Simplify.

$$-2x - 2 + 2 < 4 + 2$$
$$-2x < 6$$

Step 3 Multiply each side of the inequality by the reciprocal of the coefficient -2. Since $-\frac{1}{2}$ is a negative number, the inequality symbol must be reversed. Simplify.

$$-\tfrac{1}{2}(-2x) > -\tfrac{1}{2}(6)$$
$$x > -3$$

When an inequality contains parentheses, one of the steps in solving the inequality requires the use of the Distributive Property.

Solve: $-2(x - 7) > 3 - 4(2x - 3)$

Step 1 Use the Distributive Property to remove parentheses. Simplify.

$$-2(x - 7) > 3 - 4(2x - 3)$$
$$-2x + 14 > 3 - 8x + 12$$
$$-2x + 14 > 15 - 8x$$

Step 2 Add the opposite of the variable term $-8x$ to each side of the inequality. Simplify.

$$-2x + 8x + 14 > 15 - 8x + 8x$$
$$6x + 14 > 15$$

Step 3 Add the opposite of the constant term 14 to each side of the inequality. Simplify.

$$6x + 14 + (-14) > 15 + (-14)$$
$$6x > 1$$

Step 4 Multiply each side of the inequality by the reciprocal of the coefficient 6. Simplify.

$$\tfrac{1}{6} \cdot 6x > \tfrac{1}{6} \cdot 1$$
$$x > \tfrac{1}{6}$$

Example 1 Solve: $7x - 3 \le 3x + 17$

Solution

$$7x - 3 \le 3x + 17$$
$$7x + (-3x) - 3 \le 3x + (-3x) + 17$$
$$4x - 3 \le 17$$
$$4x - 3 + 3 \le 17 + 3$$
$$4x \le 20$$
$$\tfrac{1}{4} \cdot 4x \le \tfrac{1}{4} \cdot 20$$
$$x \le 5$$

Example 2 Solve: $5 - 4x > 9 - 8x$

Your solution

Solution on p. A58

Example 3 Solve:
$3(3 - 2x) \geq -5x - 2(3 - x)$

Solution

$$\begin{aligned} 3(3 - 2x) &\geq -5x - 2(3 - x) \\ 9 - 6x &\geq -5x - 6 + 2x \\ 9 - 6x &\geq -3x - 6 \\ 9 - 6x + 3x &\geq -3x + 3x - 6 \\ 9 - 3x &\geq -6 \\ 9 + (-9) - 3x &\geq -6 + (-9) \\ -3x &\geq -15 \\ -\tfrac{1}{3}(-3x) &\leq -\tfrac{1}{3}(-15) \\ x &\leq 5 \end{aligned}$$

Example 4 Solve:
$8 - 4(3x + 5) \leq 6(x - 8)$

Your solution

Solution on p. 458

3.2 Objective To solve application problems

Use COMPUTER TUTOR™ DISK 10

Example 5 A rectangle is 10 ft wide and $2x + 4$ ft long. Express as an integer the maximum length of the rectangle when the area is less than 200 ft². (The area of a rectangle is equal to its length times its width.)

Strategy To find the maximum length:

▷ Replace the variables in the area formula by the given values and solve for x.

▷ Replace the variable in the expression $2x + 4$ with the value found for x.

Solution

Length times width	is less than	200 ft²

$$\begin{aligned} 10(2x + 4) &< 200 \\ 20x + 40 &< 200 \\ 20x + 40 - 40 &< 200 - 40 \\ 20x &< 160 \\ \tfrac{1}{20} \cdot 20x &< \tfrac{1}{20} \cdot 160 \\ x &< 8 \end{aligned}$$

$2x + 4 < 20$

The maximum length is 19 ft.

Example 6 Company A rents cars for $8 a day and 10¢ for every mile driven. Company B rents cars for $10 a day and 8¢ per mile driven. You want to rent a car for one week. What is the maximum number of miles you can drive a Company A car if it is to cost you less than a Company B car?

Your strategy

Your solution

Solution on p. 458

3.1 Exercises

Solve.

1. $4x - 8 < 2x$
2. $7x - 4 < 3x$
3. $2x - 8 > 4x$
4. $3y + 2 > 7y$
5. $8 - 3x \le 5x$
6. $10 - 3x \le 7x$
7. $3x + 2 \ge 5x - 8$
8. $2n - 9 \ge 5n + 4$
9. $5x - 2 < 3x - 2$
10. $8x - 9 > 3x - 9$
11. $0.1(180 + x) > x$
12. $x > 0.2(50 + x)$
13. $0.15x + 55 > 0.10x + 80$
14. $-3.6b + 16 < 2.8b + 25.6$
15. $2(3x - 1) > 3x + 4$
16. $5(2x + 7) > -4x - 7$
17. $3(2x - 5) \ge 8x - 5$
18. $5x - 8 \ge 7x - 9$
19. $2(2y - 5) \le 3(5 - 2y)$
20. $2(5x - 8) \le 7(x - 3)$
21. $5(2 - x) > 3(2x - 5)$
22. $4(3d - 1) > 3(2 - 5d)$
23. $5(x - 2) > 9x - 3(2x - 4)$
24. $3x - 2(3x - 5) > 4(2x - 1)$
25. $4 - 3(3 - n) \le 3(2 - 5n)$
26. $15 - 5(3 - 2x) \le 4(x - 3)$
27. $2x - 3(x - 4) \ge 4 - 2(x - 7)$
28. $4 + 2(3 - 2y) \le 4(3y - 5) - 6y$

3.2 Application Problems

Solve.

1. Five times the sum of a number and four is less than three times the number. Find the largest integer that will satisfy the inequality.

2. The quotient of a number and two is less than twice the difference between the number and six. Find the smallest integer that will satisfy the inequality.

3. A rectangle is 8 ft wide and $3x - 2$ ft long. Express as an integer the minimum length of the rectangle when the area is greater than 104 ft^2. (The area of a rectangle is equal to its length times its width.)

4. The base of a triangle is 10 in. and the height is $2x - 4$ in. Express as an integer the maximum height of the triangle when the area is less than 80 in.^{2}. (The area of a triangle is equal to one half the base times the height.)

5. Company A rents cars for $12 a day and 12¢ for every mile driven. Company B rents cars for $30 per day with unlimited mileage. What is the maximum number of miles per day you can drive a Company A car if it is to cost you less than a Company B car?

6. Company A rents cars for $15 a day and 10¢ for every mile driven. Company B rents cars for $35 a day with unlimited mileage. You want to rent a car for one week. What is the maximum number of miles you can drive a Company A car if it is to cost you less than a Company B car?

7. Regulations require that grade A hamburger does not contain more than 20% fat. Find the maximum amount of fat that a grocer can mix with 200 lb of lean hamburger to meet the 20% regulation for grade A hamburger.

8. An account executive earns $500 per month plus an 8% commission on the amount of sales. The executive's goal is to earn a minimum of $2500 per month. Find the minimum sales that will enable the executive to earn $2500 per month.

9. Company A rents cars for $12 a day and 12¢ for every mile driven. Company B rents cars for $8 a day and 20¢ per mile driven. You want to rent a car for one week. What is the maximum number of miles you can drive a Company B car if it is to cost you less than a Company A car?

10. Company A rents cars for $10 a day and 20¢ for every mile driven. Company B rents cars for $20 a day and 10¢ for every mile driven. You want to rent a car for one week. What is the maximum number of miles you can drive a Company A car if it is to cost you less than a Company B car?

SECTION 4 Graphing Linear Inequalities

4.1 Objective To graph an inequality in two variables

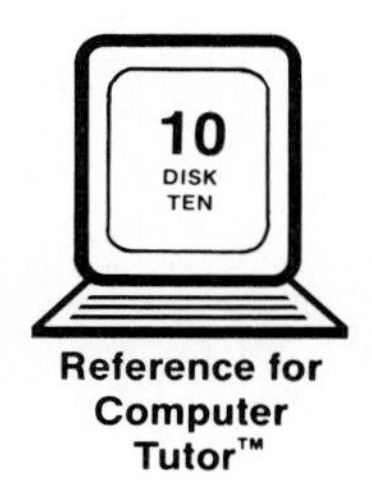

Reference for Computer Tutor™

The graph of the linear equation $y = x - 2$ separates a plane into three sets:

the set of points on the line,

the set of points above the line,

the set of points below the line.

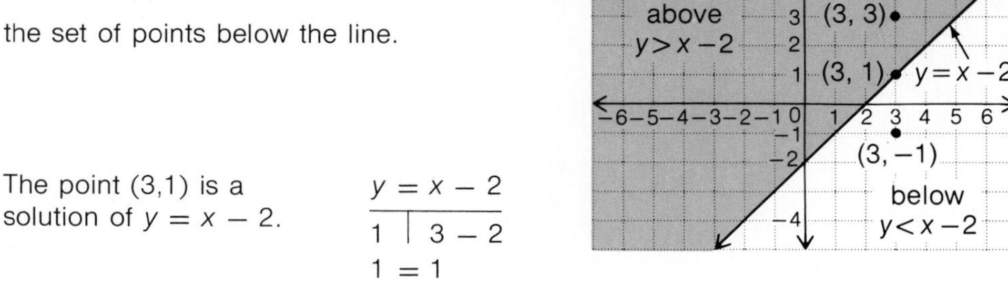

The point (3,1) is a solution of $y = x - 2$.

$$\begin{array}{c|c} \multicolumn{2}{c}{y = x - 2} \\ \hline 1 & 3 - 2 \\ \multicolumn{2}{c}{1 = 1} \end{array}$$

The point (3,3) is a solution of $y > x - 2$.

$$\begin{array}{c|c} \multicolumn{2}{c}{y > x - 2} \\ \hline 3 & 3 - 2 \\ \multicolumn{2}{c}{3 > 1} \end{array}$$

Any point above the line is a solution of $y > x - 2$.

The point (3, −1) is a solution of $y < x - 2$.

$$\begin{array}{c|c} \multicolumn{2}{c}{y < x - 2} \\ \hline -1 & 3 - 2 \\ \multicolumn{2}{c}{-1 < 1} \end{array}$$

Any point below the line is a solution of $y < x - 2$.

The solution set of $y = x - 2$ is all points on the line. The solution set of $y > x - 2$ is all points above the line. The solution set of $y < x - 2$ is all points below the line. The solution set of an inequality in two variables is a **half-plane.**

The following illustrates the procedure for graphing a linear inequality.

Graph the solution set of $2x + 3y \le 6$.

Step 1 Solve the inequality for y.

$$\begin{aligned} 2x + 3y &\le 6 \\ 2x + (-2x) + 3y &\le -2x + 6 \\ 3y &\le -2x + 6 \\ \tfrac{1}{3} \cdot 3y &\le \tfrac{1}{3}(-2x + 6) \\ y &\le -\tfrac{2}{3}x + 2 \end{aligned}$$

Step 2 Change the inequality to an equality and graph the line. If the inequality is **≥ or ≤,** the line is in the solution set and is shown by a **solid line.** If the inequality is **> or <,** the line is not a part of the solution set and is shown by a **dotted line.**

$$y = -\tfrac{2}{3}x + 2$$

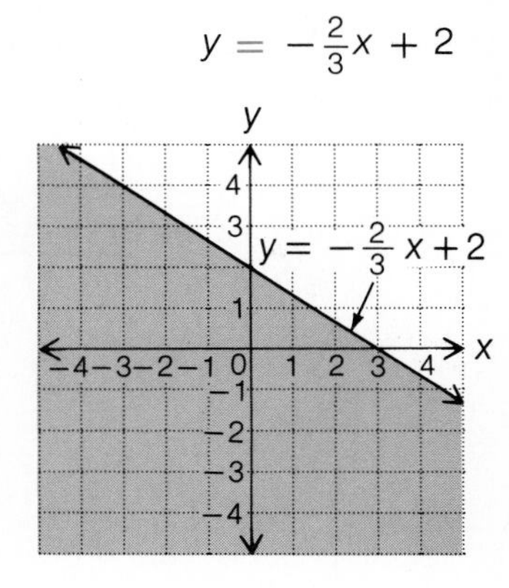

Step 3 If the inequality is **> or ≥,** shade the **upper half-plane.** If the inequality is **< or ≤,** shade the **lower half-plane.**

Example 1 Graph the solution set of $3x + y > -2$.

Solution

$$3x + y > -2$$
$$3x + (-3x) + y > -3x - 2$$
$$y > -3x - 2$$

Graph $y = -3x - 2$ as a dotted line.
Shade the upper half-plane.

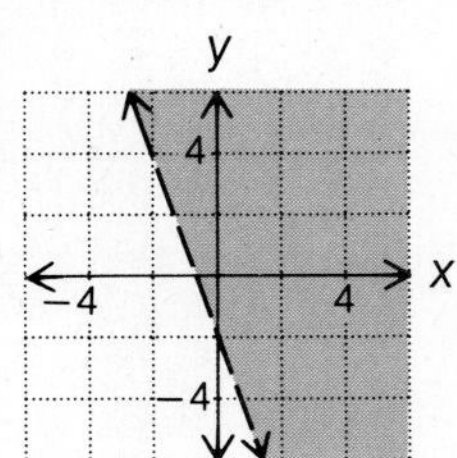

Example 2 Graph the solution set of $x - 3y < 2$.

Your solution

Example 3 Graph the solution set of $2x - y \geq 2$.

Solution

$$2x - y \geq 2$$
$$2x + (-2x) - y \geq -2x + 2$$
$$-y \geq -2x + 2$$
$$-1(-y) \leq -1(-2x + 2)$$
$$y \leq 2x - 2$$

Graph $y = 2x - 2$ as a solid line.
Shade the lower half-plane.

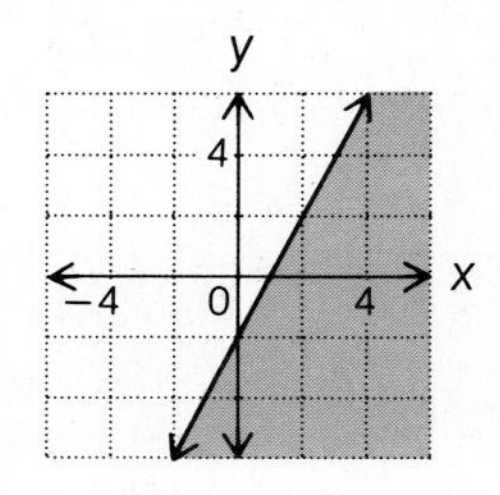

Example 4 Graph the solution set of $2x - 4y \leq 8$.

Your solution

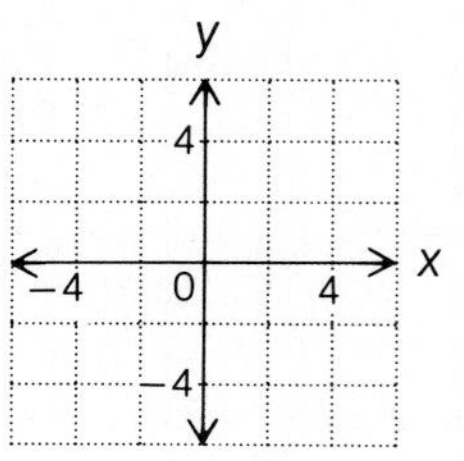

Example 5 Graph the solution set of $y > 3$.

Solution $y > 3$

Graph $y = 3$ as a dotted line.
Shade the upper half-plane.

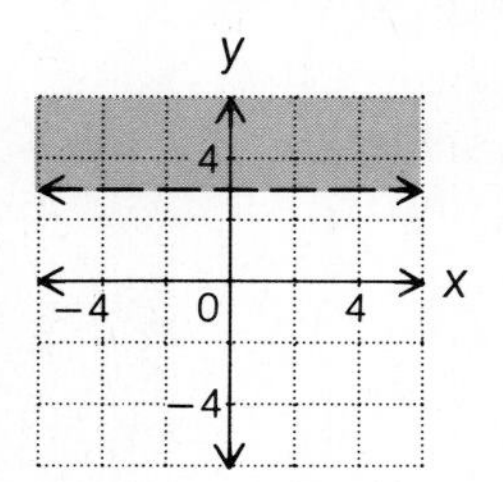

Example 6 Graph the solution set of $x < 3$.

Your solution

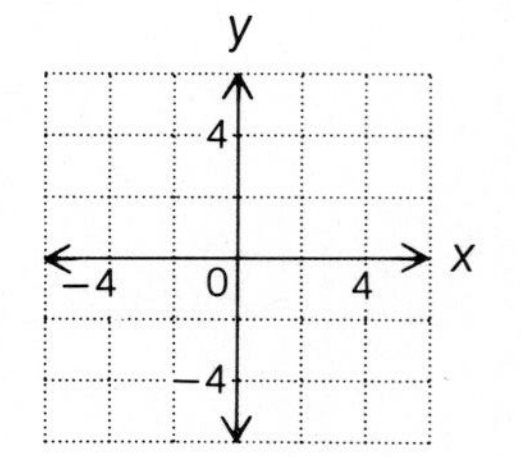

Solutions on p. A59

4.1 Exercises

Graph the solution set.

1. $x + y > 4$

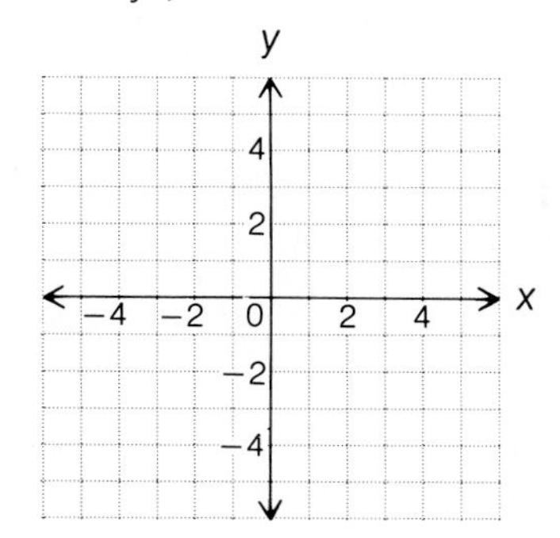

2. $x - y > -3$

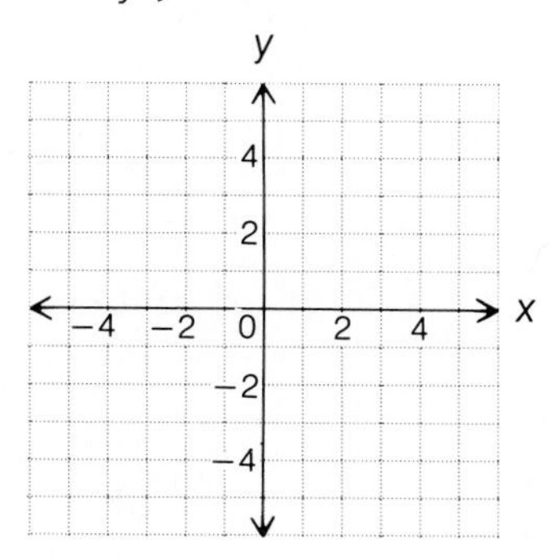

3. $2x - y < -3$

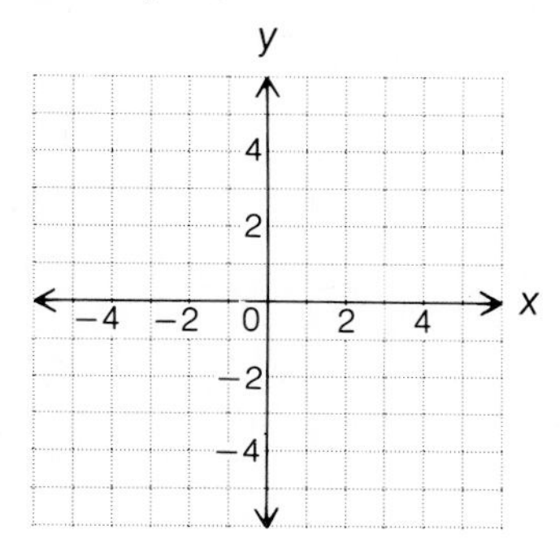

4. $3x - y < 9$

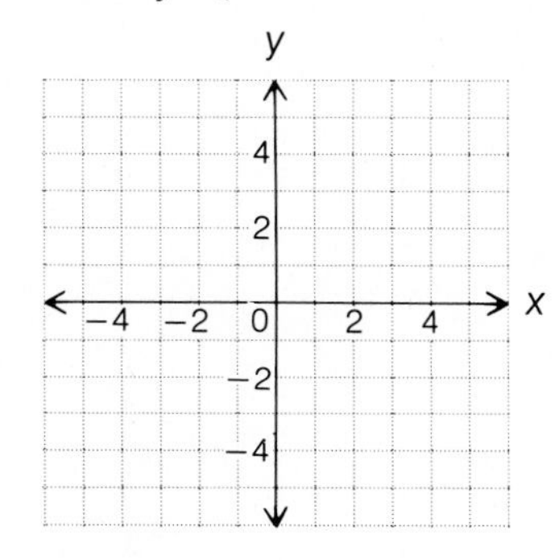

5. $2x + y \geq 4$

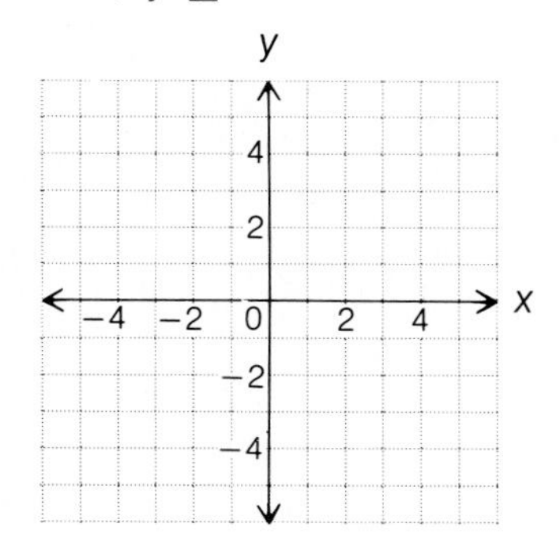

6. $3x + y \geq 6$

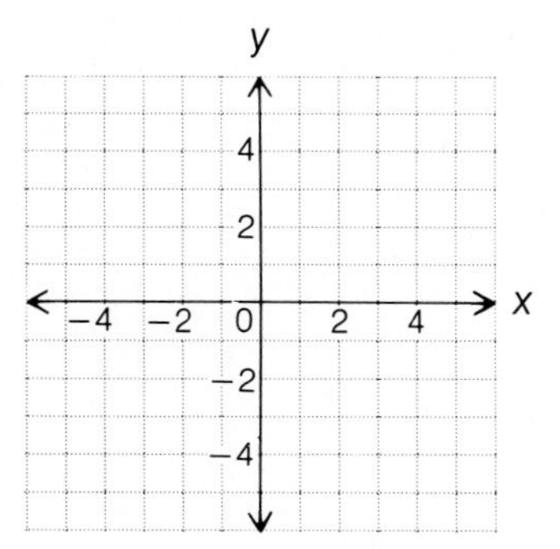

7. $y \leq -2$

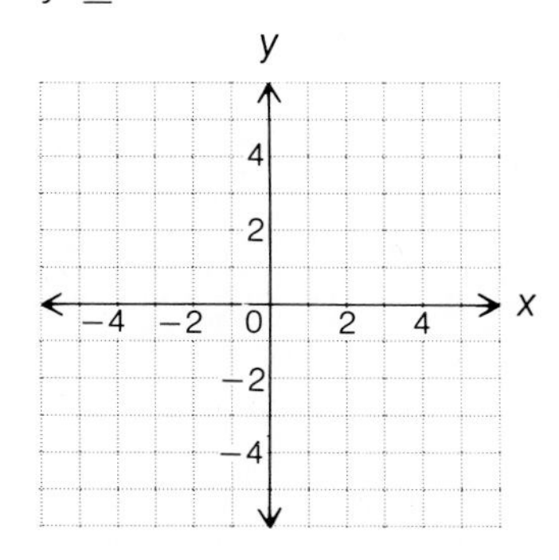

8. $y > 3$

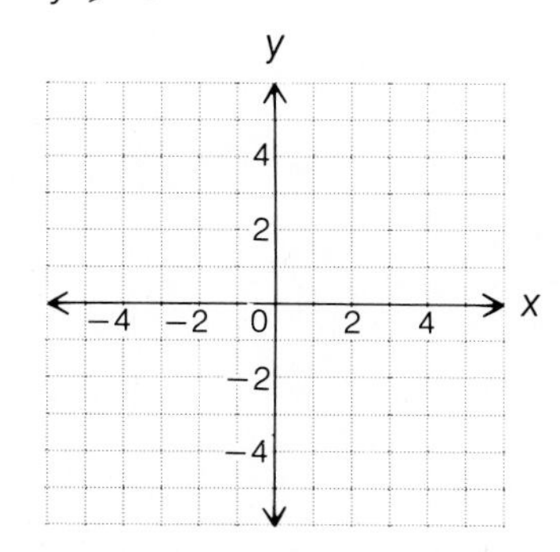

Graph the solution set.

9. $3x - 2y < 8$

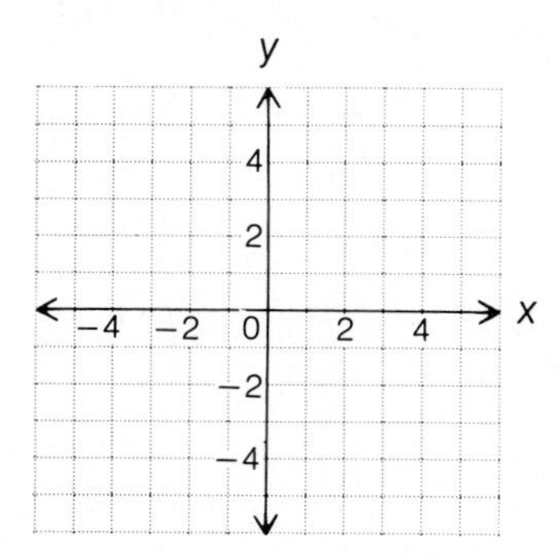

10. $5x + 4y > 4$

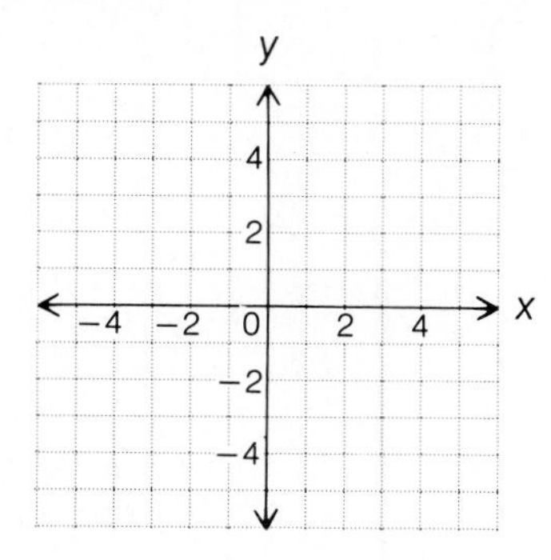

11. $-3x - 4y \geq 4$

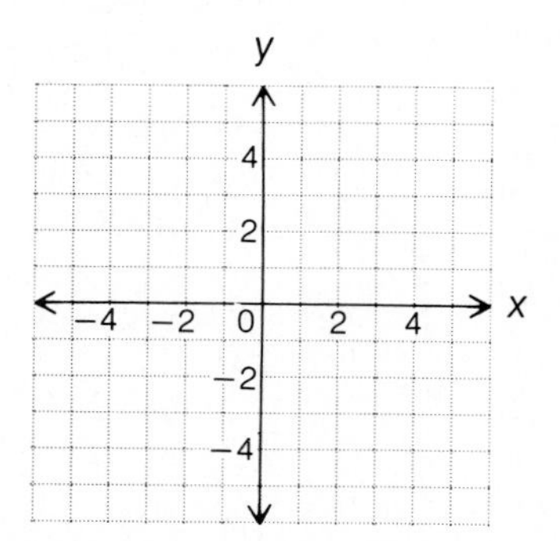

12. $-5x - 2y \geq 8$

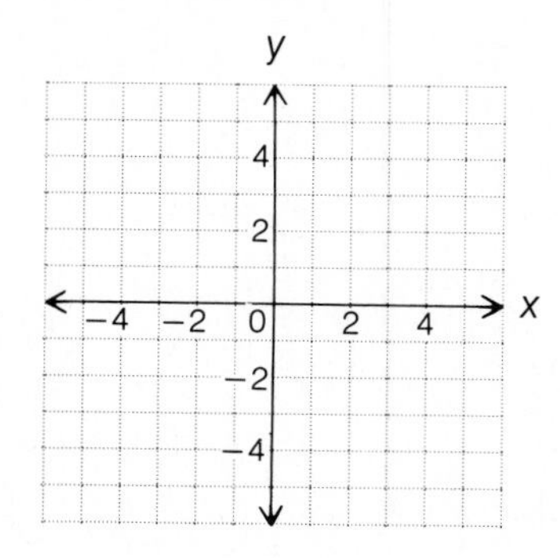

13. $6x + 5y \leq -10$

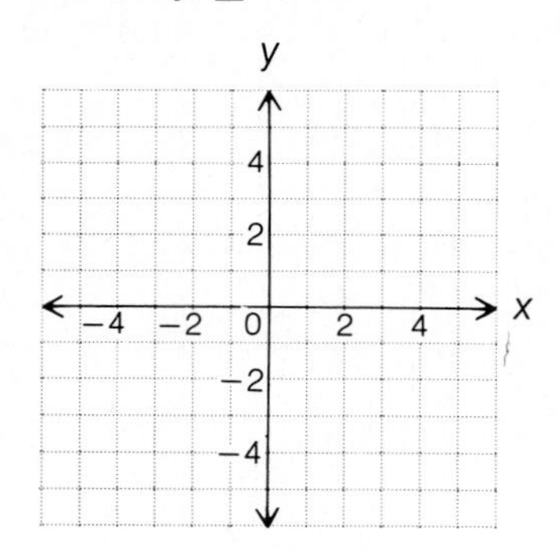

14. $2x + 2y \leq -4$

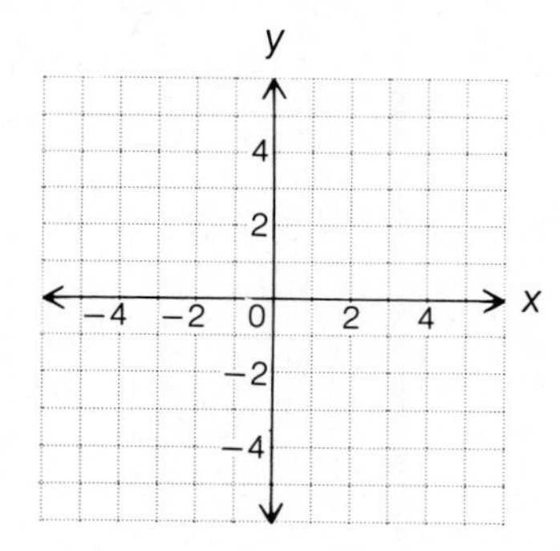

15. $-4x + 3y < -12$

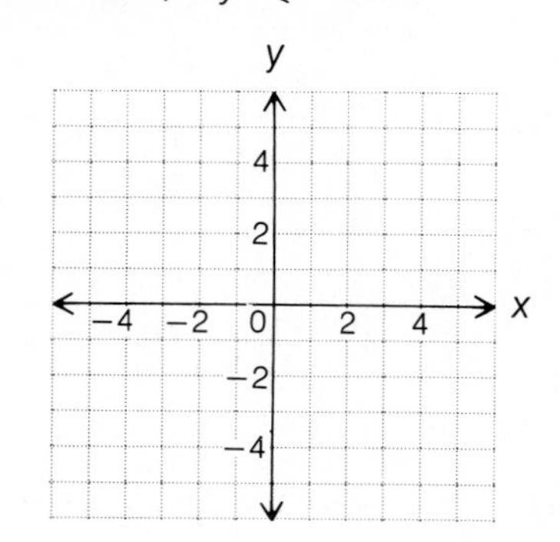

16. $-4x + 5y < 15$

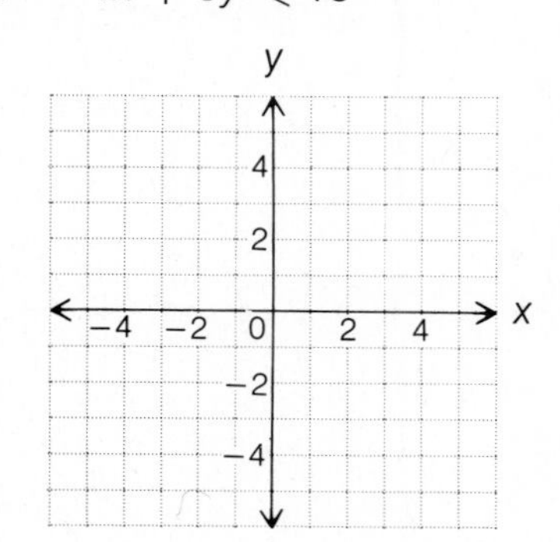

Calculators and Computers

First-Degree Inequalities in One Variable

The program FIRST-DEGREE INEQUALITIES on the Student Disk provides inequalities for you to solve. There are three levels of difficulty with the first level the easiest and the third level the most difficult.

Once you choose the level of difficulty, the program will display a problem. Using paper and pencil, solve the inequality. When you are ready, press the RETURN key and compare your solution with the one on the screen. The answers are rounded to the nearest hundredth.

After you complete a problem, you may continue practicing at the same level, return to the menu and select a different level, or quit the program.

Chapter Summary

KEY WORDS

A **set** is a collection of objects. The objects of a set are called the **elements** of the set.

The **roster method** of writing a set encloses a list of the elements in braces.

The **empty set,** or **null set,** written $\emptyset$ or $\{\ \}$, is the set which contains no elements.

The **union** of two sets, written $A \cup B$, is the set that contains all the elements of A and all of the elements of B (the elements which are in both set A and set B are listed only once).

The **intersection** of two sets, written $A \cap B$, is the set that contains the elements which are common to both A and B.

An **inequality** is an expression which contains the symbol $<$, $>$, $\leq$, or $\geq$.

The **solution set of an inequality** is a set of numbers, each element of which, when substituted for the variable, results in a true inequality. The solution set of an inequality can be graphed on the number line.

The solution set of an inequality in two variables is a **half-plane.**

ESSENTIAL RULES

Addition Property of Inequalities

The same number can be added to each side of an inequality without changing the solution set of the inequality.

If $a > b$, then $a + c > b + c$.
If $a < b$, then $a + c < b + c$.

The Addition Property of Inequalities also holds true for an inequality containing the symbol $\geq$ or $\leq$.

Multiplication Property of Inequalities

Rule 1
Each side of an inequality can be multiplied by the same **positive number** without changing the solution set of the inequality.

If $a > b$ and $c > 0$, then $ac > bc$.
If $a < b$ and $c > 0$, then $ac < bc$.

Rule 2
If each side of an inequality is multiplied by the same **negative number** and the inequality symbol is reversed, then the solution set of the inequality is not changed.

If $a > b$ and $c < 0$, then $ac < bc$.
If $a < b$ and $c < 0$, then $ac > bc$.

The Multiplication Property of Inequalities also holds true for an inequality containing the symbol $\geq$ or $\leq$.

Review/Test

SECTION 1

1.1a Use the roster method to write the set of even positive integers between 3 and 9.

1.1b Find $A \cap B$, given $A = \{6,8,10,12\}$ and $B = \{12,14,16\}$.

1.2a Use set builder notation to write the set of positive integers less than 50.

1.2b Use set builder notation to write the set of real numbers greater than -23.

1.3a Graph the solution set of $x > -2$.

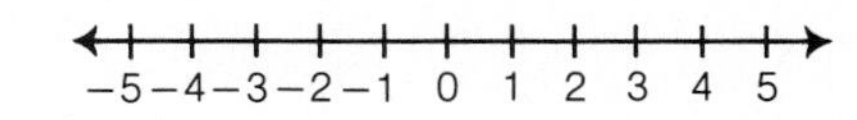

1.3b Graph the solution set of $(x < 5) \cap (x > 0)$.

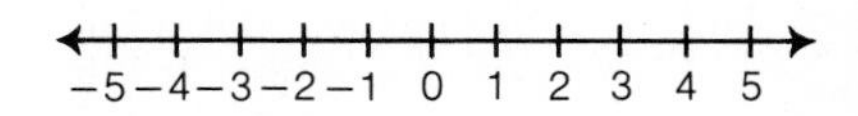

SECTION 2

2.1a Solve and graph the solution set of $4 + x < 1$.

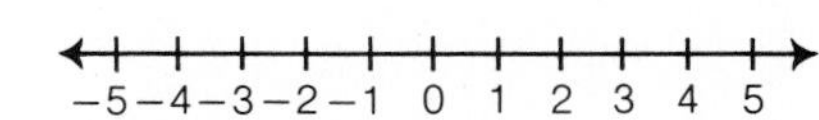

2.1b Solve: $x + \frac{1}{2} > \frac{5}{8}$

2.2a Solve and graph the solution set of $\frac{2}{3}x \geq 2$.

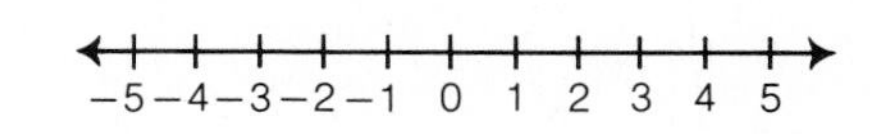

2.2b Solve: $-\frac{3}{8}x \leq 5$

Review/Test

2.3 Five more than a number is less than -3. Find the largest integer that will satisfy the inequality.

SECTION **3**

3.1a Solve:
$3(2x - 5) \geq 8x - 9$

3.1b Solve:
$6x - 3(2 - 3x) < 4(2x - 7)$

3.2 A rectangle is 15 ft long and $2x - 4$ ft wide. Express as an integer the maximum width of the rectangle when the area is less than 180 ft^2. (The area of a rectangle is equal to its length times its width.)

SECTION **4**

4.1a Graph the solution set of
$3x + y > 4$.

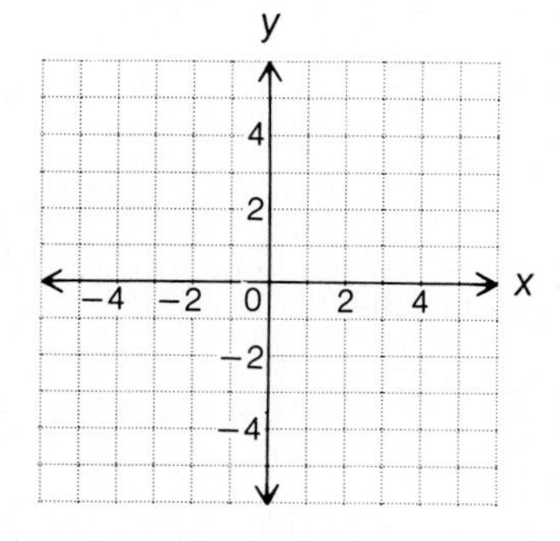

4.1b Graph the solution set of
$4x - 5y \geq 15$.

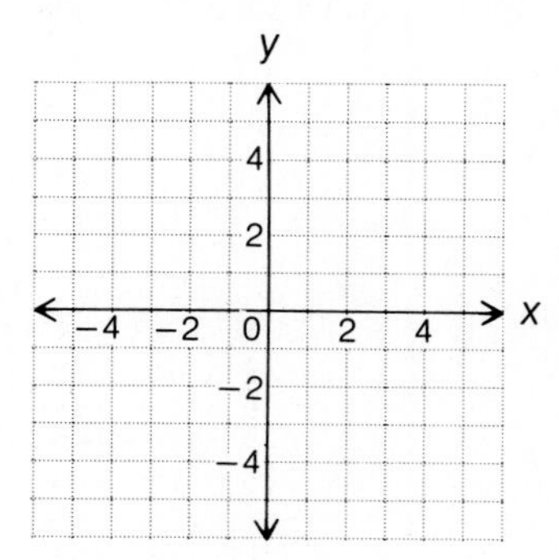

Cumulative Review/Test

1. Simplify:
$2[5a - 3(2 - 5a) - 8]$

2. Solve: $\frac{5}{8} - 4x = \frac{1}{8}$

3. Solve:
$2x - 3[x - 2(x - 3)] = 2$

4. Simplify: $(-3a)(-2a^3b^2)^2$

5. Simplify: $\frac{27a^3b^2}{(-3ab^2)^3}$

6. Simplify:
$(16x^2 - 12x - 2) \div (4x - 1)$

7. Factor $4x^2 - 21x + 5$.

8. Factor $27a^2x^2 - 3a^2$.

9. Simplify:
$\frac{x^2 - 2x}{x^2 - 2x - 8} \div \frac{x^3 - 5x^2 + 6x}{x^2 - 7x + 12}$

10. Simplify: $\frac{4a}{2a - 3} - \frac{2a}{a + 3}$

11. Solve:
$\frac{5y}{6} - \frac{5}{9} = \frac{y}{3} - \frac{5}{6}$

12. Solve $R = \frac{C - S}{t}$ for C.

13. Find the slope of the line containing the points $(2, -3)$ and $(-1, 4)$.

14. Find the equation of the line which contains the point $(1, -3)$ and has slope $-\frac{3}{2}$.

15. Solve by substitution.
$$x = 3y + 1$$
$$2x + 5y = 13$$

16. Solve by the addition method.
$$9x - 2y = 17$$
$$5x + 3y = -7$$

Cumulative Review/Test

17. Find $A \cup B$, given $A = \{0,1,2\}$ and $B = \{-2,-10\}$.

18. Use set builder notation to write the set of real numbers less than 48.

19. Graph the solution set of $(x > 1) \cup (x < -1)$.

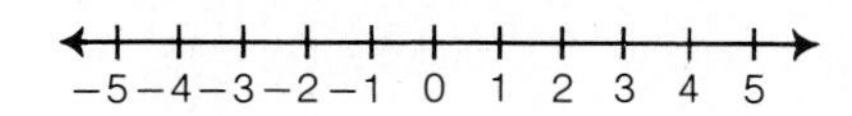

20. Graph the solution set of $\frac{3}{8}x > -\frac{3}{4}$.

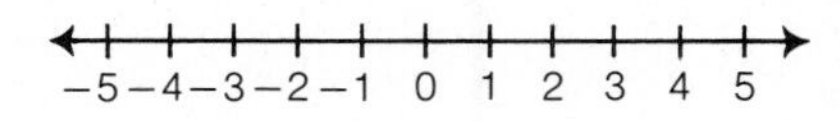

21. Solve: $-\frac{4}{5}x > 12$

22. Solve: $15 - 3(5x - 7) < 2(7 - 2x)$

23. Three fifths of a number is less than negative fifteen. Find the largest integer that will satisfy the inequality.

24. Company A rents cars for \$6 a day and 25¢ for every mile driven. Company B rents cars for \$15 a day and 10¢ per mile. You want to rent a car for 6 days. What is the maximum number of miles you can drive a Company A car if it is to cost you less than a Company B car?

25. In a lake, 100 fish are caught, tagged, and then released. Later 150 fish are caught. Three of the 150 fish are found to have tags. Estimate the number of fish in the lake.

26. A drawer contains 13¢ stamps and 18¢ stamps. The number of 13¢ stamps is two less than the number of 18¢ stamps. The total value of all the stamps is \$2.53. How many 13¢ stamps are in the drawer?

27. Graph $y = 2x - 1$.

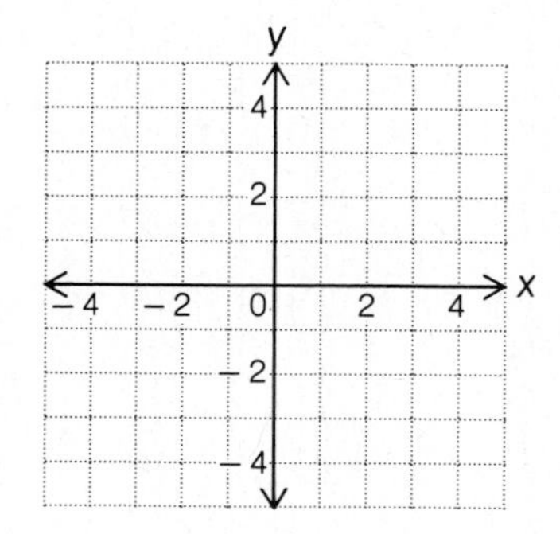

28. Graph the solution set of $6x - 3y \geq 6$.

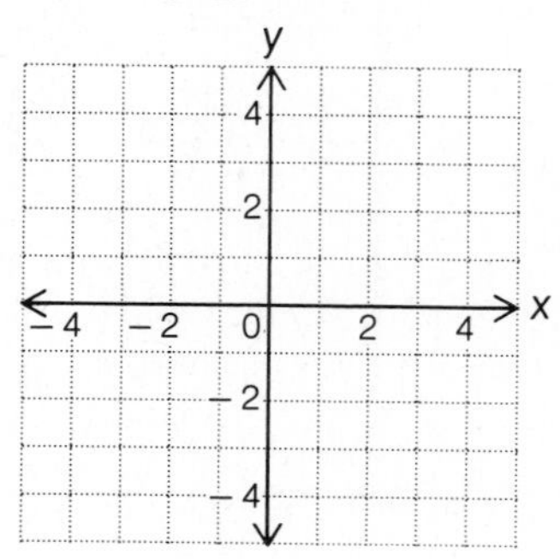

11

Radical Expressions

Objectives

- To simplify numerical radical expressions
- To simplify variable radical expressions
- To add and subtract radical expressions
- To multiply radical expressions
- To divide radical expressions
- To solve an equation containing one or more radical expressions
- To solve application problems

A Table of Square Roots

The practice of finding the square root of a number has existed for at least two thousand years. Since the process of finding a square root is tedious and time consuming, it is convenient to have tables of square roots. There is one such table in the back of this book.

But this table is not the first square root table ever written (nor is it likely to be the last). The table shown below is part of an old Babylonian clay tablet which was written around 350 B.C. It is an incomplete table of square roots written in a style called *cuneiform*.

The number base of the Babylonians was 60 instead of 10, as we use today. The symbol was used for one and ten was written as . Some examples of numbers using this system are given below.

= 9

= 40

A translation of the first couple of lines of the table are given next to that line. The number given in parentheses is the equivalent base 10 number that would be used today. You might try to translate the third line. The answer is given at the bottom of this page.

40 × 60 + 1 (= 2401)
which is the square of 49

41 × 60 + 40 (= 2500)
which is the square of 50

Answer: 43 × 60 + 21 (= 2601) is the square of 51.

SECTION 1 Introduction to Radical Expressions

1.1 Objective To simplify numerical radical expressions

Reference for Computer Tutor™

A **square root** of a positive number x is a number whose square is x.

A square root of 16 is 4 since $4^2 = 16$.
A square root of 16 is -4 since $(-4)^2 = 16$.

Every positive number has two square roots, one a positive and one a negative number. The symbol "$\sqrt{\ }$", called a **radical,** is used to indicate the positive or **principal square root** of a number. For example, $\sqrt{16} = 4$ and $\sqrt{25} = 5$. The number under the radical sign is called the **radicand.**

When the negative square root of a number is to be found, a negative sign is placed in front of the radical. For example, $-\sqrt{16} = -4$ and $-\sqrt{25} = -5$.

The square of an integer is a **perfect square.**

$$7^2 = 49$$
$$9^2 = 81$$
$$12^2 = 144$$

An integer which is a perfect square can be written as the product of prime factors, each of which has an even exponent when expressed in exponential form.

$$49 = 7 \cdot 7 = 7^2$$
$$81 = 3 \cdot 3 \cdot 3 \cdot 3 = 3^4$$
$$144 = 2 \cdot 2 \cdot 2 \cdot 2 \cdot 3 \cdot 3 = 2^4 3^2$$

To find the square root of a perfect square written in exponential form, remove the radical sign and multiply the exponent by $\frac{1}{2}$.

Simplify $\sqrt{625}$.

Write the prime factorization of the radicand in exponential form. $\sqrt{625} = \sqrt{5^4}$

Remove the radical sign and multiply the exponent by $\frac{1}{2}$. $= 5^2$

Simplify. $= 25$

If a number is not a perfect square, its square root can only be approximated, for example, $\sqrt{2}$ and $\sqrt{7}$.

$$\sqrt{2} \approx 1.4142135\ldots$$
$$\sqrt{7} \approx 2.6457513\ldots$$

These numbers are **irrational numbers.** Their decimal representations never terminate or repeat.

The approximate square roots of the positive integers up to 200 can be found in the Appendix on page A2. The square roots have been rounded to the nearest thousandth.

A radical expression is in simplest form when the radicand contains no factor which is a perfect square. The Product Property of Square Roots is used to simplify radical expressions.

The Product Property of Square Roots

If a and b are positive real numbers, then $\sqrt{ab} = \sqrt{a} \cdot \sqrt{b}$.

Simplify $\sqrt{96}$.

Write the prime factorization of the radicand in exponential form. $\sqrt{96} = \sqrt{2^5 \cdot 3}$

Write the radicand as a product of a perfect square and factors which do not contain a perfect square. $= \sqrt{2^4(2 \cdot 3)}$

Use the Product Property of Square Roots. $= \sqrt{2^4}\sqrt{2 \cdot 3}$

Simplify. $= 2^2\sqrt{2 \cdot 3}$

$= 4\sqrt{6}$

Simplify $\sqrt{-4}$.

The square root of a negative number is not a real number since the square of a real number is always positive. $\sqrt{-4}$ is not a real number.

Simplify $\sqrt{125}$. Then find the decimal approximation. Round to the nearest thousandth.

Write the prime factorization of the radicand in exponential form. $\sqrt{125} = \sqrt{5^3}$

Write the radicand as a product of a perfect square and factors which do not contain a perfect square. $= \sqrt{5^2 \cdot 5}$

Use the Product Property of Square Roots. $= \sqrt{5^2}\sqrt{5}$

Simplify. $= 5\sqrt{5}$

Replace the radical expression by the decimal approximation found on page A2. $\approx 5(2.236)$

Simplify. ≈ 11.180

Note that in the table on page A2, the decimal approximation of $\sqrt{125}$ is 11.180

Example 1

Simplify $3\sqrt{90}$.

Solution

$3\sqrt{90} = 3\sqrt{2 \cdot 3^2 \cdot 5} = 3\sqrt{3^2(2 \cdot 5)} = 3\sqrt{3^2}\sqrt{2 \cdot 5} = 3 \cdot 3\sqrt{10} = 9\sqrt{10}$

Example 2

Simplify $-5\sqrt{32}$.

Your solution

Example 3

Find the decimal approximation of $\sqrt{252}$. Use the table on page A2.

Solution

$\sqrt{252} = \sqrt{2^2 \cdot 3^2 \cdot 7} = \sqrt{2^2 \cdot 3^2}\sqrt{7} = 2 \cdot 3\sqrt{7} = 6\sqrt{7} \approx 6(2.646) \approx 15.876$

Example 4

Find the decimal approximation of $\sqrt{216}$. Use the table on page A2.

Your solution

Solutions on p. A61

1.2 Objective To simplify variable radical expressions

Reference for Computer Tutor™

Variable expressions which contain radicals do not always represent real numbers.

The variable expression at the right does not represent a real number when x is a negative number, for example, -4.

$\sqrt{x^3}$

$\sqrt{(-4)^3} = \sqrt{-64}$ Not a real number

For this reason, the variables in this unit will represent *positive* numbers unless otherwise stated.

A variable or a product of variables written in exponential form is a **perfect square** when each exponent is an even number.

To find the square root of a perfect square, remove the radical sign and multiply each exponent by $\frac{1}{2}$.

Simplify $\sqrt{a^6}$.

Remove the radical sign and multiply the exponent by $\frac{1}{2}$. $\sqrt{a^6} = a^3$

A variable radical expression is in simplest form when the radicand contains no factor which is a perfect square.

Simplify $\sqrt{x^7}$.

Write x^7 as the product of x and a perfect square. $\sqrt{x^7} = \sqrt{x^6 \cdot x}$

Use the Product Property of Square Roots. $= \sqrt{x^6}\sqrt{x}$

Simplify the perfect square. $= x^3\sqrt{x}$

Simplify $3x\sqrt{8x^3y^{13}}$.

Write the prime factorization of the coefficient of the radicand in exponential form. $3x\sqrt{8x^3y^{13}} = 3x\sqrt{2^3x^3y^{13}}$

Write the radicand as a product of a perfect square and factors which do not contain a perfect square. $= 3x\sqrt{2^2x^2y^{12}(2xy)}$

Use the Product Property of Square Roots. $= 3x\sqrt{2^2x^2y^{12}}\sqrt{2xy}$

Simplify. $= 3x \cdot 2xy^6\sqrt{2xy}$
$= 6x^2y^6\sqrt{2xy}$

Simplify $\sqrt{25(x-2)^2}$.

Write the prime factorization of the coefficient in exponential form. $\sqrt{25(x-2)^2} = \sqrt{5^2(x-2)^2}$

Note that $(x - 2)$ must be positive. $= 5(x-2)$
Simplify. $= 5x - 10$

Example 5

Simplify $\sqrt{b^{15}}$.

Solution

$\sqrt{b^{15}} = \sqrt{b^{14} \cdot b} = \sqrt{b^{14}} \cdot \sqrt{b} = b^7\sqrt{b}$

Example 6

Simplify $\sqrt{y^{19}}$.

Your solution

Example 7

Simplify $\sqrt{24x^5}$.

Solution

$\sqrt{24x^5} = \sqrt{2^3 \cdot 3 \cdot x^5} = \sqrt{2^2x^4(2 \cdot 3x)} =$
$\sqrt{2^2 \cdot x^4}\sqrt{2 \cdot 3x} = 2x^2\sqrt{6x}$

Example 8

Simplify $\sqrt{45b^7}$.

Your solution

Example 9

Simplify $2a\sqrt{18a^3b^{10}}$.

Solution

$2a\sqrt{18a^3b^{10}} = 2a\sqrt{2 \cdot 3^2 \cdot a^3 \cdot b^{10}} =$
$2a\sqrt{3^2a^2b^{10}(2a)} = 2a\sqrt{3^2a^2b^{10}}\sqrt{2a} =$
$2a \cdot 3 \cdot a \cdot b^5\sqrt{2a} = 6a^2b^5\sqrt{2a}$

Example 10

Simplify $3a\sqrt{28a^9b^{18}}$.

Your solution

Example 11

Simplify $\sqrt{16(x + 5)^2}$.

Solution

$\sqrt{16(x + 5)^2} = \sqrt{2^4(x + 5)^2} = 2^2(x + 5) =$
$4(x + 5) = 4x + 20$

Example 12

Simplify $\sqrt{25(a - 3)^2}$.

Your solution

Example 13

Simplify $\sqrt{x^2 + 10x + 25}$.

Solution

$\sqrt{x^2 + 10x + 25} = \sqrt{(x + 5)^2} = x + 5$

Example 14

Simplify $\sqrt{x^2 - 14x + 49}$.

Your solution

Solutions on p. A61

1.1 Exercises

Simplify.

1. $\sqrt{16}$	**2.** $\sqrt{64}$	**3.** $\sqrt{49}$	**4.** $\sqrt{144}$
5. $\sqrt{32}$	**6.** $\sqrt{50}$	**7.** $\sqrt{8}$	**8.** $\sqrt{12}$
9. $6\sqrt{18}$	**10.** $-3\sqrt{48}$	**11.** $5\sqrt{40}$	**12.** $2\sqrt{28}$
13. $\sqrt{15}$	**14.** $\sqrt{21}$	**15.** $\sqrt{29}$	**16.** $\sqrt{13}$
17. $-9\sqrt{72}$	**18.** $11\sqrt{80}$	**19.** $\sqrt{45}$	**20.** $\sqrt{225}$
21. $\sqrt{0}$	**22.** $\sqrt{210}$	**23.** $6\sqrt{128}$	**24.** $9\sqrt{288}$
25. $\sqrt{105}$	**26.** $\sqrt{55}$	**27.** $\sqrt{900}$	**28.** $\sqrt{300}$
29. $5\sqrt{180}$	**30.** $7\sqrt{98}$	**31.** $\sqrt{250}$	**32.** $\sqrt{120}$
33. $\sqrt{96}$	**34.** $\sqrt{160}$	**35.** $\sqrt{324}$	**36.** $\sqrt{444}$

Find the decimal approximation. Use the table on page A2.

37. $\sqrt{240}$	**38.** $\sqrt{300}$	**39.** $\sqrt{288}$	**40.** $\sqrt{600}$
41. $\sqrt{256}$	**42.** $\sqrt{324}$	**43.** $\sqrt{275}$	**44.** $\sqrt{450}$
45. $\sqrt{245}$	**46.** $\sqrt{525}$	**47.** $\sqrt{352}$	**48.** $\sqrt{363}$

1.2 Exercises

Simplify.

49. $\sqrt{x^6}$ **50.** $\sqrt{x^{12}}$ **51.** $\sqrt{y^{15}}$ **52.** $\sqrt{y^{11}}$

53. $\sqrt{a^{20}}$ **54.** $\sqrt{a^{16}}$ **55.** $\sqrt{x^4y^4}$ **56.** $\sqrt{x^{12}y^8}$

57. $\sqrt{4x^4}$ **58.** $\sqrt{25y^8}$ **59.** $\sqrt{24x^2}$ **60.** $\sqrt{x^3y^{15}}$

61. $\sqrt{x^3y^7}$ **62.** $\sqrt{a^{15}b^5}$ **63.** $\sqrt{a^3b^{11}}$ **64.** $\sqrt{24y^7}$

65. $\sqrt{60x^5}$ **66.** $\sqrt{72y^7}$ **67.** $\sqrt{49a^4b^8}$ **68.** $\sqrt{144x^2y^8}$

69. $\sqrt{18x^5y^7}$ **70.** $\sqrt{32a^5b^{15}}$ **71.** $\sqrt{40x^{11}y^7}$ **72.** $\sqrt{72x^9y^3}$

73. $\sqrt{80a^9b^{10}}$ **74.** $\sqrt{96a^5b^7}$ **75.** $2\sqrt{16a^2b^3}$ **76.** $5\sqrt{25a^4b^7}$

77. $x\sqrt{x^4y^2}$ **78.** $y\sqrt{x^3y^6}$ **79.** $4\sqrt{20a^4b^7}$ **80.** $5\sqrt{12a^3b^4}$

81. $3x\sqrt{12x^2y^7}$ **82.** $4y\sqrt{18x^5y^4}$ **83.** $2x^2\sqrt{8x^2y^3}$ **84.** $3y^2\sqrt{27x^4y^3}$

85. $\sqrt{25(a+4)^2}$ **86.** $\sqrt{81(x-y)^4}$ **87.** $\sqrt{4(x-2)^4}$

88. $\sqrt{9(x-2)^8}$ **89.** $\sqrt{x^2+4x+4}$ **90.** $\sqrt{b^2-8b+16}$

91. $\sqrt{y^2-2y+1}$ **92.** $\sqrt{a^2+6a+9}$ **93.** $\sqrt{x^2-8x+16}$

SECTION 2 Addition and Subtraction of Radical Expressions

2.1 Objective To add and subtract radical expressions

Reference for Computer Tutor™

The Distributive Property is used to simplify the sum or difference of radical expressions with like radicands.

$5\sqrt{2} + 3\sqrt{2} = (5 + 3)\sqrt{2} = 8\sqrt{2}$

$6\sqrt{2x} - 4\sqrt{2x} = (6 - 4)\sqrt{2x} = 2\sqrt{2x}$

Radical expressions which are in simplest form and have unlike radicands cannot be simplified by the Distributive Property.

$2\sqrt{3} + 4\sqrt{2}$ cannot be simplified by the Distributive Property.

Simplify $4\sqrt{8} - 10\sqrt{2}$.

Simplify each term.

$4\sqrt{8} - 10\sqrt{2} = 4\sqrt{2^3} - 10\sqrt{2}$

$= 4\sqrt{2^2 \cdot 2} - 10\sqrt{2}$ Do this step mentally.

$= 4\sqrt{2^2}\sqrt{2} - 10\sqrt{2}$

$= 4 \cdot 2\sqrt{2} - 10\sqrt{2}$

$= 8\sqrt{2} - 10\sqrt{2}$

Simplify the expression by using the Distributive Property.

$= (8 - 10)\sqrt{2}$ Do this step mentally.

$= -2\sqrt{2}$

Simplify $8\sqrt{18x} - 2\sqrt{32x}$.

Simplify each term.

$8\sqrt{18x} - 2\sqrt{32x} = 8\sqrt{2 \cdot 3^2 x} - 2\sqrt{2^5 x}$

$= 8\sqrt{3^2 \cdot 2x} - 2\sqrt{2^4 \cdot 2x}$ Do this step mentally.

$= 8\sqrt{3^2}\sqrt{2x} - 2\sqrt{2^4}\sqrt{2x}$

$= 8 \cdot 3\sqrt{2x} - 2 \cdot 2^2\sqrt{2x}$

$= 24\sqrt{2x} - 8\sqrt{2x}$

Simplify the expression by using the Distributive Property.

$= (24 - 8)\sqrt{2x}$ Do this step mentally.

$= 16\sqrt{2x}$

Example 1

Simplify $5\sqrt{2} - 3\sqrt{2} + 12\sqrt{2}$.

Solution

$5\sqrt{2} - 3\sqrt{2} + 12\sqrt{2} = 14\sqrt{2}$

Example 2

Simplify $9\sqrt{3} + 3\sqrt{3} - 18\sqrt{3}$.

Your solution

Solution on p. A61

Example 3

Simplify $3\sqrt{12} - 5\sqrt{27}$.

Solution

$3\sqrt{12} - 5\sqrt{27} = 3\sqrt{2^2 \cdot 3} - 5\sqrt{3^3} =$
$3\sqrt{2^2}\sqrt{3} - 5\sqrt{3^2}\sqrt{3} =$
$3 \cdot 2\sqrt{3} - 5 \cdot 3\sqrt{3} = 6\sqrt{3} - 15\sqrt{3} = -9\sqrt{3}$

Example 4

Simplify $2\sqrt{50} - 5\sqrt{32}$.

Your solution

Example 5

Simplify $3\sqrt{12x^3} - 2x\sqrt{3x}$.

Solution

$3\sqrt{12x^3} - 2x\sqrt{3x} = 3\sqrt{2^2 \cdot 3 \cdot x^3} - 2x\sqrt{3x} =$
$3\sqrt{2^2 \cdot x^2}\sqrt{3x} - 2x\sqrt{3x} = 3 \cdot 2 \cdot x\sqrt{3x} - 2x\sqrt{3x} =$
$6x\sqrt{3x} - 2x\sqrt{3x} = 4x\sqrt{3x}$

Example 6

Simplify $y\sqrt{28y} + 7\sqrt{63y^3}$.

Your solution

Example 7

Simplify $2x\sqrt{8y} - 3\sqrt{2x^2y} + 2\sqrt{32x^2y}$.

Solution

$2x\sqrt{8y} - 3\sqrt{2x^2y} + 2\sqrt{32x^2y} =$
$2x\sqrt{2^3y} - 3\sqrt{2x^2y} + 2\sqrt{2^5x^2y} =$
$2x\sqrt{2^2}\sqrt{2y} - 3\sqrt{x^2}\sqrt{2y} + 2\sqrt{2^4x^2}\sqrt{2y} =$
$2x \cdot 2\sqrt{2y} - 3 \cdot x\sqrt{2y} + 2 \cdot 2^2 \cdot x\sqrt{2y} =$
$4x\sqrt{2y} - 3x\sqrt{2y} + 8x\sqrt{2y} = 9x\sqrt{2y}$

Example 8

Simplify $2\sqrt{27a^5} - 4a\sqrt{12a^3} + a^2\sqrt{75a}$.

Your solution

Solutions on p. A61

2.1 Exercises

Simplify.

1. $2\sqrt{2} + \sqrt{2}$
2. $3\sqrt{5} + 8\sqrt{5}$
3. $-3\sqrt{7} + 2\sqrt{7}$
4. $4\sqrt{5} - 10\sqrt{5}$
5. $-3\sqrt{11} - 8\sqrt{11}$
6. $-3\sqrt{3} - 5\sqrt{3}$
7. $2\sqrt{x} + 8\sqrt{x}$
8. $3\sqrt{y} + 2\sqrt{y}$
9. $8\sqrt{y} - 10\sqrt{y}$
10. $-5\sqrt{2a} + 2\sqrt{2a}$
11. $-2\sqrt{3b} - 9\sqrt{3b}$
12. $-7\sqrt{5a} - 5\sqrt{5a}$
13. $3x\sqrt{2} - x\sqrt{2}$
14. $2y\sqrt{3} - 9y\sqrt{3}$
15. $2a\sqrt{3a} - 5a\sqrt{3a}$
16. $-5b\sqrt{3x} - 2b\sqrt{3x}$
17. $3\sqrt{xy} - 8\sqrt{xy}$
18. $-4\sqrt{xy} + 6\sqrt{xy}$
19. $\sqrt{45} + \sqrt{125}$
20. $\sqrt{32} - \sqrt{98}$
21. $2\sqrt{2} + 3\sqrt{8}$
22. $4\sqrt{128} - 3\sqrt{32}$
23. $5\sqrt{18} - 2\sqrt{75}$
24. $5\sqrt{75} - 2\sqrt{18}$
25. $5\sqrt{4x} - 3\sqrt{9x}$
26. $-3\sqrt{25y} + 8\sqrt{49y}$
27. $3\sqrt{3x^2} - 5\sqrt{27x^2}$
28. $-2\sqrt{8y^2} + 5\sqrt{32y^2}$
29. $2x\sqrt{xy^2} - 3y\sqrt{x^2y}$
30. $4a\sqrt{b^2a} - 3b\sqrt{a^2b}$
31. $3x\sqrt{12x} - 5\sqrt{27x^3}$
32. $2a\sqrt{50a} + 7\sqrt{32a^3}$
33. $4y\sqrt{8y^3} - 7\sqrt{18y^5}$
34. $2a\sqrt{8ab^2} - 2b\sqrt{2a^3}$
35. $b^2\sqrt{a^5b} + 3a^2\sqrt{ab^5}$
36. $y^2\sqrt{x^5y} + x\sqrt{x^3y^5}$

Simplify.

37. $4\sqrt{2} - 5\sqrt{2} + 8\sqrt{2}$

38. $3\sqrt{3} + 8\sqrt{3} - 16\sqrt{3}$

39. $5\sqrt{x} - 8\sqrt{x} + 9\sqrt{x}$

40. $\sqrt{x} - 7\sqrt{x} + 6\sqrt{x}$

41. $8\sqrt{2} - 3\sqrt{y} - 8\sqrt{2}$

42. $8\sqrt{3} - 5\sqrt{2} - 5\sqrt{3}$

43. $8\sqrt{8} - 4\sqrt{32} - 9\sqrt{50}$

44. $2\sqrt{12} - 4\sqrt{27} + \sqrt{75}$

45. $-2\sqrt{3} + 5\sqrt{27} - 4\sqrt{45}$

46. $-2\sqrt{8} - 3\sqrt{27} + 3\sqrt{50}$

47. $4\sqrt{75} + 3\sqrt{48} - \sqrt{99}$

48. $2\sqrt{75} - 5\sqrt{20} + 2\sqrt{45}$

49. $\sqrt{25x} - \sqrt{9x} + \sqrt{16x}$

50. $\sqrt{4x} - \sqrt{100x} - \sqrt{49x}$

51. $3\sqrt{3x} + \sqrt{27x} - 8\sqrt{75x}$

52. $5\sqrt{5x} + 2\sqrt{45x} - 3\sqrt{80x}$

53. $2a\sqrt{75b} - a\sqrt{20b} + 4a\sqrt{45b}$

54. $2b\sqrt{75a} - 5b\sqrt{27a} + 2b\sqrt{20a}$

55. $x\sqrt{3y^2} - 2y\sqrt{12x^2} + xy\sqrt{3}$

56. $a\sqrt{27b^2} + 3b\sqrt{147a^2} - ab\sqrt{3}$

57. $3\sqrt{ab^3} + 4a\sqrt{ab} - 5b\sqrt{4ab}$

58. $5\sqrt{a^3b} + a\sqrt{4ab} - 3\sqrt{49a^3b}$

59. $3a\sqrt{2ab^2} - \sqrt{a^2b^2} + 4b\sqrt{3a^2b}$

60. $2\sqrt{4a^2b^2} - 3a\sqrt{9ab^2} + 4b\sqrt{a^2b}$

SECTION 3 Multiplication and Division of Radical Expressions

3.1 Objective To multiply radical expressions

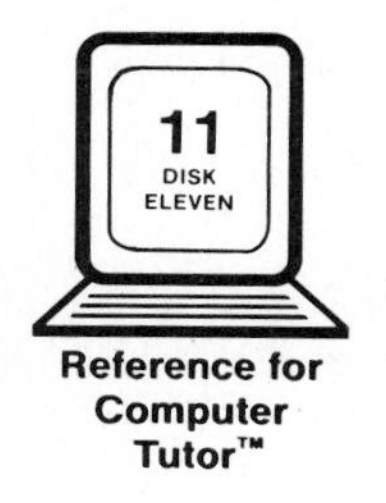

Reference for Computer Tutor™

The Product Property of Square Roots can also be used to multiply variable radical expressions. $\sqrt{2x}\sqrt{3y} = \sqrt{2x \cdot 3y} = \sqrt{6xy}$

Simplify $(\sqrt{x})^2$.

Multiply the radicands. $(\sqrt{x})^2 = \sqrt{x}\sqrt{x} = \sqrt{x \cdot x}$ Do this step mentally.

Simplify. $= \sqrt{x^2}$

$= x$

Note: For $a > 0$, $\mathbf{(\sqrt{a})^2 = \sqrt{a^2} = a}$

Simplify $\sqrt{2x^2}\sqrt{32x^5}$.

Use the Product Property of Square Roots. $\sqrt{2x^2}\sqrt{32x^5} = \sqrt{2x^2 \cdot 32x^5}$ Do this step mentally.

Multiply the radicands. $= \sqrt{64x^7}$

Simplify. $= \sqrt{2^6x^7}$

$= \sqrt{2^6x^6}\sqrt{x}$

$= 2^3x^3\sqrt{x}$

$= 8x^3\sqrt{x}$

Simplify $\sqrt{2x}(x + \sqrt{2x})$.

Use the Distributive Property to remove parentheses. $\sqrt{2x}(x + \sqrt{2x}) = \sqrt{2x}(x) + \sqrt{2x}\sqrt{2x}$ Do this step mentally.

$= x\sqrt{2x} + \sqrt{4x^2}$

$= x\sqrt{2x} + \sqrt{2^2x^2}$

$= x\sqrt{2x} + 2x$

Simplify $(\sqrt{2} - 3x)(\sqrt{2} + x)$.

Use the FOIL method to remove parentheses. $(\sqrt{2} - 3x)(\sqrt{2} + x) = \sqrt{2 \cdot 2} + x\sqrt{2} - 3x\sqrt{2} - 3x^2$

$= \sqrt{2^2} + (x - 3x)\sqrt{2} - 3x^2$

$= 2 - 2x\sqrt{2} - 3x^2$

The expressions $a + b$ and $a - b$, which are the sum and difference of two terms, are called **conjugates** of each other. Conjugates differ only in the sign of the second term.

Simplify $(2 + \sqrt{7})(2 - \sqrt{7})$.

The product of conjugates of the form $(a + b)(a - b) = a^2 - b^2$. $(2 + \sqrt{7})(2 - \sqrt{7}) = 2^2 - \sqrt{7^2}$

$= 4 - 7$

$= -3$

Simplify $(3 + \sqrt{y})(3 - \sqrt{y})$.

The product of conjugates of the form $(a + b)(a - b) = a^2 - b^2$. $(3 + \sqrt{y})(3 - \sqrt{y}) = 3^2 - \sqrt{y^2}$

$= 9 - y$

Example 1

Simplify $\sqrt{3x^4}\sqrt{2x^2y}\sqrt{6xy^2}$.

Solution

$\sqrt{3x^4}\sqrt{2x^2y}\sqrt{6xy^2} = \sqrt{36x^7y^3} =$
$\sqrt{2^2 3^2 x^7 y^3} = \sqrt{2^2 3^2 x^6 y^2}\sqrt{xy} =$
$2 \cdot 3x^3y\sqrt{xy} = 6x^3y\sqrt{xy}$

Example 2

Simplify $\sqrt{5a}\sqrt{15a^3b^4}\sqrt{3b^5}$.

Your solution

Example 3

Simplify $\sqrt{3ab}(\sqrt{3a} + \sqrt{9b})$.

Solution

$\sqrt{3ab}(\sqrt{3a} + \sqrt{9b}) = \sqrt{3^2a^2b} + \sqrt{3^3ab^2} =$
$\sqrt{3^2a^2}\sqrt{b} + \sqrt{3^2b^2}\sqrt{3a} = 3a\sqrt{b} + 3b\sqrt{3a}$

Example 4

Simplify $\sqrt{5x}(\sqrt{5x} - \sqrt{25y})$.

Your solution

Example 5

Simplify $(\sqrt{a} - \sqrt{b})(\sqrt{a} + \sqrt{b})$.

Solution

$(\sqrt{a} - \sqrt{b})(\sqrt{a} + \sqrt{b}) = \sqrt{a^2} - \sqrt{b^2} = a - b$

Example 6

Simplify $(2\sqrt{x} + 7)(2\sqrt{x} - 7)$.

Your solution

Example 7

Simplify $(2\sqrt{x} - \sqrt{y})(5\sqrt{x} - 2\sqrt{y})$.

Solution

$(2\sqrt{x} - \sqrt{y})(5\sqrt{x} - 2\sqrt{y}) =$
$10\sqrt{x^2} - 4\sqrt{xy} - 5\sqrt{xy} + 2\sqrt{y^2} =$
$10x - 9\sqrt{xy} + 2y$

Example 8

Simplify $(3\sqrt{x} - \sqrt{y})(5\sqrt{x} - 2\sqrt{y})$.

Your solution

3.2 Objective To divide radical expressions

Reference for Computer Tutor™

The Quotient Property of Square Roots

The square root of a quotient is equal to the quotient of the square roots.

If a and b are positive real numbers and $b \neq 0$, then $\sqrt{\frac{a}{b}} = \frac{\sqrt{a}}{\sqrt{b}}$ and $\frac{\sqrt{a}}{\sqrt{b}} = \sqrt{\frac{a}{b}}$.

Simplify $\sqrt{\frac{4x^2}{z^6}}$.

Rewrite the radical expression as the quotient of the square roots.

$$\sqrt{\frac{4x^2}{z^6}} = \frac{\sqrt{4x^2}}{\sqrt{z^6}}$$

Do this step mentally.

Simplify.

$$= \frac{\sqrt{2^2x^2}}{\sqrt{z^6}} = \frac{2x}{z^3}$$

Simplify $\sqrt{\dfrac{24x^3y^7}{3x^7y^2}}$.

Simplify the radicand. $\sqrt{\dfrac{24x^3y^7}{3x^7y^2}} = \sqrt{\dfrac{8y^5}{x^4}}$

Rewrite the radical expression as the quotient of the square roots. $= \dfrac{\sqrt{8y^5}}{\sqrt{x^4}}$ Do this step mentally.

Simplify.

$$= \frac{\sqrt{2^3y^5}}{\sqrt{x^4}}$$

$$= \frac{\sqrt{2^2y^4}\sqrt{2y}}{\sqrt{x^4}}$$

$$= \frac{2y^2\sqrt{2y}}{x^2}$$

Simplify $\dfrac{\sqrt{4x^2y}}{\sqrt{xy}}$.

Use the Quotient Property of Square Roots. $\dfrac{\sqrt{4x^2y}}{\sqrt{xy}} = \sqrt{\dfrac{4x^2y}{xy}}$

Simplify the radicand. $= \sqrt{4x}$

Simplify the radical expression.

$$= \sqrt{2^2}\sqrt{x}$$

$$= 2\sqrt{x}$$

A radical expression is not considered to be in simplest form if a radical remains in the denominator. The procedure used to remove a radical from the denominator is called **rationalizing the denominator.**

Simplify $\dfrac{2}{\sqrt{3}}$.

Multiply the expression by $\dfrac{\sqrt{3}}{\sqrt{3}}$, which equals 1. $\dfrac{2}{\sqrt{3}} = \dfrac{2}{\sqrt{3}} \cdot \dfrac{\sqrt{3}}{\sqrt{3}}$

The radicand in the denominator is a perfect square. $= \dfrac{2\sqrt{3}}{\sqrt{3^2}}$ Do this step mentally.

Simplify.
The radical has been removed from the denominator. $= \dfrac{2\sqrt{3}}{3}$

Simplify $\dfrac{1}{\sqrt{y}+3}$.

Multiply the numerator and denominator by $\sqrt{y}-3$, the conjugate of $\sqrt{y}+3$. $\dfrac{1}{\sqrt{y}+3} = \dfrac{1}{\sqrt{y}+3} \cdot \dfrac{\sqrt{y}-3}{\sqrt{y}-3}$

Simplify. $= \dfrac{\sqrt{y}-3}{\sqrt{y^2}-3^2}$ Do this step mentally.

$$= \frac{\sqrt{y}-3}{y-9}$$

Simplify $\frac{\sqrt{2} + \sqrt{18y^2}}{\sqrt{2}}$.

Divide each term in the numerator by the denominator.

$$\frac{\sqrt{2} + \sqrt{18y^2}}{\sqrt{2}} = \frac{\sqrt{2}}{\sqrt{2}} + \frac{\sqrt{18y^2}}{\sqrt{2}}$$

Use the Quotient Property of Square Roots.

$$= 1 + \sqrt{\frac{18y^2}{2}}$$

Simplify.

$$= 1 + \sqrt{9y^2}$$
$$= 1 + \sqrt{3^2y^2}$$
$$= 1 + 3y$$

Example 9

Simplify $\frac{\sqrt{4x^2y^5}}{\sqrt{3x^4y}}$.

Solution

$$\frac{\sqrt{4x^2y^5}}{\sqrt{3x^4y}} = \sqrt{\frac{2^2x^2y^5}{3x^4y}} = \sqrt{\frac{2^2y^4}{3x^2}} = \frac{2y^2}{x\sqrt{3}} =$$
$$\frac{2y^2}{x\sqrt{3}} \cdot \frac{\sqrt{3}}{\sqrt{3}} = \frac{2y^2\sqrt{3}}{3x}$$

Example 10

Simplify $\frac{\sqrt{15x^6y^7}}{\sqrt{3x^7y^9}}$.

Your solution

Example 11

Simplify $\frac{\sqrt{2}}{\sqrt{2} - \sqrt{x}}$.

Solution

$$\frac{\sqrt{2}}{\sqrt{2} - \sqrt{x}} = \frac{\sqrt{2}}{\sqrt{2} - \sqrt{x}} \cdot \frac{\sqrt{2} + \sqrt{x}}{\sqrt{2} + \sqrt{x}} = \frac{2 + \sqrt{2x}}{2 - x}$$

Example 12

Simplify $\frac{\sqrt{y}}{\sqrt{y} + 3}$.

Your solution

Example 13

Simplify $\frac{\sqrt{20} - 2\sqrt{125}}{\sqrt{5}}$.

Solution

$$\frac{\sqrt{20} - 2\sqrt{125}}{\sqrt{5}} = \frac{\sqrt{20}}{\sqrt{5}} - \frac{2\sqrt{125}}{\sqrt{5}} =$$
$$\sqrt{\frac{20}{5}} - 2\sqrt{\frac{125}{5}} = \sqrt{4} - 2\sqrt{25} =$$
$$\sqrt{2^2} - 2\sqrt{5^2} = 2 - 2 \cdot 5 = 2 - 10 = -8$$

Example 14

Simplify $\frac{\sqrt{27x^3} - 3\sqrt{12x}}{\sqrt{3x}}$.

Your solution

Solutions on p. A62

3.1 Exercises

Simplify.

1. $\sqrt{5}\cdot\sqrt{5}$

2. $\sqrt{11}\cdot\sqrt{11}$

3. $\sqrt{3}\cdot\sqrt{12}$

4. $\sqrt{2}\cdot\sqrt{8}$

5. $\sqrt{x}\cdot\sqrt{x}$

6. $\sqrt{y}\cdot\sqrt{y}$

7. $\sqrt{xy^3}\cdot\sqrt{x^5y}$

8. $\sqrt{a^3b^5}\cdot\sqrt{ab^5}$

9. $\sqrt{3a^2b^5}\cdot\sqrt{6ab^7}$

10. $\sqrt{5x^3y}\cdot\sqrt{10x^2y}$

11. $\sqrt{6a^3b^2}\cdot\sqrt{24a^5b}$

12. $\sqrt{8ab^5}\cdot\sqrt{12a^7b}$

13. $\sqrt{2}(\sqrt{2}-\sqrt{3})$

14. $3(\sqrt{12}-\sqrt{3})$

15. $\sqrt{x}(\sqrt{x}-\sqrt{y})$

16. $\sqrt{b}(\sqrt{a}-\sqrt{b})$

17. $\sqrt{5}(\sqrt{10}-\sqrt{x})$

18. $\sqrt{6}(\sqrt{y}-\sqrt{18})$

19. $\sqrt{8}(\sqrt{2}-\sqrt{5})$

20. $\sqrt{10}(\sqrt{20}-\sqrt{a})$

21. $(\sqrt{x}-3)^2$

22. $(2\sqrt{a}-y)^2$

23. $\sqrt{3a}(\sqrt{3a}-\sqrt{3b})$

24. $\sqrt{5x}(\sqrt{10x}-\sqrt{x})$

25. $\sqrt{2ac}\cdot\sqrt{5ab}\cdot\sqrt{10cb}$

26. $\sqrt{3xy}\cdot\sqrt{6x^3y}\cdot\sqrt{2y^2}$

27. $(3\sqrt{x}-2y)(5\sqrt{x}-4y)$

28. $(5\sqrt{x}+2\sqrt{y})(3\sqrt{x}-\sqrt{y})$

29. $(\sqrt{x}-\sqrt{y})(\sqrt{x}+\sqrt{y})$

30. $(\sqrt{3x}+y)(\sqrt{3x}-y)$

31. $(2\sqrt{x}+\sqrt{y})(5\sqrt{x}+4\sqrt{y})$

32. $(5\sqrt{x}-2\sqrt{y})(3\sqrt{x}-4\sqrt{y})$

3.2 Exercises

Simplify.

33. $\dfrac{\sqrt{32}}{\sqrt{2}}$

34. $\dfrac{\sqrt{45}}{\sqrt{5}}$

35. $\dfrac{\sqrt{98}}{\sqrt{2}}$

36. $\dfrac{\sqrt{48}}{\sqrt{3}}$

37. $\dfrac{\sqrt{27a}}{\sqrt{3a}}$

38. $\dfrac{\sqrt{72x^5}}{\sqrt{2x}}$

39. $\dfrac{\sqrt{15x^3y}}{\sqrt{3xy}}$

40. $\dfrac{\sqrt{40x^5y^2}}{\sqrt{5xy}}$

41. $\dfrac{\sqrt{2a^5b^4}}{\sqrt{98ab^4}}$

42. $\dfrac{\sqrt{48x^5y^2}}{\sqrt{3x^3y}}$

43. $\dfrac{1}{\sqrt{3}}$

44. $\dfrac{1}{\sqrt{8}}$

45. $\dfrac{3}{\sqrt{x}}$

46. $\dfrac{4}{\sqrt{2x}}$

47. $\dfrac{\sqrt{8x^2y}}{\sqrt{2x^4y^2}}$

48. $\dfrac{\sqrt{9xy^2}}{\sqrt{27x}}$

49. $\dfrac{\sqrt{4x^2y}}{\sqrt{3xy^3}}$

50. $\dfrac{\sqrt{16x^3y^2}}{\sqrt{8x^3y}}$

51. $\dfrac{1}{\sqrt{2}-3}$

52. $\dfrac{5}{\sqrt{7}-3}$

53. $\dfrac{3}{5+\sqrt{5}}$

54. $\dfrac{7}{\sqrt{2}-7}$

55. $\dfrac{\sqrt{xy}}{\sqrt{x}-\sqrt{y}}$

56. $\dfrac{\sqrt{x}}{\sqrt{x}-\sqrt{y}}$

57. $\dfrac{3\sqrt{2}-8\sqrt{2}}{\sqrt{2}}$

58. $\dfrac{5\sqrt{3}-2\sqrt{3}}{2\sqrt{3}}$

59. $\dfrac{2\sqrt{8}+3\sqrt{2}}{\sqrt{32}}$

60. $\dfrac{6x\sqrt{5}-3\sqrt{125}}{\sqrt{45}}$

61. $\dfrac{3x\sqrt{x^3}-4\sqrt{x^5}}{\sqrt{x}}$

62. $\dfrac{7a\sqrt{a^5}+5a^3\sqrt{a}}{\sqrt{9a}}$

SECTION 4 Solving Equations Containing Radical Expressions

4.1 Objective To solve an equation containing one or more radical expressions

Reference for Computer Tutor™

An equation that contains a variable expression in a radicand is a **radical equation.**

$$\left.\begin{aligned}\sqrt{x} &= 4\\ \sqrt{x+2} &= \sqrt{x-7}\end{aligned}\right\}\text{ Radical Equations}$$

The following property of equality is used to solve radical equations.

Property of Squaring Both Sides of an Equation

If two numbers are equal, then the squares of the numbers are equal.

If a and b are real numbers and $a = b$, then $a^2 = b^2$.

Solve $\sqrt{x-2} - 7 = 0$.

Rewrite the equation with the radical on one side of the equation and the constant on the other side.

$$\sqrt{x-2} - 7 = 0$$
$$\sqrt{x-2} = 7$$

Square both sides of the equation.

$$(\sqrt{x-2})^2 = 7^2$$

Solve the resulting equation.

$$x - 2 = 49$$
$$x = 51$$

The solution is 51.

Check the solution.
When squaring both sides of an equation, the resulting equation may have a solution which is not a solution of the original equation.

Check:
$$\sqrt{x-2} - 7 = 0$$
$$\sqrt{51-2} - 7 = 0$$
$$\sqrt{49} - 7 = 0$$
$$\sqrt{7^2} - 7 = 0$$
$$7 - 7 = 0$$
$$0 = 0 \quad \text{A true equation}$$

Example 1

Solve: $\sqrt{3x} + 2 = 5$

Solution

$$\sqrt{3x} + 2 = 5$$
$$\sqrt{3x} = 3$$
$$(\sqrt{3x})^2 = 3^2$$
$$3x = 9$$
$$x = 3$$

Check:
$$\sqrt{3x} + 2 = 5$$
$$\sqrt{3 \cdot 3} + 2 = 5$$
$$\sqrt{3^2} + 2 = 5$$
$$3 + 2 = 5$$
$$5 = 5$$

The solution is 3.

Example 2

Solve: $\sqrt{4x} + 3 = 7$

Your solution

Solution on p. A62

Example 3

Solve: $0 = 3 - \sqrt{2x - 3}$

Solution

$$
\begin{aligned}
0 &= 3 - \sqrt{2x - 3} \\
-3 &= -\sqrt{2x - 3} \\
(-3)^2 &= (-\sqrt{2x - 3})^2 \\
9 &= 2x - 3 \\
12 &= 2x \\
6 &= x
\end{aligned}
$$

Check:

$$
\begin{aligned}
0 &= 3 - \sqrt{2x - 3} \\
0 &= 3 - \sqrt{2 \cdot 6 - 3} \\
0 &= 3 - \sqrt{12 - 3} \\
0 &= 3 - \sqrt{9} \\
0 &= 3 - \sqrt{3^2} \\
0 &= 3 - 3 \\
0 &= 0
\end{aligned}
$$

The solution is 6.

Example 4

Solve: $\sqrt{3x - 2} - 5 = 0$

Your solution

Example 5

Solve: $\sqrt{2x + 1} = \sqrt{3x - 4}$

Solution

$$
\begin{aligned}
\sqrt{2x + 1} &= \sqrt{3x - 4} \\
(\sqrt{2x + 1})^2 &= (\sqrt{3x - 4})^2 \\
2x + 1 &= 3x - 4 \\
2x &= 3x - 5 \\
-x &= -5 \\
x &= 5
\end{aligned}
$$

Check:

$$
\begin{aligned}
\sqrt{2x + 1} &= \sqrt{3x - 4} \\
\sqrt{2 \cdot 5 + 1} &= \sqrt{3 \cdot 5 - 4} \\
\sqrt{10 + 1} &= \sqrt{15 - 4} \\
\sqrt{11} &= \sqrt{11}
\end{aligned}
$$

The solution is 5.

Example 6

Solve: $\sqrt{4x + 3} = \sqrt{x + 12}$

Your solution

Solutions on pp. A62–A63

4.2 Objective To solve application problems

Reference for Computer Tutor™

A right triangle contains one 90° angle. The side opposite the 90° angle is called the **hypotenuse.** The other two sides are called legs.

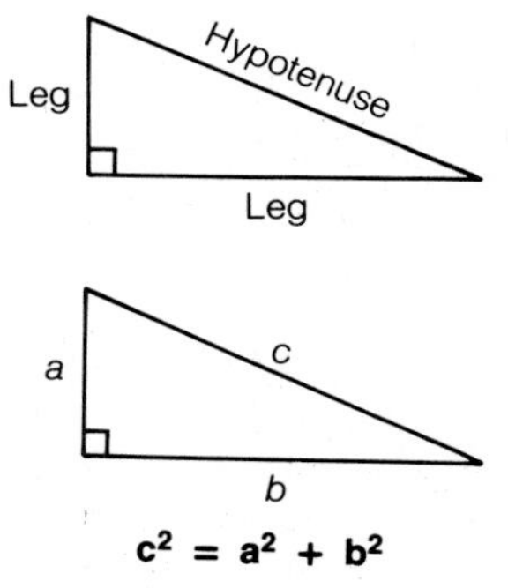

Pythagoras, a Greek mathematician, discovered that the square of the hypotenuse of a right triangle is equal to the sum of the squares of the two legs. This is called the **Pythagorean Theorem.**

Using this theorem, the hypotenuse of a right triangle can be found when the two legs are known. Use the formula

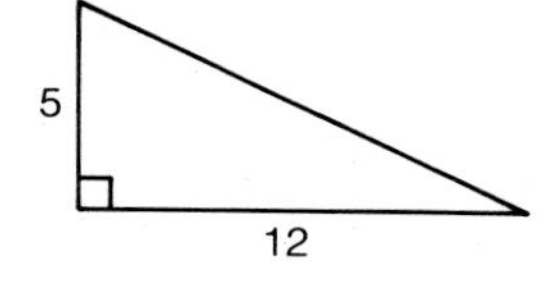

$$
\begin{aligned}
\text{Hypotenuse} &= \sqrt{(\text{leg})^2 + (\text{leg})^2} \\
c &= \sqrt{a^2 + b^2} \\
&= \sqrt{(5)^2 + (12)^2} \\
&= \sqrt{25 + 144} \\
&= \sqrt{169} \\
&= 13
\end{aligned}
$$

The leg of a right triangle can be found when one leg and the hypotenuse are known. Use the formula

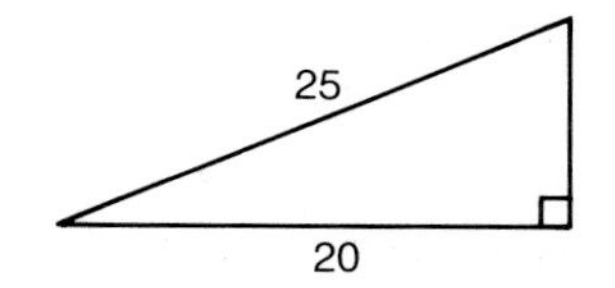

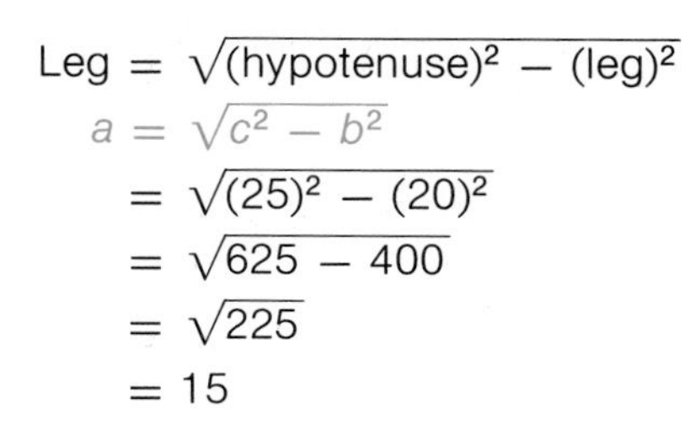

$$
\begin{aligned}
\text{Leg} &= \sqrt{(\text{hypotenuse})^2 - (\text{leg})^2} \\
a &= \sqrt{c^2 - b^2} \\
&= \sqrt{(25)^2 - (20)^2} \\
&= \sqrt{625 - 400} \\
&= \sqrt{225} \\
&= 15
\end{aligned}
$$

Examples 7 and 8 illustrate the use of the Pythagorean Theorem. Examples 9 and 10 illustrate other applications of radical equations.

Example 7

A guy wire is attached to a point 20 m above the ground on a telephone pole. The wire is anchored to the ground at a point 8 m from the base of the pole. Find the length of the guy wire.

Strategy

To find the length of the guy wire, use the Pythagorean Theorem. One leg is the distance from the bottom of the wire to the base of the telephone pole. The other leg is the distance from the top of the wire to the base of the telephone pole. The guy wire is the hypotenuse. Solve the Pythagorean Theorem for the hypotenuse.

Solution

$c = \sqrt{a^2 + b^2}$
$c = \sqrt{(20)^2 + (8)^2}$
$c = \sqrt{400 + 64}$
$c = \sqrt{464}$
$c \approx 21.541$

The guy wire has a length of 21.541 m.

Example 8

A ladder 8 ft long is resting against a building. How high on the building will the ladder reach when the bottom of the ladder is 3 ft from the building?

Your strategy

Your solution

Example 9

How far would a submarine periscope have to be above the water to locate a ship 4 mi away? The equation for the distance in miles that the lookout can see is $d = 1.4\sqrt{h}$, where h is the height in feet above the surface of the water. Round to the nearest hundredth.

Strategy

To find the height above the water, replace d in the equation with the given value and solve for h.

Solution

$1.4\sqrt{h} = d$
$1.4\sqrt{h} = 4$
$\sqrt{h} = \frac{4}{1.4}$
$(\sqrt{h})^2 = \left(\frac{4}{1.4}\right)^2$
$h = \frac{16}{1.96}$
$h \approx 8.16$

The periscope must be 8.16 ft above the water.

Example 10

Find the length of a pendulum that makes one swing in 2.5 s. The equation for the time for one swing is $T = 2\pi\sqrt{\frac{L}{32}}$, where T is the time in seconds and L is the length in feet. Use 3.14 for π. Round to the nearest hundredth.

Your strategy

Your solution

Solutions on p. A63

4.1 Exercises

Solve and check.

1. $\sqrt{x} = 5$
2. $\sqrt{y} = 7$
3. $\sqrt{a} = 12$
4. $\sqrt{a} = 9$
5. $\sqrt{5x} = 5$
6. $\sqrt{3x} = 4$
7. $\sqrt{4x} = 8$
8. $\sqrt{6x} = 3$
9. $\sqrt{2x} - 4 = 0$
10. $3 - \sqrt{5x} = 0$
11. $\sqrt{4x} + 5 = 2$
12. $\sqrt{3x} + 9 = 4$
13. $\sqrt{3x - 2} = 4$
14. $\sqrt{5x + 6} = 1$
15. $\sqrt{2x + 1} = 7$
16. $\sqrt{5x + 4} = 3$
17. $0 = 2 - \sqrt{3 - x}$
18. $0 = 5 - \sqrt{10 + x}$
19. $\sqrt{5x + 2} = 0$
20. $\sqrt{3x - 7} = 0$
21. $\sqrt{3x} - 6 = -4$
22. $\sqrt{5x} + 8 = 23$
23. $0 = \sqrt{3x - 9} - 6$
24. $0 = \sqrt{2x + 7} - 3$
25. $\sqrt{5x - 1} = \sqrt{3x + 9}$
26. $\sqrt{3x + 4} = \sqrt{12x - 14}$
27. $\sqrt{5x - 3} = \sqrt{4x - 2}$
28. $\sqrt{5x - 9} = \sqrt{2x - 3}$
29. $\sqrt{x^2 - 5x + 6} = \sqrt{x^2 - 8x + 9}$
30. $\sqrt{x^2 - 2x + 4} = \sqrt{x^2 + 5x - 12}$

4.2 Application Problems

Solve.

1. A rope used to support a 4 ft tent pole is attached to the top of the pole and anchored in the ground at a point 2 ft from the base of the pole. Find the length of the rope.

2. A wooden beam 12 ft long is resting against a wall. The beam touches the wall at a point 8 ft above the ground. Find the distance from the base of the wall to the base of the beam.

3. How far would a submarine periscope have to be above the water to locate a ship 5 mi away? The equation for the distance in miles that the lookout can see is $d = 1.4\sqrt{h}$, where h is the height in feet above the surface of the water. Round to the nearest hundredth.

4. How far would a submarine periscope have to be above the water to locate a ship 6 mi away? The equation for the distance in miles that the lookout can see is $d = 1.4\sqrt{h}$, where h is the height in feet above the surface of the water. Round to the nearest hundredth.

5. An object is dropped from a high building. Find the distance the object has fallen when the speed reaches 64 ft/s. The equation for the distance is $v = \sqrt{64d}$, where v is the speed of the object and d is the distance.

6. An object is dropped from a plane. Find the distance the object has fallen when the speed reaches 512 ft/s. The equation for the distance is $v = \sqrt{64d}$, where v is the speed of the object and d is the distance.

7. A stone is dropped from a bridge and hits the water 1.5 s later. How high is the bridge? The equation for the distance an object falls in T seconds is given by $T = \sqrt{\frac{d}{16}}$, where d is the distance in feet. Round to the nearest hundredth.

8. A stone is dropped into a mine shaft and hits the bottom 3 s later. How deep is the mine shaft? The equation for the distance an object falls in T seconds is given by $T = \sqrt{\frac{d}{16}}$, where d is the distance in feet. Round to the nearest hundredth.

9. Find the length of a pendulum that makes one swing in 2 s. The equation for the time of one swing of a pendulum is given by $T = 2\pi\sqrt{\frac{L}{32}}$, where T is the time in seconds and L is the length in feet. Use 3.14 for π. Round to the nearest hundredth.

10. Find the length of a pendulum that makes one swing in 1.5 s. The equation for the time of one swing of a pendulum is given by $T = 2\pi\sqrt{\frac{L}{32}}$, where T is the time in seconds and L is the length in feet. Use 3.14 for π. Round to the nearest hundredth.

Calculators and Computers

Simplifying Radical Expressions

Chapter 11 presents simplification of numerical and variable radical expressions and operations with radical expressions (addition, subtraction, multiplication, and division). These concepts are followed by solving equations containing one or more radical expressions and solving application problems which involve radical expressions.

Just as expressions with the same variable part can be added or subtracted, expressions with like radicands can be added or subtracted.

$$2x + 3x = 5x \qquad 2\sqrt{x} + 3\sqrt{x} = 5\sqrt{x}$$

If the variable parts or the radicands are unlike, the expressions cannot be added or subtracted.

$$2x + 3y = 2x + 3y \qquad 2\sqrt{x} + 3\sqrt{y} = 2\sqrt{x} + 3\sqrt{y}$$

Expressions with like or unlike variable parts or radicands can be multiplied.

$$(2x)(3x) = 6x^2 \qquad (2\sqrt{x})(3\sqrt{x}) = 6\sqrt{x^2} = 6x$$
$$(2x)(3y) = 6xy \qquad (2\sqrt{x})(3\sqrt{y}) = 6\sqrt{xy}$$

A computer program can be written to solve radical expressions. One such program is on the Student Disk.

The program RADICAL EXPRESSIONS on the Student Disk will allow you to practice simplifying radical expressions. The program will display a radical expression. Then, using pencil and paper, simplify the expression. When you are finished, press the RETURN key. The correct solution will be displayed on the screen.

After you complete a problem, you have the opportunity to continue to practice or to quit the program. You will press the letter 'C' to continue or the letter 'Q' to quit.

Chapter Summary

KEY WORDS

A **square root** of a positive number x is a number whose square is x.

The **principal square root** of a number is the positive square root.

The symbol $\sqrt{\ }$ is called a radical and is used to indicate the principal square root of a number.

The **radicand** is the number under the radical sign.

The square of an integer is a **perfect square.**

If a number is not a perfect square, its square root can only be approximated. These numbers are **irrational numbers.** Their decimal representations never terminate or repeat.

Conjugates are binomial expressions which differ only in the sign of the second term. (The expressions $a + b$ and $a - b$ are conjugates.)

Rationalizing the denominator is the procedure used to remove a radical from the denominator of a fraction.

A **radical equation** is an equation that contains a variable expression in a radicand.

ESSENTIAL RULES

The Product Property of Square Roots

If a and b are positive real numbers, then $\sqrt{ab} = \sqrt{a}\sqrt{b}$.

The Quotient Property of Square Roots

If a and b are positive real numbers and $b \neq 0$, then $\sqrt{\frac{a}{b}} = \frac{\sqrt{a}}{\sqrt{b}}$ and $\frac{\sqrt{a}}{\sqrt{b}} = \sqrt{\frac{a}{b}}$.

Property of Squaring Both Sides of An Equation

If a and b are real numbers and $a = b$, then $a^2 = b^2$.

Pythagorean Theorem

$c^2 = a^2 + b^2$

Review/Test

SECTION 1

1.1a Simplify $\sqrt{45}$.

1.1b Simplify $\sqrt{75}$.

1.1c Find the decimal approximation of $\sqrt{125}$. Use the table on page A2.

1.2a Simplify $\sqrt{121x^8y^2}$.

1.2b Simplify $\sqrt{72x^7y^2}$.

1.2c Simplify $\sqrt{32a^5b^{11}}$.

SECTION 2

2.1a Simplify $8\sqrt{y} - 3\sqrt{y}$.

2.1b Simplify $5\sqrt{8} - 3\sqrt{50}$.

2.1c Simplify $3\sqrt{8y} - 2\sqrt{72x} + 5\sqrt{18y}$.

2.1d Simplify $2x\sqrt{3xy^3} - 2y\sqrt{12x^3y} - 3xy\sqrt{xy}$.

SECTION 3

3.1a Simplify $\sqrt{8x^3y}\sqrt{10xy^4}$.

3.1b Simplify $\sqrt{3x^2y}\sqrt{6xy^2}\sqrt{2x}$.

Review/Test

3.1c Simplify $\sqrt{a}(\sqrt{a} - \sqrt{b})$.

3.1d Simplify $(\sqrt{y} - 3)(\sqrt{y} + 5)$.

3.2a Simplify $\frac{\sqrt{162}}{\sqrt{2}}$.

3.2b Simplify $\frac{\sqrt{98a^6b^4}}{\sqrt{2a^3b^2}}$.

3.2c Simplify $\frac{2}{\sqrt{3} - 1}$.

3.2d Simplify $\frac{3\sqrt{x^3} - 4\sqrt{9x}}{3\sqrt{x}}$.

SECTION **4**

4.1a Solve: $\sqrt{5x - 6} = 7$

4.1b Solve: $\sqrt{9x} + 3 = 18$

4.2a The square root of the sum of two consecutive odd integers is equal to 10. Find the larger integer.

4.2b Find the length of a pendulum that makes one swing in 3 s. The equation for the time of one swing of a pendulum is given by $T = 2\pi\sqrt{\frac{L}{32}}$, where T is the time in seconds and L is the length in feet. Use 3.14 for π. Round to the nearest hundredth.

Cumulative Review/Test

1. Simplify:
$\left(\frac{2}{3}\right)^2 \cdot \left(\frac{3}{4} - \frac{3}{2}\right) + \left(\frac{1}{2}\right)^2$

2. Simplify:
$-3[x - 2(3 - 2x) - 5x] + 2x$

3. Solve:
$2x - 4[3x - 2(1 - 3x)] = 2(3 - 4x)$

4. Simplify $(-3x^2y)(-2x^3y^4)$.

5. Simplify:
$\frac{12b^4 - 6b^2 + 2}{-6b^2}$

6. Factor $12x^3y^2 - 9x^2y^3$.

7. Factor $2a^3 - 16a^2 + 30a$.

8. Simplify:
$\frac{3x^3 - 6x^2}{4x^2 + 4x} \cdot \frac{3x - 9}{9x^3 - 45x^2 + 54x}$

9. Simplify:
$\frac{x + 2}{x - 4} - \frac{6}{(x - 4)(x - 3)}$

10. Solve:
$\frac{x}{2x - 5} - 2 = \frac{3x}{2x - 5}$

11. Find the equation of the line which contains the point $(-2, -3)$ and has slope $\frac{1}{2}$.

12. Solve by substitution.
$4x - 3y = 1$
$2x + y = 3$

13. Solve by the addition method.
$5x + 4y = 7$
$3x - 2y = 13$

14. Solve: $3(x - 7) \geq 5x - 12$

15. Simplify: $\sqrt{108}$

16. Simplify: $3\sqrt{32} - 2\sqrt{128}$

Cumulative Review/Test

17. Simplify:
$2a\sqrt{2ab^3} + b\sqrt{8a^3b} - 5ab\sqrt{ab}$

18. Simplify:
$\sqrt{2a^9b}\sqrt{98ab^3}\sqrt{2a}$

19. Simplify: $\sqrt{3}(\sqrt{6} - \sqrt{x^2})$

20. Simplify: $\frac{\sqrt{320}}{\sqrt{5}}$

21. Simplify: $\frac{3}{2 - \sqrt{5}}$

22. Solve: $\sqrt{3x - 2} - 4 = 0$

23. The selling price for a book is \$29.40. The markup rate used by the bookstore is 20%. Find the cost of the book.

24. How many ounces of pure water must be added to 40 oz of a 12% salt solution to make a salt solution which is 5% salt?

25. The sum of two numbers is twenty-one. The product of the two numbers is one hundred four. Find the two numbers.

26. A small water pipe takes twice as long to fill a tank as does a larger water pipe. With both pipes open it takes 16 h to fill the tank. Find the time it would take the small pipe working alone to fill the tank.

27. Solve by graphing.
$3x - 2y = 8$
$4x + 5y = 3$

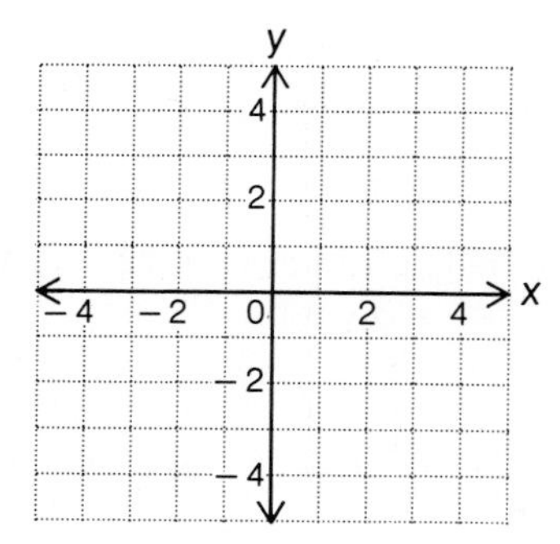

28. Graph the solution set of
$3x + y < 2$.

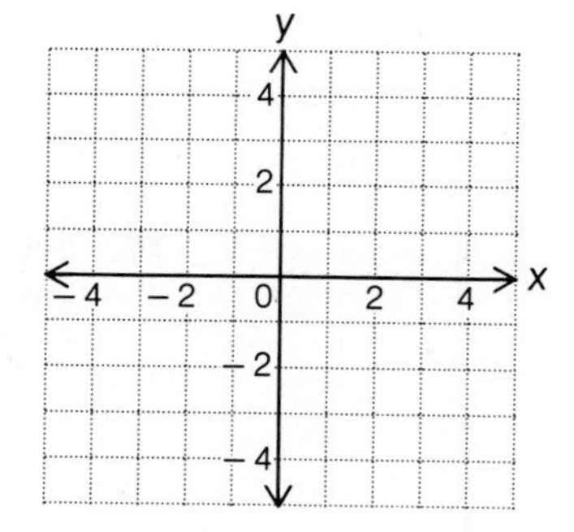

29. The square root of the sum of two consecutive integers is equal to 9. Find the smaller integer.

30. A stone is dropped from a building and hits the ground 5 s later. How high is the building? The equation for the distance an object falls in T seconds is given by $T = \sqrt{\frac{d}{16}}$, where d is the distance in feet. Round to the nearest hundredth.

12

Quadratic Equations

Objectives

- **To solve a quadratic equation by factoring**
- **To solve a quadratic equation by taking square roots**
- **To solve a quadratic equation by completing the square**
- **To solve a quadratic equation by using the quadratic formula**
- **To graph a quadratic equation of the form $y = ax^2 + bx + c$**
- **To solve application problems**

Algebraic Symbolism

The way in which an algebraic expression or equation is written has gone through several stages of development. First there was the *rhetoric,* which was in vogue until the late 13th century. In this method, an expression would be written out in sentences. The word 'res' was used to represent an unknown.

Rhetoric: From the additive *res* in the additive *res* results in a square *res*. From the three in an additive x come three additive *res* and from the subtractive four in the additive *res* come subtractive four *res*. From three in subtractive four comes subtractive twelve.

Modern: $(x + 3)(x - 4) = x^2 - x - 12$

The second stage was *syncoptic,* which was a shorthand in which abbreviations were used for words.

Syncoptic: *a* 6 in *b* quad − *c* plano 4 in *b* + *b* cub
Modern: $6ab^2 - 4cb + b^3$

The current modern stage, called the *symbolic stage,* began with the use of exponents rather than words to symbolize exponential expressions. This occurred near the beginning of the 17th century with the publication of the book La Geometrie by René Descartes. Modern notation is still evolving as mathematicians continue to search for convenient methods to symbolize a concept.

SECTION 1 Solving Quadratic Equations by Factoring or by Taking Square Roots

1.1 Objective To solve a quadratic equation by factoring

Reference for Computer Tutor™

An equation of the form $ax^2 + bx + c = 0$, where a, b, and c are constants and $a > 0$, is a **quadratic equation.**

$4x^2 - 3x + 1 = 0$, $a = 4$, $b = -3$, $c = 1$

$3x^2 - 4 = 0$, $a = 3$, $b = 0$, $c = -4$

A quadratic equation is also called a **second-degree equation.**

A quadratic equation is in **standard form** when the polynomial is in descending order and equal to zero.

Recall that the Principle of Zero Products states that if the product of two factors is zero, then at least one of the factors must be zero.

If $a \cdot b = 0$, then $a = 0$ or $b = 0$.

The Principle of Zero Products can be used in solving quadratic equations.

Solve by factoring: $2x^2 - x = 1$

$$2x^2 - x = 1$$

Write the equation in standard form.

$$2x^2 - x - 1 = 0$$

Factor.

$$(2x + 1)(x - 1) = 0$$

Let each factor equal zero.

$$2x + 1 = 0 \qquad x - 1 = 0$$

Rewrite each equation in the form *variable = constant*.

$$2x = -1 \qquad x = 1$$

$$x = -\frac{1}{2}$$

Write the solutions.

The solutions are $-\frac{1}{2}$ and 1.

Check:

$$2x^2 - x = 1 \qquad 2x^2 - x = 1$$

$$2\left(-\frac{1}{2}\right)^2 - \left(-\frac{1}{2}\right) = 1 \qquad 2(1)^2 - 1 = 1$$

$$2 \cdot \frac{1}{4} + \frac{1}{2} = 1 \qquad 2 \cdot 1 - 1 = 1$$

$$\frac{1}{2} + \frac{1}{2} = 1 \qquad 2 - 1 = 1$$

$$1 = 1 \qquad 1 = 1$$

Example 1 Solve by factoring: $x^2 + 10x + 25 = 0$

Solution

$$x^2 + 10x + 25 = 0$$

$$(x + 5)(x + 5) = 0$$

$$x + 5 = 0 \qquad x + 5 = 0$$

$$x = -5 \qquad x = -5$$

-5 is a double root of the equation.

The solution is -5.

Example 2 Solve by factoring: $2x^2 = (x + 2)(x + 3)$

Your solution

Solution on p. A65

1.2 Objective To solve a quadratic equation by taking square roots

Reference for Computer Tutor™

The quadratic equation $x^2 = 25$ can be read "the square of a number equals 25." The solution is the positive or the negative square root of 25, 5 or -5.

$$x^2 = 25 \qquad x^2 = 25$$
$$5^2 \mid 25 \qquad (-5)^2 \mid 25$$
$$25 = 25 \qquad 25 = 25$$

5 is a solution. -5 is a solution.

The solution can be found by taking the square root of each side of the equation and writing the positive and the negative square roots of the number.
$x = \pm 5$ means $x = 5$ or $x = -5$.

$$x^2 = 25$$
$$\sqrt{x^2} = \sqrt{25}$$
$$x = \pm\sqrt{25} = \pm 5$$

The solutions are 5 and -5.

Solve by taking square roots: $3x^2 = 36$

$$3x^2 = 36$$

Solve for x^2.

$$x^2 = 12$$

Take the square root of each side of the equation.

$$\sqrt{x^2} = \sqrt{12}$$

Simplify.

$$x = \pm\sqrt{12} = \pm 2\sqrt{3}$$

Write the solutions.

The solutions are $2\sqrt{3}$ and $-2\sqrt{3}$.

Check:

$$3x^2 = 36 \qquad 3x^2 = 36$$
$$3(2\sqrt{3})^2 = 36 \qquad 3(-2\sqrt{3})^2 = 36$$
$$3(12) = 36 \qquad 3(12) = 36$$
$$36 = 36 \qquad 36 = 36$$

An equation containing the square of a binomial can be solved by taking square roots.

Solve by taking square roots: $2(x-1)^2 - 36 = 0$

$$2(x-1)^2 - 36 = 0$$

Solve for $(x-1)^2$.

$$2(x-1)^2 = 36$$
$$(x-1)^2 = 18$$

Take the square root of each side of the equation.

$$\sqrt{(x-1)^2} = \sqrt{18}$$

Simplify.

$$x - 1 = \pm\sqrt{18} = \pm 3\sqrt{2}$$

Solve for x.

$$x - 1 = 3\sqrt{2} \qquad x - 1 = -3\sqrt{2}$$
$$x = 1 + 3\sqrt{2} \qquad x = 1 - 3\sqrt{2}$$

Write the solutions.

The solutions are $1 + 3\sqrt{2}$ and $1 - 3\sqrt{2}$.

$1 + 3\sqrt{2}$ and $1 - 3\sqrt{2}$ check as solutions.

Example 3 Solve by taking square roots: $x^2 + 16 = 0$

Solution

$$x^2 + 16 = 0$$
$$x^2 = -16$$
$$\sqrt{x^2} = \sqrt{-16}$$

$\sqrt{-16}$ is not a real number.

The equation has no real number solution.

Example 4 Solve by taking square roots: $x^2 + 81 = 0$

Your solution

Solution on p. A65

1.1 Exercises

Solve by factoring.

1. $x^2 + 2x - 15 = 0$
2. $t^2 + 3t - 10 = 0$
3. $z^2 - 4z + 3 = 0$
4. $s^2 - 5s + 4 = 0$
5. $p^2 + 3p + 2 = 0$
6. $v^2 + 6v + 5 = 0$
7. $x^2 - 6x + 9 = 0$
8. $y^2 - 8y + 16 = 0$
9. $12y^2 + 8y = 0$
10. $6x^2 - 9x = 0$
11. $r^2 - 10 = 3r$
12. $t^2 - 12 = 4t$
13. $3v^2 - 5v + 2 = 0$
14. $2p^2 - 3p - 2 = 0$
15. $3s^2 + 8s = 3$
16. $3x^2 + 5x = 12$
17. $9z^2 = 12z - 4$
18. $6r^2 = 12 - r$
19. $4t^2 = 4t + 3$
20. $5y^2 + 11y = 12$
21. $4v^2 - 4v + 1 = 0$
22. $9s^2 - 6s + 1 = 0$
23. $x^2 - 9 = 0$
24. $t^2 - 16 = 0$
25. $4y^2 - 1 = 0$
26. $9z^2 - 4 = 0$
27. $x + 15 = x(x - 1)$
28. $p + 18 = p(p - 2)$
29. $r^2 - r - 2 = (2r - 1)(r - 3)$
30. $s^2 + 5s - 4 = (2s + 1)(s - 4)$
31. $x^2 + x + 5 = (3x + 2)(x - 4)$

1.2 Exercises

Solve by taking square roots.

32. $x^2 = 36$

33. $y^2 = 49$

34. $v^2 - 1 = 0$

35. $z^2 - 64 = 0$

36. $4x^2 - 49 = 0$

37. $9w^2 - 64 = 0$

38. $9y^2 = 4$

39. $4z^2 = 25$

40. $16v^2 - 9 = 0$

41. $25x^2 - 64 = 0$

42. $y^2 + 81 = 0$

43. $z^2 + 49 = 0$

44. $w^2 - 24 = 0$

45. $v^2 - 48 = 0$

46. $(x - 1)^2 = 36$

47. $(y + 2)^2 = 49$

48. $2(x + 5)^2 = 8$

49. $4(z - 3)^2 = 100$

50. $9(x - 1)^2 - 16 = 0$

51. $4(y + 3)^2 - 81 = 0$

52. $49(v + 1)^2 - 25 = 0$

53. $81(y - 2)^2 - 64 = 0$

54. $(x - 4)^2 - 20 = 0$

55. $(y + 5)^2 - 50 = 0$

56. $(x + 1)^2 + 36 = 0$

57. $(y - 7)^2 + 49 = 0$

58. $2\left(z - \frac{1}{2}\right)^2 = 12$

59. $3\left(v + \frac{3}{4}\right)^2 = 36$

SECTION 2 Solving Quadratic Equations by Completing the Square

2.1 Objective To solve a quadratic equation by completing the square

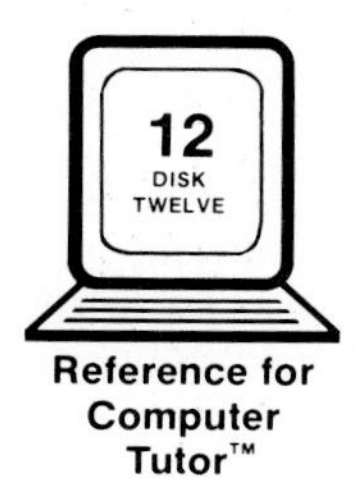

Reference for Computer Tutor™

Recall that a perfect square trinomial is the square of a binomial.

Perfect Square Trinomial		Square of a Binomial
$x^2 + 6x + 9$	=	$(x + 3)^2$
$x^2 - 10x + 25$	=	$(x - 5)^2$
$x^2 + 8x + 16$	=	$(x + 4)^2$

For each perfect square trinomial, the square of $\frac{1}{2}$ of the coefficient of x equals the constant term.

$x^2 + 6x + 9, \quad \left(\frac{1}{2} \cdot 6\right)^2 = 9$

$x^2 - 10x + 25, \quad \left[\frac{1}{2}(-10)\right]^2 = 25$

$x^2 + 8x + 16, \quad \left(\frac{1}{2} \cdot 8\right)^2 = 16$

$$\left(\frac{1}{2} \text{ coefficient of } x\right)^2 = \text{constant term}$$

This relationship can be used to write the constant term for a perfect square trinomial. Adding to a binomial the constant term which makes it a perfect square trinomial is called **completing the square.**

Complete the square on $x^2 - 8x$. Write the resulting perfect square trinomial as the square of a binomial.

Find the constant term. $\left[\frac{1}{2}(-8)\right]^2 = 16$

Complete the square on $x^2 - 8x$ by adding the constant term. $x^2 - 8x + 16$

Write the resulting perfect square trinomial as the square of a binomial. $x^2 - 8x + 16 = (x - 4)^2$

Complete the square on $y^2 + 5y$. Write the resulting perfect square trinomial as the square of a binomial.

Find the constant term. $\left(\frac{1}{2} \cdot 5\right)^2 = \left(\frac{5}{2}\right)^2 = \frac{25}{4}$

Complete the square on $y^2 + 5y$ by adding the constant term. $y^2 + 5y + \frac{25}{4}$

Write the resulting perfect square trinomial as the square of a binomial. $y^2 + 5y + \frac{25}{4} = \left(y + \frac{5}{2}\right)^2$

A quadratic equation which cannot be solved by factoring can be solved by completing the square. Add to both sides of the equation the term which completes the square. Rewrite the quadratic equation in the form $(x + a)^2 = b$. Take the square root of each side of the equation and then solve for x.

Solve by completing the square: $x^2 - 6x - 3 = 0$

$$x^2 - 6x - 3 = 0$$

Add the opposite of the constant term to each side of the equation.

$$x^2 - 6x = 3$$

Find the constant term which completes the square on $x^2 - 6x$.

$$\left[\frac{1}{2}(-6)\right]^2 = 9$$

Do this step mentally.

Add this term to each side of the equation.

$$x^2 - 6x + 9 = 3 + 9$$

Factor the perfect square trinomial.

$$(x - 3)^2 = 12$$

Take the square root of each side of the equation.

$$\sqrt{(x - 3)^2} = \sqrt{12}$$

Simplify.

$$x - 3 = \pm\sqrt{12} = \pm 2\sqrt{3}$$

Solve for x.

$$x - 3 = 2\sqrt{3} \qquad x - 3 = -2\sqrt{3}$$
$$x = 3 + 2\sqrt{3} \qquad x = 3 - 2\sqrt{3}$$

Write the solution.

The solutions are $3 + 2\sqrt{3}$ and $3 - 2\sqrt{3}$.

Check:

$$x^2 - 6x - 3 = 0$$
$$(3 + 2\sqrt{3})^2 - 6(3 + 2\sqrt{3}) - 3 = 0$$
$$9 + 12\sqrt{3} + 12 - 18 - 12\sqrt{3} - 3 = 0$$
$$0 = 0$$

$$x^2 - 6x - 3 = 0$$
$$(3 - 2\sqrt{3})^2 - 6(3 - 2\sqrt{3}) - 3 = 0$$
$$9 - 12\sqrt{3} + 12 - 18 + 12\sqrt{3} - 3 = 0$$
$$0 = 0$$

Solve by completing the square: $2x^2 - x - 1 = 0$

$$2x^2 - x - 1 = 0$$

Add the opposite of the constant term to each side of the equation.

$$2x^2 - x = 1$$

To complete the square, the coefficient of the x^2 term must be 1. Multiply each term by the reciprocal of the coefficient of x^2.

$$\frac{1}{2}(2x^2 - x) = \frac{1}{2} \cdot 1$$

$$x^2 - \frac{1}{2}x = \frac{1}{2}$$

Find the term which completes the square on $x^2 - \frac{1}{2}x$.

$$\left[\frac{1}{2}\left(-\frac{1}{2}\right)\right]^2 = \left(-\frac{1}{4}\right)^2 = \frac{1}{16}$$

Do this step mentally.

Add this term to each side of the equation.

$$x^2 - \frac{1}{2}x + \frac{1}{16} = \frac{1}{2} + \frac{1}{16}$$

Factor the perfect square trinomial.

$$\left(x - \frac{1}{4}\right)^2 = \frac{9}{16}$$

Take the square root of each side of the equation.

$$\sqrt{\left(x - \frac{1}{4}\right)^2} = \sqrt{\frac{9}{16}}$$

Simplify.

$$x - \frac{1}{4} = \pm\frac{3}{4}$$

Solve for x.

$$x - \frac{1}{4} = \frac{3}{4} \qquad x - \frac{1}{4} = -\frac{3}{4}$$

$$x = 1 \qquad x = -\frac{1}{2}$$

The solutions are 1 and $-\frac{1}{2}$.

1 and $-\frac{1}{2}$ check as solutions.

Example 1 Solve by completing the square: $2x^2 - 4x - 1 = 0$

Solution

$$2x^2 - 4x - 1 = 0$$

$$2x^2 - 4x = 1$$

$$\frac{1}{2}(2x^2 - 4x) = \frac{1}{2} \cdot 1$$

$$x^2 - 2x = \frac{1}{2}$$

$$x^2 - 2x + 1 = \frac{1}{2} + 1 \quad \text{Complete the square.}$$

$$(x - 1)^2 = \frac{3}{2}$$

$$\sqrt{(x - 1)^2} = \sqrt{\frac{3}{2}}$$

$$x - 1 = \pm\sqrt{\frac{3}{2}} = \pm\frac{\sqrt{6}}{2}$$

$$x - 1 = \frac{\sqrt{6}}{2} \qquad x - 1 = -\frac{\sqrt{6}}{2}$$

$$x = 1 + \frac{\sqrt{6}}{2} \qquad x = 1 - \frac{\sqrt{6}}{2}$$

$$= \frac{2 + \sqrt{6}}{2} \qquad = \frac{2 - \sqrt{6}}{2}$$

The solutions are $\frac{2 + \sqrt{6}}{2}$ and $\frac{2 - \sqrt{6}}{2}$.

Example 2 Solve by completing the square: $3x^2 - 6x - 2 = 0$

Your solution

Solution on p. A66

Example 3 Solve by completing the square:
$x^2 + 4x + 5 = 0$

Solution

$$x^2 + 4x + 5 = 0$$
$$x^2 + 4x = -5$$

Complete the square.

$$x^2 + 4x + 4 = -5 + 4$$
$$(x + 2)^2 = -1$$
$$\sqrt{(x + 2)^2} = \sqrt{-1}$$

$\sqrt{-1}$ is not a real number.

The quadratic equation has no real number solution.

Example 4 Solve by completing the square:
$x^2 + 6x + 12 = 0$

Your solution

Example 5 Solve by completing the square:
$x^2 + 6x + 4 = 0$
Approximate the solutions. Use the Table of Square Roots on page A2.

Solution

$$x^2 + 6x + 4 = 0$$
$$x^2 + 6x = -4$$

Complete the square.

$$x^2 + 6x + 9 = -4 + 9$$
$$(x + 3)^2 = 5$$
$$\sqrt{(x + 3)^2} = \sqrt{5}$$
$$x + 3 = \pm\sqrt{5}$$

$$x + 3 = \sqrt{5} \qquad x + 3 = -\sqrt{5}$$
$$x = -3 + \sqrt{5} \qquad x = -3 - \sqrt{5}$$
$$\approx -3 + 2.236 \qquad \approx -3 - 2.236$$
$$\approx -0.764 \qquad \approx -5.236$$

The solutions are approximately -0.764 and -5.236.

Example 6 Solve by completing the square:
$x^2 + 8x + 8 = 0$
Approximate the solutions. Use the Table of Square Roots on page A2.

Your solution

Solutions on p. A66

2.1 Exercises

Solve by completing the square.

1. $x^2 + 2x - 3 = 0$
2. $y^2 + 4y - 5 = 0$
3. $z^2 - 6z - 16 = 0$
4. $w^2 + 8w - 9 = 0$
5. $x^2 = 4x - 4$
6. $z^2 = 8z - 16$
7. $v^2 - 6v + 13 = 0$
8. $x^2 + 4x + 13 = 0$
9. $y^2 + 5y + 4 = 0$
10. $v^2 - 5v - 6 = 0$
11. $w^2 + 7w = 8$
12. $y^2 + 5y = -4$
13. $v^2 + 4v + 1 = 0$
14. $y^2 - 2y - 5 = 0$
15. $x^2 + 6x = 5$
16. $w^2 - 8w = 3$
17. $z^2 = 2z + 1$
18. $y^2 = 10y - 20$
19. $p^2 + 3p = 1$
20. $r^2 + 5r = 2$
21. $t^2 - 3t = -2$
22. $z^2 - 5z = -3$
23. $v^2 + v - 3 = 0$
24. $x^2 - x = 1$
25. $y^2 = 7 - 10y$
26. $v^2 = 14 + 16v$
27. $r^2 - 3r = 5$
28. $s^2 + 3s = -1$
29. $t^2 - t = 4$
30. $y^2 + y - 4 = 0$
31. $x^2 - 3x + 5 = 0$
32. $z^2 + 5z + 7 = 0$
33. $2t^2 - 3t + 1 = 0$
34. $2x^2 - 7x + 3 = 0$
35. $2r^2 + 5r = 3$
36. $2y^2 - 3y = 9$

Solve by completing the square.

37. $2s^2 = 7s - 6$

38. $2x^2 = 3x + 20$

39. $2v^2 = v + 1$

40. $2z^2 = z + 3$

41. $3r^2 + 5r = 2$

42. $3t^2 - 8t = 3$

43. $3y^2 + 8y + 4 = 0$

44. $3z^2 - 10z - 8 = 0$

45. $4x^2 + 4x - 3 = 0$

46. $4v^2 + 4v - 15 = 0$

47. $6s^2 + 7s = 3$

48. $6z^2 = z + 2$

49. $6p^2 = 5p + 4$

50. $6t^2 = t - 2$

51. $4v^2 - 4v - 1 = 0$

52. $2s^2 - 4s - 1 = 0$

53. $4z^2 - 8z = 1$

54. $3r^2 - 2r = 2$

55. $3y - 6 = (y - 1)(y - 2)$

56. $7s + 55 = (s + 5)(s + 4)$

57. $4p + 2 = (p - 1)(p + 3)$

58. $v - 10 = (v + 3)(v - 4)$

Solve by completing the square. Approximate the solutions to the nearest thousandth. Use the Table of Square Roots on page A2.

59. $y^2 + 3y = 5$

60. $w^2 + 5w = 2$

61. $2z^2 - 3z = 7$

62. $2x^2 + 3x = 11$

63. $4x^2 + 6x - 1 = 0$

64. $4x^2 + 2x - 3 = 0$

SECTION 3 Solving Quadratic Equations by Using the Quadratic Formula

3.1 Objective To solve a quadratic equation by using the quadratic formula

Reference for Computer Tutor™

Any quadratic equation can be solved by completing the square. Applying this method to the standard form of a quadratic equation produces a formula that can be used to solve any quadratic equation.

Solve $ax^2 + bx + c = 0$ by completing the square.

$$ax^2 + bx + c = 0$$

Add the opposite of the constant term to each side of the equation.

$$ax^2 + bx + c + (-c) = 0 + (-c)$$

$$ax^2 + bx = -c$$

Multiply each side of the equation by the reciprocal of a, the coefficient of x^2.

$$\frac{1}{a}(ax^2 + bx) = \frac{1}{a}(-c)$$

$$x^2 + \frac{b}{a}x = -\frac{c}{a}$$

Complete the square by adding $\left(\frac{1}{2} \cdot \frac{b}{a}\right)^2$ to each side of the equation.

$$x^2 + \frac{b}{a}x + \left(\frac{1}{2} \cdot \frac{b}{a}\right)^2 = \left(\frac{1}{2} \cdot \frac{b}{a}\right)^2 - \frac{c}{a}$$

$$x^2 + \frac{b}{a}x + \frac{b^2}{4a^2} = \frac{b^2}{4a^2} - \frac{c}{a}$$

Simplify the right side of the equation.

$$x^2 + \frac{b}{a}x + \frac{b^2}{4a^2} = \frac{b^2}{4a^2} - \left(\frac{c}{a} \cdot \frac{4a}{4a}\right)$$

$$x^2 + \frac{b}{a}x + \frac{b^2}{4a^2} = \frac{b^2}{4a^2} - \frac{4ac}{4a^2}$$

$$x^2 + \frac{b}{a}x + \frac{b^2}{4a^2} = \frac{b^2 - 4ac}{4a^2}$$

Factor the perfect square trinomial on the left side of the equation.

$$\left(x + \frac{b}{2a}\right)^2 = \frac{b^2 - 4ac}{4a^2}$$

Take the square root of each side of the equation.

$$\sqrt{\left(x + \frac{b}{2a}\right)^2} = \sqrt{\frac{b^2 - 4ac}{4a^2}}$$

$$\left(x + \frac{b}{2a}\right) = \pm\frac{\sqrt{b^2 - 4ac}}{2a}$$

Solve for x.

$$x + \frac{b}{2a} = \frac{\sqrt{b^2 - 4ac}}{2a} \qquad x + \frac{b}{2a} = -\frac{\sqrt{b^2 - 4ac}}{2a}$$

$$x = -\frac{b}{2a} + \frac{\sqrt{b^2 - 4ac}}{2a} \qquad x = -\frac{b}{2a} - \frac{\sqrt{b^2 - 4ac}}{2a}$$

$$= \frac{-b + \sqrt{b^2 - 4ac}}{2a} \qquad = \frac{-b - \sqrt{b^2 - 4ac}}{2a}$$

The Quadratic Formula

The solution of $ax^2 + bx + c = 0$ is $x = \frac{-b + \sqrt{b^2 - 4ac}}{2a}$ or $x = \frac{-b - \sqrt{b^2 - 4ac}}{2a}$.

The quadratic formula is frequently written in the form $\mathbf{x = \frac{-b \pm \sqrt{b^2 - 4ac}}{2a}}$.

Solve by using the quadratic formula: $2x^2 = 4x - 1$

$$2x^2 = 4x - 1$$

Write the equation in standard form.

$$2x^2 - 4x + 1 = 0$$

$a = 2$, $b = -4$, and $c = 1$.

Replace a, b, and c in the quadratic formula by their values.

$$x = \frac{-b \pm \sqrt{b^2 - 4ac}}{2a}$$

$$= \frac{-(-4) \pm \sqrt{(-4)^2 - 4\cdot 2\cdot 1}}{2\cdot 2}$$

Simplify.

$$= \frac{4 \pm \sqrt{16 - 8}}{4} = \frac{4 \pm \sqrt{8}}{4}$$

$$= \frac{4 \pm 2\sqrt{2}}{4} = \frac{2 \pm \sqrt{2}}{2}$$

Write the solutions.

The solutions are $\frac{2 + \sqrt{2}}{2}$ and $\frac{2 - \sqrt{2}}{2}$.

$\frac{2 + \sqrt{2}}{2}$ and $\frac{2 - \sqrt{2}}{2}$ check as solutions.

Example 1 Solve by using the quadratic formula: $2x^2 - 3x + 1 = 0$

Solution $2x^2 - 3x + 1 = 0$

$a = 2$, $b = -3$, $c = 1$

$$x = \frac{-(-3) \pm \sqrt{(-3)^2 - 4(2)(1)}}{2\cdot 2}$$

$$= \frac{3 \pm \sqrt{9 - 8}}{4} = \frac{3 \pm \sqrt{1}}{4} = \frac{3 \pm 1}{4}$$

$$x = \frac{3 + 1}{4} = \frac{4}{4} = 1 \qquad x = \frac{3 - 1}{4} = \frac{2}{4} = \frac{1}{2}$$

The solutions are 1 and $\frac{1}{2}$.

Example 2 Solve by using the quadratic formula: $3x^2 + 4x - 4 = 0$

Your solution

Example 3 Solve by using the quadratic formula: $2x^2 = 8x - 5$

Solution

$$2x^2 = 8x - 5$$

$$2x^2 - 8x + 5 = 0$$

$a = 2$, $b = -8$, $c = 5$

$$x = \frac{-(-8) \pm \sqrt{(-8)^2 - 4(2)(5)}}{2\cdot 2}$$

$$= \frac{8 \pm \sqrt{64 - 40}}{4} = \frac{8 \pm \sqrt{24}}{4}$$

$$= \frac{8 \pm 2\sqrt{6}}{4} = \frac{4 \pm \sqrt{6}}{2}$$

The solutions are $\frac{4 + \sqrt{6}}{2}$ and $\frac{4 - \sqrt{6}}{2}$.

Example 4 Solve by using the quadratic formula: $x^2 + 2x = 1$

Your solution

3.1 Exercises

Solve by using the quadratic formula.

1. $x^2 - 4x - 5 = 0$

2. $y^2 + 3y + 2 = 0$

3. $z^2 - 2z - 15 = 0$

4. $v^2 + 5v + 4 = 0$

5. $z^2 + 6z - 7 = 0$

6. $s^2 + 3s - 10 = 0$

7. $t^2 + t - 6 = 0$

8. $x^2 - x - 2 = 0$

9. $y^2 = 2y + 3$

10. $w^2 = 3w + 18$

11. $r^2 = 5 - 4r$

12. $z^2 = 3 - 2z$

13. $2y^2 - y - 1 = 0$

14. $2t^2 - 5t + 3 = 0$

15. $w^2 + 3w + 5 = 0$

16. $x^2 - 2x + 6 = 0$

17. $p^2 - p = 0$

18. $2v^2 + v = 0$

19. $4t^2 - 9 = 0$

20. $4s^2 - 25 = 0$

21. $4y^2 + 4y = 15$

22. $4r^2 + 4r = 3$

23. $3t^2 = 7t + 6$

24. $3x^2 = 10x + 8$

25. $5z^2 + 11z = 12$

26. $4v^2 = v + 3$

27. $6s^2 - s - 2 = 0$

28. $6y^2 + 5y - 4 = 0$

29. $2x^2 + x + 1 = 0$

30. $3r^2 - r + 2 = 0$

31. $t^2 - 2t = 5$

32. $y^2 - 4y = 6$

33. $t^2 + 6t - 1 = 0$

Solve by using the quadratic formula.

34. $z^2 + 4z + 1 = 0$ **35.** $w^2 = 4w + 9$ **36.** $y^2 = 8y + 3$

37. $4t^2 - 4t - 1 = 0$ **38.** $4x^2 - 8x - 1 = 0$ **39.** $v^2 + 6v + 1 = 0$

40. $s^2 + 4s - 8 = 0$ **41.** $4t^2 - 12t - 15 = 0$ **42.** $4w^2 - 20w + 5 = 0$

43. $9y^2 + 6y - 1 = 0$ **44.** $9s^2 - 6s - 2 = 0$ **45.** $4p^2 + 4p + 1 = 0$

46. $9z^2 + 12z + 4 = 0$ **47.** $2x^2 = 4x - 5$ **48.** $3r^2 = 5r - 6$

49. $4p^2 + 16p = -11$ **50.** $4y^2 - 12y = -1$ **51.** $4x^2 = 4x + 11$

52. $4s^2 + 12s = 3$ **53.** $9v^2 = -30v - 23$ **54.** $9t^2 = 30t + 17$

Solve by using the quadratic formula. Approximate the solutions to the nearest thousandth. Use the Table of Square Roots on page A2.

55. $x^2 - 2x - 21 = 0$ **56.** $y^2 + 4y - 11 = 0$ **57.** $s^2 - 6s - 13 = 0$

58. $w^2 + 8w - 15 = 0$ **59.** $2p^2 - 7p - 10 = 0$ **60.** $3t^2 - 8t - 1 = 0$

61. $4z^2 + 8z - 1 = 0$ **62.** $4x^2 + 7x + 1 = 0$ **63.** $5v^2 - v - 5 = 0$

SECTION 4 Graphing Quadratic Equations in Two Variables

4.1 Objective To graph a quadratic equation of the form $y = ax^2 + bx + c$

Reference for Computer Tutor™

An equation of the form $y = ax^2 + bx + c$ is a **quadratic equation in two variables.** Examples of quadratic equations in two variables are shown at the right.

$y = 3x^2 - x + 1$

$y = -x^2 - 3$

$y = 2x^2 - 5x$

The graph of a quadratic equation in two variables is a **parabola.** The graph is "cup shaped" and opens either up or down. The graphs of two parabolas are shown below.

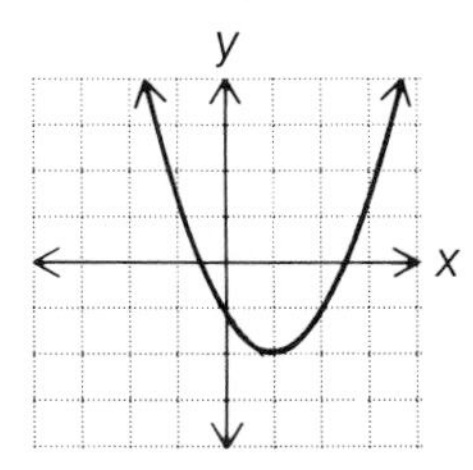

Parabola which opens up

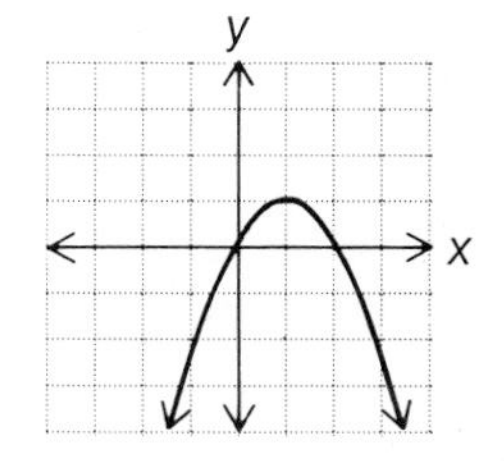

Parabola which opens down

Graph $y = x^2 - 2x - 3$.

Find several solutions of the equation. Since the graph is not a straight line, several solutions must be found in order to determine the cup shape.
Display the ordered pair solutions in a table.

x	y
0	−3
1	−4
−1	0
2	−3
3	0

Graph the ordered pair solutions on a rectangular coordinate system.
Draw a parabola through the points.

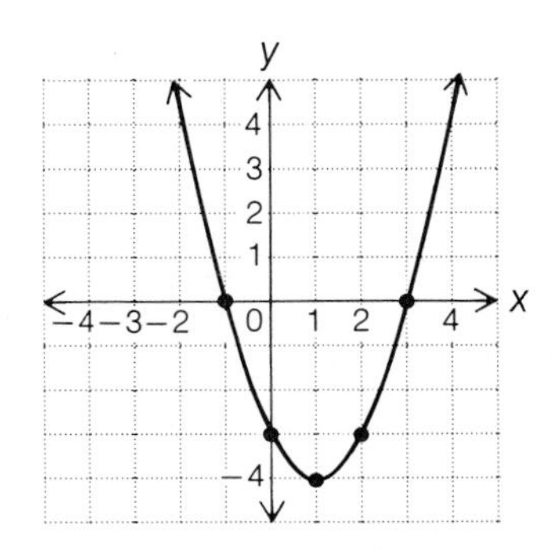

Graph $y = -2x^2 + 1$.

Find enough solutions of the equation to determine the cup shape.
Display the ordered pair solutions in a table.

x	y
0	1
1	−1
−1	−1
2	−7
−2	−7

Graph the ordered pair solutions on a rectangular coordinate system.
Draw a parabola through the points.

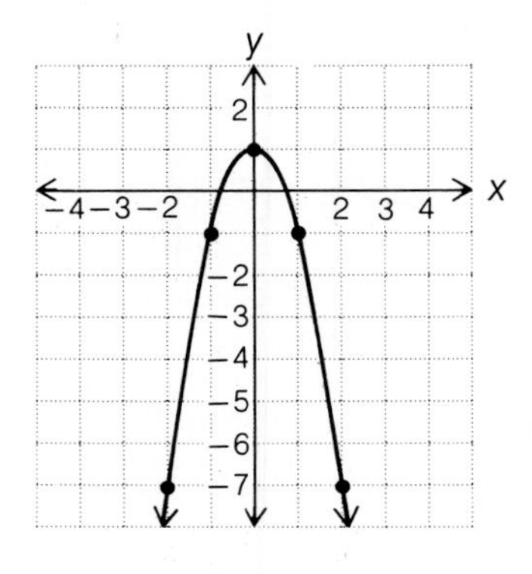

Note in the first example above that the coefficient of x^2 is **positive** and the graph **opens up.** In the second example, the coefficient of x^2 is **negative** and the graph **opens down.**

Example 1 Graph $y = x^2 - 2x$.

Solution

x	y
0	0
1	−1
−1	3
2	0
3	3

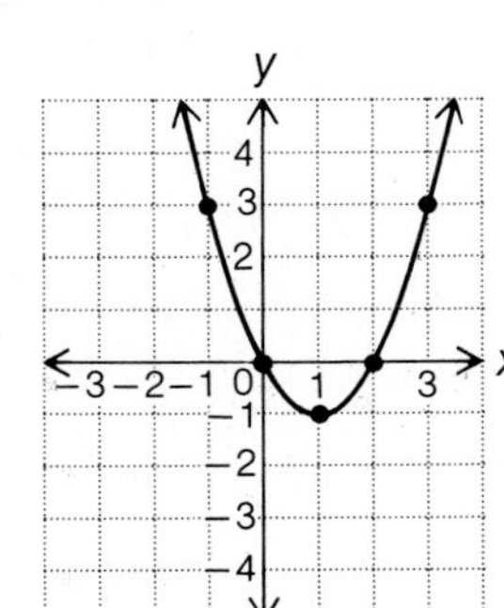

Example 2 Graph $y = x^2 + 2$.

Your solution

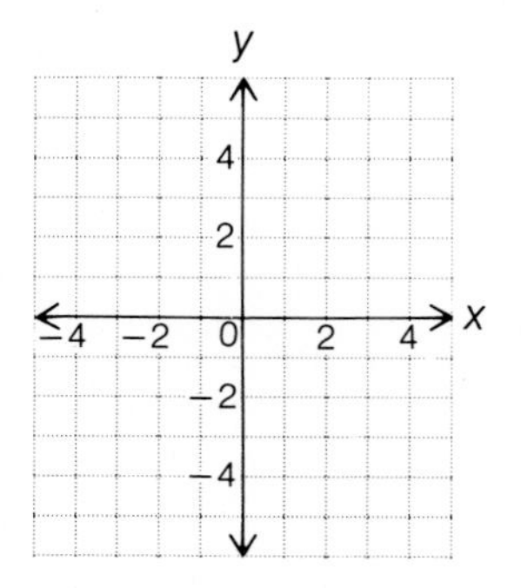

Example 3 Graph $y = x^2 - 4x + 4$.

Solution

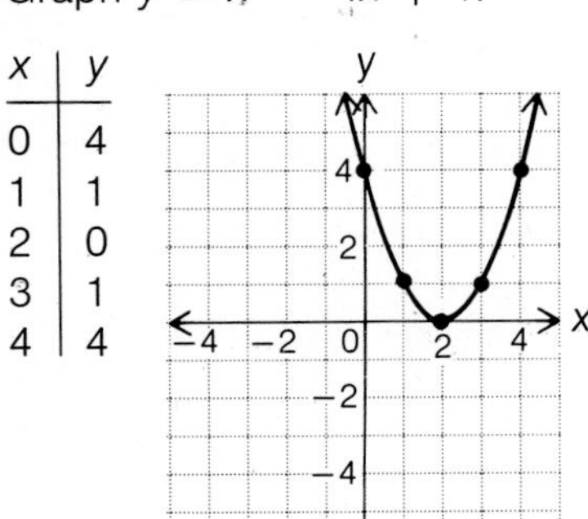

x	y
0	4
1	1
2	0
3	1
4	4

Example 4 Graph $y = x^2 + 2x + 1$.

Your solution

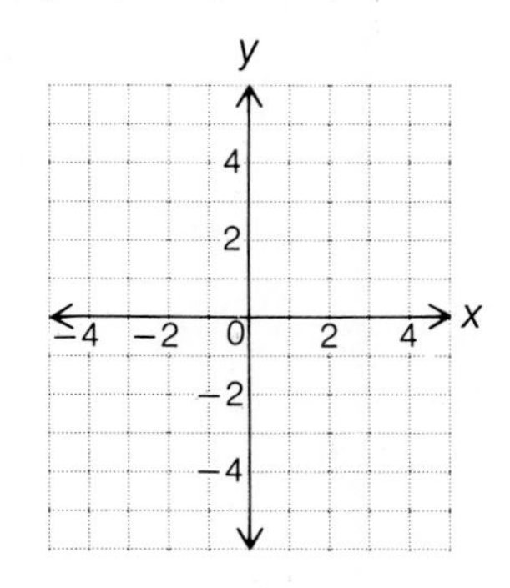

Solutions on p. A67

4.1 Exercises

Graph.

1. $y = x^2$

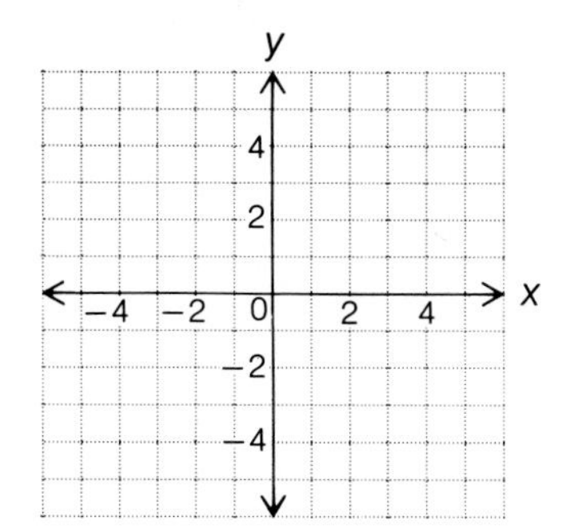

2. $y = -x^2$

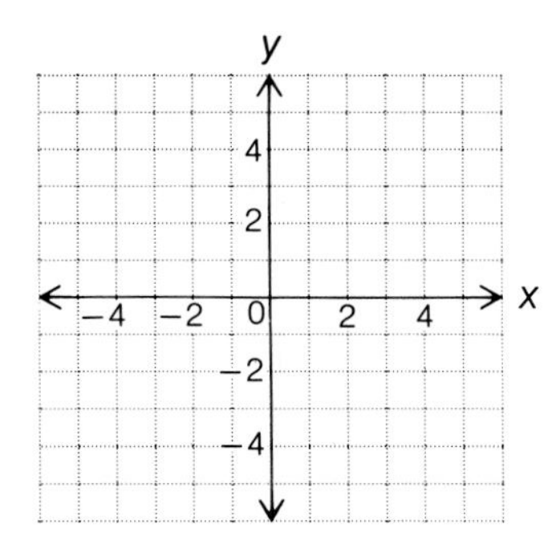

3. $y = -x^2 + 1$

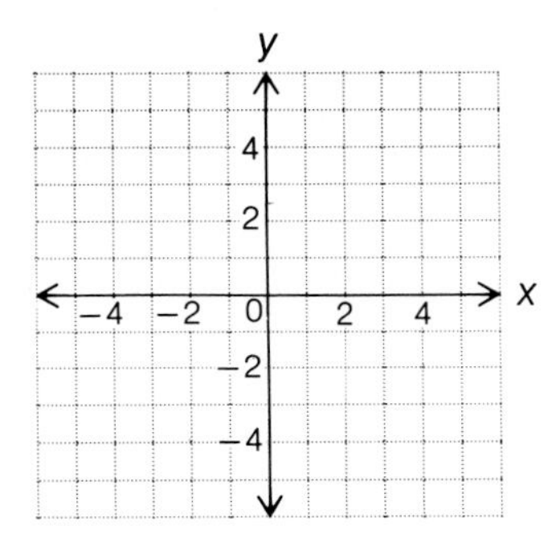

4. $y = x^2 - 1$

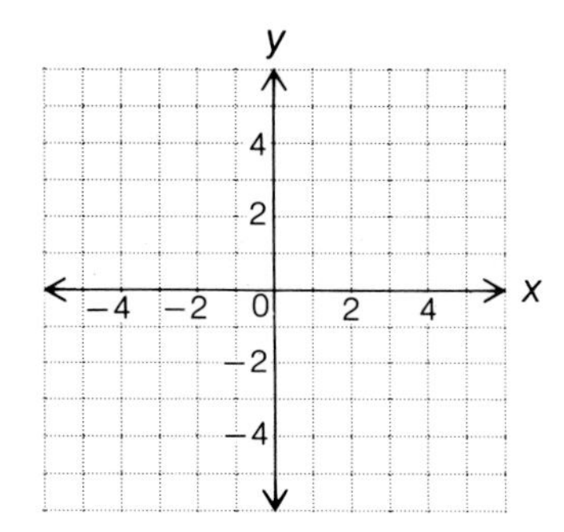

5. $y = 2x^2$

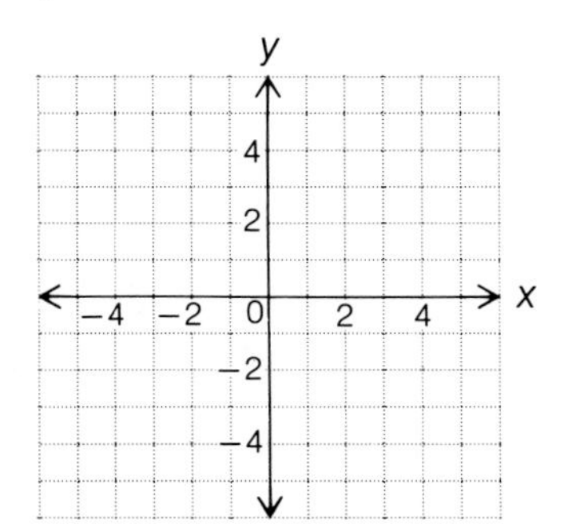

6. $y = \frac{1}{2}x^2$

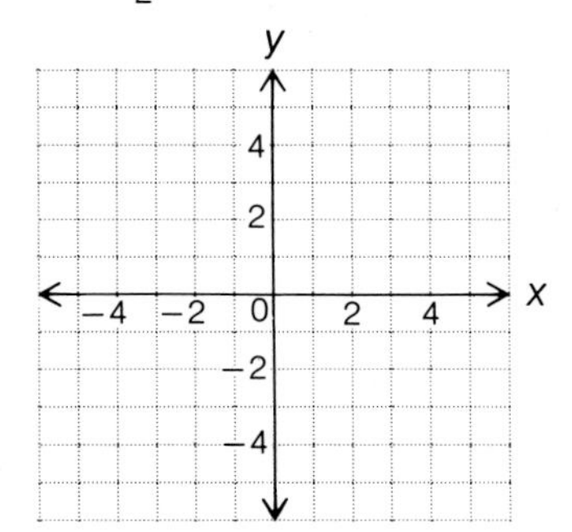

7. $y = -\frac{1}{2}x^2 + 1$

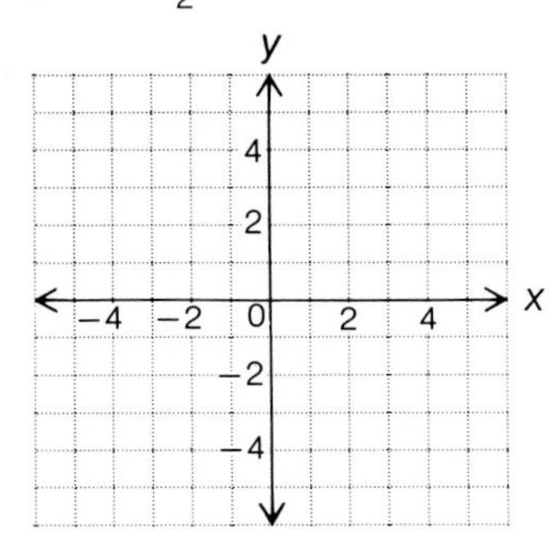

8. $y = 2x^2 - 1$

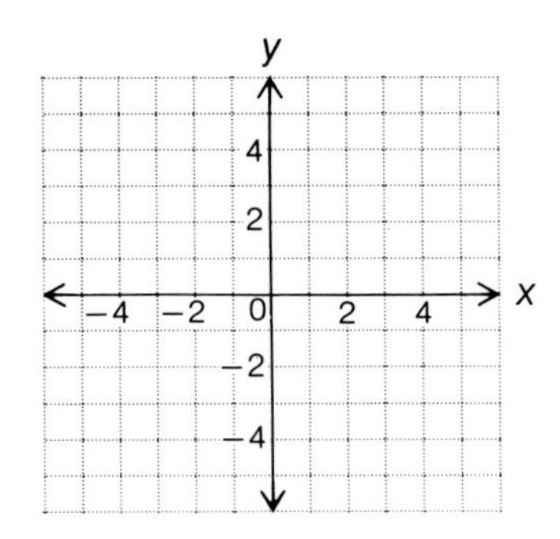

Graph.

9. $y = x^2 - 4x$

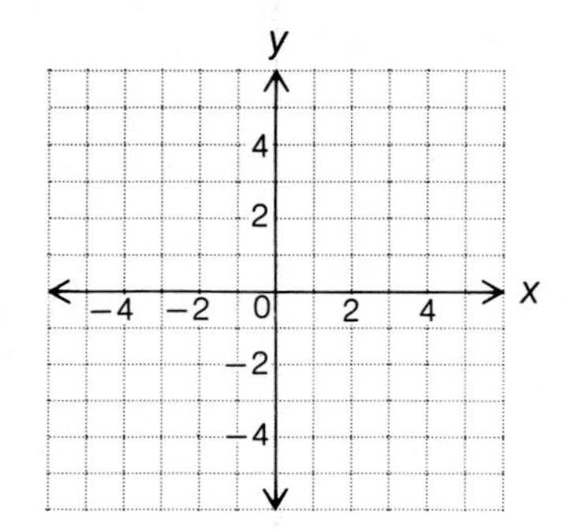

10. $y = x^2 + 4x$

11. $y = x^2 - 2x + 5$

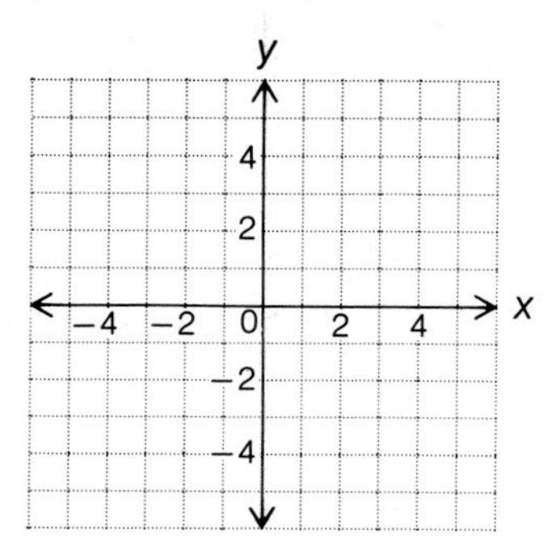

12. $y = x^2 - 4x + 2$

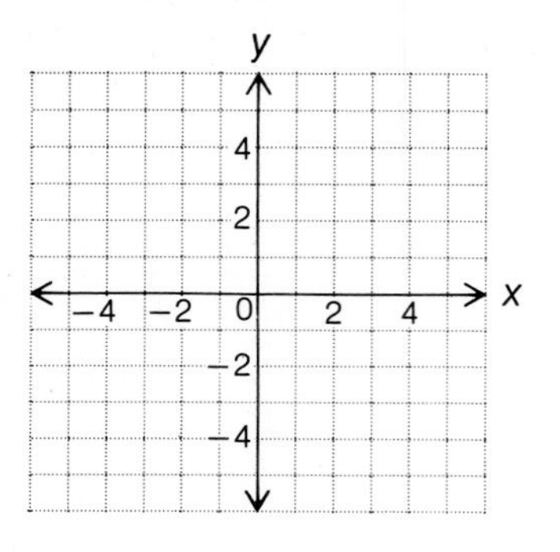

13. $y = -x^2 + 2x + 3$

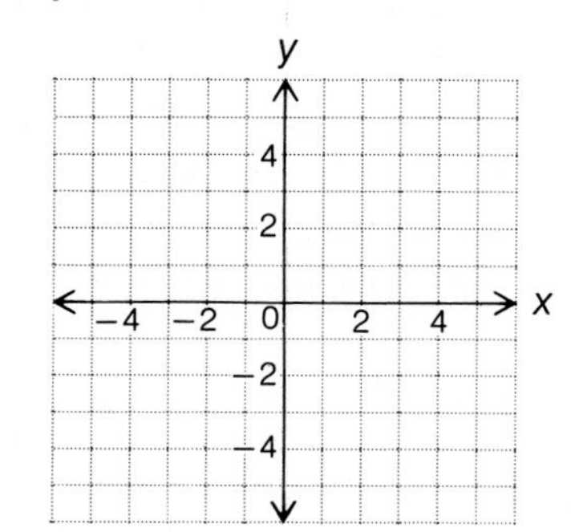

14. $y = -x^2 - 2x + 3$

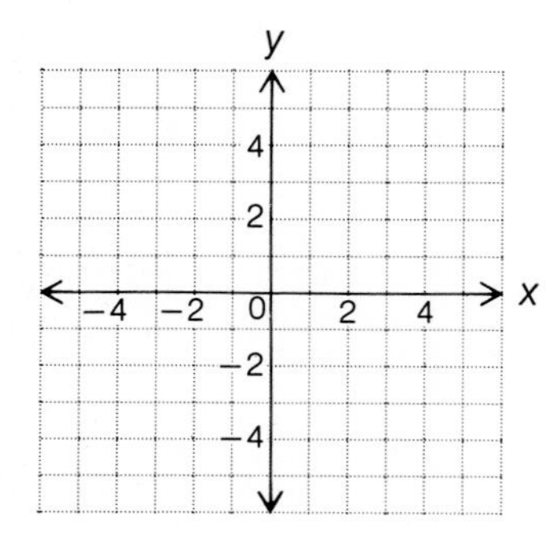

15. $y = -x^2 + 4x - 4$

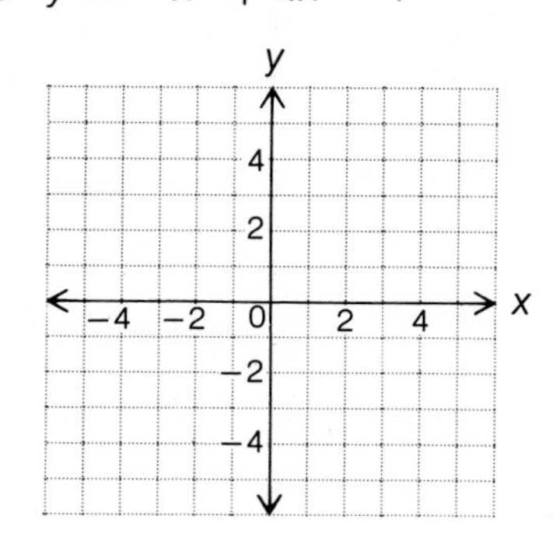

16. $y = -x^2 + 6x - 9$

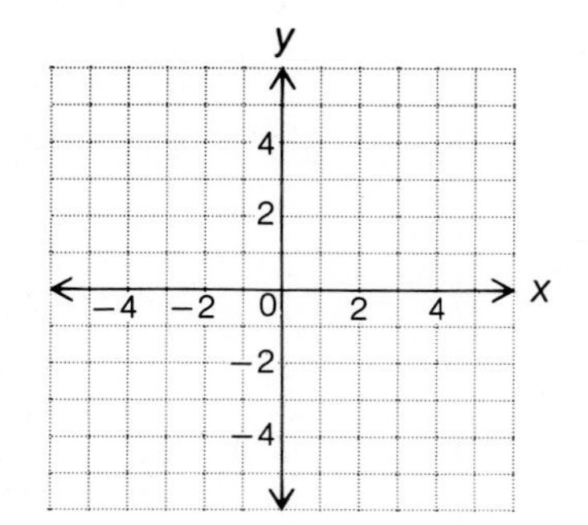

SECTION 5 Application Problems

5.1 Objective To solve application problems

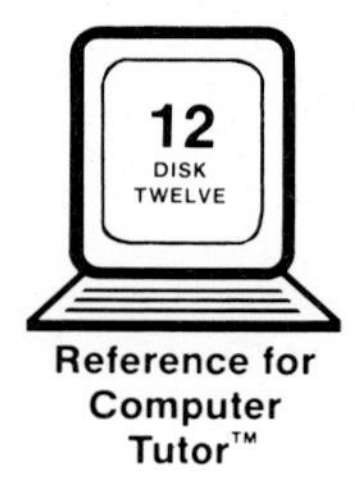

Reference for Computer Tutor™

The application problems in this section are varieties of those problems solved earlier in the text. Each of the strategies for the problems in this section will result in a quadratic equation.

In 5 h, two campers rowed 12 mi down a stream and then rowed back to their campsite. The rate of the stream's current was 1 mph. Find the rate at which the campers rowed.

STRATEGY FOR SOLVING AN APPLICATION PROBLEM

▷ Determine the type of problem. For example, is it a distance-rate problem, a geometry problem, a work problem, or an age problem?

The problem is a distance-rate problem.

▷ Choose a variable to represent the unknown quantity. Write numerical or variable expressions for all the remaining quantities. These results can be recorded in a table.

The unknown rate of the campers: r

	Distance	÷	Rate	=	Time
Downstream	12	÷	$r + 1$	=	$\frac{12}{r+1}$
Upstream	12	÷	$r - 1$	=	$\frac{12}{r-1}$

▷ Determine how the quantities are related. If necessary, review the strategies presented in Chapter 4.

The total time of the trip was 5 h.

$$\frac{12}{r+1} + \frac{12}{r-1} = 5$$

$$(r+1)(r-1)\left(\frac{12}{r+1} + \frac{12}{r-1}\right) = (r+1)(r-1)5$$

$$(r-1)12 + (r+1)12 = (r^2 - 1)5$$

$$12r - 12 + 12r + 12 = 5r^2 - 5$$

$$24r = 5r^2 - 5$$

$$0 = 5r^2 - 24r - 5$$

$$0 = (5r+1)(r-5)$$

$$5r + 1 = 0 \qquad r - 5 = 0$$

$$5r = -1 \qquad r = 5$$

$$r = -\frac{1}{5}$$

The solution $r = -\frac{1}{5}$ is not possible, since the rate cannot be a negative number.

The rowing rate was 5 mph.

Example 1

A painter and the painter's apprentice working together can paint a room in 2 h. The apprentice working alone requires 3 more hours to paint the room than the painter requires working alone. How long does it take the painter working alone to paint the room?

Strategy

▷ This is a work problem.

▷ Time for the painter to paint the room: t
Time for the apprentice to paint the room: $t + 3$

	Rate	Time	Part
Painter	$\frac{1}{t}$	2	$\frac{2}{t}$
Apprentice	$\frac{1}{t+3}$	2	$\frac{2}{t+3}$

▷ The sum of the parts of the task completed must equal 1.

Solution

$$\frac{2}{t} + \frac{2}{t+3} = 1$$
$$t(t+3)\left(\frac{2}{t} + \frac{2}{t+3}\right) = t(t+3) \cdot 1$$
$$(t+3)2 + t(2) = t(t+3)$$
$$2t + 6 + 2t = t^2 + 3t$$
$$0 = t^2 - t - 6$$
$$0 = (t-3)(t+2)$$

$$t - 3 = 0 \qquad t + 2 = 0$$
$$t = 3 \qquad t = -2$$

The solution $t = -2$ is not possible.

The time is 3 h.

Example 2

The length of a rectangle is 2 m more than the width. The area is 15 m². Find the width.

Your strategy

Your solution

Solution on p. A68

5.1 Application Problems

Solve.

1. The length of a rectangle is twice the width. The area of the rectangle is 32 ft^2. Find the length and width. (Area $= l \cdot w$)

2. The height of a triangle is four times the length of the base. The area of the triangle is 18 m^2. Find the height and the length of the base of the triangle. $\left(\text{Area} = \frac{1}{2}bh\right)$

3. The height of a triangle is 2 m more than twice the length of the base. The area of the triangle is 20 m^2. Find the height and the length of the base of the triangle.

4. The length of a rectangle is 1 ft more than twice the width. The area of the rectangle is 120 ft^2. Find the length and width of the rectangle.

5. The sum of the squares of two consecutive positive odd integers is thirty-four. Find the two integers.

6. The difference between the squares of two consecutive positive even integers is twenty-eight. Find the two integers.

7. The sum of the squares of three consecutive integers is two. Find the three integers.

8. The sum of the squares of three consecutive even integers is eight. Find the three integers.

9. An integer plus twice the square of the integer is 21. Find the integer.

10. Twice the sum of three times an integer and the square of the integer is 36. Find the integer.

11. One car is two years older than a second car. Two years ago the product of their ages was 24. Find the present ages of the two cars.

12. One coin is twice the age of a second coin. One year ago the product of their ages was 10. Find the present ages of the coins.

Solve.

13. One stamp is three times the age of a second stamp. Eight years ago the product of their ages was 19. Find the present ages of the stamps.

14. One child is twice the age of a second child. Three years ago the product of the sum of their ages and the difference between their ages was 45. Find the present ages of the children.

15. A small pipe takes 8 h longer to fill a tank than a larger pipe. Working together, the pipes can fill the tank in 3 h. How long would it take each pipe working alone to fill the tank?

16. One painter takes 6 h longer to paint a room than does a second painter. Working together, the painters can paint the room in 4 h. How long would it take each painter working alone to paint the room?

17. One photocopy machine takes 16 min longer to reproduce a report than does a second machine. Working together, it takes 6 min to reproduce the report. How long would it take each machine working alone to reproduce the report?

18. A water tank has two drains. One drain takes 21 min longer to empty the tank than does the second drain. With both drains open, the tank empties in 10 min. How long would it take each drain working alone to empty the tank?

19. A motorboat traveled 24 mi at a constant rate before reducing the speed by 2 mph. Another 20 mi was traveled at the reduced speed. The total time for the 44-mile trip was 4 h. Find the rate of the boat during the first 24 mi.

20. A motorist traveled 120 mi at a constant rate before increasing the speed by 10 mph. Another 100 mi was driven at the increased speed. The total time for the 220-mile trip was 5 h. Find the rate during the first 120 mi.

21. It took a motorboat one more hour to travel 48 mi against the current than it did to go 48 mi with the current. The rate of the current was 2 mph. Find the rate of the boat in calm water.

22. It took a small plane one more hour to fly 240 mi against the wind than it did to fly the same distance with the wind. The rate of the wind was 20 mph. Find the rate of the plane in calm air.

Calculators and Computers

Checking Solutions to Quadratic Equations

A calculator can be used to check solutions to quadratic equations. Here are some examples:

Solve and check: $x^2 - 6x - 16 = 0$

Solve by factoring. The solutions are 8 and -2.

To check the solutions, the [x^2] key will be used. The method of evaluating polynomials given in Chapter 5 could also be used.

To check the solutions, replace x by 8. $8^2 - 6 \cdot 8 - 16 \stackrel{?}{=} 0$

Enter the following keystrokes: 8 [x^2] [−] [(] 6 [×] 8 [)] [−] 16 [=]

The result in the display should be zero. The solution is correct. The solution -2 can be checked in a similar manner.

One note about the calculation—the parentheses keys are used to ensure that multiplication is completed before subtraction.

Solve and check: $2x^2 - x - 9 = 0$

Use the quadratic formula. The solutions are $\frac{1 + \sqrt{73}}{4}$ and $\frac{1 - \sqrt{73}}{4}$.

To check the solutions first evaluate the expression $\frac{1 + \sqrt{73}}{4}$ and store the result in the calculator's memory.

1 [+] 73 [√] [÷] 4 [=] [M+]

Now replace x in the equation by the solution and determine if the left and right sides of the equation are equal. The [MR] key recalls the solution from memory.

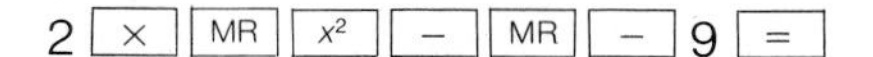

Is the result in the display zero? Probably not! Nonetheless the answer is very close to zero. The result in our display was 9.000000 − 10. The −10 at the end of the display means that the decimal point should be moved ten places to the left. That makes the number 0.0000000009, which is indeed close to zero.

The reason the answer was not exactly zero is that $\sqrt{73}$ is an irrational number and therefore has an infinitely long decimal representation. The calculator, on the other hand, can store only 8 or 9 places past the decimal point. Thus the calculator is using only an approximation of $\sqrt{73}$ so that when checking the solution the result is not exactly zero.

The solution $\frac{1 - \sqrt{73}}{4}$ can be checked in a similar manner.

Chapter Summary

KEY WORDS

A **quadratic equation** is an equation which can be written in the form $ax^2 + bx + c = 0$, where a, b, and c are constants and $a > 0$. A quadratic equation is also called a **second-degree equation.**

A quadratic equation is in **standard form** when the polynomial is in descending order and equal to zero.

The graph of an equation of the form $y = ax^2 + bx + c$ is a **parabola.**

ESSENTIAL RULES

The Quadratic Formula

$$x = \frac{-b \pm \sqrt{b^2 - 4ac}}{2a}$$

Review/Test

SECTION 1

1.1 Solve by factoring:
$3x^2 + 7x = 20$

1.2 Solve by taking square roots:
$3(x + 4)^2 - 60 = 0$

SECTION 2

2.1a Solve by completing the square:
$x^2 + 4x - 16 = 0$

2.1b Solve by completing the square:
$x^2 + 3x = 8$

2.1c Solve by completing the square:
$2x^2 - 6x + 1 = 0$

2.1d Solve by completing the square:
$2x^2 + 8x = 3$

SECTION 3

3.1a Solve by using the quadratic formula: $x^2 + 4x + 2 = 0$

3.1b Solve by using the quadratic formula: $x^2 - 3x = 6$

Review/Test

3.1c Solve by using the quadratic formula: $2x^2 - 5x - 3 = 0$

3.1d Solve by using the quadratic formula: $3x^2 - x = 1$

SECTION 4

4.1 Graph $y = x^2 + 2x - 4$.

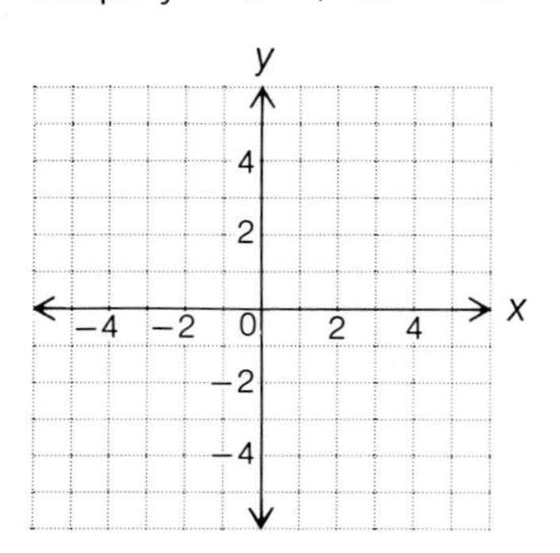

SECTION 5

5.1a The length of a rectangle is 2 ft less than twice the width. The area of the rectangle is 40 ft^2. Find the length and width of the rectangle.

5.1b It took a motorboat one hour more to travel 60 mi against a current than it did to go 60 mi with the current. The rate of the current was 1 mph. Find the rate of the boat in calm water.

Cumulative Review/Test

1. Simplify:
$2x - 3[2x - 4(3 - 2x) + 2] - 3$

2. Solve: $-\frac{3}{5}x = -\frac{9}{10}$

3. Solve:
$2x - 3(4x - 5) = -3x - 6$

4. Simplify:
$(2a^2b)^2(-3a^4b^2)$

5. Simplify: $(x^2 - 8) \div (x - 2)$

6. Factor $3x^3 + 2x^2 - 8x$.

7. Simplify:
$\frac{3x^2 - 6x}{4x - 6} \div \frac{2x^2 + x - 6}{6x^3 - 24x}$

8. Simplify:
$\frac{x}{2(x - 1)} - \frac{1}{(x - 1)(x + 1)}$

9. Simplify: $\dfrac{1 - \frac{7}{x} + \frac{12}{x^2}}{2 - \frac{1}{x} - \frac{15}{x^2}}$

10. Find the x- and y-intercepts for the line $4x - 3y = 12$.

11. Find the equation of the line which contains the point $(-3,2)$ and has slope $-\frac{4}{3}$.

12. Solve by substitution.
$3x - y = 5$
$y = 2x - 3$

13. Solve by the addition method.
$3x + 2y = 2$
$5x - 2y = 14$

14. Solve:
$2x - 3(2 - 3x) > 2x - 5$

15. Simplify:
$(\sqrt{a} - \sqrt{2})(\sqrt{a} + \sqrt{2})$

16. Simplify: $\frac{\sqrt{108a^7b^3}}{\sqrt{3a^4b}}$

Cumulative Review/Test

17. Simplify: $\dfrac{\sqrt{12x^2} - \sqrt{27}}{\sqrt{3}}$

18. Solve: $3 = 8 - \sqrt{5x}$

19. Solve by factoring:
$6x^2 - 17x = -5$

20. Solve by taking square roots:
$2(x - 5)^2 = 36$

21. Solve by completing the square:
$3x^2 + 7x = -3$

22. Solve by using the quadratic formula: $2x^2 - 3x - 2 = 0$

23. Find the selling price per pound of a mixture made from 20 lb of cashews which cost \$3.50 per lb and 50 lb of peanuts which cost \$1.75 per pound.

24. A stock investment of 100 shares paid a dividend of \$215. At this rate, how many additional shares are required to earn a dividend of \$752.50?

25. A 720-mile trip from one city to another takes 3 h when a plane is flying with the wind. The return trip, against the wind, takes 4.5 h. Find the rate of the plane in still air and the rate of the wind.

26. A student received a 70, a 91, an 85, and a 77 on four tests in a math class. What scores on the last test will enable the student to receive a minimum of 400 points?

27. The sum of the squares of three consecutive odd integers is 83. Find the middle odd integer.

28. A jogger ran 7 mi at a constant rate and then reduced the rate by 3 mph. An additional 8 mi was run at the reduced rate. The total time spent jogging the 15 mi was 3 h. Find the rate for the last 8 mi.

29. Graph $2x - 3y > 6$.

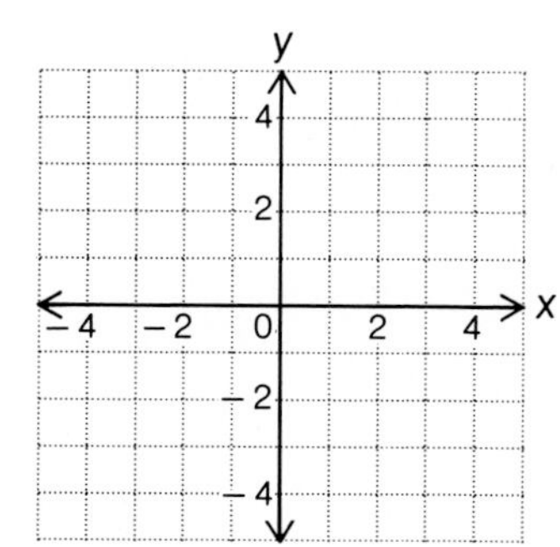

30. Graph $y = x^2 - 2x - 3$.

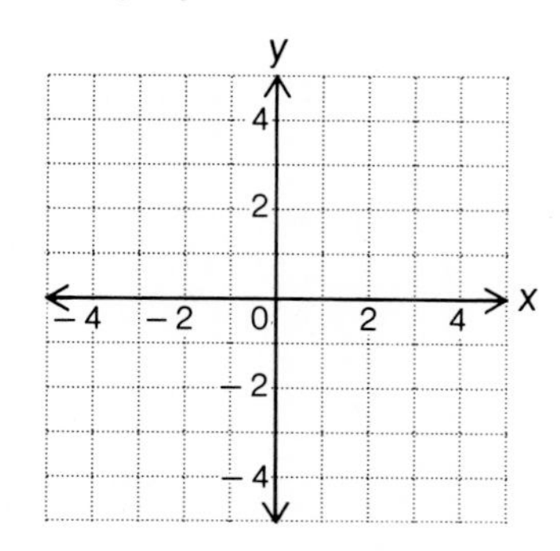

Cumulative Review/Final Exam

1. Evaluate $|-10|$.

2. Subtract: $8 - (-10) - 4$

3. Simplify $-2^3 \cdot (-3)^2$.

4. Simplify $8 + \frac{16 - 4}{3 - (-3)} \div 2$.

5. Evaluate $\frac{-a^2 - b}{a - b}$ when $a = 2$ and $b = -3$.

6. Simplify $-2a - (-3b) - 4a + 6$.

7. Simplify $(-10y)\left(-\frac{3}{5}\right)$.

8. Simplify $-3[3x - 3(2x - 4) - 3]$.

9. Solve: $-\frac{5}{6}x = 40$

10. Solve: $3x - 2(4 - 3x) = -3(2 - x)$

11. Write $\frac{3}{8}$ as a percent.

12. 22% of what is 9.9?

13. Simplify $(3x^2 - 2x + 4) - (5x^2 - 4x - 8)$.

14. Simplify $(-2x^2y)^5$.

15. Simplify $(-2x^2 + 2x - 1)(x - 3)$.

16. Simplify $\frac{(-6xy^2)^3}{(9x^2y)^2}$.

Cumulative Review/Final Exam

17. Simplify: $\frac{16x^2 - 8x + 2}{-2x^2}$

18. Simplify:
$(6x^2 - 3x + 1) \div (2x - 3)$

19. Simplify: $(-2x^{-2}y^2)(3x^2y^{-2})$

20. Factor $-3y^4 + 6y^3 - 21y^2$.

21. Factor $x^2 - 8x - 9$.

22. Factor $6x^2 - x - 12$.

23. Factor $-6x^3 - 21x^2 - 18x$.

24. Factor $49y^2 - 1$.

25. Factor $x(y - 1) - 2(y - 1)$.

26. Factor $12x^2 - 27x^2y^2$.

27. Solve: $x(x + 1) = 12$

28. Simplify:
$\frac{x^2 - 7x + 12}{x^2 - 4x} \cdot \frac{2x^2 + x}{2x^2 - 5x - 3}$

29. Simplify: $\frac{3}{2x - 3} - \frac{2x}{x + 4}$

30. Simplify: $\frac{y - \frac{3}{2y - 1}}{1 - \frac{2}{2y - 1}}$

31. Solve:
$\frac{3x}{2x - 3} - 4 = \frac{2}{2x - 3}$

32. Solve $L = a(1 + ct)$ for t.

Cumulative Review/Final Exam

33. Find the slope of the line containing the points $(-2,4)$ and $(-1,1)$.

34. Find the equation of the line which contains the point $(0,-2)$ and has slope $-\frac{3}{4}$.

35. Solve by substitution.
$$2x - 3y = 7$$
$$3x + y = 5$$

36. Solve by the addition method.
$$x - 2y = -7$$
$$3x + 4y = -1$$

37. Solve: $-\frac{2}{3}x < -\frac{4}{9}$

38. Solve:
$3(3y - 2) + 2 < -3(1 - 2y)$

39. Simplify $\sqrt{81a^4}$.

40. Simplify
$2\sqrt{27a} - 5\sqrt{49b} + 8\sqrt{48a}$.

41. Simplify $\frac{3}{\sqrt{2} - 1}$.

42. Solve: $\sqrt{3x} - 2 = 1$

43. Solve by factoring:
$2x^2 - x = 3$

44. Solve by using the quadratic formula: $3x^2 - 2x = 3$

45. Translate and simplify "the difference between twice a number and four times the sum of the number and two."

46. Due to depreciation, the value of a car is now \$4560. This is five eighths of its original value. Find the original value.

47. The manufacturer's cost for a bicycle is \$125. The manufacturer then sells the bicycle for \$195. What is the markup rate?

48. An investment of \$6000 is made at an annual simple interest rate of 7.5%. How much additional money must be invested at 9.5% so that the total interest earned is \$735?

Cumulative Review/Final Exam

49. Find the selling price per pound of a mixture of coffee made from 60 lb of coffee which cost \$3.20 per pound and 20 lb of coffee which cost \$8.40 per pound.

50. Eight grams of sugar are added to a 60-gram serving of a breakfast cereal which is 15% sugar. What is the percent concentration of sugar in the resulting mixture?

51. A 620-mile, 4-hour plane trip was flown at two speeds. For the first part of the trip, the average speed was 180 mph. For the remainder of the trip, the average speed was 140 mph. For how long did the plane fly at each speed?

52. The perimeter of a rectangle is 80 ft. The length of the rectangle is 5 ft less than twice the width. Find the length and width of the rectangle.

53. A coin bank contains 28 coins in dimes and quarters. The coins have a total value of \$4.45. Find the number of dimes and quarters in the bank.

54. The width of a rectangle is 4 in. less than the length. The area of the rectangle is 320 in.^{2}. Find the length and width of the rectangle.

55. A pre-election survey showed that 4 out of every 7 voters would vote in an election. At this rate, how many people would be expected to vote in a city of 133,000?

56. A business report for a company can be printed in 30 min using one computer. A second computer can print the report in 42 min. How long will it take to print the report with both computers operating?

57. With the wind, a plane flies 700 mi in 2.5 h. Against the wind, the plane requires 3.5 h to fly the same distance. Find the rate of the plane in calm air and the rate of the wind.

58. It took a plane two more hours to fly 900 mi against the wind than it did to fly the same distance with the wind. The rate of the wind was 60 mph. Find the rate of the plane in calm air.

59. Solve by graphing: $2x + 3y = -4$
$3x - 2y = 7$

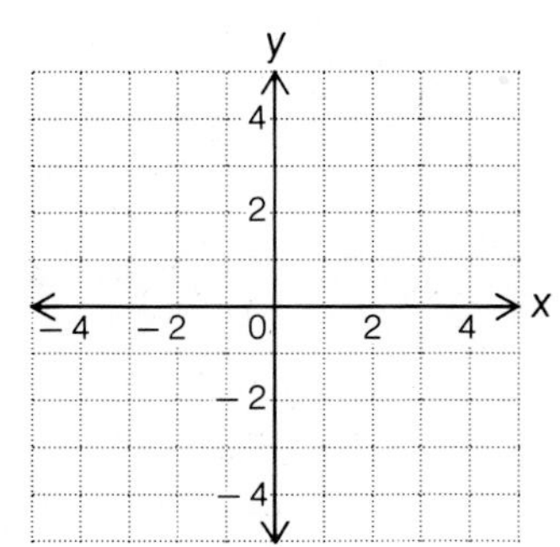

60. Graph $y = -x^2 + 2x + 3$.

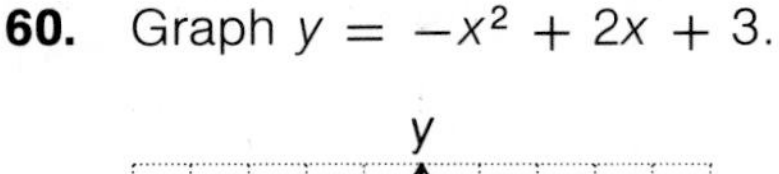

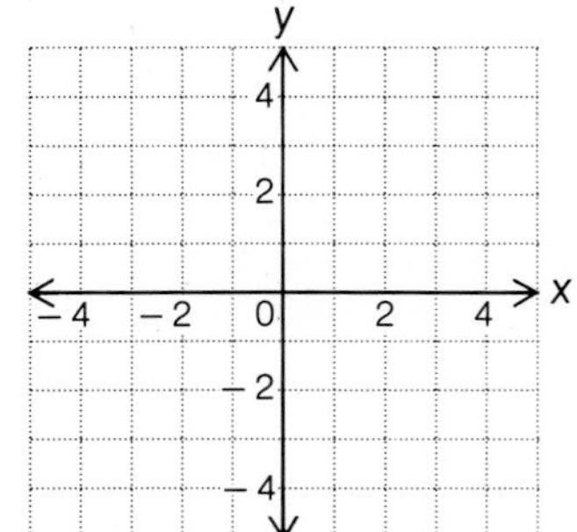

Appendix

Table of Square Roots

Decimal approximations have been rounded to the nearest thousandth.

Number	Square Root	Number	Square Root	Number	Square Root	Number	Square Root
1	1	51	7.141	101	10.050	151	12.288
2	1.414	52	7.211	102	10.100	152	12.329
3	1.732	53	7.280	103	10.149	153	12.369
4	2	54	7.348	104	10.198	154	12.410
5	2.236	55	7.416	105	10.247	155	12.450
6	2.449	56	7.483	106	10.296	156	12.490
7	2.646	57	7.550	107	10.344	157	12.530
8	2.828	58	7.616	108	10.392	158	12.570
9	3	59	7.681	109	10.440	159	12.610
10	3.162	60	7.746	110	10.488	160	12.649
11	3.317	61	7.810	111	10.536	161	12.689
12	3.464	62	7.874	112	10.583	162	12.728
13	3.606	63	7.937	113	10.630	163	12.767
14	3.742	64	8	114	10.677	164	12.806
15	3.873	65	8.062	115	10.724	165	12.845
16	4	66	8.124	116	10.770	166	12.884
17	4.123	67	8.185	117	10.817	167	12.923
18	4.243	68	8.246	118	10.863	168	12.961
19	4.359	69	8.307	119	10.909	169	13
20	4.472	70	8.367	120	10.954	170	13.038
21	4.583	71	8.426	121	11	171	13.077
22	4.690	72	8.485	122	11.045	172	13.115
23	4.796	73	8.544	123	11.091	173	13.153
24	4.899	74	8.602	124	11.136	174	13.191
25	5	75	8.660	125	11.180	175	13.229
26	5.099	76	8.718	126	11.225	176	13.267
27	5.196	77	8.775	127	11.269	177	13.304
28	5.292	78	8.832	128	11.314	178	13.342
29	5.385	79	8.888	129	11.358	179	13.379
30	5.477	80	8.944	130	11.402	180	13.416
31	5.568	81	9	131	11.446	181	13.454
32	5.657	82	9.055	132	11.489	182	13.491
33	5.745	83	9.110	133	11.533	183	13.528
34	5.831	84	9.165	134	11.576	184	13.565
35	5.916	85	9.220	135	11.619	185	13.601
36	6	86	9.274	136	11.662	186	13.638
37	6.083	87	9.327	137	11.705	187	13.675
38	6.164	88	9.381	138	11.747	188	13.711
39	6.245	89	9.434	139	11.790	189	13.748
40	6.325	90	9.487	140	11.832	190	13.784
41	6.403	91	9.539	141	11.874	191	13.820
42	6.481	92	9.592	142	11.916	192	13.856
43	6.557	93	9.644	143	11.958	193	13.892
44	6.633	94	9.695	144	12	194	13.928
45	6.708	95	9.747	145	12.042	195	13.964
46	6.782	96	9.798	146	12.083	196	14
47	6.856	97	9.849	147	12.124	197	14.036
48	6.928	98	9.899	148	12.166	198	14.071
49	7	99	9.950	149	12.207	199	14.107
50	7.071	100	10	150	12.247	200	14.142

Table of Properties

Properties of Real Numbers

Associative Property of Addition

If a, b, and c are real numbers, then $(a + b) + c = a + (b + c)$.

Associative Property of Multiplication

If a, b, and c are real numbers, then $(a \cdot b) \cdot c = a \cdot (b \cdot c)$.

Commutative Property of Addition

If a and b are real numbers, then $a + b = b + a$.

Commutative Property of Multiplication

If a and b are real numbers, then $a \cdot b = b \cdot a$.

Addition Property of Zero

If a is a real number, then $a + 0 = 0 + a = a$.

Multiplication Property of One

If a is a real number, then $a \cdot 1 = 1 \cdot a = a$.

Inverse Property of Addition

If a is a real number, then

$a + (-a) = (-a) + a = 0$.

Inverse Property of Multiplication

If a is a real number and $a \neq 0$, then

$a \cdot \frac{1}{a} = \frac{1}{a} \cdot a = 1$.

Distributive Property

If a, b, and c are real numbers, then $a(b + c) = ab + ac$.

Properties of Equations

Addition Property of Equations

If $a = b$, then $a + c = b + c$.

Multiplication Property of Equations

If $a = b$ and $c \neq 0$, then $a \cdot c = b \cdot c$.

Properties of Exponents

If m and n are integers, then

$x^m \cdot x^n = x^{m+n}$.

If m and n are integers and $x \neq 0$, then

$\frac{x^m}{x^n} = x^{m-n}$.

If m and n are integers, then $(x^m)^n = x^{m \cdot n}$.

If m, n, and p are integers, then $(x^m \cdot y^n)^p = x^{m \cdot p}y^{n \cdot p}$.

If x is a real number and $x \neq 0$, then

$x^0 = 1$.

If n is a positive integer and $x \neq 0$, then

$x^{-n} = \frac{1}{x^n}$.

Property of Zero Products

If $a \cdot b = 0$, then $a = 0$ or $b = 0$.

Properties of Radical Expressions

If a and b are positive real numbers,

then $\sqrt{ab} = \sqrt{a} \cdot \sqrt{b}$.

If a and b are positive real numbers and $b \neq 0$,

then $\sqrt{\frac{a}{b}} = \frac{\sqrt{a}}{\sqrt{b}}$.

Properties of Inequalities

Addition Property of Inequalities

If $a > b$, then $a + c > b + c$.
If $a < b$, then $a + c < b + c$.

Multiplication Property of Inequalities

If $a > b$ and $c > 0$, then $ac > bc$.
If $a < b$ and $c > 0$, then $ac < bc$.
If $a > b$ and $c < 0$, then $ac < bc$.
If $a < b$ and $c < 0$, then $ac > bc$.

Property of Squaring Both Sides of an Equation

If a and b are real numbers and $a = b$, then $a^2 = b^2$.

Table of Symbols

Symbol	Meaning
$+$	add
$-$	subtract
$\cdot$, x, $(a)(b)$	multiply
$\frac{a}{b}$, $\div$	divide
$(\ )$	parentheses, a grouping symbol
$[\ ]$	brackets, a grouping symbol
π	pi, a number approximately equal to $\frac{22}{7}$ or 3.14
$-a$	the opposite, or additive inverse, of a
$\frac{1}{a}$	the reciprocal, or multiplicative inverse, of a
$=$	is equal to
$\approx$	is approximately equal to
$\neq$	is not equal to
$<$	is less than
$\leq$	is less than or equal to
$>$	is greater than
$\geq$	is greater than or equal to
(a, b)	an ordered pair whose first component is a and whose second component is b
$^\circ$	degree (for angles)
$\sqrt{a}$	the principal square root of a
$\varnothing$, $\{\ \}$	the empty set
$\lvert a \rvert$	the absolute value of a
$\cup$	union of two sets
$\cap$	intersection of two sets
$\in$	is an element of (for sets)

Table of Measurement Abbreviations

U.S. Customary System

Length

in.	inch
ft	feet
yd	yard
mi	mile

Capacity

oz	ounce
c	cup
qt	quart
gal	gallon

Weight

oz	ounce
lb	pound

Area

in.^{2}	square inches
ft^2	square feet

Metric System

Length

mm	millimeter (0.001 m)
cm	centimeter (0.01 m)
dm	decimeter (0.1 m)
m	meter
dam	decameter (10 m)
hm	hectometer (100 m)
km	kilometer (1000 m)

Capacity

ml	milliliter (0.001 L)
cl	centiliter (0.01 L)
dl	deciliter (0.1 L)
L	liter
dal	decaliter (10 L)
hl	hectoliter (100 L)
kl	kiloliter (1000 L)

Weight/Mass

mg	milligram (0.001 g)
cg	centigram (0.01 g)
dg	decigram (0.1 g)
g	gram
dag	decagram (10 g)
hg	hectogram (100 g)
kg	kilogram (1000 g)

Area

cm^2	square centimeters
m^2	square meters

Time

h	hours
min	minutes
s	seconds

Answers to Chapter 1

SECTION 1 Pages 17–18

Example 2 a. $5 > -13$ b. $-8 > -22$

Example 4 a. 9 b. −62

Example 6 $|-5| = 5$ $-|-9| = -9$

Pages 19–20

1. $3 < 5$ **3.** $-2 > -5$ **5.** $-16 < 1$ **7.** $3 > -7$ **9.** $-11 < -8$ **11.** $-1 > -6$ **13.** $0 > -3$ **15.** $6 > -8$ **17.** $-14 < 16$ **19.** $35 > 28$ **21.** $-42 < 27$ **23.** $21 > -34$ **25.** $-27 > -39$ **27.** $-87 < 63$ **29.** $68 > -79$ **31.** $-62 > -84$ **33.** $94 > 83$ **35.** $59 > -67$ **37.** $-93 < -55$ **39.** $-88 < 57$ **41.** $0 < 129$ **43.** $-131 < 101$ **45.** $-194 < -180$ **47.** −16 **49.** 3 **51.** −45 **53.** 88 **55.** 2 **57.** 6 **59.** 5 **61.** 1 **63.** −8 **65.** 0 **67.** 19 **69.** 22 **71.** −20 **73.** −18 **75.** 23 **77.** −27 **79.** −41 **81.** 25 **83.** 30 **85.** −34 **87.** −45 **89.** 36 **91.** 61 **93.** −52 **95.** −93 **97.** 119

SECTION 2 Pages 21–24

Example 2
$-154 + (-37)$
-191

Example 4
$-36 + 17 + (-21)$
$-19 + (-21)$
-40

Example 6
$-5 + (-2) + 9 + (-3)$
$-7 + 9 + (-3)$
$2 + (-3)$
-1

Example 8
$-8 - 14$
$-8 + (-14)$
-22

Example 10
$3 - (-4) - 15$
$3 + 4 + (-15)$
$7 + (-15)$
-8

Example 12
$4 - (-3) - 12 - (-7) - 20$
$4 + 3 + (-12) + 7 + (-20)$
$7 + (-12) + 7 + (-20)$
$-5 + 7 + (-20)$
$2 + (-20)$
-18

Example 14
$17 - 10 - 2 - (-6) - 9$
$17 + (-10) + (-2) + 6 + (-9)$
$7 + (-2) + 6 + (-9)$
$5 + 6 + (-9)$
$11 + (-9)$
2

Pages 25–26

1. −2 **3.** 20 **5.** −11 **7.** −9 **9.** −3 **11.** 1 **13.** −5 **15.** −30 **17.** 9 **19.** 1 **21.** −10 **23.** −28 **25.** −41 **27.** −392 **29.** 8 **31.** −7 **33.** −9 **35.** 9 **37.** −3 **39.** 18 **41.** −9 **43.** 11 **45.** 0 **47.** 11 **49.** 2 **51.** −138 **53.** −8 **55.** −337

SECTION 3 Pages 27–30

Example 2
$(-3) \cdot 4 \cdot (-5)$
$(-12) \cdot (-5)$
60

Example 4
$-38 \cdot 51$
-1938

Example 6

$-6 \cdot 8 \cdot (-11) \cdot 3$
$-48 \cdot (-11) \cdot 3$
$528 \cdot 3$
1584

Example 8

$-7(-8)(9)(-2)$
$56(9)(-2)$
$504(-2)$
-1008

Example 10

$(-135) \div (-9)$
15

Example 12

$-72 \div 4$
-18

Example 14

$84 \div (-6)$
-14

Example 16

Strategy To find the difference, subtract the average temperature throughout the earth's stratosphere ($-70°$) from the average temperature on earth's surface ($57°$).

Solution $57 - (-70)$
$57 + 70$
127

The difference is 127°F.

Example 18

Strategy To find the average daily low temperature:
- ▷ Add the seven temperature readings.
- ▷ Divide by 7.

Solution $-6 + (-7) + 1 + 0 + (-5) + (-10) + (-1)$
$-13 + 1 + 0 + (-5) + (-10) + (-1)$
$-12 + 0 + (-5) + (-10) + (-1)$
$-12 + (-5) + (-10) + (-1)$
$-17 + (-10) + (-1)$
$-27 + (-1)$
-28

$-28 \div 7 = -4$

The average daily low temperature was $-4°$.

Pages 31–33

1. 42 **3.** −24 **5.** 6 **7.** 18 **9.** −20 **11.** −16 **13.** 25 **15.** 0 **17.** −72 **19.** −102 **21.** 140 **23.** −228 **25.** −320 **27.** −156 **29.** −70 **31.** 162 **33.** 120 **35.** 36 **37.** 192 **39.** −108 **41.** −2100 **43.** 0 **45.** −251,636 **47.** −2 **49.** 8 **51.** 0 **53.** −9 **55.** −9 **57.** 9 **59.** −24 **61.** −12 **63.** 31 **65.** 17 **67.** 15 **69.** −13 **71.** −18 **73.** 19 **75.** 13 **77.** −19 **79.** 17 **81.** 26 **83.** 23 **85.** 25 **87.** −34 **89.** 11 **91.** −13 **93.** 13 **95.** 12 **97.** −14 **99.** 290

Page 34

1. The temperature was −1°C. **3.** Your score was −15 points. **5.** The difference in elevation is 20,602 ft. **7.** The difference in elevation is 30,314 ft. **9.** The average daily low temperature was −3°.

SECTION 4 Pages 35–40

Example 2

$$\begin{array}{r} 0.16 \\ 25\overline{)4.00} \\ -2\,5 \\ \hline 1\,50 \\ -1\,50 \\ \hline 0 \end{array}$$

$\frac{4}{25} = 0.16$

Example 4

$$\begin{array}{r} 0.444 \\ 9\overline{)4.000} \\ -3\,6 \\ \hline 40 \\ -36 \\ \hline 40 \\ -36 \\ \hline 4 \end{array}$$

$\frac{4}{9} = 0.\overline{4}$

Example 6

The LCM of 9 and 12 is 36.

$$\frac{5}{9} - \frac{11}{12} = \frac{20}{36} - \frac{33}{36} = \frac{20}{36} + \frac{-33}{36} =$$

$$\frac{20-33}{36} = \frac{-13}{36} = -\frac{13}{36}$$

Example 8

The LCM of 8, 6 and 2 is 24.

$$-\frac{7}{8} - \frac{5}{6} + \frac{1}{2} = \frac{-21}{24} - \frac{20}{24} + \frac{12}{24} =$$

$$\frac{-21}{24} + \frac{-20}{24} + \frac{12}{24} = \frac{-21-20+12}{24} =$$

$$\frac{-29}{24} = -1\frac{5}{24}$$

Example 10

$$\begin{array}{r} 3.907 \\ 4.9 \\ +\ 6.63 \\ \hline 15.437 \end{array}$$

Example 12

$$\begin{array}{r} 67.910 \\ -\ 16.127 \\ \hline 51.783 \end{array}$$

$16.127 - 67.91 = -51.783$

Example 14

$2.7 + (-9.44) + 6.2$

$-6.74 + 6.2$

-0.54

Example 16

The product is negative.

$$-\frac{7}{12} \times \frac{9}{14} = -\frac{7 \cdot 9}{12 \cdot 14} =$$

$$-\frac{\overset{1}{\cancel{7}} \cdot \overset{1}{\cancel{3}} \cdot 3}{2 \cdot 2 \cdot \underset{1}{\cancel{3}} \cdot 2 \cdot \underset{1}{\cancel{7}}} = -\frac{3}{8}$$

Example 18

The quotient is positive.

$$-\frac{3}{8} \div -\frac{5}{12} = \frac{3}{8} \times \frac{12}{5} =$$

$$\frac{3 \cdot 12}{8 \cdot 5} = \frac{3 \cdot \overset{1}{\cancel{2}} \cdot \overset{1}{\cancel{2}} \cdot 3}{\underset{1}{\cancel{2}} \cdot \underset{1}{\cancel{2}} \cdot 2 \cdot 5} = \frac{9}{10}$$

Example 20

$$\begin{array}{r} 5.44 \\ \times\ 3.8 \\ \hline 4352 \\ 1632 \\ \hline 20.672 \end{array}$$

$-5.44 \times 3.8 = -20.672$

Example 22

$3.44 \times (-1.7) \times 0.6$

$(-5.848) \times 0.6$

-3.5088

Example 24

$$\begin{array}{r} 0.231 \\ 1.7\overline{)\,0.3.940} \\ -3\,4 \\ \hline 54 \\ -51 \\ \hline 30 \\ -17 \\ \hline 13 \end{array}$$

$-0.394 \div 1.7 \approx -0.23$

Pages 41–44

1. $0.\overline{3}$ **3.** 0.25 **5.** 0.4 **7.** $0.1\overline{6}$ **9.** 0.125 **11.** $0.\overline{2}$ **13.** $0.\overline{45}$ **15.** $0.58\overline{3}$ **17.** $0.2\overline{6}$ **19.** 0.5625 **21.** $0.3\overline{8}$ **23.** 0.05 **25.** 0.24 **27.** $0.2\overline{3}$ **29.** 0.225 **31.** $0.36\overline{1}$ **33.** $0.6\overline{81}$ **35.** $0.458\overline{3}$ **37.** $0.\overline{15}$ **39.** $0.\overline{081}$ **41.** $1\frac{1}{12}$ **43.** $-\frac{5}{24}$ **45.** $-\frac{19}{24}$ **47.** $\frac{5}{26}$ **49.** $\frac{7}{24}$ **51.** 0 **53.** $-\frac{7}{16}$ **55.** $\frac{11}{24}$ **57.** 1 **59.** $1\frac{3}{8}$ **61.** $\frac{169}{315}$ **63.** -38.8 **65.** -6.192 **67.** 13.355 **69.** 4.676 **71.** -10.03 **73.** -37.19 **75.** -17.5 **77.** 853.2594 **79.** $-\frac{3}{8}$ **81.** $\frac{1}{10}$ **83.** $-\frac{4}{9}$ **85.** $-\frac{7}{30}$ **87.** $\frac{2}{27}$ **89.** $-1\frac{1}{9}$ **91.** $-4\frac{19}{32}$ **93.** $3\frac{1}{8}$ **95.** $\frac{2}{3}$ **97.** 4.164 **99.** 4.347 **101.** -4.028 **103.** -2.22 **105.** 12.26448 **107.** -274.44 **109.** -2.59 **111.** -5.11 **113.** -2060.55 **115.** 4781.29

SECTION 5 Pages 45–48

Example 2 $-6^3 = -(6 \cdot 6 \cdot 6) = -216$

Example 4 $(-3)^4 = (-3)(-3)(-3)(-3) = 81$

Example 6 $(3^3)(-2)^3 = (3)(3)(3) \cdot (-2)(-2)(-2) =$
$27(-8) = -216$

Example 8 $\left(-\frac{2}{5}\right)^2 = \left(-\frac{2}{5}\right)\left(-\frac{2}{5}\right) = \frac{4}{25}$

Example 10 $-3(0.3)^3 = -3(0.3)(0.3)(0.3) =$
$-0.9(0.3)(0.3) = -0.27(0.3) = -0.081$

Example 12
$18 - 5[8 - 2(2 - 5)] \div 10$
$18 - 5[8 - 2(-3)] \div 10$
$18 - 5[8 + 6] \div 10$
$18 - 5[14] \div 10$
$18 - 70 \div 10$
$18 - 7$
11

Example 14
$36 \div (8 - 5)^2 - (-3)^2 \cdot 2$
$36 \div (3)^2 - (-3)^2 \cdot 2$
$36 \div 9 - 9 \cdot 2$
$4 - 9 \cdot 2$
$4 - 18$
-14

Example 16
$(6.97 - 4.72)^2 \times 4.5 \div 0.05$
$(2.25)^2 \times 4.5 \div 0.05$
$5.0625 \times 4.5 \div 0.05$
$22.78125 \div 0.05$
455.625

Example 18
$\frac{5}{8} \div \left(\frac{1}{3} - \frac{3}{4}\right) + \frac{7}{12}$
$\frac{5}{8} \div \left(\frac{-5}{12}\right) + \frac{7}{12}$
$\frac{5}{8} \cdot \left(-\frac{12}{5}\right) + \frac{7}{12}$
$-\frac{3}{2} + \frac{7}{12}$
$-\frac{18}{12} + \frac{7}{12}$
$-\frac{11}{12}$

Pages 49–50

1. 36 **3.** -49 **5.** 9 **7.** 81 **9.** $\frac{1}{4}$ **11.** 0.09 **13.** 12 **15.** 0.216 **17.** -12 **19.** 16 **21.** -864 **23.** -1008 **25.** 3 **27.** $-77{,}760$ **29.** 9 **31.** 12 **33.** 1 **35.** 13 **37.** -36 **39.** 13 **41.** 4 **43.** 15 **45.** -1 **47.** 4 **49.** 0.51 **51.** 1.7 **53.** $\frac{1}{4}$ **55.** $\frac{17}{48}$

REVIEW/TEST Pages 53–54

1.1a $-2 > -40$ **1.1b** $-1 < 0$ **1.2a** 4 **1.2b** -4 **2.1a** -3 **2.1b** -16 **2.2a** -14 **2.2b** 4 **3.1a** -48 **3.1b** 90 **3.2a** -9 **3.2b** 17 **3.3a** The temperature is 7°C. **3.3b** The average low temperature was −2°. **4.1a** 0.85 **4.1b** $0.\overline{7}$ **4.2a** $\frac{1}{15}$ **4.2b** -6.881 **4.3a** $-\frac{1}{2}$ **4.3b** -5.3578 **5.1a** -108 **5.1b** 12 **5.2a** 9 **5.2b** 8

CUMULATIVE REVIEW/TEST Pages 55–56

1. $-8 < -4$ (1.1.1) **2.** $1 > -1$ (1.1.1) **3.** 2 (1.1.2) **4.** -3 (1.1.2) **5.** -2 (1.2.1) **6.** -8 (1.2.1) **7.** -11 (1.2.1) **8.** -14 (1.2.2) **9.** 18 (1.2.2) **10.** -13 (1.2.2) **11.** -54 (1.3.1) **12.** 72 (1.3.1) **13.** -8 (1.3.2) **14.** 12 (1.3.2) **15.** The temperature was -7°C. (1.3.3) **16.** The average daily high temperature was -4°. (1.3.3) **17.** 0.35 (1.4.1) **18.** $0.\overline{63}$ (1.4.1) **19.** $-\frac{11}{24}$ (1.4.2) **20.** $\frac{1}{24}$ (1.4.2) **21.** 3.561 (1.4.2) **22.** $-\frac{7}{6}$ (1.4.3) **23.** $\frac{1}{8}$ (1.4.3) **24.** -11.5 (1.4.3) **25.** -4 (1.5.1) **26.** $-\frac{3}{4}$ (1.5.1) **27.** $-\frac{9}{8}$ (1.5.1) **28.** 16 (1.5.1) **29.** 12 (1.5.2) **30.** 8 (1.5.2) **31.** $-\frac{75}{176}$ (1.5.2) **32.** $\frac{27}{50}$ (1.5.2)

Note: The numbers in parentheses following the answers in the Cumulative Review/Tests are a reference to the objective which corresponds with that problem. The first number of the reference indicates the chapter and the second and third numbers indicate the objective. For example, the reference (1.2.1) stands for Chapter 1, Objective 2.1. This notation will be used for all cumulative tests throughout the text.

Answers to Chapter 2

SECTION 1 Pages 59–60

Example 2 -4

Example 4
$2xy + y^2$
$2(-4)(2) + (2)^2$
$2(-4)(2) + 4$
$-8(2) + 4$
$-16 + 4$
-12

Example 6
$\frac{a^2 + b^2}{a + b}$
$\frac{(5)^2 + (-3)^2}{5 + (-3)}$
$\frac{25 + 9}{5 + (-3)}$
$\frac{34}{2} = 17$

Example 8
$x^3 - 2(x + y) + z^2$
$(2)^3 - 2[2 + (-4)] + (-3)^2$
$(2)^3 - 2(-2) + (-3)^2$
$8 - 2(-2) + 9$
$8 + 4 + 9$
$12 + 9$
21

Pages 61–62

1. $2x^2$, $5x$, -8 **3.** 6, $-a^4$ **5.** $7x^2y$, $6xy^2$ **7.** 1, -9 **9.** 1, -4, -1 **11.** 10 **13.** 32 **15.** 21 **17.** 16 **19.** -9 **21.** 3 **23.** -7 **25.** 13 **27.** -15 **29.** 41 **31.** 1 **33.** 5 **35.** 1 **37.** 57 **39.** 5 **41.** 12 **43.** 6 **45.** 10 **47.** 8 **49.** -3 **51.** -2 **53.** -22 **55.** 4 **57.** 20 **59.** 24 **61.** 4.96 **63.** -5.68

SECTION 2 Pages 63–68

Example 2
$3a - 2b - 5a + 6b$
$-2a + 4b$

Example 4
$-3y^2 + 7 + 8y^2 - 14$
$5y^2 - 7$

Example 6
$-5(4y^2)$
$-20y^2$

Example 8
$-7(-2a)$
$14a$

Example 10
$(-5x)(-2)$
$10x$

Example 12
$5(3 + 7b)$
$15 + 35b$

Example 14
$(3a - 1)5$
$15a - 5$

Example 16
$-8(-2a + 7b)$
$16a - 56b$

Example 18
$-(5x - 12)$
$-5x + 12$

Example 20
$3(-a^2 - 6a + 7)$
$-3a^2 - 18a + 21$

Example 22
$3y - 2(y - 7x)$
$3y - 2y + 14x$
$y + 14x$

Example 24
$-2(x - 2y) + 4(x - 3y)$
$-2x + 4y + 4x - 12y$
$2x - 8y$

Example 26
$-5(-2y - 3x) + 4y$
$10y + 15x + 4y$
$14y + 15x$

Example 28
$3y - 2[x - 4(2 - 3y)]$
$3y - 2[x - 8 + 12y]$
$3y - 2x + 16 - 24y$
$-21y - 2x + 16$

Pages 69–72

1. $14x$ **3.** $5a$ **5.** $-6y$ **7.** $-3b-7$ **9.** $5a$ **11.** $-2ab$ **13.** $5xy$ **15.** 0 **17.** $-\frac{5}{6}x$ **19.** $-\frac{1}{24}x^2$
21. $11x$ **23.** $7a$ **25.** $-14x^2$ **27.** $-x+3y$ **29.** $17x-3y$ **31.** $-2a-6b$ **33.** $-3x-8y$
35. $-4x^2-2x$ **37.** $12x$ **39.** $-21a$ **41.** $6y$ **43.** $8x$ **45.** $-6a$ **47.** $12b$ **49.** $-15x^2$ **51.** x^2
53. a **55.** x **57.** n **59.** x **61.** y **63.** $3x$ **65.** $-2x$ **67.** $-8a^2$ **69.** $8y$ **71.** $4y$ **73.** $-2x$
75. $6a$ **77.** $-x-7$ **79.** $10x-35$ **81.** $-5a-80$ **83.** $-15y+35$ **85.** $20-14b$ **87.** $-35+50x$
89. $18x^2+12x$ **91.** $10x-35$ **93.** $-14x+49$ **95.** $-30x^2-15$ **97.** $-24y^2+96$ **99.** $5x^2+5y^2$
101. $-4x^2+20y^2$ **103.** $3x^2+6x-18$ **105.** $-2y^2+4y-8$ **107.** $-2a^2-4a+6$
109. $10x^2+15x-35$ **111.** $6x^2+3xy-9y^2$ **113.** $-3a^2-5a+4$ **115.** $-2x-16$ **117.** $-9-12y$
119. $7n-7$ **121.** $-2x+41$ **123.** $3y-3$ **125.** $-a-7b$ **127.** $-4x+24$ **129.** $-2x-16$
131. $-3x+21$ **133.** $-7x+24$ **135.** $-x+50$ **137.** $20x-41y$

SECTION 3

Pages 73–76

Example 2 the difference between twice n and one third of n

$2n-\frac{1}{3}n$

Example 4 the quotient of 7 less than b and 15

$\frac{b-7}{15}$

Example 6 the unknown number: n
the cube of the number: n^3
the total of ten and the cube of the number: $10+n^3$

$n(10+n^3)$

Example 8 the first integer: x
the second integer: $x+1$
the third integer: $x+2$

$x+(x+1)+(x+2)$

Example 10 the unknown number: x
the difference between the number and sixty: $x-60$

$5(x-60)$

Example 12 the unknown number: x
three eighths of the number: $\frac{3}{8}x$
five sixths of the number: $\frac{5}{6}x$

$\frac{3}{8}x-\frac{5}{6}x$

$\frac{9}{24}x-\frac{20}{24}x$

$-\frac{11}{24}x$

Example 14 the first consecutive integer: x
the second consecutive integer: $x+1$
the third consecutive integer: $x+2$

$x+(x+1)+(x+2)$
$3x+3$

Pages 77–80

1. $8+y$ **3.** $t+10$ **5.** $z+14$ **7.** x^2-20 **9.** $\frac{3}{4}n+12$ **11.** $8+\frac{n}{4}$ **13.** $3(y+7)$ **15.** $t(t+16)$
17. $\frac{1}{2}x^2+15$ **19.** $5n^3+n^2$ **21.** $r-\frac{r}{3}$ **23.** $x^2-(x+17)$ **25.** $9(z+4)$ **27.** $12-x$ **29.** $\frac{2}{3}x$ **31.** $\frac{2x}{9}$
33. $11x-8$ **35.** $(x+2)-9$ **37.** $\frac{7}{5+x}$ **39.** $5+\frac{1}{2}(x+3)$ **41.** $x(2x-4)$ **43.** $(x-5)x$ **45.** $\frac{2x+5}{x}$
47. $x-(3x-8)$ **49.** $x+3x$; $4x$ **51.** $(x+6)+5$; $x+11$ **53.** $x-(x+10)$; -10 **55.** $\frac{1}{6}x+\frac{4}{9}x$; $\frac{11}{18}x$
57. $\frac{x}{3}+x$; $\frac{4x}{3}$ **59.** $14\left(\frac{1}{7}x\right)$; $2x$ **61.** $10x-2x$; $8x$ **63.** $(x-6)+16$; $x+10$ **65.** $4x-x$; $3x$ **67.** $15x-5x$; $10x$ **69.** $(9-x)-12$; $-3-x$ **71.** $10-\left(\frac{2}{3}x+4\right)$; $6-\frac{2}{3}x$ **73.** $5[x+(x+1)]$; $10x+5$
75. $x+(x+1)+6$; $2x+7$ **77.** $2[x+(x+1)+(x+2)]$; $6x+6$ **79.** x^2+3+2x^2; $3x^2+3$
81. $(x+7)+(2-2x)$; $-x+9$ **83.** $(6+x)+2(x-3)$; $3x$ **85.** $x+4(x-9)$; $5x-36$

REVIEW/TEST Pages 83–84

1.1a 22 **1.1b** 3 **2.1a** $5x$ **2.1b** y^2 **2.1c** $-9x - 7y$ **2.1d** $-2x - 5y$ **2.2a** $2x$ **2.2b** $3x$ **2.2c** $36y$ **2.2d** $-10a$ **2.3a** $15 - 35b$ **2.3b** $-4x + 8$ **2.3c** $-6x^2 + 21y^2$ **2.3d** $-10x^2 + 15x - 30$ **2.4a** $-x + 6$ **2.4b** $7x + 38$ **2.4c** $-7x + 33$ **2.4d** $2x + y$ **3.1a** $b - 7b$ **3.1b** $10(x - 3)$ **3.2a** $x + 2x^2$ **3.2b** $\frac{6}{x} - 3$ **3.3a** $8[x + (x + 1)]$; $16x + 8$ **3.3b** $11 + 2(x + 4)$; $19 + 2x$

CUMULATIVE REVIEW/TEST Pages 85–86

1. -7 (1.2.1) **2.** 5 (1.2.2) **3.** 24 (1.3.1) **4.** -5 (1.3.2) **5.** 1.25 (1.4.1) **6.** $\frac{11}{48}$ (1.4.2) **7.** $\frac{1}{6}$ (1.4.3) **8.** $\frac{1}{4}$ (1.4.3) **9.** $\frac{8}{3}$ (1.5.1) **10.** -5 (1.5.2) **11.** $\frac{53}{48}$ (1.5.2) **12.** 16 (2.1.1) **13.** $5x^2$ (2.2.1) **14.** $-7a - 10b$ (2.2.1) **15.** $6a$ (2.2.2) **16.** $30b$ (2.2.2) **17.** $24 - 6x$ (2.2.3) **18.** $6y - 18$ (2.2.3) **19.** $-8x^2 + 12y^2$ (2.2.3) **20.** $-9y^2 + 9y + 21$ (2.2.3) **21.** $-7x + 14$ (2.2.4) **22.** $5x - 43$ (2.2.4) **23.** $17x - 24$ (2.2.4) **24.** $-3x + 21y$ (2.2.4) **25.** $\frac{1}{2}b + b$ (2.3.1) **26.** $\frac{10}{y - 2}$ (2.3.1) **27.** $8 - \frac{n}{12}$ (2.3.2) **28.** $n(n + 2)$ (2.3.2) **29.** $4[n + (n + 1)]$; $8n + 4$ (2.3.3) **30.** $(3 + n)5 + 12$; $27 + 5n$ (2.3.3)

Answers to Chapter 3

SECTION 1 Pages 89–92

Example 2

$$5 - 4x = 8x + 2$$

$5 - 4\left(\frac{1}{4}\right)$	$8\left(\frac{1}{4}\right) + 2$
$5 - 1$	$2 + 2$

$$4 = 4$$

Yes, $\frac{1}{4}$ is a solution.

Example 4

$$10x - x^2 = 3x - 10$$

$10(5) - (5)^2$	$3(5) - 10$
$50 - 25$	$15 - 10$

$$25 \neq 5$$

No, 5 is not a solution.

Example 6

$$\frac{1}{2} = x - \frac{2}{3}$$

$$\frac{1}{2} + \frac{2}{3} = x - \frac{2}{3} + \frac{2}{3}$$

$$\frac{7}{6} = x$$

The solution is $\frac{7}{6}$.

Example 8

$$-\frac{2}{5}x = 6$$

$$\left(-\frac{5}{2}\right)\left(-\frac{2}{5}x\right) = \left(-\frac{5}{2}\right)(6)$$

$$x = -15$$

The solution is -15.

Example 10

$$4x - 8x = 16$$

$$-4x = 16$$

$$\left(-\frac{1}{4}\right)(-4x) = \left(-\frac{1}{4}\right)(16)$$

$$x = -4$$

The solution is -4.

Pages 93–96

1. Yes **3.** No **5.** No **7.** Yes **9.** Yes **11.** Yes **13.** No **15.** No **17.** Yes **19.** No **21.** Yes **23.** No **25.** No **27.** Yes **29.** Yes **31.** $x = 2$ **33.** $b = 15$ **35.** $a = 6$ **37.** $m = -6$ **39.** $n = 3$ **41.** $b = 0$ **43.** $a = -2$ **45.** $z = -7$ **47.** $m = -7$ **49.** $x = -12$ **51.** $b = 2$ **53.** $x = 6$ **55.** $x = -5$ **57.** $m = 15$ **59.** $w = 9$ **61.** $b = 14$ **63.** $a = 19$ **65.** $c = -1$ **67.** $x = 1$ **69.** $a = -\frac{2}{3}$ **71.** $b = -\frac{1}{2}$ **73.** $n = \frac{4}{15}$ **75.** $c = \frac{5}{12}$ **77.** $w = 1.869$ **79.** $t = 0.884$ **81.** $x = 7.251$ **83.** $y = 7$ **85.** $a = -7$ **87.** $m = -4$ **89.** $n = 5$ **91.** $t = 9$ **93.** $x = -8$ **95.** $x = 0$ **97.** $x = -7$ **99.** $y = 8$ **101.** $t = -7$ **103.** $x = 12$ **105.** $b = -18$ **107.** $x = 15$ **109.** $m = -20$ **111.** $x = -12$ **113.** $x = 15$ **115.** $x = 25$ **117.** $x = \frac{8}{3}$ **119.** $y = \frac{1}{3}$ **121.** $m = \frac{15}{7}$ **123.** $n = 4$ **125.** $y = 3$ **127.** $x = 4.745$ **129.** $a = 2.06$ **131.** $x = -2.13$

SECTION 2

Pages 97–98

Example 2

$$5x + 7 = 10$$
$$5x + 7 + (-7) = 10 + (-7)$$
$$5x = 3$$
$$\frac{1}{5} \cdot 5x = \frac{1}{5} \cdot 3$$
$$x = \frac{3}{5}$$

The solution is $\frac{3}{5}$.

Example 4

$$2 = 11 + 3x$$
$$2 + (-11) = 11 + (-11) + 3x$$
$$-9 = 3x$$
$$\frac{1}{3}(-9) = \frac{1}{3}(3x)$$
$$-3 = x$$

The solution is -3.

Example 6

$$x - 5 + 4x = 25$$
$$5x - 5 = 25$$
$$5x - 5 + 5 = 25 + 5$$
$$5x = 30$$
$$\frac{1}{5} \cdot 5x = \frac{1}{5} \cdot 30$$
$$x = 6$$

The solution is 6.

Example 8

Strategy To find the depth, replace P with the given value and solve for D.

Solution

$$P = 15 + \frac{1}{2}D$$
$$45 = 15 + \frac{1}{2}D$$
$$45 + (-15) = 15 + (-15) + \frac{1}{2}D$$
$$30 = \frac{1}{2}D$$
$$2(30) = 2 \cdot \frac{1}{2}D$$
$$60 = D$$

The depth is 60 ft.

Pages 99–101

1. $x = 3$ **3.** $a = 6$ **5.** $x = -1$ **7.** $x = -3$ **9.** $d = 2$ **11.** $c = -2$ **13.** $w = 2$ **15.** $t = 2$ **17.** $a = 5$ **19.** $b = -3$ **21.** $x = 6$ **23.** $x = 2$ **25.** $x = 3$ **27.** $a = 1$ **29.** $b = 6$ **31.** $m = -7$ **33.** $y = 0$ **35.** $c = 2$ **37.** $x = \frac{6}{7}$ **39.** $a = \frac{2}{3}$ **41.** $x = \frac{13}{9}$ **43.** $n = -1$ **45.** $x = \frac{3}{4}$ **47.** $x = \frac{4}{9}$ **49.** $x = \frac{1}{3}$ **51.** $w = -\frac{1}{2}$ **53.** $b = -\frac{3}{4}$ **55.** $x = -\frac{1}{7}$ **57.** $a = \frac{1}{3}$ **59.** $x = -\frac{1}{6}$ **61.** $a = 1$ **63.** $x = 1$ **65.** $x = 0$ **67.** $x = \frac{13}{10}$ **69.** $a = \frac{2}{5}$ **71.** $x = -\frac{4}{3}$ **73.** $x = -\frac{3}{2}$ **75.** $m = 18$ **77.** $n = 8$ **79.** $b = -16$ **81.** $y = 25$ **83.** $c = 21$ **85.** $w = 15$ **87.** $x = -16$ **89.** $x = -21$ **91.** $x = \frac{15}{2}$ **93.** $y = -\frac{18}{5}$ **95.** $y = 2$ **97.** $z = 3$ **99.** $b = 1$ **101.** $m = -2$ **103.** $y = -0.74$ **105.** $x = 0.15$

Page 102

1. The time is 2 s. **3.** The length is 30 ft. **5.** The height is 5 in. **7.** The magnification is 11. **9.** The height is 7 ft.

SECTION 3 Pages 103–106

Example 2

$$\begin{aligned} 5x + 4 &= 6 + 10x \\ 5x + (-10x) + 4 &= 6 + 10x + (-10x) \\ -5x + 4 &= 6 \\ -5x + 4 + (-4) &= 6 + (-4) \\ -5x &= 2 \\ \left(-\tfrac{1}{5}\right)(-5x) &= -\tfrac{1}{5}\cdot 2 \\ x &= -\tfrac{2}{5} \end{aligned}$$

The solution is $-\frac{2}{5}$.

Example 4

$$\begin{aligned} 5x - 10 - 3x &= 6 - 4x \\ 2x - 10 &= 6 - 4x \\ 2x + 4x - 10 &= 6 - 4x + 4x \\ 6x - 10 &= 6 \\ 6x - 10 + 10 &= 6 + 10 \\ 6x &= 16 \\ \tfrac{1}{6}\cdot 6x &= \tfrac{1}{6}\cdot 16 \\ x &= \tfrac{8}{3} \end{aligned}$$

The solution is $\frac{8}{3}$.

Example 6

$$\begin{aligned} 5x - 4(3 - 2x) &= 2(3x - 2) + 6 \\ 5x - 12 + 8x &= 6x - 4 + 6 \\ 13x - 12 &= 6x + 2 \\ 13x + (-6x) - 12 &= 6x + (-6x) + 2 \\ 7x - 12 &= 2 \\ 7x - 12 + 12 &= 2 + 12 \\ 7x &= 14 \\ \tfrac{1}{7}\cdot 7x &= \tfrac{1}{7}\cdot 14 \\ x &= 2 \end{aligned}$$

The solution is 2.

Example 8

$$\begin{aligned} -2[3x - 5(2x - 3)] &= 3x - 8 \\ -2[3x - 10x + 15] &= 3x - 8 \\ -2[-7x + 15] &= 3x - 8 \\ 14x - 30 &= 3x - 8 \\ 14x + (-3x) - 30 &= 3x + (-3x) - 8 \\ 11x - 30 &= -8 \\ 11x - 30 + 30 &= -8 + 30 \\ 11x &= 22 \\ \tfrac{1}{11}\cdot 11x &= \tfrac{1}{11}\cdot 22 \\ x &= 2 \end{aligned}$$

The solution is 2.

Example 10

Strategy To find the location of the fulcrum when the system balances, replace the variables F_1, F_2, and d in the lever system equation by the given values and solve for x.

Solution

$$\begin{aligned} F_1\cdot x &= F_2\cdot(d - x) \\ 45x &= 80(25 - x) \\ 45x &= 2000 - 80x \\ 45x + 80x &= 2000 - 80x + 80x \\ 125x &= 2000 \\ \tfrac{1}{125}\cdot 125x &= \tfrac{1}{125}\cdot 2000 \\ x &= 16 \end{aligned}$$

The fulcrum is 16 ft from the 45-pound force.

Pages 107–109

1. $x = 2$ **3.** $m = 3$ **5.** $x = 3$ **7.** $y = -1$ **9.** $x = -1$ **11.** $x = 2$ **13.** $x = -2$ **15.** $b = -3$ **17.** $x = -8$ **19.** $y = 0$ **21.** $x = -1$ **23.** $x = -3$ **25.** $x = -1$ **27.** $m = 4$ **29.** $x = -2$ **31.** $n = 3$ **33.** $x = -6$ **35.** $a = -2$ **37.** $b = \frac{2}{3}$ **39.** $x = \frac{5}{6}$ **41.** $n = -\frac{2}{3}$ **43.** $y = 3.5$ **45.** $x = 2.45$ **47.** $y = 1$ **49.** $x = 4$ **51.** $m = -1$ **53.** $b = 1$ **55.** $x = -1$ **57.** $y = 2$ **59.** $x = -4$ **61.** $n = -6$ **63.** $x = \frac{5}{6}$ **65.** $a = -\frac{7}{10}$ **67.** $x = \frac{1}{4}$ **69.** $x = 5$ **71.** $x = \frac{20}{3}$ **73.** $x = 2$ **75.** $y = 2.5$

Page 110

1. The fulcrum is 12 ft from the 26-pound force. **3.** A 400-pound force must be applied to the other end. **5.** The break-even point is 200 power saws. **7.** The break-even point is 40 popcorn poppers. **9.** To break even, 250 cassettes must be sold.

SECTION 4 Pages 111–114

Example 2 The unknown number: n

four less than one third of a number	equals	five minus two thirds of the number

$$\frac{1}{3}x - 4 = 5 - \frac{2}{3}x$$
$$\frac{1}{3}x + \frac{2}{3}x - 4 = 5 - \frac{2}{3}x + \frac{2}{3}x$$
$$x - 4 = 5$$
$$x - 4 + 4 = 5 + 4$$
$$x = 9$$

The number is 9.

Example 4 The unknown number: n

two times the difference between a number and eight	is equal to	the sum of six times the number and eight

$$2(n - 8) = 6n + 8$$
$$2n - 16 = 6n + 8$$
$$2n + (-6n) - 16 = 6n + (-6n) + 8$$
$$-4n - 16 = 8$$
$$-4n - 16 + 16 = 8 + 16$$
$$-4n = 24$$
$$-\frac{1}{4}(-4n) = -\frac{1}{4}(24)$$
$$n = -6$$

The number is -6.

Example 6 The smaller number: n
The larger number: $12 - n$

the total of three times the smaller and six	amounts to	seven less than the product of four and the larger

$$3n + 6 = 4(12 - n) - 7$$
$$3n + 6 = 48 - 4n - 7$$
$$3n + 6 = 41 - 4n$$
$$3n + 4n + 6 = 41 - 4n + 4n$$
$$7n + 6 = 41$$
$$7n + 6 + (-6) = 41 + (-6)$$
$$7n = 35$$
$$\frac{1}{7} \cdot 7n = \frac{1}{7} \cdot 35$$
$$n = 5$$

$$12 - n = 12 - 5 = 7$$

The smaller number is 5.
The larger number is 7.

Example 8

Strategy To find the number of carbon atoms in a butane molecule, write and solve an equation using c to represent the number of carbon atoms in a butane molecule.

Solution

8	is	twice the number of carbon atoms in a butane molecule

$$8 = 2c$$
$$\frac{1}{2}\cdot 8 = \frac{1}{2}\cdot 2c$$
$$4 = c$$

There are 4 carbon atoms in a butane molecule.

Example 10

Strategy To find the Fahrenheit temperature, write and solve an equation using F to represent the Fahrenheit temperature.

Solution

20	is	$\frac{5}{9}$ of the difference between the Fahrenheit temperature and 32

$$20 = \frac{5}{9}(F - 32)$$
$$20 = \frac{5}{9}F - \frac{160}{9}$$
$$20 + \frac{160}{9} = \frac{5}{9}F - \frac{160}{9} + \frac{160}{9}$$
$$\frac{340}{9} = \frac{5}{9}F$$
$$\frac{9}{5}\cdot\frac{340}{9} = \frac{9}{5}\cdot\frac{5}{9}F$$
$$68 = F$$

The Fahrenheit temperature is 68°.

Example 12

Strategy To find the length of the wire which produces an A note, write and solve an equation using A to represent the length of the wire.

Solution

10	is	6 in. less than $\frac{1}{2}$ the length of the wire

$$10 = \frac{1}{2}A - 6$$
$$10 + 6 = \frac{1}{2}A - 6 + 6$$
$$16 = \frac{1}{2}A$$
$$2\cdot 16 = 2\cdot\frac{1}{2}A$$
$$32 = A$$

The length of the wire which produces an A note is 32 in.

Example 14

Strategy To find the number of color TV's made each day, write and solve an equation, using x to represent the number of color TV's and $140 - x$ to represent the number of black and white TV's.

Solution

three times the number of black and white TV's	equals	20 less than the number of color TV's

$$3(140 - x) = x - 20$$
$$420 - 3x = x - 20$$
$$420 - 3x + (-x) = x + (-x) - 20$$
$$420 - 4x = -20$$
$$420 + (-420) - 4x = -20 + (-420)$$
$$-4x = -440$$
$$-\frac{1}{4}(-4x) = -\frac{1}{4}(-440)$$
$$x = 110$$

There are 110 color TV's made each day.

Pages 115–116

1. $x + 12 = 10; x = -2$ **3.** $\frac{2}{3}x = 6; x = 9$ **5.** $3(x + 4) = 15; x = 1$ **7.** $12 - 5x = 7; x = 1$
9. $3x + x = 12; x = 3$ **11.** $2x + (x + 3) = 15; x = 4$ **13.** $15 = 4x - 1; x = 4$ **15.** $3(x + 2) = 15; x = 3$
17. $4x = 2x + 10; x = 5$ **19.** $x + 4 = 3x - 8; x = 6$ **21.** $4(3x - 1) = 2x + 6; x = 1$
23. $2(25 - x) = 3x; x = 10; 25 - x = 15$ **25.** $3x + 1 = 2(10 - x) - 4; x = 3; 10 - x = 7$
27. $(16 - x) + 12 = 4x + 3; x = 5; 16 - x = 11$

Pages 117–118

1. $52 = x + 23$. The cost is \$29. **3.** $3600 = \frac{3}{5}x$. The original value was \$6000. **5.** $36{,}000 = x + 3x$. \$9000 was spent for radio advertising and \$27,000 was spent for television advertising. **7.** $2100 = 2x + x$. During the holiday season, 1400 part-time employees are employed. **9.** $20 = x + x + 2x$. There are 5 oxygen atoms, 5 carbon atoms, and 10 hydrogen atoms. **11.** $9536 = 48x + 600$. The monthly payment is \$186.17. **13.** $570 = 3x + 30$. The pressure is 180 lb/in.^{2} **15.** $2000 = 2x - 400$. The drive shaft speed is 1200 rpm. **17.** $155 = 80 + 25x$. There were 3 h of labor. **19.** $180 = 90 + 2x + x$. The second and third angles measure 30° and 60°. **21.** $3(42 - x) = 4x$. The pieces measure 18 in. and 24 in. **23.** $3x = 2(12 - x) + 1$. The pieces measure 5 ft and 7 ft.

REVIEW/TEST Pages 121–122

1.1a No **1.1b** Yes **1.2** $x = -5$ **1.3** $x = -12$ **2.1a** $x = -3$ **2.1b** $x = 5$ **2.2** 200 calculators were produced. **3.1a** $x = -5$ **3.1b** $x = -\frac{1}{2}$ **3.2a** $x = -\frac{1}{3}$ **3.2b** $x = \frac{12}{11}$ **3.3** The final temperature is 60°C. **4.1a** $3x - 15 = 27$; $x = 14$ **4.1b** $5x + 6 = 3(x + 12)$; $x = 15$ **4.1c** 8 and 10 **4.2a** The time is 7 h. **4.2b** The pieces measured 6 ft and 12 ft.

CUMULATIVE REVIEW/TEST Pages 123–124

1. 6 (1.2.2) **2.** -48 (1.3.1) **3.** $-\frac{19}{48}$ (1.4.2) **4.** -2 (1.4.3) **5.** 54 (1.5.1) **6.** 24 (1.5.2) **7.** 6 (2.1.1) **8.** $-17x$ (2.2.1) **9.** $-5a - 2b$ (2.2.1) **10.** $2x$ (2.2.2) **11.** $36y$ (2.2.2) **12.** $2x^2 + 6x - 4$ (2.2.3) **13.** $-4x + 14$ (2.2.4) **14.** $6x - 34$ (2.2.4) **15.** Yes (3.1.1) **16.** No (3.1.1) **17.** $x = -5$ (3.1.2) **18.** $x = -25$ (3.1.3) **19.** $x = -3$ (3.2.1) **20.** $x = 3$ (3.2.1) **21.** $x = 13$ (3.3.2) **22.** $x = 2$ (3.3.2) **23.** $x = -3$ (3.3.1) **24.** $x = \frac{1}{2}$ (3.3.1) **25.** 250 cameras were produced. (3.2.2) **26.** The final temperature is 60°C. (3.3.3) **27.** $12 - 5n = -18$; $n = 6$ (3.4.1) **28.** $6n + 13 = 3n - 5$; $n = -6$ (3.4.1) **29.** The area of the garage is 600 ft^2. (3.4.2) **30.** The pieces measure 5 ft and 11 ft. (3.4.2)

Answers to Chapter 4

SECTION 1 Pages 127–128

Example 2 $125\% = 125\left(\frac{1}{100}\right) = \frac{125}{100} = 1\frac{1}{4}$

$125\% = 125(0.01) = 1.25$

Example 4 $16\frac{2}{3}\% = 16\frac{2}{3}\left(\frac{1}{100}\right) = \frac{50}{3}\left(\frac{1}{100}\right) = \frac{1}{6}$

Example 6 $0.5\% = 0.5(0.01) = 0.005$

Example 8 $0.043 = 0.043(100\%) = 4.3\%$

Example 10 $2.57 = 2.57(100\%) = 257\%$

Example 12 $\frac{5}{9} = \frac{5}{9}(100\%) = \frac{500}{9}\% \approx 55.6\%$

Example 14 $\frac{9}{16} = \frac{9}{16}(100\%) = \frac{900}{16}\% = 56\frac{1}{4}\%$

Pages 129–130

1. $\frac{3}{4}$; 0.75 **3.** $\frac{1}{2}$; 0.50 **5.** $\frac{16}{25}$; 0.64 **7.** $1\frac{1}{4}$; 1.25 **9.** $\frac{19}{100}$; 0.19 **11.** $\frac{1}{20}$; 0.05 **13.** $4\frac{1}{2}$; 4.50 **15.** $\frac{2}{25}$; 0.08
17. $\frac{1}{9}$ **19.** $\frac{1}{8}$ **21.** $\frac{5}{16}$ **23.** $\frac{1}{400}$ **25.** $\frac{23}{400}$ **27.** $\frac{1}{16}$ **29.** 0.073 **31.** 0.158 **33.** 0.003 **35.** 0.0915
37. 0.1823 **39.** 0.0015 **41.** 15% **43.** 5% **45.** 17.5% **47.** 115% **49.** 62% **51.** 316.5%
53. 0.8% **55.** 6.5% **57.** 54% **59.** 33.3% **61.** 45.5% **63.** 87.5% **65.** 166.7% **67.** 128.6%
69. 34% **71.** $37\frac{1}{2}\%$ **73.** $35\frac{5}{7}\%$ **75.** $18\frac{3}{4}\%$ **77.** 125% **79.** $155\frac{5}{9}\%$

SECTION 2 Pages 131–132

Example 2

$$P \times B = A$$
$$P(60) = 27$$
$$P(60)\left(\frac{1}{60}\right) = 27\left(\frac{1}{60}\right)$$
$$P = 0.45$$

The percent is 45%.

Example 4

Strategy To find the percent of defective wheel bearings, solve the basic percent equation using $A = 6$ and $B = 200$. The percent is unknown.

Solution

$$P \times B = A$$
$$P(200) = 6$$
$$P(200)\left(\frac{1}{200}\right) = 6\left(\frac{1}{200}\right)$$
$$P = 0.03$$

3% of the wheel bearings were defective.

Example 6

Strategy To find the percent decrease, solve the basic percent equation using $B = 2500$ and $A = 500$. The percent is unknown.

Solution

$$P \times B = A$$
$$P(2500) = 500$$
$$P(2500)\left(\frac{1}{2500}\right) = 500\left(\frac{1}{2500}\right)$$
$$P = 0.2$$

The percent decrease is 20%.

Page 133

1. 24% **3.** 7.2 **5.** 400 **7.** 9 **9.** 25% **11.** 5 **13.** 200% **15.** 400 **17.** 7.7 **19.** 200 **21.** 400 **23.** 20 **25.** 80.34%

Page 134

1. 20% of the total budget is for materials. **3.** 6.4 gal of the gasoline are used efficiently. **5.** The value of the car today is $4000. **7.** The percent decrease is 40%. **9.** There were 270,000 take-offs and landings last year. **11.** 28.75% (or $28\frac{3}{4}$%) of the people voted in the election.

SECTION 3 Pages 135–136

Example 2

Strategy Given: $C = \$40$

$S = \$60$

Unknown: r

Use the equation $S = C + rC$.

Solution

$$S = C + rC$$
$$60 = 40 + 40r$$
$$20 = 40r$$
$$0.5 = r$$

The markup rate is 50%.

Example 4

Strategy Given: $R = \$27.60$

$S = \$20.70$

Unknown: r

Use the equation $S = R - rR$.

Solution

$$S = R - rR$$
$$20.70 = 27.60 - 27.60r$$
$$-6.90 = -27.60r$$
$$0.25 = r$$

The discount rate is 25%.

Pages 137–138

1. The selling price is $16.20. **3.** The cost is $50. **5.** The markup rate is 60%. **7.** The selling price is $3.05. **9.** The markup rate is 75%. **11.** The markup rate is 56.25%. **13.** The sale price is $22.40. **15.** The regular price is $32.50. **17.** The discount rate is 25%. **19.** The regular price is $300. **21.** The discount rate is 26%. **23.** The discount rate is 23.3%.

SECTION 4 Pages 139–140

Example 2

Strategy

▷ Additional amount: x

Principal	Rate	Interest
5000	0.08	0.08(5000)
x	0.14	$0.14x$
$5000 + x$	0.11	$0.11(5000 + x)$

▷ The sum of the interest earned by the two investments equals the interest earned on the total investment.

Solution

$$0.08(5000) + 0.14x = 0.11(5000 + x)$$
$$400 + 0.14x = 550 + 0.11x$$
$$400 + 0.03x = 550$$
$$0.03x = 150$$
$$x = 5000$$

$5000 more must be invested at 14%.

Pages 141–142

1. The amount invested at 8% is \$2000. The amount invested at 12% is \$3000. **3.** \$4000 more must be invested at 8%. **5.** The amount invested at 12% is \$1600. The amount invested at 8% is \$2400. **7.** \$2500 was deposited in the 12% account. **9.** \$8000 more must be invested at 8%. **11.** The amount invested at 7.5% is \$15,000. The amount invested at 11.25% is \$10,000. **13.** The total amount invested is \$24,000. **15.** The total amount to be invested is \$50,000.

SECTION 5 Pages 143–146

Example 2

Strategy

▷ Pounds of \$.55 fertilizer: x

	Amount	Cost	Value
\$.80 fertilizer	20	\$.80	0.80(20)
\$.55 fertilizer	x	\$.55	$0.55x$
\$.75 fertilizer	$20 + x$	\$.75	$0.75(20 + x)$

▷ The sum of the values before mixing equals the value after mixing.

Solution

$$\begin{aligned} 0.80(20) + 0.55x &= 0.75(20 + x) \\ 16 + 0.55x &= 15 + 0.75x \\ 16 - 0.20x &= 15 \\ -0.20x &= -1 \\ x &= 5 \end{aligned}$$

5 lb of the \$.55 fertilizer must be added.

Example 4

Strategy

▷ Liters of water: x

	Amount	Percent	Quantity
Water	x	0	$0x$
12%	5	0.12	5(0.12)
8%	$x + 5$	0.08	$0.08(x + 5)$

▷ The sum of the quantities before mixing is equal to the quantity after mixing.

Solution

$$\begin{aligned} 0x + 5(0.12) &= 0.08(x + 5) \\ 0.60 &= 0.08x + 0.40 \\ 0.20 &= 0.08x \\ 2.5 &= x \end{aligned}$$

The pharmacist adds 2.5 L of water to the 12% solution to get an 8% solution.

Pages 147–150

1. To make the mixture, 60 lb of the \$2.50 hamburger and 20 lb of the \$3.10 hamburger were used. **3.** 16 oz of pure gold must be used. **5.** The selling price of the mixture is \$2.75 per pound. **7.** 25 gal of cranberry juice should be mixed with 75 gal of apple juice. **9.** 36 lb of walnuts must be used. **11.** The selling price of the mixture is \$3.70 per pound. **13.** 17.25 lb of the \$4.20 cheese must be used. **15.** 1500 bushels of soybeans and 3500 bushels of corn were used. **17.** The selling price is \$3 per ounce. **19.** 1.5 L of the \$80 face cream must be used. **21.** 40 gal of the 21% butterfat and 20 gal of the 15% butterfat must be used. **23.** 10 g of pure acid must be used. **25.** The percent concentration of the resulting solution is 20%. **27.** 40 L of the 85% maple syrup must be mixed with 110 L of pure maple syrup. **29.** 20 lb of oats must be added. **31.** The percent concentration of sugar in the mixture is 44%. **33.** 300 lb of 20% polyester should be woven with 300 lb of 50% polyester. **35.** 20 lb of the 40% wheat flour were used. **37.** The 300-pound alloy contains 10% tin. **39.** 16.67 oz of the 5% solution should be mixed with 33.33 oz of the 8% solution.

SECTION 6 Pages 151–152

Example 2

Strategy

▷ Rate of the first train: r

Rate of the second train: $2r$

	Rate	Time	Distance
1st train	r	3	$3r$
2nd train	$2r$	3	$3(2r)$

▷ The sum of the distances traveled by each train equals 288 mi.

Solution

$3r + 3(2r) = 288$

$3r + 6r = 288$

$9r = 288$

$r = 32$

$2r = 2(32) = 64$

The first train is traveling at 32 mph.
The second train is traveling at 64 mph.

Example 4

Strategy

▷ Time spent flying out: t

Time spent flying back: $5 - t$

	Rate	Time	Distance
Out	150	t	$150t$
Back	100	$5 - t$	$100(5 - t)$

▷ The distance out equals the distance back.

Solution

$150t = 100(5 - t)$

$150t = 500 - 100t$

$250t = 500$

$t = 2$ (The time out was 2 h.)

The distance $= 150t = 150(2) = 300$ mi.

The parcel of land was 300 mi away.

Pages 153–154

1. The first cyclist is riding at a rate of 8 mph. The second cyclist is riding at a rate of 16 mph. **3.** In 2 h the cabin cruiser will be alongside the motorboat. **5.** The distance to the resort is 150 mi. **7.** The length of the track is 120 m. **9.** The car overtakes the cyclist 48 mi from the starting point. **11.** For 3 h the car traveled at 45 mph, and for 2 h the car traveled at 30 mph. **13.** The average speed on the winding road was 32 mph. **15.** The corporate offices are 120 mi from the airport. **17.** The car is traveling at 50 mph. **19.** The cyclists will meet after 1.5 $\left(\text{or } 1\frac{1}{2}\right)$ h.

SECTION 7 Pages 155–156

Example 2

Strategy

▷ Width of the rectangle: w

Length of the rectangle: $w + 3$

▷ Use the equation for the perimeter of a rectangle.

Solution

$2l + 2w = p$

$2(w + 3) + 2w = 34$

$2w + 6 + 2w = 34$

$4w + 6 = 34$

$4w = 28$

$w = 7$

The width of the rectangle is 7 m.

Example 4

Strategy

▷ Measure of the first angle: $2x$

Measure of the second angle: x

Measure of the third angle: $x - 4$

▷ Use the equation $A + B + C = 180°$.

Solution

$A + B + C = 180$

$2x + x + (x - 4) = 180$

$4x - 4 = 180$

$4x = 184$

$x = 46$

$2x = 2(46) = 92$

$x - 4 = 46 - 4 = 42$

The measure of the first angle is 92°.
The measure of the second angle is 46°.
The measure of the third angle is 42°.

Pages 157–158

1. The length is 15 m. The width is 10 m. **3.** The length is 100 cm. The width is 25 cm. **5.** The lengths of the sides are 4 m, 4 m, and 2 m. **7.** The width is 8 m. The length is 13 m. **9.** The measures of the sides are 20 cm, 40 cm, and 50 cm. **11.** The length is 14.48 m. The width is 13.64 m. **13.** The measure of each angle is 60°. **15.** The measure of each angle is 45°. **17.** The measure of the angles are 42°, 42°, and 96°. **19.** The measures of the angles are 30°, 60°, and 90°. **21.** The measures of the angles are 37°, 111°, and 32°. **23.** The measures of the angles are 38°, 76°, and 66°.

SECTION 8 Pages 159–162

Example 2

Strategy

▷ First consecutive integer: n
Second consecutive integer: $n + 1$
Third consecutive integer: $n + 2$
▷ The sum of the three integers is -6.

Solution

$$\begin{aligned} n + (n + 1) + (n + 2) &= -6 \\ 3n + 3 &= -6 \\ 3n &= -9 \\ n &= -3 \end{aligned}$$

$n + 1 = -3 + 1 = -2$

$n + 2 = -3 + 2 = -1$

The three consecutive integers are -3, -2, and -1.

Example 4

Strategy

▷ Number of dimes: x
Number of nickels: $4x$
Number of quarters: $x + 5$

Coin	Number	Value	Total Value
Dime	x	10	$10x$
Nickel	$4x$	5	$5(4x)$
Quarter	$x + 5$	25	$25(x + 5)$

▷ The sum of the total values of each denomination of coin equals the total value of all the coins (675 cents).

Solution

$$\begin{aligned} 10x + 5(4x) + 25(x + 5) &= 675 \\ 10x + 20x + 25x + 125 &= 675 \\ 55x + 125 &= 675 \\ 55x &= 550 \\ x &= 10 \end{aligned}$$

$4x = 4(10) = 40$

$x + 5 = 10 + 5 = 15$

The bank contains 10 dimes, 40 nickels, and 15 quarters.

Example 6

Strategy

▷ The number of years ago: x

	Present age	Past age
Half dollar	25	$25 - x$
Dime	15	$15 - x$

▷ At a past age, the half dollar was twice as old as the dime.

Solution

$25 - x = 2(15 - x)$
$25 - x = 30 - 2x$
$25 + x = 30$
$x = 5$

Five years ago the half dollar was twice as old as the dime.

Pages 163–166

1. The integers are 15, 16, and 17. **3.** The integers are 20, 22, and 24. **5.** The integers are 15, 17, and 19. **7.** The integers are 4 and 6. **9.** The integers are 3 and 5. **11.** The integers are −8, −7, and −6. **13.** The integers are 11, 13, and 15. **15.** The integers are −1, 1, and 3. **17.** The coin purse contains 12 nickels and 4 dimes. **19.** Eight 15¢ stamps and two 25¢ stamps were sold. **21.** There are 15 dimes and 20 quarters in the bank. **23.** There were 6 twenty-dollar bills and 3 ten-dollar bills. **25.** There are 5 pennies in the bank. **27.** There are fifteen 2¢ stamps, seven 5¢ stamps, and fourteen 7¢ stamps in the collection. **29.** There are 30 nickels, 15 dimes, and 33 quarters in the bank. **31.** In 7 years the autographed first edition will be three times as old as the reprint. **33.** The present age of the nickel is 16 years and the present age of the dime is 40 years. **35.** In 25 years, the crystal vase will be three times as old as the porcelain vase. **37.** Sixty-five years ago the butterchurn was twice the age of the ice box. **39.** The present age of the antique is 47 years, and the present age of the replica is 2 years. **41.** The oil painting is 17 years old and the watercolor is 3 years old. **43.** The 5¢ coin is 4 years old, and the 10¢ coin is 8 years old.

REVIEW/TEST Pages 169–170

1.1a $\frac{3}{5}$; 0.60 **1.1b** $\frac{5}{8}$ **1.2a** 37.5% **1.2b** $87\frac{1}{2}\%$ **2.1a** 6.4 **2.1b** 125% **2.2** The value of the computer last year was $3000. **3.1** The cost is $200. **3.2** The discount rate is 20%. **4.1** $1500 should be invested at 7%, and $5500 should be invested at 9%. **5.1** The merchant should use 8 lb of the $7 coffee and 4 lb of the $4 coffee. **5.2** 20 gal of the 15% solution must be used. **6.1** The first plane is traveling at a rate of 225 mph. The second plane is traveling at a rate of 125 mph. **7.1** The length is 14 m. The width is 5 m. **7.2** The measures of the angles are 48°, 33°, and 99°. **8.1** The integers are 5, 7, and 9. **8.2** There are 15 nickels and 35 quarters in the bank. **8.3** In 10 years the 5¢ stamp will be three times the age of the 20¢ stamp.

CUMULATIVE REVIEW/TEST Pages 171–172

1. 6 (1.2.2) **2.** $-\frac{1}{6}$ (1.5.1) **3.** $-\frac{11}{6}$ (1.5.2) **4.** -21 (2.1.1) **5.** $6x + 4y$ (2.2.1) **6.** $-15 + 10x + 25x^3$ (2.2.3) **7.** $3x - 6$ (2.2.4) **8.** No (3.1.1) **9.** $x = 3$ (3.1.2) **10.** $x = -16$ (3.1.3) **11.** $x = 1$ (3.2.1) **12.** $x = -16$ (3.3.2) **13.** $\frac{4}{5}$ (4.1.1) **14.** $\frac{1}{6}$ (4.1.1) **15.** The two numbers are 4 and 11. (3.4.1) **16.** The bill included a charge for 5 hours of labor. (3.4.2) **17.** 7.5% (4.1.2) **18.** 8% (4.1.2) **19.** 20 (4.2.1) **20.** 40 (4.2.1) **21.** 20% of the libraries had the reference book. (4.2.2) **22.** $2000 more must be deposited into the account. (4.4.1) **23.** The markup rate is 75%. (4.3.1) **24.** The sale price is $79.20. (4.3.2) **25.** 60 grams of gold alloy must be mixed. (4.5.1) **26.** 30 ounces of pure water must be added. (4.5.2) **27.** The measure of the first side is 22 ft. (4.7.1) **28.** The measure of one of the equal angles is 47°. (4.7.2) **29.** The middle integer is 14. (4.8.1) **30.** There are 7 dimes in the bank. (4.8.2)

Answers to Chapter 5

SECTION 1 Pages 175–176

Example 2

$$\begin{array}{r} 2x^2 + 4x - 3 \\ +\ 5x^2 - 6x \quad\ \\ \hline 7x^2 - 2x - 3 \end{array}$$

Example 4 $(-3x^2 + 2y^2) + (-8x^2 + 9xy)$
$-11x^2 + 9xy + 2y^2$

Example 6

$$\begin{array}{r} 8y^2 - 4xy + x^2 \\ -\ 2y^2 - xy + 5x^2 \\ \hline \end{array} =$$

$$\begin{array}{r} 8y^2 - 4xy + x^2 \\ +\ -2y^2 + xy - 5x^2 \\ \hline 6y^2 - 3xy - 4x^2 \end{array}$$

Example 8 $(-3a^2 - 4a + 2) - (5a^3 + 2a - 6)$
$(-3a^2 - 4a + 2) + (-5a^3 - 2a + 6)$
$-5a^3 - 3a^2 - 6a + 8$

Pages 177–178

1. $-2x^2 + 3x$ **3.** $y^2 - 8$ **5.** $5x^2 + 7x + 20$ **7.** $x^3 + 2x^2 - 6x - 6$ **9.** $2a^3 - 3a^2 - 11a + 2$
11. $5x^2 + 8x$ **13.** $7x^2 + xy - 4y^2$ **15.** $3a^2 - 3a + 17$ **17.** $5x^3 + 10x^2 - x - 4$ **19.** $3r^3 + 2r^2 - 11r + 7$
21. $-2x^3 + 3x^2 + 10x + 11$ **23.** $4x$ **25.** $3y^2 - 4y - 2$ **27.** $-7x - 7$ **29.** $4x^3 + 3x^2 + 3x + 1$
31. $y^3 - y^2 + 6y - 6$ **33.** $-y^2 - 13xy$ **35.** $2x^2 - 3x - 1$ **37.** $-2x^3 + x^2 + 2$ **39.** $3a^3 - 2$
41. $4y^3 - 2y^2 + 2y - 4$

SECTION 2 Pages 179–180

Example 2 $(3x^2)(6x^3) = (3 \cdot 6)(x^2 \cdot x^3) =$
$18x^5$

Example 4 $(-3xy^2)(-4x^2y^3) =$
$[(-3)(-4)](x \cdot x^2)(y^2 \cdot y^3) =$
$12x^3y^5$

Example 6 $(3x)(2x^2y)^3 =$
$(3x)(2^3x^6y^3) = (3x)(8x^6y^3) =$
$(3 \cdot 8)(x \cdot x^6)y^3 = 24x^7y^3$

Example 8 $(3x^2)^2(-2xy^2)^3 =$
$(3^2x^4)[(-2)^3x^3y^6] =$
$(9x^4)(-8x^3y^6) =$
$[9(-8)](x^4 \cdot x^3)y^6 =$
$-72x^7y^6$

Pages 181–182

1. $2x^2$ **3.** $12x^2$ **5.** $6a^7$ **7.** x^3y^5 **9.** $-10x^9y$ **11.** x^7y^8 **13.** $-6x^3y^5$ **15.** x^4y^5z **17.** $a^3b^5c^4$
19. $-a^5b^8$ **21.** $-6a^5b$ **23.** $40y^{10}z^6$ **25.** $-20a^2b^3$ **27.** $x^3y^5z^3$ **29.** $-12a^{10}b^7$ **31.** $-36a^3b^2c^3$
33. 81 **35.** -27 **37.** -512 **39.** y^8 **41.** y^{15} **43.** $-x^6$ **45.** $27y^3$ **47.** $9y^6$ **49.** $x^{15}y^{20}$
51. $16a^4b^{12}$ **53.** $16a^{12}b^2$ **55.** $-54y^{13}$ **57.** a^6b^4 **59.** $x^{13}y^5$ **61.** $192x^6y^{10}$ **63.** $24x^6y^4$ **65.** $9a^4b^{10}$
67. $-24a^3b^8$ **69.** $729a^9b^6$

SECTION 3 Pages 183–186

Example 2 $(-2y + 3)(-4y) = 8y^2 - 12y$

Example 4 $-a^2(3a^2 + 2a - 7) =$
$-3a^4 - 2a^3 + 7a^2$

Example 6

$$\begin{array}{r} 2y^3 + 2y^2 - 3 \\ \times\ 3y\ \ - 1 \\ \hline -2y^3 - 2y^2 \quad + 3 \\ 6y^4 + 6y^3 \qquad - 9y \\ \hline 6y^4 + 4y^3 - 2y^2 - 9y + 3 \end{array}$$

Example 8 $(4y - 5)(2y - 3) =$
$8y^2 - 12y - 10y + 15 =$
$8y^2 - 22y + 15$

Example 10 $(3b + 2)(3b - 5) =$
$9b^2 - 15b + 6b - 10 =$
$9b^2 - 9b - 10$

Example 12 $(2a + 5c)(2a - 5c) = 4a^2 - 25c^2$

Example 14 $(3x + 2y)^2 = 9x^2 + 12xy + 4y^2$

Example 16

Strategy To find the area, replace the variable r in the equation $A = \pi r^2$ by the given value and solve for A.

Solution $A = \pi r^2$
$A = 3.14(x - 4)^2$
$A = 3.14(x^2 - 8x + 16)$
$A = 3.14x^2 - 25.12x + 50.24$
The area is $3.14x^2 - 25.12x + 50.24$.

5

Pages 187–190

1. $x^2 - 2x$ **3.** $-x^2 - 7x$ **5.** $3a^3 - 6a^2$ **7.** $-5x^4 + 5x^3$ **9.** $-3x^5 + 7x^3$ **11.** $12x^3 - 6x^2$
13. $6x^2 - 12x$ **15.** $3x^2 + 4x$ **17.** $-x^3y + xy^3$ **19.** $2x^4 - 3x^2 + 2x$ **21.** $2a^3 + 3a^2 + 2a$
23. $3x^6 - 3x^4 - 2x^2$ **25.** $-6y^4 - 12y^3 + 14y^2$ **27.** $-2a^3 - 6a^2 + 8a$ **29.** $6y^4 - 3y^3 + 6y^2$
31. $x^3y - 3x^2y^2 + xy^3$ **33.** $x^3 + 4x^2 + 5x + 2$ **35.** $a^3 - 6a^2 + 13a - 12$ **37.** $-2b^3 + 7b^2 + 19b - 20$
39. $-6x^3 + 31x^2 - 41x + 10$ **41.** $x^3 - 3x^2 + 5x - 15$ **43.** $x^4 - 4x^3 - 3x^2 + 14x - 8$
45. $15y^3 - 16y^2 - 70y + 16$ **47.** $5a^4 - 20a^3 - 5a^2 + 22a - 8$ **49.** $y^4 + 4y^3 + y^2 - 5y + 2$
51. $x^2 + 4x + 3$ **53.** $a^2 + a - 12$ **55.** $y^2 - 5y - 24$ **57.** $y^2 - 10y + 21$ **59.** $2x^2 + 15x + 7$
61. $3x^2 + 11x - 4$ **63.** $4x^2 - 31x + 21$ **65.** $3y^2 - 2y - 16$ **67.** $9x^2 + 54x + 77$ **69.** $21a^2 - 83a + 80$
71. $15b^2 + 47b - 78$ **73.** $2a^2 + 7ab + 3b^2$ **75.** $6a^2 + ab - 2b^2$ **77.** $2x^2 - 3xy - 2y^2$
79. $10x^2 + 29xy + 21y^2$ **81.** $6a^2 - 25ab + 14b^2$ **83.** $2a^2 - 11ab - 63b^2$ **85.** $100a^2 - 100ab + 21b^2$
87. $15x^2 + 56xy + 48y^2$ **89.** $14x^2 - 97xy - 60y^2$ **91.** $56x^2 - 61xy + 15y^2$ **93.** $y^2 - 25$ **95.** $4x^2 - 9$
97. $x^2 + 2x + 1$ **99.** $9a^2 - 30a + 25$ **101.** $9x^2 - 49$ **103.** $4a^2 + 4ab + b^2$ **105.** $x^2 - 4xy + 4y^2$
107. $16 - 9y^2$ **109.** $25x^2 + 20xy + 4y^2$

Page 190

1. The area is $8x^2 - 12x$. **3.** The area is $x^2 + 4x + 4$. **5.** The area is $3.14x^2 + 18.84x + 28.26$.

SECTION 4 Pages 191–194

Example 2 $\frac{42y^{12}}{-14y^{17}} = -\frac{\overset{1}{\cancel{2}} \cdot 3 \cdot \overset{1}{\cancel{7}}y^{12}}{\underset{1}{\cancel{2}} \cdot \underset{1}{\cancel{7}}y^{17}} = -\frac{3}{y^5}$

Example 4 $\frac{12r^4s^2}{-8r^3s} = -\frac{\overset{1}{\cancel{2}} \cdot \overset{1}{\cancel{2}} \cdot 3r^4s^2}{\underset{1}{\cancel{2}} \cdot \underset{1}{\cancel{2}} \cdot 2r^3s} = -\frac{3rs}{2}$

Example 6 $\frac{(2x^2y)^3}{-4xy^5} = -\frac{2^3x^6y^3}{4xy^5} =$

$-\frac{\overset{1}{\cancel{2}} \cdot \overset{1}{\cancel{2}} \cdot 2x^6y^3}{\underset{1}{\cancel{2}} \cdot \underset{1}{\cancel{2}}xy^5} = -\frac{2x^5}{y^2}$

Example 8 $\frac{4x^3y + 8x^2y^2 - 4xy^3}{2xy} =$

$\frac{4x^3y}{2xy} + \frac{8x^2y^2}{2xy} - \frac{4xy^3}{2xy} =$

$2x^2 + 4xy - 2y^2$

Example 10 $\frac{24x^2y^2 - 18xy + 6y}{6xy} =$

$\frac{24x^2y^2}{6xy} - \frac{18xy}{6xy} + \frac{6y}{6xy} =$

$4xy - 3 + \frac{1}{x}$

Example 12

$$\begin{array}{r} x^2 + 2x - 1 \\ 2x - 3\,\overline{)\,2x^3 + x^2 - 8x - 3} \\ \underline{2x^3 - 3x^2} \qquad\qquad \\ 4x^2 - 8x \qquad \\ \underline{4x^2 - 6x} \qquad \\ -2x - 3 \\ \underline{-2x + 3} \\ -6 \end{array}$$

$(2x^3 + x^2 - 8x - 3) \div (2x - 3) =$

$x^2 + 2x - 1 - \frac{6}{2x - 3}$

Pages 195–198

1. $3x$ **3.** $-x$ **5.** $4x^3$ **7.** $-\frac{4}{x^2}$ **9.** $\frac{a}{b^4}$ **11.** y^3 **13.** $\frac{3}{5}$ **15.** $\frac{24}{b}$ **17.** $-\frac{3}{5ab^2}$ **19.** $-\frac{4b}{9}$ **21.** $-\frac{2x^2y^2}{11z^5}$ **23.** $-8a^3b^4$ **25.** $\frac{4a^2}{9b^3}$ **27.** $\frac{x^2y^2}{z^3}$ **29.** $\frac{a^2}{b}$ **31.** $-\frac{a^2}{c^2}$ **33.** y^4 **35.** $y + 1$ **37.** $2b - 5$ **39.** $6y + 4$ **41.** $12x - 7$ **43.** $5y - 3$ **45.** $-y + 9$ **47.** $a^2 - 5a + 7$ **49.** $a^6 - 5a^3 - 3a$ **51.** $xy - 3$ **53.** $-2x^2 + 3$ **55.** $8y + 2 - \frac{3}{y}$ **57.** $2y - 6 + \frac{9}{y}$ **59.** $2a + 1 - 3b$ **61.** $a - 3 + 6b$ **63.** $x + 5$ **65.** $b - 7$ **67.** $y - 5$ **69.** $2y - 7$ **71.** $2y + 6 + \frac{25}{y - 3}$ **73.** $x - 2 + \frac{8}{x + 2}$ **75.** $3y - 5 + \frac{20}{2y + 4}$ **77.** $6x - 12 + \frac{19}{x + 2}$ **79.** $b - 5 - \frac{24}{b - 3}$ **81.** $3x + 17 + \frac{64}{x - 4}$ **83.** $5y + 3 + \frac{1}{2y + 3}$ **85.** $4a + 1$ **87.** $2a + 9 + \frac{33}{3a - 1}$ **89.** $x^2 - 5x + 2$ **91.** $2a^2 + a + 1 + \frac{6}{2a + 3}$ **93.** $2b^2 - 3b + 4 - \frac{17}{2b + 3}$ **95.** $5x^2 + 3x + 3 + \frac{6}{x - 2}$ **97.** $x^2 + 5$

SECTION 5 Pages 199–200

Example 2 $\frac{2^{-2}}{2^3} = 2^{-5} = \frac{1}{2^5} = \frac{1}{32}$

Example 4 $(-2x^2)(x^{-3}y^{-4})^{-2} =$

$(-2x^2)(x^6y^8) = -2x^8y^8$

Example 6 $\frac{(3x^{-2}y)^3}{9xy^0} = \frac{3^3x^{-6}y^3}{9xy^0} =$

$\frac{\overset{1}{\cancel{3}} \cdot \overset{1}{\cancel{3}} \cdot 3x^{-6}y^3}{\underset{1}{\cancel{3}} \cdot \underset{1}{\cancel{3}}xy^0} = 3x^{-7}y^3 = \frac{3y^3}{x^7}$

Pages 201–202

1. $\frac{1}{5^2} = \frac{1}{25}$ **3.** $\frac{1}{7^1} = \frac{1}{7}$ **5.** $\frac{1}{3^3} = \frac{1}{27}$ **7.** $\frac{1}{2^6} = \frac{1}{64}$ **9.** $\frac{1}{x^2}$ **11.** $\frac{1}{a^6}$ **13.** $\frac{x^2}{y^3}$ **15.** $\frac{1}{xy^2}$ **17.** x
19. $\frac{1}{a^7}$ **21.** $\frac{1}{x^4}$ **23.** $\frac{1}{a^8}$ **25.** $\frac{y}{x^3}$ **27.** $\frac{b}{a^2}$ **29.** $\frac{1}{a^4}$ **31.** $\frac{1}{a^6}$ **33.** x^6 **35.** a^{18} **37.** 1 **39.** $\frac{y^4}{x^4}$
41. $\frac{x}{y^5}$ **43.** $\frac{1}{x^4y^3}$ **45.** $\frac{y^2}{x^4}$ **47.** x^9y^9 **49.** $-\frac{8x^3}{y^6}$ **51.** $\frac{16x^4}{y^6}$ **53.** $\frac{9}{x^2y^4}$ **55.** $\frac{2}{x^4}$ **57.** $-\frac{5}{a^8}$ **59.** $-\frac{a^5}{8b^4}$
61. $\frac{10y^3}{x^4}$ **63.** $\frac{1}{a^5b^6}$ **65.** $\frac{1}{4x^3}$ **67.** $\frac{16}{3a^5}$ **69.** $\frac{2y}{x^3}$ **71.** $\frac{1}{x^3}$ **73.** $\frac{1}{x^{12}y^{12}}$

REVIEW/TEST Pages 205–206

1.1 $3x^3 + 6x^2 - 8x + 3$ **1.2** $-5a^3 + 3a^2 - 4a + 3$ **2.1a** a^4b^7 **2.1b** $-6x^3y^6$ **2.2a** x^8y^{12} **2.2b** $-8a^6b^3$ **3.1a** $4x^3 - 6x^2$ **3.1b** $6y^4 - 9y^3 + 18y^2$ **3.2a** $x^3 - 7x^2 + 17x - 15$ **3.2b** $-4x^4 + 8x^3 - 3x^2 - 14x + 21$ **3.3a** $a^2 + 3ab - 10b^2$ **3.3b** $10x^2 - 43xy + 28y^2$ **3.4a** $16y^2 - 9$ **3.4b** $4x^2 - 20x + 25$ **3.5** The area is $3.14x^2 - 31.4x + 78.5$. **4.1a** $-\frac{4}{x^6}$ **4.1b** $\frac{9y^6}{x}$ **4.2a** $4x^4 - 2x^2 + 5$ **4.2b** $4x - 1 + \frac{3}{x^2}$ **4.3a** $x + 7$ **4.3b** $2x + 3 + \frac{2}{2x - 3}$ **5.1a** $\frac{a^4}{b^6}$ **5.1b** $-\frac{6b}{a}$

5

CUMULATIVE REVIEW/TEST Pages 207–208

1. $\frac{5}{144}$ (1.4.2) **2.** $\frac{5}{3}$ (1.5.1) **3.** $\frac{25}{11}$ (1.5.2) **4.** $-\frac{22}{9}$ (2.1.1) **5.** $5x - 3xy$ (2.2.1) **6.** $-9x$ (2.2.2) **7.** $-18x + 12$ (2.2.4) **8.** $x = -16$ (3.1.3) **9.** $x = -16$ (3.3.1) **10.** $x = 15$ (3.3.2) **11.** 22% (4.2.1) **12.** $4b^3 - 4b^2 - 8b - 4$ (5.1.1) **13.** $3y^3 + 2y^2 - 10y$ (5.1.2) **14.** a^9b^{15} (5.2.2) **15.** $-8x^3y^6$ (5.2.1) **16.** $6y^4 + 8y^3 - 16y^2$ (5.3.1) **17.** $10a^3 - 39a^2 + 20a - 21$ (5.3.2) **18.** $15b^2 - 31b + 14$ (5.3.3) **19.** $9b^2 + 12b + 4$ (5.3.4) **20.** $\frac{1}{2b^2}$ (5.4.1) **21.** $6a - 4 + \frac{2}{a^2}$ (5.4.2) **22.** $a - 7$ (5.4.3) **23.** $\frac{6y^2}{x^6}$ (5.5.1) **24.** $8n - 2n = 18; n = 3$ (3.4.1) **25.** The selling price is \$43.20. (4.3.1) **26.** The resulting mixture is 28% orange juice. (4.5.2) **27.** The car overtakes the cyclist 25 mi from the starting point. (4.6.1) **28.** The length is 15 m. The width is 6 m. (4.7.1) **29.** Twenty years ago, the gold coin was twice the age of the silver coin. (4.8.3) **30.** The area is $4x^2 + 12x + 9$. (5.3.5)

Answers to Chapter 6

SECTION 1 Pages 211–212

Example 2

$12x^3y^6 = 2 \cdot 2 \cdot 3 \cdot x^3y^6$

$15x^2y^3 = 3 \cdot 5 \cdot x^2 \cdot y^3$

The GCF is $3x^2y^3$.

Example 4

$14a^2 = 2 \cdot 7 \cdot a^2$

$21a^4b = 3 \cdot 7 \cdot a^4b$

The GCF is $7a^2$.

$14a^2 - 21a^4b = 7a^2(2) + 7a^2(-3a^2b) =$
$7a^2(2 - 3a^2b)$

Example 6

$6x^4y^2 = 2 \cdot 3 \cdot x^4 \cdot y^2$

$9x^3y^2 = 3 \cdot 3 \cdot x^3 \cdot y^2$

$12x^2y^4 = 2 \cdot 2 \cdot 3 \cdot x^2 \cdot y^4$

The GCF is $3x^2y^2$.

$6x^4y^2 - 9x^3y^2 + 12x^2y^4 =$
$3x^2y^2(2x^2) + 3x^2y^2(-3x) + 3x^2y^2(4y^2) =$
$3x^2y^2(2x^2 - 3x + 4y^2)$

Pages 213–214

1. x^3 **3.** xy^4 **5.** xy^4z^2 **7.** ab^2c^3 **9.** $3x^2$ **11.** $2a$ **13.** $7a^3$ **15.** There is no common factor other than 1. **17.** $3a^2b^2$ **19.** ab **21.** $2x$ **23.** $3x$ **25.** $5(a + 1)$ **27.** $8(2 - a^2)$ **29.** $4(2x + 3)$ **31.** $6(5a - 1)$ **33.** $x(7x - 3)$ **35.** $a^2(3 + 5a^3)$ **37.** $y(14y + 11)$ **39.** $2x(x^3 - 2)$ **41.** $2x^2(5x^2 - 6)$ **43.** $4a^5(2a^3 - 1)$ **45.** $xy(xy - 1)$ **47.** $3xy(xy^3 - 2)$ **49.** $xy(x - y^2)$ **51.** There is no common factor other than 1. **53.** $6b^2(a^2b - 2)$ **55.** $a(a^2 - 3a + 5)$ **57.** $5(x^2 - 3x + 7)$ **59.** $3x(x^2 + 2x + 3)$ **61.** $2x^2(x^2 - 2x + 3)$ **63.** $2x(x^2 + 3x - 7)$ **65.** $y^3(2y^2 - 3y + 7)$ **67.** $xy(x^2 - 3xy + 7y^2)$ **69.** $5y(y^2 + 2y - 5)$ **71.** $3b^2(a^2 - 3a + 5)$

SECTION 2 Pages 215–218

Example 2

$(x - \blacksquare)(x - \blacksquare)$

Factors	Sums
−1, −20	−21
−2, −10	−12
−4, −5	−9

$(x - 4)(x - 5)$

$x^2 - 9x + 20 = (x - 4)(x - 5)$

Example 4

$(x + \blacksquare)(x - \blacksquare)$

Factors	Sums
+1, −18	−17
−1, +18	17
+2, −9	−7
−2, +9	7
+3, −6	−3
−3, +6	3

$(x + 6)(x - 3)$

$x^2 + 3x - 18 = (x + 6)(x - 3)$

Example 6
The GCF is $3b$.

$3a^2b - 18ab - 81b = 3b(a^2 - 6a - 27)$

Factor the trinomial.

$3b(a + \quad)(a - \quad)$

Factors	Sums
+1, −27	−26
−1, +27	26
+3, −9	−6
−3, +9	6

$3b(a + 3)(a - 9)$

$3a^2b - 18ab - 81b = 3b(a + 3)(a - 9)$

Example 8
The GCF is 3.

$3x^2 - 9xy - 12y^2 = 3(x^2 - 3xy - 4y^2)$

Factor the trinomial.

$3(x + \quad y)(x - \quad y)$

Factors	Sums
+1, −4	−3
−1, +4	3
+2, −2	0

$3(x + y)(x - 4y)$

$3x^2 - 9xy - 12y^2 = 3(x + y)(x - 4y)$

Pages 219–222

1. $(x + 1)(x + 2)$ **3.** $(x + 1)(x - 2)$ **5.** $(a + 4)(a - 3)$ **7.** $(a - 2)(a - 1)$ **9.** $(a + 2)(a - 1)$ **11.** $(b - 3)(b - 3)$ **13.** $(b + 8)(b - 1)$ **15.** $(y + 11)(y - 5)$ **17.** $(y - 2)(y - 3)$ **19.** $(z - 5)(z - 9)$ **21.** $(z + 8)(z - 20)$ **23.** $(p + 3)(p + 9)$ **25.** $(x + 10)(x + 10)$ **27.** $(b + 4)(b + 5)$ **29.** $(x + 3)(x - 14)$ **31.** $(b + 4)(b - 5)$ **33.** $(y + 3)(y - 17)$ **35.** $(p + 3)(p - 7)$ **37.** Irreducible over the integers. **39.** $(x - 5)(x - 15)$ **41.** $(x - 7)(x - 8)$ **43.** $(x + 8)(x - 7)$ **45.** $(a + 3)(a - 24)$ **47.** $(a - 3)(a - 12)$ **49.** $(z + 8)(z - 17)$ **51.** $(c + 9)(c - 10)$ **53.** $(z + 4)(z + 11)$ **55.** $(c + 2)(c + 17)$ **57.** $(x + 8)(x - 12)$ **59.** $(x - 8)(x - 14)$ **61.** $(b + 15)(b - 7)$ **63.** $(a + 3)(a - 12)$ **65.** $(b - 6)(b - 17)$ **67.** $(a + 3)(a + 24)$ **69.** $(x + 12)(x + 13)$ **71.** $(x + 6)(x - 16)$ **73.** $2(x + 1)(x + 2)$ **75.** $3(a + 3)(a - 2)$ **77.** $a(b + 5)(b - 3)$ **79.** $x(y - 2)(y - 3)$ **81.** $z(z - 3)(z - 4)$ **83.** $3y(y - 2)(y - 3)$ **85.** $3(x + 4)(x - 3)$ **87.** $5(z + 4)(z - 7)$ **89.** $2a(a + 8)(a - 4)$ **91.** $(x - 2y)(x - 3y)$ **93.** $(a - 4b)(a - 5b)$ **95.** $(x + 4y)(x - 7y)$ **97.** Irreducible over the integers. **99.** $z^2(z - 5)(z - 7)$ **101.** $b^2(b - 10)(b - 12)$ **103.** $2y^2(y + 3)(y - 16)$ **105.** $x^2(x + 8)(x - 1)$ **107.** $4y(x + 7)(x - 2)$ **109.** $8(y - 1)(y - 3)$ **111.** $c(c + 3)(c + 10)$ **113.** $3x(x - 3)(x - 9)$ **115.** $(x - 3y)(x - 5y)$ **117.** $(a - 6b)(a - 7b)$ **119.** $(y + z)(y + 7z)$ **121.** $3y(x + 21)(x - 1)$ **123.** $3x(x - 3)(x + 4)$ **125.** $4z(z + 11)(z - 3)$ **127.** $4x(x + 3)(x - 1)$ **129.** $5(p + 12)(p - 7)$ **131.** $p^2(p + 12)(p - 3)$ **133.** $(t - 5s)(t - 7s)$ **135.** $(a + 3b)(a - 11b)$ **137.** $5x^2(x - 2)(x - 4)$ **139.** $15a(b + 4)(b - 1)$ **141.** $3y(x + 15)(x - 3)$

SECTION 3 Pages 223–226

Example 2
$(\blacksquare x + \blacksquare)(\blacksquare x - \blacksquare)$ or $(\blacksquare x - \blacksquare)(\blacksquare x + \blacksquare)$

Factors of 2: 1,2 Factors of -3: $+1, -3$
$-1, +3$

Trial Factors	Middle Term
$(1x + 1)(2x - 3)$	$-3x + 2x = -x$
$(1x - 3)(2x + 1)$	$x - 6x = -5x$
$(1x - 1)(2x + 3)$	$3x - 2x = x$
$(1x + 3)(2x - 1)$	$-x + 6x = 5x$

$(x + 1)(2x - 3)$

$2x^2 - x - 3 = (x + 1)(2x - 3)$

Example 4
The GCF is $2a^2$.

$4a^2b^2 + 26a^2b - 14a^2 = 2a^2(2b^2 + 13b - 7)$

Factor the trinomial.

$2a^2(\blacksquare b + \blacksquare)(\blacksquare b - \blacksquare)$ or $2a^2(\blacksquare b - \blacksquare)(\blacksquare b + \blacksquare)$

Factors of 2: 1,2 Factors of -7: $+1, -7$
$-1, +7$

Trial Factors	Middle Term
$(1b + 1)(2b - 7)$	$-7b + 2b = -5b$
$(1b - 7)(2b + 1)$	$b - 14b = -13b$
$(1b - 1)(2b + 7)$	$7b - 2b = 5b$
$(1b + 7)(2b - 1)$	$-b + 14b = 13b$

$2a^2(b + 7)(2b - 1)$

$4a^2b^2 + 26a^2b - 14a^2 = 2a^2(b + 7)(2b - 1)$

Example 6
The GCF is $3y$.

$12y + 12y^2 - 45y^3 = 3y(4 + 4y - 15y^2)$

Factor the trinomial.

$3y(\blacksquare + \blacksquare y)(\blacksquare - \blacksquare y)$ or $3y(\blacksquare - \blacksquare y)(\blacksquare + \blacksquare y)$

Factors of 4: 1,4 Factors of -15: $+1, -15$
2,2 $-1, +15$
$+3, -5$
$-3, +5$

Trial Factors	Middle Term
$(1 + 1y)(4 - 15y)$	$-15y + 4y = -11y$
$(1 - 15y)(4 + 1y)$	$y - 60y = -59y$
$(1 - 1y)(4 + 15y)$	$15y - 4y = 11y$
$(1 + 15y)(4 - 1y)$	$-y + 60y = 59y$
$(1 + 3y)(4 - 5y)$	$-5y + 12y = 7y$
$(1 - 5y)(4 + 3y)$	$3y - 20y = -17y$
$(1 - 3y)(4 + 5y)$	$5y - 12y = -7y$
$(1 + 5y)(4 - 3y)$	$-3y + 20y = 17y$
$(2 + 1y)(2 - 15y)$	$-30y + 2y = -28y$
$(2 - 1y)(2 + 15y)$	$30y - 2y = 28y$
$(2 + 3y)(2 - 5y)$	$-10y + 6y = -4y$
$(2 - 3y)(2 + 5y)$	$10y - 6y = 4y$

$3y(2 - 3y)(2 + 5y)$

$12y + 12y^2 - 45y^3 = 3y(2 - 3y)(2 + 5y)$

Pages 227–230

1. $(x + 1)(2x + 1)$ **3.** $(y + 3)(2y + 1)$ **5.** $(a - 1)(2a - 1)$ **7.** $(b - 5)(2b - 1)$ **9.** $(x + 1)(2x - 1)$ **11.** $(x - 3)(2x + 1)$ **13.** $(t + 2)(2t - 5)$ **15.** $(p - 5)(3p - 1)$ **17.** $(3y - 1)(4y - 1)$ **19.** Irreducible over the integers. **21.** $(2t - 1)(3t - 4)$ **23.** $(x + 4)(8x + 1)$ **25.** Irreducible over the integers. **27.** $(3y + 1)(4y + 5)$ **29.** $(a + 7)(7a - 2)$ **31.** $(b - 4)(3b - 4)$ **33.** $(z - 14)(2z + 1)$ **35.** $(p + 8)(3p - 2)$ **37.** $(2x - 3)(3x - 4)$ **39.** $(b + 7)(5b - 2)$ **41.** $(2a - 3)(3a + 8)$ **43.** $(z + 2)(4z + 3)$ **45.** $(2p + 5)(11p - 2)$ **47.** $(y + 1)(8y + 9)$ **49.** $(3t + 1)(6t - 5)$ **51.** $(b + 12)(6b - 1)$ **53.** $(3x + 2)(3x + 2)$ **55.** $(2b - 3)(3b - 2)$ **57.** $(3b + 5)(11b - 7)$ **59.** $(3y - 4)(6y - 5)$ **61.** $(3a + 7)(5a - 3)$ **63.** $(2y - 5)(4y - 3)$ **65.** $(2z + 3)(4z - 5)$ **67.** Irreducible over the integers. **69.** $(2z - 5)(5z - 2)$ **71.** $(6z + 5)(6z + 7)$ **73.** $2(x + 1)(2x + 1)$ **75.** $5(y - 1)(3y - 7)$ **77.** $x(x - 5)(2x - 1)$ **79.** $b(a - 4)(3a - 4)$ **81.** Irreducible over the integers. **83.** $(x + y)(3x - 2y)$ **85.** $(a + 2b)(3a - b)$ **87.** $(y - 2z)(4y - 3z)$ **89.** $(3 - x)(4 + x)$ **91.** $(4 + z)(7 - z)$ **93.** $(8 + x)(1 - x)$ **95.** $3(x + 5)(3x - 4)$ **97.** $4(4y - 1)(5y - 1)$ **99.** $z(2z + 3)(4z + 1)$ **101.** $y(2x - 5)(3x + 2)$ **103.** $4(2x - 3)(3x - 2)$ **105.** $a^2(7a - 1)(5a + 2)$ **107.** $5(3b - 2)(b - 7)$ **109.** $(x - 7y)(3x - 5y)$ **111.** $3(8y - 1)(9y + 1)$ **113.** $(1 - x)(21 + x)$ **115.** $(3a - 2b)(5a + 7b)$ **117.** $z(3 - z)(11 + z)$ **119.** $2x(x + 1)(5x + 1)$ **121.** $5(t + 2)(2t - 5)$ **123.** $p(p - 5)(3p - 1)$ **125.** $2(z + 4)(13z - 3)$ **127.** $2y(y - 4)(5y - 2)$ **129.** $yz(z + 2)(4z - 3)$ **131.** $b^2(4b + 5)(5b + 4)$ **133.** $2a(2a - 3)(3a + 8)$ **135.** $p^2(3 - p)(12 + p)$ **137.** $y(2x - 7y)(4x - 5y)$ **139.** $xy(3x + 2)(3x + 2)$ **141.** $ab(2a - b)(a - 5b)$

SECTION 4

Pages 231–234

Example 2

$25a^2 - b^2 = (5a)^2 - b^2 = (5a + b)(5a - b)$

Example 4

$n^8 - 36 = (n^4)^2 - 6^2 = (n^4 + 6)(n^4 - 6)$

Example 6

$\sqrt{a^2} = a$ $\quad 2(10a) = 20a$

$\sqrt{100} = 10$

The trinomial is a perfect square.

$a^2 + 20a + 100 = (a + 10)^2$

Example 8

$\sqrt{25a^2} = 5a$ $\quad 2(5a \cdot 3b) = 30ab$

$\sqrt{9b^2} = 3b$

The trinomial is a perfect square.

$25a^2 - 30ab + 9b^2 = (5a - 3b)^2$

Example 10

$5x(2x + 3) - 4(2x + 3) = (2x + 3)(5x - 4)$

Example 12

$2y(5x - 2) - 3(2 - 5x) =$

$2y(5x - 2) + 3(5x - 2) = (5x - 2)(2y + 3)$

Example 14

The GCF is $3x$.

$12x^3 - 75x = 3x(4x^2 - 25)$

Factor the difference of two squares.

$3x(2x + 5)(2x - 5)$

Example 16

The common binomial factor is $b - 7$.

$a^2(b - 7) + (7 - b) =$

$a^2(b - 7) - (b - 7) = (b - 7)(a^2 - 1)$

Factor the difference of two squares.

$(b - 7)(a + 1)(a - 1)$

Example 18

The GCF is $4x$.

$4x^3 + 28x^2 - 120x = 4x(x^2 + 7x - 30)$

Factor the trinomial.

$4x(x + 10)(x - 3)$

6

Pages 235–238

1. $(x + 2)(x - 2)$ **3.** $(a + 9)(a - 9)$ **5.** $(2x + 1)(2x - 1)$ **7.** $(x^3 + 3)(x^3 - 3)$ **9.** $(5x + 1)(5x - 1)$ **11.** $(1 + 7x)(1 - 7x)$ **13.** Irreducible over the integers. **15.** $(x^2 + y)(x^2 - y)$ **17.** $(3x + 4y)(3x - 4y)$ **19.** $(xy + 2)(xy - 2)$ **21.** $(4 + xy)(4 - xy)$ **23.** $(y + 7)^2$ **25.** Irreducible over the integers. **27.** $(x - 6)^2$ **29.** $(x + 3y)^2$ **31.** $(5x + 1)^2$ **33.** $(3a + 1)^2$ **35.** $(2a - 5)^2$ **37.** $(3a - 7)^2$ **39.** $(2a - 3b)^2$ **41.** $(2y - 9z)^2$ **43.** $(a + b)(x + 2)$ **45.** $(b + 2)(x - y)$ **47.** $(x - 3)(z - 1)$ **49.** $(b - 2c)(x + y)$ **51.** $(x - 2)(a - 5)$ **53.** $(y - 2)(b - 2a)$ **55.** $(y - 3)(b - 3)$ **57.** $(x - y)(a + 2)$ **59.** $5(x - 1)(x + 1)$ **61.** $x(x + 2)^2$ **63.** $x^2(x + 7)(x - 5)$ **65.** $5(b + 3)(b + 12)$ **67.** Irreducible over the integers. **69.** $2y(x - 3)(x + 11)$ **71.** $x(x^2 - 6x - 5)$ **73.** $3(y^2 - 12)$ **75.** $(2a + 1)(10a + 1)$ **77.** $y^2(x - 8)(x + 1)$ **79.** $5(2a - 3b)(a + b)$ **81.** $2(5 - x)(5 + x)$ **83.** $b^2(a - 5)^2$ **85.** $ab(3a - b)(4a + b)$ **87.** $3a(2a - 1)^2$ **89.** $3(81 + a^2)$ **91.** $2a(2a - 5)(3a - 4)$ **93.** $a(2a + 5)^2$ **95.** $3b(3a - 1)^2$ **97.** $6(4 + x)(2 - x)$ **99.** $x^2(x - y)(x + y)$ **101.** $2a(3a + 2)^2$ **103.** $b(2 - 3a)(1 + 2a)$ **105.** $8x(3y + 1)^2$ **107.** $y^2(5 + x)(3 - x)$ **109.** $3(x - 3y)(x + 3y)$ **111.** $y(y - 3)(y + 3)$ **113.** $x^2y^2(3x - 5y)(5x + 4y)$ **115.** $2(x - 1)(a + b)$ **117.** $(x - 2)(x - 1)(x + 1)$ **119.** $(x - 2)(x + 2)(a + b)$ **121.** $(2 - x)(2 + x)(x - 5)$ **123.** $(x + 2)(x - 2)^2$

SECTION 5 Pages 239–242

Example 2

$2x(x + 7) = 0$

$2x = 0 \qquad x + 7 = 0$

$x = 0 \qquad x = -7$

The solutions are 0 and -7.

Example 4

$4x^2 - 9 = 0$

$(2x - 3)(2x + 3) = 0$

$2x - 3 = 0 \qquad 2x + 3 = 0$

$2x = 3 \qquad 2x = -3$

$x = \frac{3}{2} \qquad x = -\frac{3}{2}$

The solutions are $\frac{3}{2}$ and $-\frac{3}{2}$

Example 6

$(x + 2)(x - 7) = 52$

$x^2 - 5x - 14 = 52$

$x^2 - 5x - 66 = 0$

$(x + 6)(x - 11) = 0$

$x + 6 = 0 \qquad x - 11 = 0$

$x = -6 \qquad x = 11$

The solutions are -6 and 11.

Example 8

Strategy
First positive consecutive integer: n
Second positive consecutive integer: $n + 1$

The sum of the squares of two positive consecutive integers is 61.

Solution
$n^2 + (n + 1)^2 = 61$
$n^2 + n^2 + 2n + 1 = 61$
$2n^2 + 2n + 1 = 61$
$2n^2 + 2n - 60 = 0$
$2(n^2 + n - 30) = 0$
$2(n - 5)(n + 6) = 0$

$n - 5 = 0$ $\quad$ $n + 6 = 0$
$n = 5$ $\quad$ $n = -6$

Since -6 is not a positive integer, it is not a solution.

$n = 5$
$n + 1 = 5 + 1 = 6$

The two integers are 5 and 6.

Example 10

Strategy
Width $= x$
Length $= 2x + 4$

The area of a rectangle is 96 in.^{2}.
Use the equation $A = l \cdot w$.

Solution
$A = l \cdot w$
$96 = (2x + 4)x$
$96 = 2x^2 + 4x$
$0 = 2x^2 + 4x - 96$
$0 = 2(x^2 + 2x - 48)$
$0 = 2(x + 8)(x - 6)$

$x + 8 = 0$ $\quad$ $x - 6 = 0$
$x = -8$ $\quad$ $x = 6$

Since the width cannot be a negative number, -8 is not a solution.

$x = 6$
$2x + 4 = 2(6) + 4 = 12 + 4 = 16$

The width is 6 in.
The length is 16 in.

6

Pages 243–244

1. The solutions are -3 and -2. **3.** The solutions are 7 and 3. **5.** The solutions are 0 and 5. **7.** The solutions are 0 and 9. **9.** The solutions are 0 and $-\frac{3}{2}$. **11.** The solutions are 0 and $\frac{2}{3}$. **13.** The solutions are -2 and 5. **15.** The solutions are 9 and -9. **17.** The solutions are $\frac{7}{2}$ and $-\frac{7}{2}$. **19.** The solutions are $\frac{1}{3}$ and $-\frac{1}{3}$. **21.** The solutions are -2 and -4. **23.** The solutions are -7 and 2. **25.** The solutions are 2 and 3. **27.** The solutions are -7 and 3. **29.** The solutions are $-\frac{1}{2}$ and 5. **31.** The solutions are $-\frac{1}{3}$ and $-\frac{1}{2}$. **33.** The solutions are 0 and 3. **35.** The solutions are 0 and 7. **37.** The solutions are -1 and -4. **39.** The solutions are 2 and 3. **41.** The solutions are $\frac{1}{2}$ and -4. **43.** The solutions are $\frac{1}{3}$ and 4. **45.** The solutions are 3 and 9. **47.** The solutions are 9 and -2. **49.** The solutions are -1 and -2. **51.** The solutions are -9 and 5. **53.** The solutions are -7 and 4. **55.** The solutions are -2 and -3. **57.** The solutions are -5 and -8. **59.** The solutions are 1 and 3. **61.** The solutions are 5 and -12. **63.** The solutions are $-\frac{3}{2}$ and -2. **65.** The solutions are $-\frac{1}{2}$ and -4.

Pages 245–246

1. The number is 6. **3.** The numbers are 2 and 4. **5.** The numbers are 6 and 7. **7.** The numbers are 3 and 7. **9.** The number is 9 or -15. **11.** The numbers are 14 and 15. **13.** The height is 5 ft and the length is 20 ft. **15.** The width is 10 in. and the length is 30 in. **17.** The width is 5 in. and the length is 15 in. **19.** The length of a side of the original square is 10 in. **21.** The time is 4 s. **23.** The radius of the original circle is 3.8078556 in.

REVIEW/TEST Pages 249–250

1.1 $4ab^3$ **1.2** $2x(3x^2 - 4x + 5)$ **2.1a** $(p + 2)(p + 3)$ **2.1b** $(a - 3)(a - 16)$ **2.1c** $(x - 3)(x + 5)$ **2.1d** $(x + 3)(x - 12)$ **2.2a** $5(x^2 - 9x - 3)$ **2.2b** $2y^2(y - 8)(y + 1)$ **3.1a** Irreducible over the integers. **3.1b** $(2x + 1)(3x + 8)$ **3.2a** $4(2x - 3)(x + 4)$ **3.2b** $3y^2(2x^2 + 3x + 4)$ **4.1a** $(b + 4)(b - 4)$ **4.1b** $(2x - 7y)(2x + 7y)$ **4.2a** $(p + 6)^2$ **4.2b** $(2a - 3b)^2$ **4.3a** $(x - 2)(a + b)$ **4.3b** $(p + 1)(x - 1)$ **4.4a** $3(a - 5)(a + 5)$ **4.4b** $3(x + 2y)^2$ **5.1a** The solutions are -7 and $\frac{3}{2}$. **5.1b** The solutions are 3 and 5. **5.2** The width is 6 cm and the length is 15 cm.

CUMULATIVE REVIEW/TEST Pages 251–252

1. 7 (1.2.1) **2.** 4 (1.5.2) **3.** -7 (2.1.1) **4.** $15x^2$ (2.2.2) **5.** 12 (2.2.4) **6.** $x = \frac{2}{3}$ (3.1.3) **7.** $x = \frac{7}{4}$ (3.3.1) **8.** $x = 3$ (3.3.2) **9.** 45 (4.2.1) **10.** $9a^6b^4$ (5.2.2) **11.** $x^3 - 3x^2 - 6x + 8$ (5.3.2) **12.** $4x + 8 + \frac{21}{2x - 3}$ (5.4.3) **13.** $\frac{y^6}{x^8}$ (5.5.1) **14.** $6xy^2$ (6.1.1) **15.** $5xy^2(3 - 4y^2)$ (6.1.2) **16.** $(x - 7y)(x + 2y)$ (6.2.2) **17.** $(p - 10)(p + 1)$ (6.2.1) **18.** $3a(2a + 5)(3a + 2)$ (6.3.2) **19.** $(6a - 7b)(6a + 7b)$ (6.4.1) **20.** $(2x + 7y)^2$ (6.4.2) **21.** $(3x - 2)(3x + 7)$ (6.3.1) **22.** $2(3x - 4y)^2$ (6.4.4) **23.** $(x - 3)(3y - 2)$ (6.4.3) **24.** The solutions are $\frac{2}{3}$ and -7. (6.5.1) **25.** The pieces measure 4 ft and 6 ft. (3.4.2) **26.** The discount rate is 40%. (4.3.2) **27.** \$6500 more must be invested. (4.4.1) **28.** The distance to the resort was 168 mi. (4.6.1) **29.** The integers are 10, 12, and 14. (4.8.1) **30.** The length is 12 in. (6.5.2)

Answers to Chapter 7

SECTION 1 Pages 255–258

Example 2

$$\frac{6x^5y}{12x^2y^3} = \frac{\overset{1}{\cancel{2}} \cdot \overset{1}{\cancel{3}} \cdot x^5y}{\underset{1}{\cancel{2}} \cdot 2 \cdot \underset{1}{\cancel{3}} \cdot x^2y^3} = \frac{x^3}{2y^2}$$

Example 4

$$\frac{x^2 + 2x - 24}{16 - x^2} = \frac{\overset{-1}{\cancel{(x-4)}}(x+6)}{\underset{1}{\cancel{(4-x)}}(4+x)} = -\frac{x+6}{x+4}$$

Example 6

$$\frac{x^2 + 4x - 12}{x^2 - 3x + 2} = \frac{\overset{1}{\cancel{(x-2)}}(x+6)}{(x-1)\underset{1}{\cancel{(x-2)}}} = \frac{x+6}{x-1}$$

Example 8

$$\frac{12x^2 + 3x}{10x - 15} \cdot \frac{8x - 12}{9x + 18} =$$

$$\frac{3x(4x+1)}{5(2x-3)} \cdot \frac{4(2x-3)}{9(x+2)} =$$

$$\frac{\overset{1}{\cancel{3}}x(4x+1) \cdot 2 \cdot 2\overset{1}{\cancel{(2x-3)}}}{5\underset{1}{\cancel{(2x-3)}} \cdot \underset{1}{\cancel{3}} \cdot 3(x+2)} = \frac{4x(4x+1)}{15(x+2)}$$

Example 10

$$\frac{x^2 + 2x - 15}{9 - x^2} \cdot \frac{x^2 - 3x - 18}{x^2 - 7x + 6} =$$

$$\frac{(x-3)(x+5)}{(3-x)(3+x)} \cdot \frac{(x+3)(x-6)}{(x-1)(x-6)} =$$

$$\frac{\overset{-1}{\cancel{(x-3)}}(x+5) \cdot \overset{1}{\cancel{(x+3)}}\overset{1}{\cancel{(x-6)}}}{\underset{1}{\cancel{(3-x)}}\underset{1}{\cancel{(3+x)}} \cdot (x-1)\underset{1}{\cancel{(x-6)}}} = -\frac{x+5}{x-1}$$

Example 12

$$\frac{a^2}{4bc^2 - 2b^2c} \div \frac{a}{6bc - 3b^2} =$$

$$\frac{a^2}{4bc^2 - 2b^2c} \cdot \frac{6bc - 3b^2}{a} =$$

$$\frac{a^2 \cdot 3\overset{1}{\cancel{b}}\overset{1}{\cancel{(2c-b)}}}{2\underset{1}{\cancel{b}}c\underset{1}{\cancel{(2c-b)}} \cdot a} = \frac{3a}{2c}$$

Example 14

$$\frac{3x^2 + 26x + 16}{3x^2 - 7x - 6} \div \frac{2x^2 + 9x - 5}{x^2 + 2x - 15} =$$

$$\frac{3x^2 + 26x + 16}{3x^2 - 7x - 6} \cdot \frac{x^2 + 2x - 15}{2x^2 + 9x - 5} =$$

$$\frac{\overset{1}{\cancel{(3x+2)}}(x+8)}{\underset{1}{\cancel{(3x+2)}}\underset{1}{\cancel{(x-3)}}} \cdot \frac{\overset{1}{\cancel{(x+5)}}\overset{1}{\cancel{(x-3)}}}{(2x-1)\underset{1}{\cancel{(x+5)}}} = \frac{x+8}{2x-1}$$

Pages 259–262

1. $\frac{3}{4x}$ **3.** $\frac{1}{x+3}$ **5.** -1 **7.** $\frac{2}{3y}$ **9.** $\frac{-3}{4x}$ **11.** $\frac{a}{b}$ **13.** $-\frac{2}{x}$ **15.** $\frac{y-2}{y-3}$ **17.** $\frac{x+5}{x+4}$ **19.** $\frac{x+4}{x-3}$ **21.** $-\frac{x+2}{x+5}$ **23.** $\frac{2(x+2)}{x+3}$ **25.** $\frac{2x-1}{2x+3}$ **27.** $-\frac{x+7}{x+6}$ **29.** $\frac{5ab^2}{12x^2y}$ **31.** $\frac{4x^3y^3}{3a^2}$ **33.** $\frac{3}{4}$ **35.** ab^2 **37.** $\frac{x^2(x-1)}{y(x+3)}$ **39.** $\frac{y(x-1)}{x^2(x+10)}$ **41.** $-ab^2$ **43.** $\frac{x+5}{x+4}$ **45.** 1 **47.** $-\frac{n-10}{n-7}$ **49.** $\frac{x(x+2)}{2(x-1)}$ **51.** $-\frac{x+2}{x-6}$ **53.** $\frac{x+5}{x-12}$ **55.** $\frac{3y+2}{3y+1}$ **57.** $-\frac{3x-5}{4x-5}$ **59.** $\frac{7a^3y^2}{40bx}$ **61.** $\frac{4}{3}$ **63.** $\frac{3a}{2}$ **65.** $\frac{x^2(x+4)}{y^2(x+2)}$ **67.** $\frac{x(x-2)}{y(x-6)}$ **69.** $-\frac{3by}{ax}$ **71.** $\frac{(x-3)(x+6)}{(x+7)(x-6)}$ **73.** 1 **75.** $-\frac{x+8}{x-4}$ **77.** $\frac{2n+1}{2n-3}$ **79.** $-\frac{3x+1}{2x-3}$ **81.** $\frac{(3x+2)(5-4x)(5x-4)}{(2x-3)(4x+5)(3x-5)}$

SECTION 2 Pages 263–264

Example 2

$8uv^2 = 2 \cdot 2 \cdot 2 \cdot u \cdot v \cdot v$ $\quad 12uw = 2 \cdot 2 \cdot 3 \cdot u \cdot w$

$\text{LCM} = 2 \cdot 2 \cdot 2 \cdot 3 \cdot u \cdot v \cdot v \cdot w = 24uv^2w$

Example 4

$m^2 - 6m + 9 = (m - 3)(m - 3)$

$m^2 - 2m - 3 = (m + 1)(m - 3)$

$\text{LCM} = (m - 3)(m - 3)(m + 1)$

Example 6

The LCM is $36xy^2z$.

$\frac{x-3}{4xy^2} = \frac{x-3}{4xy^2} \cdot \frac{9z}{9z} = \frac{9xz - 27z}{36xy^2z}$

$\frac{2x+1}{9y^2z} = \frac{2x+1}{9y^2z} \cdot \frac{4x}{4x} = \frac{8x^2 + 4x}{36xy^2z}$

Example 8

The LCM is $(x - 5)(x + 5)(x - 2)$.

$\frac{2x}{25 - x^2} = \frac{2x}{-(x^2 - 25)} = -\frac{2x}{x^2 - 25}$

$\frac{x+4}{x^2 - 3x - 10} = \frac{x+4}{(x+2)(x-5)} \cdot \frac{x+5}{x+5} = \frac{x^2 + 9x + 20}{(x+2)(x-5)(x+5)}$

$\frac{2x}{25 - x^2} = -\frac{2x}{(x-5)(x+5)} \cdot \frac{x+2}{x+2} = -\frac{2x^2 + 4x}{(x+2)(x-5)(x+5)}$

Pages 265–266

1. $24x^3y^2$ **3.** $30x^4y^2$ **5.** $8x^2(x + 2)$ **7.** $6x^2y(x + 4)$ **9.** $36x(x + 2)^2$ **11.** $6(x + 1)^2$ **13.** $(x - 1)(x + 2)(x + 3)$ **15.** $(x - 5)(2x + 3)^2$ **17.** $(x - 1)(x - 2)$ **19.** $(x - 3)(x + 2)(x + 4)$ **21.** $(x + 4)(x + 1)(x - 7)$ **23.** $(x - 6)(x + 6)(x + 4)$ **25.** $(x + 3)(x - 8)(x - 10)$ **27.** $(x + 2)(x - 3)(3x - 2)$ **29.** $(x + 2)(x - 3)$ **31.** $(x - 5)(x + 1)$ **33.** $(x - 1)(x - 2)(x - 3)(x - 6)$ **35.** $\frac{5}{ab^2}, \frac{6b}{ab^2}$ **37.** $\frac{15y^2}{18x^2y}, \frac{14x}{18x^2y}$ **39.** $\frac{ay + 5a}{y^2(y + 5)}, \frac{6y}{y^2(y + 5)}$ **41.** $\frac{a^2y + 7a^2}{y(y + 7)^2}, \frac{ay}{y(y + 7)^2}$ **43.** $\frac{b}{y(y - 4)}, -\frac{b^2y}{y(y - 4)}$ **45.** $-\frac{3y - 21}{(y - 7)^2}, \frac{2}{(y - 7)^2}$ **47.** $\frac{2y^2}{y^2(y - 3)}, \frac{3}{y^2(y - 3)}$ **49.** $\frac{x^3 + 4x^2}{(2x - 1)(x + 4)}, \frac{2x^2 + x - 1}{(2x - 1)(x + 4)}$ **51.** $\frac{3x^2 + 15x}{(x - 5)(x + 5)}, \frac{4}{(x - 5)(x + 5)}$ **53.** $\frac{x - 3}{(3x - 2)(x + 2)}, \frac{6x - 4}{(3x - 2)(x + 2)}$ **55.** $\frac{x^2 - 1}{(x + 5)(x - 3)(x + 1)}, \frac{x^2 - 3x}{(x + 5)(x - 3)(x + 1)}$ **57.** $-\frac{2x^2 - 6x}{(x - 3)(x - 5)(x + 2)}, \frac{x^2 + 4x + 4}{(x - 3)(x - 5)(x + 2)}$ **59.** $\frac{x^2 - 6x - 7}{(x + 5)(x - 7)}, \frac{x^2 + 7x + 10}{(x + 5)(x - 7)}, \frac{-3}{(x + 5)(x - 7)}$

SECTION 3 Pages 267–270

Example 2

$\frac{3}{xy} + \frac{12}{xy} = \frac{3 + 12}{xy} = \frac{15}{xy}$

Example 4

$\frac{2x^2}{x^2 - x - 12} - \frac{7x + 4}{x^2 - x - 12} = \frac{2x^2 - (7x + 4)}{x^2 - x - 12} =$

$\frac{2x^2 - 7x - 4}{x^2 - x - 12} = \frac{(2x + 1)\overset{1}{\cancel{(x - 4)}}}{(x + 3)\underset{1}{\cancel{(x - 4)}}} = \frac{2x + 1}{x + 3}$

Example 6

$\frac{x^2 - 1}{x^2 - 8x + 12} - \frac{2x + 1}{x^2 - 8x + 12} + \frac{x}{x^2 - 8x + 12} =$

$\frac{(x^2 - 1) - (2x + 1) + x}{x^2 - 8x + 12} = \frac{x^2 - 1 - 2x - 1 + x}{x^2 - 8x + 12} =$

$\frac{x^2 - x - 2}{x^2 - 8x + 12} = \frac{(x + 1)\overset{1}{\cancel{(x - 2)}}}{\underset{1}{\cancel{(x - 2)}}(x - 6)} = \frac{x + 1}{x - 6}$

Example 8

The LCM of the denominators is $24y$.

$\frac{z}{8y} = \frac{z}{8y} \cdot \frac{3}{3} = \frac{3z}{24y}$ $\quad \frac{4z}{3y} = \frac{4z}{3y} \cdot \frac{8}{8} = \frac{32z}{24y}$

$\frac{5z}{4y} = \frac{5z}{4y} \cdot \frac{6}{6} = \frac{30z}{24y}$

$\frac{z}{8y} - \frac{4z}{3y} + \frac{5z}{4y} = \frac{3z}{24y} - \frac{32z}{24y} + \frac{30z}{24y} =$

$\frac{3z - 32z + 30z}{24y} = \frac{z}{24y}$

Example 10

The LCM is $x - 2$.

$$\frac{5x}{x-2} = \frac{5x}{x-2}\cdot\frac{1}{1} = \frac{5x}{x-2}$$

$$\frac{3}{2-x} = \frac{3}{-(x-2)}\cdot\frac{-1}{-1} = \frac{-3}{x-2}$$

$$\frac{5x}{x-2} - \frac{3}{2-x} = \frac{5x}{x-2} - \frac{-3}{x-2} =$$

$$\frac{5x-(-3)}{x-2} = \frac{5x+3}{x-2}$$

Example 12

The LCM is $(3x - 1)(x + 4)$.

$$\frac{4x}{3x-1} = \frac{4x}{3x-1}\cdot\frac{x+4}{x+4} = \frac{4x^2+16x}{(3x-1)(x+4)}$$

$$\frac{9}{x+4} = \frac{9}{x+4}\cdot\frac{3x-1}{3x-1} = \frac{27x-9}{(3x-1)(x+4)}$$

$$\frac{4x}{3x-1} - \frac{9}{x+4} =$$

$$\frac{4x^2+16x}{(3x-1)(x+4)} - \frac{27x-9}{(3x-1)(x+4)} =$$

$$\frac{4x^2+16x-(27x-9)}{(3x-1)(x+4)} = \frac{4x^2+16x-27x+9}{(3x-1)(x+4)} =$$

$$\frac{4x^2-11x+9}{(3x-1)(x+4)}$$

Example 14

The LCM is $(x + 5)(x - 5)$.

$$\frac{2x-1}{x^2-25} = \frac{2x-1}{(x+5)(x-5)}$$

$$\frac{2}{5-x} = \frac{2}{-(x-5)}\cdot\frac{-1\cdot(x+5)}{-1\cdot(x+5)} = \frac{-2(x+5)}{(x+5)(x-5)}$$

$$\frac{2x-1}{x^2-25} + \frac{2}{5-x} =$$

$$\frac{2x-1}{(x+5)(x-5)} + \frac{-2(x+5)}{(x+5)(x-5)} =$$

$$\frac{2x-1+(-2)(x+5)}{(x+5)(x-5)} = \frac{2x-1-2x-10}{(x+5)(x-5)} =$$

$$\frac{-11}{(x+5)(x-5)} = -\frac{11}{(x+5)(x-5)}$$

Example 16

The LCM is $(3x + 2)(x - 1)$.

$$\frac{2x-3}{3x^2-x-2} = \frac{2x-3}{(3x+2)(x-1)}$$

$$\frac{5}{3x+2} = \frac{5}{3x+2}\cdot\frac{x-1}{x-1} = \frac{5x-5}{(3x+2)(x-1)}$$

$$\frac{1}{x-1} = \frac{1}{x-1}\cdot\frac{3x+2}{3x+2} = \frac{3x+2}{(3x+2)(x-1)}$$

$$\frac{2x-3}{3x^2-x-2} + \frac{5}{3x+2} - \frac{1}{x-1} =$$

$$\frac{2x-3}{(3x+2)(x-1)} + \frac{5x-5}{(3x+2)(x-1)} - \frac{3x+2}{(3x+2)(x-1)} =$$

$$\frac{(2x-3)+(5x-5)-(3x+2)}{(3x+2)(x-1)} =$$

$$\frac{2x-3+5x-5-3x-2}{(3x+2)(x-1)} =$$

$$\frac{4x-10}{(3x+2)(x-1)} = \frac{2(2x-5)}{(3x+2)(x-1)}$$

Pages 271–274

1. $\frac{11}{y^2}$ **3.** $-\frac{7}{x+4}$ **5.** $\frac{8x}{2x+3}$ **7.** $\frac{5x+7}{x-3}$ **9.** $\frac{2x-5}{x+9}$ **11.** $\frac{-3x-4}{2x+7}$ **13.** $\frac{1}{x+5}$ **15.** $\frac{1}{x-6}$ **17.** $\frac{3}{2y-1}$ **19.** $\frac{1}{x-5}$ **21.** $\frac{4y+5x}{xy}$ **23.** $\frac{19}{2x}$ **25.** $\frac{5}{12x}$ **27.** $\frac{19x-12}{6x^2}$ **29.** $\frac{52y-35x}{20xy}$ **31.** $\frac{13x+2}{15x}$ **33.** $\frac{7}{24}$ **35.** $\frac{x+90}{45x}$ **37.** $\frac{x^2+2x+2}{2x^2}$ **39.** $\frac{2x^2+3x-10}{4x^2}$ **41.** $\frac{-3x^2+16x+2}{12x^2}$ **43.** $\frac{x^2-x-2}{x^2y}$ **45.** $\frac{16xy-12y+6x^2+3x}{12x^2y^2}$ **47.** $\frac{3xy-6y-2x^2-14x}{24x^2y}$ **49.** $\frac{9x+2}{(x-2)(x+3)}$ **51.** $\frac{2(x+23)}{(x-7)(x+3)}$ **53.** $\frac{2x^2-5x+1}{(x+1)(x-3)}$ **55.** $\frac{4x^2-34x+5}{(2x-1)(x-6)}$ **57.** $\frac{2a-5}{a-7}$ **59.** $\frac{4x+9}{(x-3)(x+3)}$ **61.** $\frac{-x+9}{(x+2)(x-3)}$ **63.** $\frac{14}{(x-5)(x-5)}$ **65.** $\frac{-2(x+7)}{(x+6)(x-7)}$ **67.** $\frac{x-4}{x-6}$ **69.** $\frac{2x+1}{x-1}$ **71.** $\frac{-3(x^2+8x+25)}{(x-3)(x+7)}$

7

SECTION 4

Pages 275–276

Example 2

The LCM of 3, x, 9, and x^2 is $9x^2$.

$$\frac{\frac{1}{3}-\frac{1}{x}}{\frac{1}{9}-\frac{1}{x^2}} = \frac{\frac{1}{3}-\frac{1}{x}}{\frac{1}{9}-\frac{1}{x^2}}\cdot\frac{9x^2}{9x^2} = \frac{\frac{1}{3}\cdot 9x^2-\frac{1}{x}\cdot 9x^2}{\frac{1}{9}\cdot 9x^2-\frac{1}{x^2}\cdot 9x^2} =$$

$$\frac{3x^2-9x}{x^2-9} = \frac{3x\overset{1}{\cancel{(x-3)}}}{\underset{1}{\cancel{(x-3)}}(x+3)} = \frac{3x}{x+3}$$

Example 4

The LCM of x and x^2 is x^2.

$$\frac{1+\frac{4}{x}+\frac{3}{x^2}}{1+\frac{10}{x}+\frac{21}{x^2}} = \frac{1+\frac{4}{x}+\frac{3}{x^2}}{1+\frac{10}{x}+\frac{21}{x^2}}\cdot\frac{x^2}{x^2} =$$

$$\frac{1\cdot x^2+\frac{4}{x}\cdot x^2+\frac{3}{x^2}\cdot x^2}{1\cdot x^2+\frac{10}{x}\cdot x^2+\frac{21}{x^2}\cdot x^2} = \frac{x^2+4x+3}{x^2+10x+21} =$$

$$\frac{(x+1)\overset{1}{\cancel{(x+3)}}}{\underset{1}{\cancel{(x+3)}}(x+7)} = \frac{x+1}{x+7}$$

Example 6

The LCM is $x-5$.

$$\frac{x+3-\frac{20}{x-5}}{x+8+\frac{30}{x-5}} = \frac{x+3-\frac{20}{x-5}}{x+8+\frac{30}{x-5}}\cdot\frac{x-5}{x-5} =$$

$$\frac{(x+3)(x-5)-\frac{20}{x-5}\cdot(x-5)}{(x+8)(x-5)+\frac{30}{x-5}\cdot(x-5)} =$$

$$\frac{x^2-2x-15-20}{x^2+3x-40+30} = \frac{x^2-2x-35}{x^2+3x-10} =$$

$$\frac{\overset{1}{\cancel{(x+5)}}(x-7)}{(x-2)\underset{1}{\cancel{(x+5)}}} = \frac{x-7}{x-2}$$

Pages 277–278

1. $\frac{x}{x-3}$ **3.** $\frac{2}{3}$ **5.** $\frac{y+3}{y-4}$ **7.** $\frac{2(2x+13)}{5x+36}$ **9.** $\frac{x+2}{x+3}$ **11.** $\frac{x-6}{x+5}$ **13.** $-\frac{x-2}{x+1}$ **15.** $x-1$ **17.** $\frac{1}{2x-1}$
19. $\frac{x-3}{x+5}$ **21.** $\frac{x-7}{x-8}$ **23.** $\frac{2y-1}{2y+1}$ **25.** $\frac{x-2}{2x-5}$ **27.** $-\frac{x+1}{4x-3}$ **29.** $\frac{x+1}{2(5x-2)}$ **31.** $\frac{b+11}{4b-21}$

SECTION 5

Pages 279–280

Example 2

$$\frac{x}{x+6} = \frac{3}{x} \quad \text{The LCM is } x(x+6).$$

$$\frac{x\overset{1}{\cancel{(x+6)}}}{1}\cdot\frac{x}{\underset{1}{\cancel{x+6}}} = \frac{x(x+6)}{1}\cdot\frac{3}{x}$$

$$x^2 = (x+6)3$$
$$x^2 = 3x+18$$
$$x^2-3x-18 = 0$$
$$(x+3)(x-6) = 0$$
$$x+3 = 0 \qquad x-6 = 0$$
$$x = -3 \qquad x = 6$$

Both -3 and 6 check as solutions.
The solutions are -3 and 6.

Example 4

$$\frac{5x}{x+2} = 3-\frac{10}{x+2} \quad \text{The LCM is } x+2.$$

$$\frac{(x+2)}{1}\cdot\frac{5x}{x+2} = \frac{(x+2)}{1}\left(3-\frac{10}{x+2}\right)$$

$$\frac{\overset{1}{\cancel{x+2}}}{1}\cdot\frac{5x}{\underset{1}{\cancel{x+2}}} = \frac{x+2}{1}\cdot 3-\frac{\overset{1}{\cancel{x+2}}}{1}\cdot\frac{10}{\underset{1}{\cancel{x+2}}}$$

$$5x = (x+2)3-10$$
$$5x = 3x+6-10$$
$$5x = 3x-4$$
$$2x = -4$$
$$x = -2$$

-2 does not check as a solution.
The equation has no solution.

Pages 281–282

1. The solution is 3. **3.** The solution is 1. **5.** The solution is 9. **7.** The solution is 1. **9.** The solution is $\frac{1}{4}$. **11.** The solution is 1. **13.** The solution is -3. **15.** The solution is $\frac{1}{2}$. **17.** The solution is 8. **19.** The solution is 5. **21.** The solution is -1. **23.** The solution is 5. **25.** The equation has no solution. **27.** The solutions are 2 and 4. **29.** The solutions are $-\frac{3}{2}$ and 4. **31.** The solution is 3. **33.** The solution is 4. **35.** The solution is -1.

SECTION **6** Pages 283–284

Example 2

$$\frac{2}{x+3} = \frac{6}{5x+5}$$

$$\frac{(x+3)(5x+5)}{1} \cdot \frac{2}{x+3} = \frac{(x+3)(5x+5)}{1} \cdot \frac{6}{5x+5}$$

$$\frac{\overset{1}{\cancel{(x+3)}}(5x+5)}{1} \cdot \frac{2}{\underset{1}{\cancel{x+3}}} = \frac{(x+3)\overset{1}{\cancel{(5x+5)}}}{1} \cdot \frac{6}{\underset{1}{\cancel{5x+5}}}$$

$$\begin{aligned} (5x+5)2 &= (x+3)6 \\ 10x + 10 &= 6x + 18 \\ 4x + 10 &= 18 \\ 4x &= 8 \\ x &= 2 \end{aligned}$$

The solution is 2.

Example 4

Strategy
To find the total area that 256 ceramic tiles will cover, write and solve a proportion using x to represent the number of square feet that 256 tiles will cover.

Solution

$$\frac{9}{16} = \frac{x}{256}$$

$$256\left(\frac{9}{16}\right) = 256\left(\frac{x}{256}\right)$$

$$144 = x$$

A 144-square-foot area can be tiled using 256 ceramic tiles.

Example 6

Strategy
To find the additional amount of medication required for a 200-pound adult, write and solve a proportion using x to represent the additional medication. Then $3 + x$ is the total amount required for a 200-pound adult.

Solution

$$\frac{150}{3} = \frac{200}{3+x}$$

$$\frac{50}{1} = \frac{200}{3+x}$$

$$(3+x) \cdot 50 = (3+x) \cdot \frac{200}{3+x}$$

$$\begin{aligned} (3+x) \cdot 50 &= 200 \\ 150 + 50x &= 200 \\ 50x &= 50 \\ x &= 1 \end{aligned}$$

One additional ounce is required for a 200-pound adult.

7

Page 285

1. The solution is 9. **3.** The solution is 12. **5.** The solution is 7. **7.** The solution is 6. **9.** The solution is 1. **11.** The solution is -6. **13.** The solution is 4. **15.** The solution is $-\frac{2}{3}$.

Pages 285–286

1. 10 lb of salt are required for 25 gal of water. **3.** 10,000 people are expected to vote. **5.** 150 ft^2 of decking can be made from 36 pieces of lumber. **7.** The license fee is \$90. **9.** There are 240 deer in the preserve. **11.** To earn a dividend of \$186, 50 additional shares are needed. **13.** $1\frac{1}{2}$ additional gal of fruit punch are necessary.

SECTION 7 Pages 287–288

Example 2

$$5x - 2y = 10$$
$$5x + (-5x) - 2y = (-5x) + 10$$
$$-2y = -5x + 10$$
$$-\tfrac{1}{2}(-2y) = -\tfrac{1}{2}(-5x + 10)$$
$$y = \tfrac{5}{2}x - 5$$

Example 4

$$s = \frac{A + L}{2}$$
$$2 \cdot s = 2\left(\frac{A + L}{2}\right)$$
$$2s = A + L$$
$$2s + (-A) = A + (-A) + L$$
$$2s - A = L$$

Example 6

$$S = a + (n - 1)d$$
$$S = a + nd - d$$
$$S + (-a) = a + (-a) + nd - d$$
$$S - a = nd - d$$
$$S - a + d = nd - d + d$$
$$S - a + d = nd$$
$$\tfrac{1}{d}(S - a + d) = \tfrac{1}{d}(nd)$$
$$\frac{S - a + d}{d} = n$$

Example 8

$$S = C + rC$$
$$S = (1 + r)C$$
$$\frac{1}{1 + r} \cdot S = \frac{1}{1 + r}(1 + r)C$$
$$\frac{S}{1 + r} = C$$

Pages 289–290

1. $y = -3x + 10$ **3.** $y = 4x - 3$ **5.** $y = -\frac{3}{2}x + 3$ **7.** $y = \frac{2}{5}x - 2$ **9.** $y = -\frac{2}{7}x + 2$ **11.** $y = -\frac{1}{3}x + 2$ **13.** $y = \frac{1}{4}x - 3$ **15.** $y = \frac{7}{2}x - 7$ **17.** $y = 3x + 7$ **19.** $x = -3y + 6$ **21.** $x = \frac{1}{3}y + 1$ **23.** $x = -\frac{5}{2}y + 5$ **25.** $x = 2y - 1$ **27.** $x = -\frac{4}{5}y - 4$ **29.** $x = \frac{2}{3}y + 5$ **31.** $h = \frac{2A}{b}$ **33.** $t = \frac{d}{r}$ **35.** $T = \frac{PV}{nR}$ **37.** $l = \frac{P - 2w}{2}$ **39.** $b_1 = \frac{2A - hb_2}{h}$ **41.** $h = \frac{3V}{A}$ **43.** $S = C - Rt$ **45.** $P = \frac{A}{1 + rt}$ **47.** $W = \frac{A}{S + 1}$

SECTION 8 Pages 291–294

Example 2

Strategy

▷ Time for one printer to complete the job: t

	Rate	Time	Part
1st printer	$\frac{1}{t}$	2	$\frac{2}{t}$
2nd printer	$\frac{1}{t}$	5	$\frac{5}{t}$

▷ The sum of the parts of the task completed must equal 1.

Solution

$$\frac{2}{t} + \frac{5}{t} = 1$$
$$t\left(\frac{2}{t} + \frac{5}{t}\right) = t \cdot 1$$
$$2 + 5 = t$$
$$7 = t$$

Working alone, one printer takes 7 h to print the payroll.

Example 4

Strategy

▷ Rate sailing across the lake: r
Rate sailing back: $3r$

	Distance	Rate	Time
Across	6	r	$\frac{6}{r}$
Back	6	$3r$	$\frac{6}{3r}$

▷ The total time for the trip was 2 h.

Solution

$$\frac{6}{r} + \frac{6}{3r} = 2$$
$$3r\left(\frac{6}{r} + \frac{6}{3r}\right) = 3r(2)$$
$$3r \cdot \frac{6}{r} + 3r \cdot \frac{6}{3r} = 6r$$
$$18 + 6 = 6r$$
$$24 = 6r$$
$$4 = r$$

The rate across the lake was 4 km/h.

Pages 295–298

1. It will take 2 h to fill the fountain with both sprinklers operating. **3.** With both members working together, it would take 12 min to mow the lawn. **5.** It will take 30 min to print the report with both computers operating. **7.** It would take the new machine 12 h to complete the task. **9.** The assistant working alone would take 90 min. **11.** It would take the cold water faucet 10 min to fill the tub. **13.** It would take the apprentice 3 h to complete the installation. **15.** It would take the second welder 2 h to complete the welds. **17.** Working alone the smaller pipe would take $14\frac{2}{3}$ h to fill the pool. **19.** Working alone the small compressor would take 12 h to cool the storage locker. **21.** The freight train travels at 30 mph and the express train travels at 50 mph. **23.** The rate of the twin-engine plane is 200 mph. **25.** The rate of the prop plane was 150 mph and the rate of the jet was 600 mph. **27.** The rate of the plane was 210 mph. **29.** The rate of the freight train was 30 mph, and the rate of the passenger train was 60 mph. **31.** The rate of the wind is 20 mph. **33.** The rate of the wind is 20 mph. **35.** The rate of the balloon on the first leg of the trip was 5 mph. **37.** The rate of the current is 3 mph. **39.** The rate of the plane for the first 200 mi was 100 mph.

REVIEW/TEST Pages 301–302

1.1a $\frac{2x^3}{3y^3}$ **1.1b** $-\frac{x+5}{x+1}$ **1.2** $\frac{x+1}{x^3(x-2)}$ **1.3** $\frac{x+5}{x+4}$ **2.1** $3(2x-1)(x+1)$ **2.2** $\frac{3x+6}{x(x-2)(x+2)}, \frac{x^2}{x(x-2)(x+2)}$ **3.1** $\frac{2}{x+5}$ **3.2** $\frac{5}{(2x-1)(3x+1)}$ **4.1** $\frac{x-3}{x-2}$ **5.1a** The solution is 2. **5.1b** No solution **6.1** The solution is -1. **6.2** An additional 2 lb of salt are required. **7.1a** $y = \frac{3}{8}x - 2$ **7.1b** $t = \frac{d-s}{r}$ **8.1** It would take 4 h to fill the pool with both pipes turned on. **8.2** The rate of the wind is 20 mph.

7

CUMULATIVE REVIEW/TEST Pages 303–304

1. $\frac{31}{30}$ (1.5.2) **2.** 21 (2.1.1) **3.** $5x - 2y$ (2.2.1) **4.** $-8x + 26$ (2.2.4) **5.** $x = -\frac{9}{2}$ (3.2.1)
6. $x = -12$ (3.2.2) **7.** 10 (4.2.1) **8.** a^3b^7 (5.2.1) **9.** $a^2 + ab - 12b^2$ (5.3.3) **10.** $3b^3 - b + 2$ (5.4.2)
11. $x^2 + 2x + 4$ (5.4.3) **12.** $(4x + 1)(3x - 1)$ (6.3.1) **13.** $(y - 6)(y - 1)$ (6.2.1)
14. $a(2a - 3)(a + 5)$ (6.3.2) **15.** $4(b + 5)(b - 5)$ (6.4.4) **16.** The solutions are -3 and $\frac{5}{2}$. (6.5.1)
17. $\frac{2x^3}{3y^5}$ (7.1.1) **18.** $-\frac{x - 2}{x + 5}$ (7.1.1) **19.** 1 (7.1.3) **20.** $\frac{3}{2x^2 + x - 1}$ (7.3.2) **21.** $\frac{x + 3}{x + 5}$ (7.4.1)
22. $x = 4$ (7.5.1) **23.** $x = 3$ (7.6.1) **24.** $t = \frac{f - v}{a}$ (7.7.1) **25.** $5n - 13 = -8$; $n = 1$ (3.4.1)
26. The 120 gram alloy is 70% silver. (4.5.2) **27.** The base is 10 in. and the height is 6 in. (6.5.2)
28. A policy of $5000 would cost $64 more. (7.6.2) **29.** Working together, it would take the pipes 6 min to fill the tank. (7.8.1) **30.** The rate of the current is 2 mph. (7.8.2)

Answers to Chapter 8

SECTION 1 Pages 307–310

Example 2

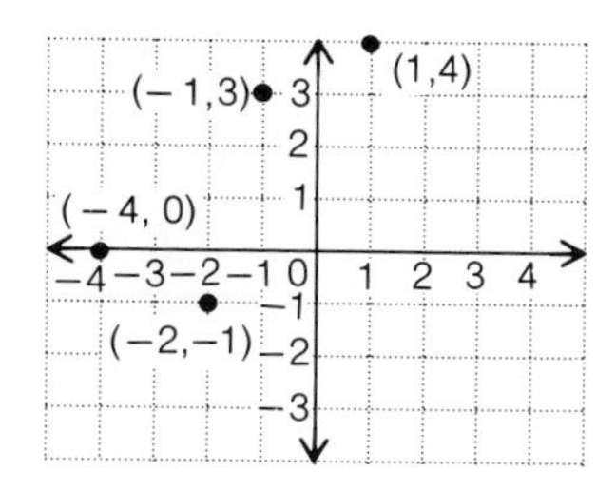

Example 4

$A(4,2)$
$B(-3,4)$
$C(-3,0)$
$D(0,0)$

Example 6

a) Abscissa of point A: 2
Abscissa of point C: -3

b) Ordinate of point B: -2
Ordinate of point D: 0

Example 8

$$y = -\tfrac{1}{2}x - 3$$

$$\begin{array}{c|l} -4 & -\tfrac{1}{2}(2) - 3 \\ & -1 - 3 \\ & -4 \end{array}$$

$-4 = -4$

Yes, $(2, -4)$ is a solution of $y = -\tfrac{1}{2}x - 3$.

Example 10

$$\begin{aligned} y &= -\tfrac{1}{4}x + 1 \\ &= -\tfrac{1}{4}(4) + 1 \\ &= -1 + 1 \\ &= 0 \end{aligned}$$

The ordered pair solution is $(4,0)$.

Example 12

Strategy Graph the ordered pairs $(1,8)$, $(3,10)$, and $(4,11)$.

Solution

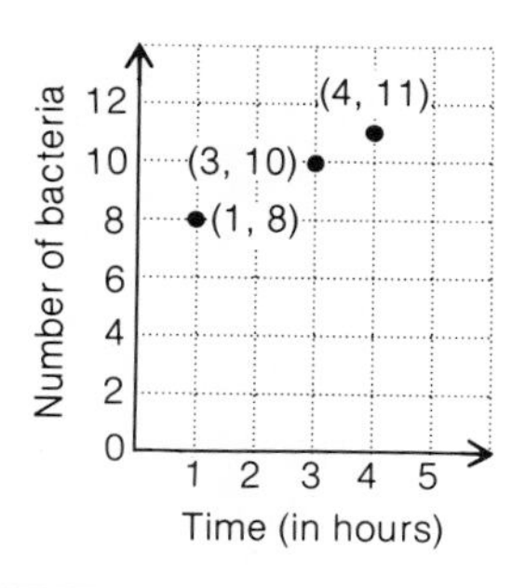

Pages 311–313

1.

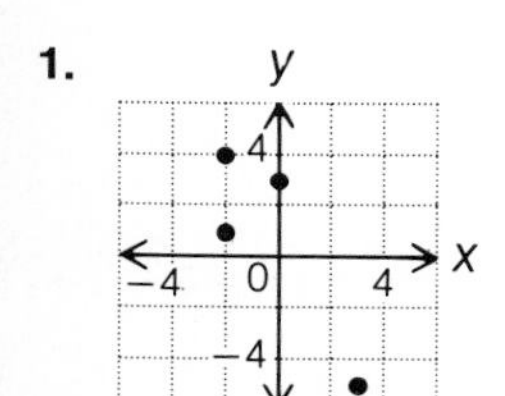

3.

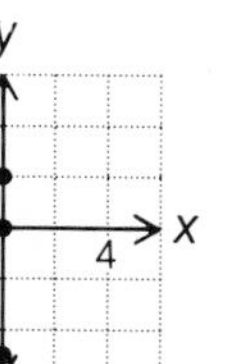

5.

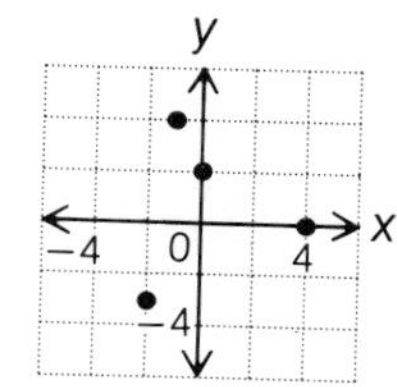

7. A is (2,3), B is (4,0), C is (−4,1) and D is (−2,−2). **9.** A is (−2,5), B is (3,4), C is (0,0) and D is (−3,−2). **11. a)** The abscissa of point A is 2. The abscissa of point C is −4. **b)** The ordinate of point B is 1. The ordinate of point D is −3. **13.** Yes, (3,4) is a solution of $y = -x + 7$. **15.** No, (−1,2) is not a solution of $y = \frac{1}{2}x - 1$. **17.** No, (4,1) is not a solution of $y = \frac{1}{4}x + 1$. **19.** Yes, (0,4) is a solution of $y = \frac{3}{4}x + 4$. **21.** No, (0,0) is not a solution of $y = 3x + 2$. **23.** The ordered pair solution is (3,7). **25.** The ordered pair solution is (6,3). **27.** The ordered pair solution is (0,1). **29.** The ordered pair solution is (−5,0).

Page 314

1.

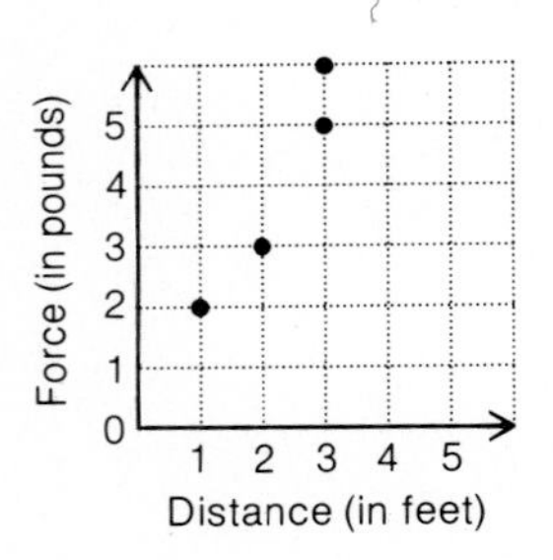

3.

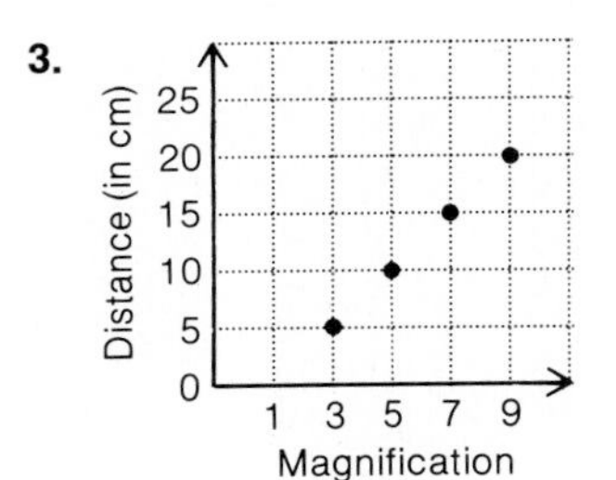

SECTION 2 Pages 315–318

Example 2

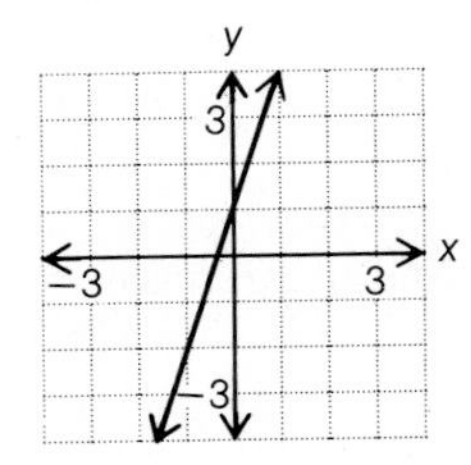

Example 4

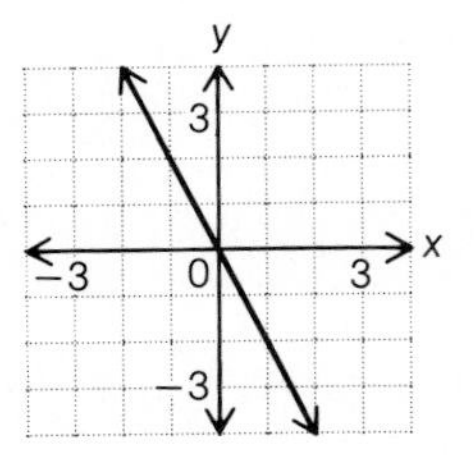

Example 6

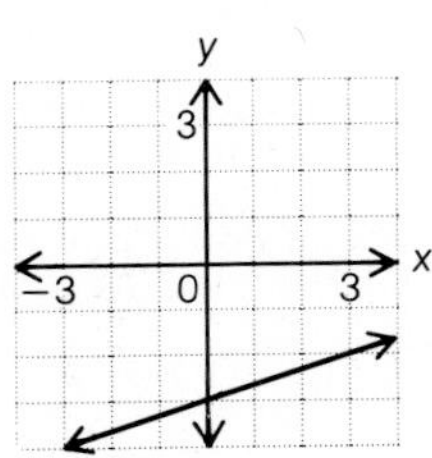

Example 8 $5x - 2y = 10$

$-2y = -5x + 10$

$y = \frac{5}{2}x - 5$

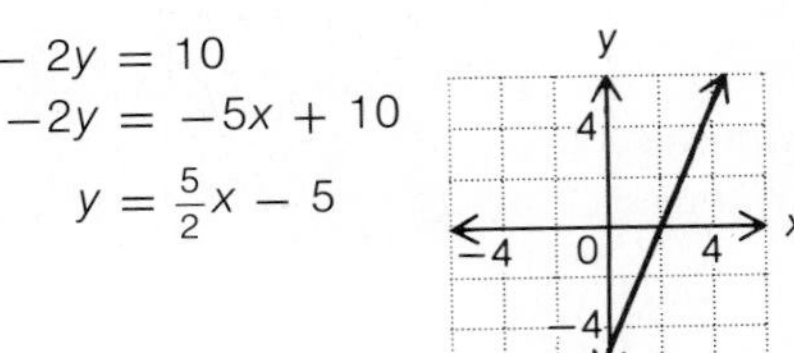

Example 10 $x - 3y = 9$

$-3y = -x + 9$

$y = \frac{1}{3}x - 3$

Example 12

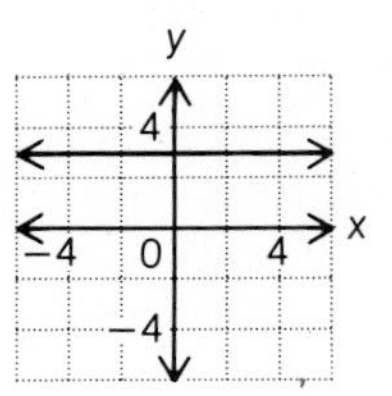

Example 14

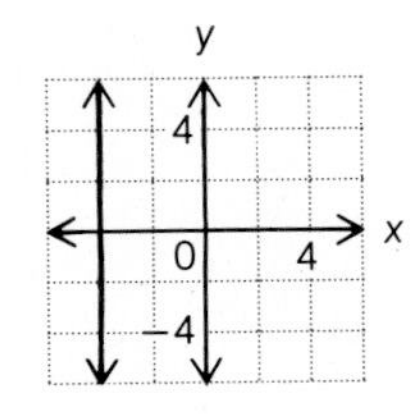

Pages 319–322

1.
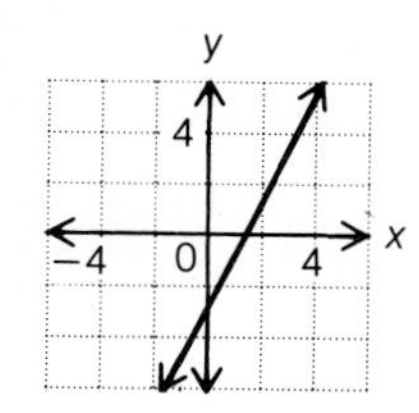

3.
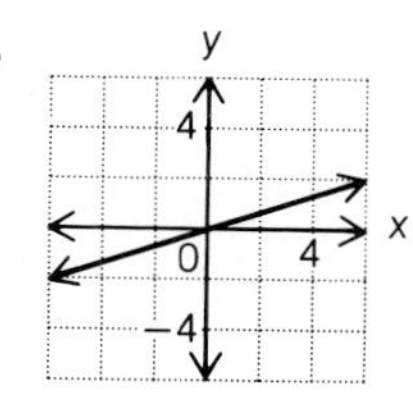

5.
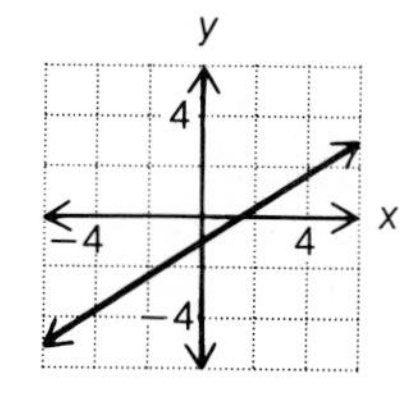

7.
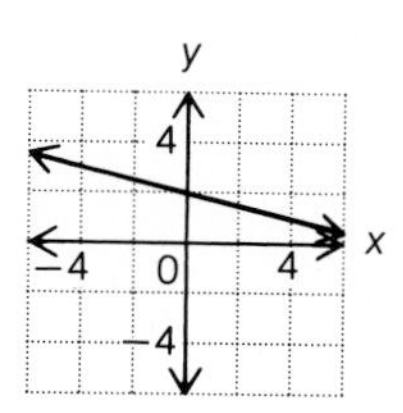

9.
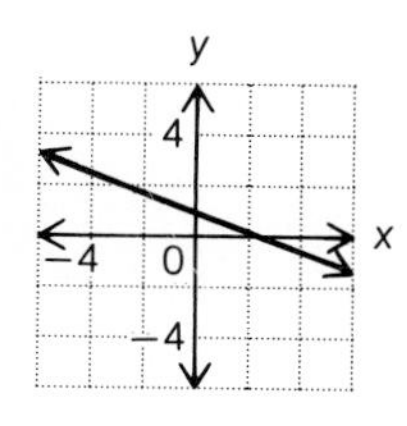

11.
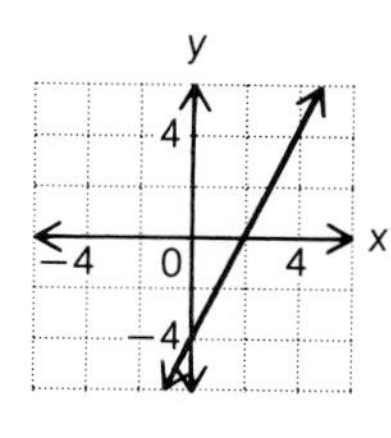

13.
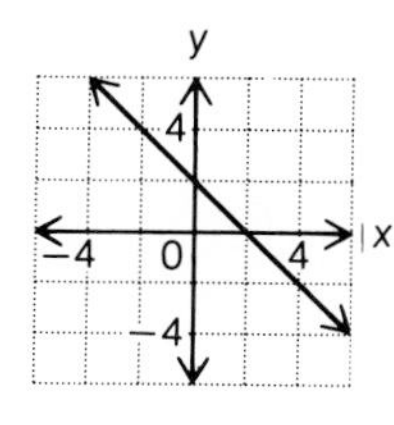

15.
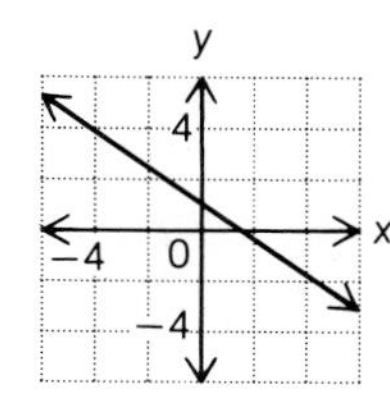

17.
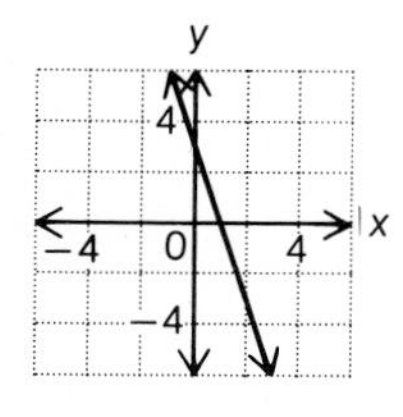

19.
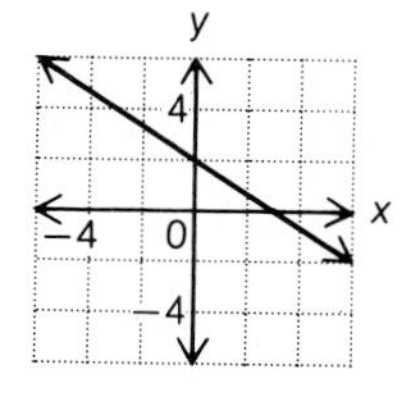

21.
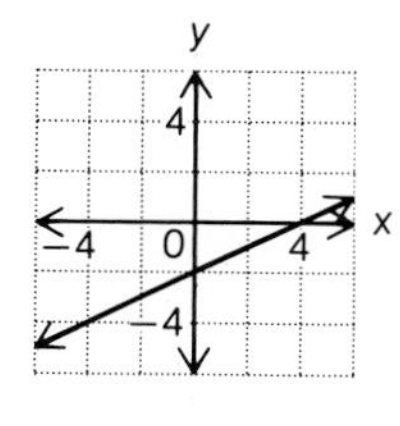

23.
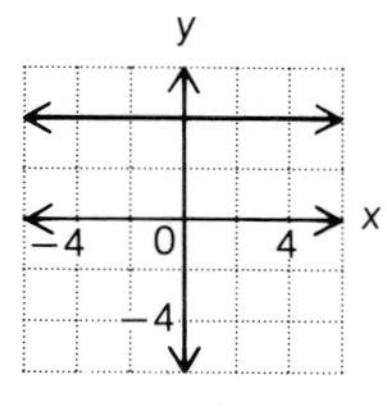

25.
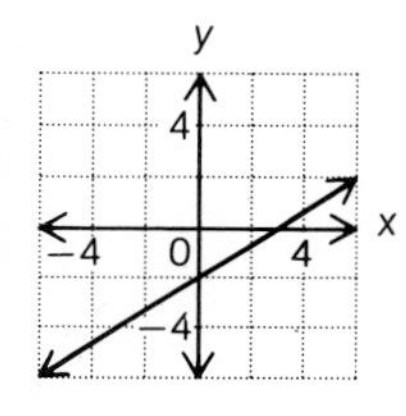

27.
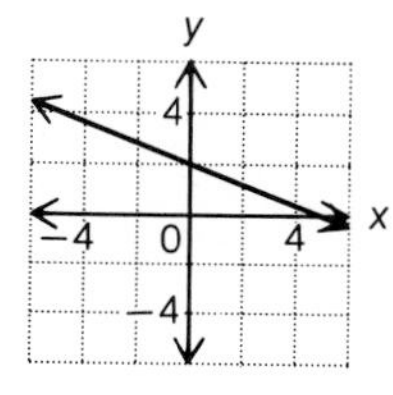

29.
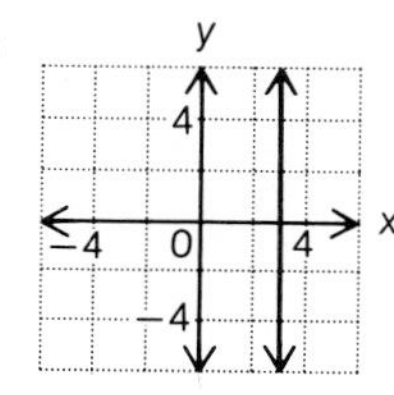

31.
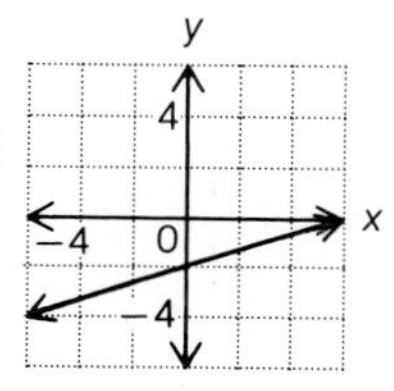

SECTION 3

Pages 323–328

Example 2

x-intercept:

$4x - y = 4$
$4x - 0 = 4$
$4x = 4$
$x = 1$
(1,0)

y-intercept:

$4x - y = 4$
$4(0) - y = 4$
$-y = 4$
$y = -4$
(0, −4)

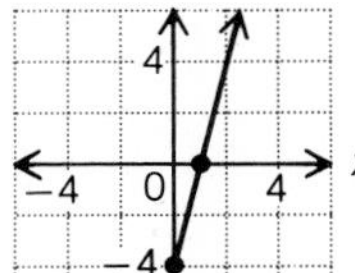

Example 4

x-intercept:

$y = 3x - 6$
$0 = 3x - 6$
$-3x = -6$
$x = 2$
(2,0)

y-intercept:

$y = 3x - 6$
$b = -6$
(0, −6)

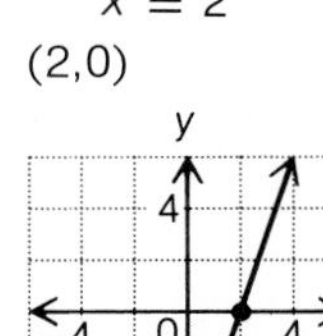
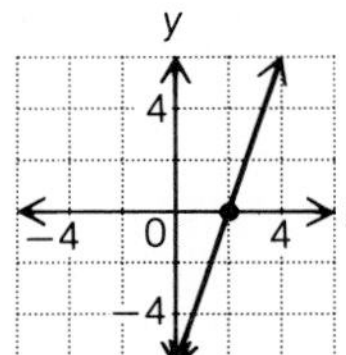

Example 6

Let $P_1 = (-1,2)$ and $P_2 = (1,3)$.

$$m = \frac{y_2 - y_1}{x_2 - x_1} = \frac{3 - 2}{1 - (-1)} = \frac{1}{2}$$

The slope is $\frac{1}{2}$.

Example 8

Let $P_1 = (1,2)$ and $P_2 = (4,-5)$.

$$m = \frac{y_2 - y_1}{x_2 - x_1} = \frac{-5 - 2}{4 - 1} = \frac{-7}{3}$$

The slope is $-\frac{7}{3}$.

Example 10 Let $P_1 = (2,3)$ and $P_2 = (2,7)$.

$$m = \frac{y_2 - y_1}{x_2 - x_1} = \frac{7-3}{2-2} = \frac{4}{0}$$

The line has no slope.

Example 12 Let $P_1 = (1,-3)$ and $P_2 = (-5,-3)$.

$$m = \frac{y_2 - y_1}{x_2 - x_1} = \frac{-3-(-3)}{-5-1} = \frac{0}{-6} = 0$$

The line has zero slope.

Example 14 y-intercept $= (0,b) = (0,-1)$

$m = -\frac{1}{4}$

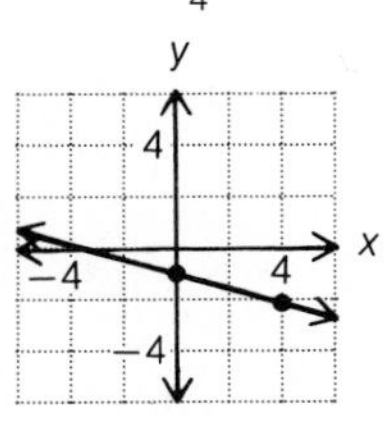

Example 16 y-intercept $= (0,b) = (0,0)$

$m = -\frac{3}{5}$

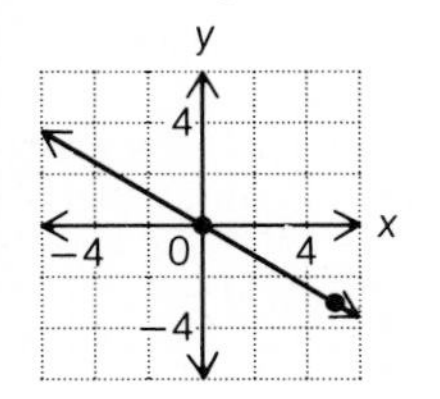

Example 18 Solve the equation for y.

$$x - 2y = 4$$
$$-2y = -x + 4$$
$$y = \frac{1}{2}x - 2$$

y-intercept $= (0,b) = (0,-2)$

$m = \frac{1}{2}$

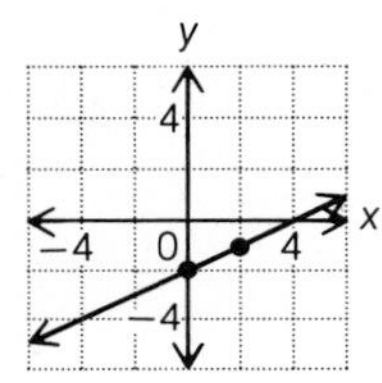

Pages 329–332

1. The x-intercept is (3,0), and the y-intercept is (0,−3). **3.** The x-intercept is (2,0), and the y-intercept is (0,−6). **5.** The x-intercept is (10,0), and the y-intercept is (0,−2). **7.** The x-intercept is (−4,0) and the y-intercept is (0,12). **9.** The x-intercept and the y-intercept are both (0,0). **11.** The x-intercept is (6,0), and the y-intercept is (0,3).

13. x-intercept: (2,0)
y-intercept: (0,5)

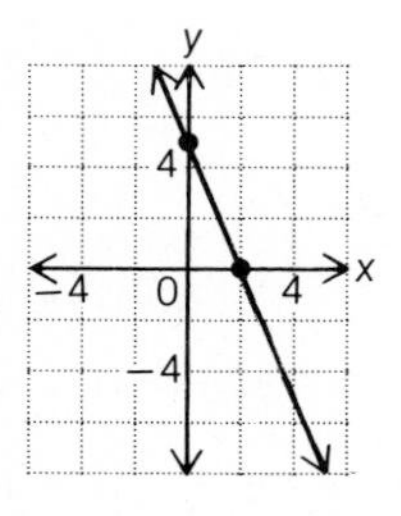

15. x-intercept: (4,0)
y-intercept: (0,−3)

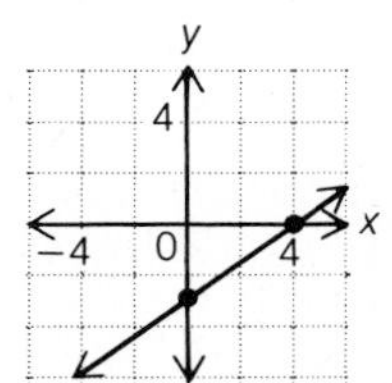

17. The slope is −2. **19.** The slope is $\frac{1}{3}$. **21.** The slope is $-\frac{5}{2}$. **23.** The slope is $-\frac{1}{2}$. **25.** The slope is −1. **27.** The line has no slope. **29.** The line has zero slope. **31.** The slope is $-\frac{1}{3}$. **33.** The line has zero slope. **35.** The slope is −5. **37.** The line has no slope. **39.** The slope is $-\frac{2}{3}$.

41.
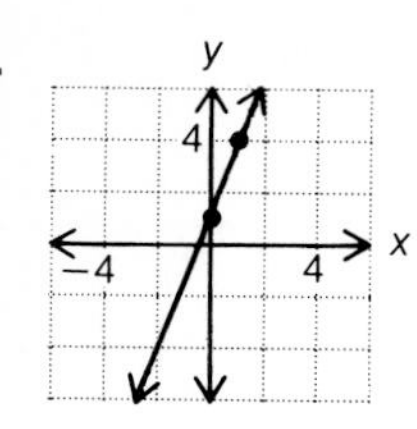

43.
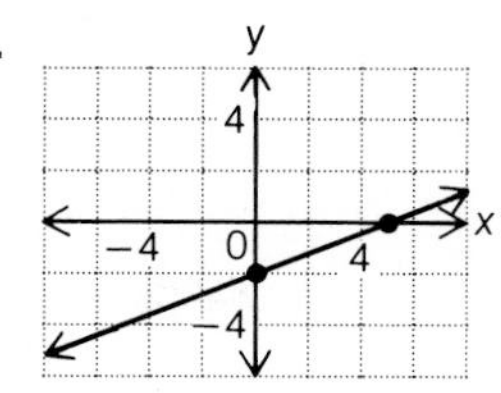

45.
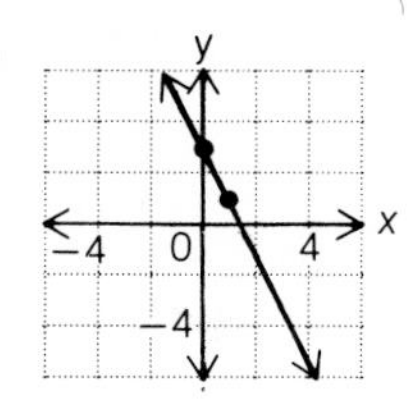

47.
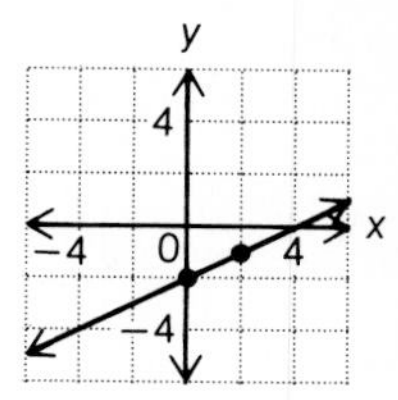

49.
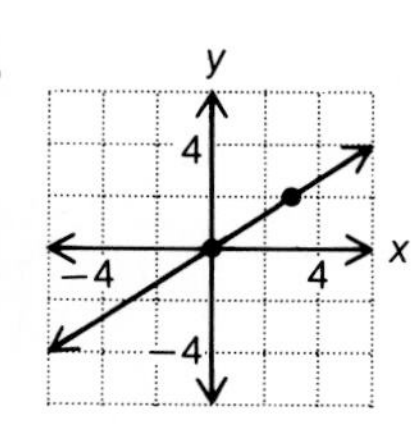

51.
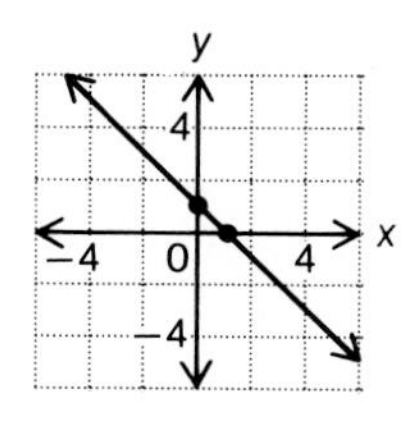

53.
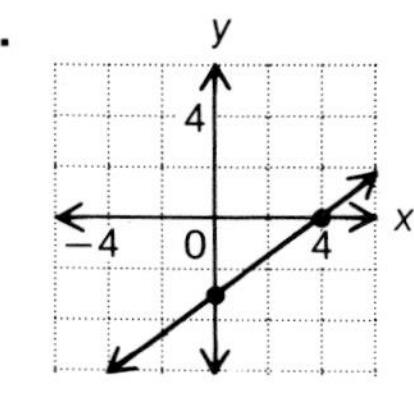

55.
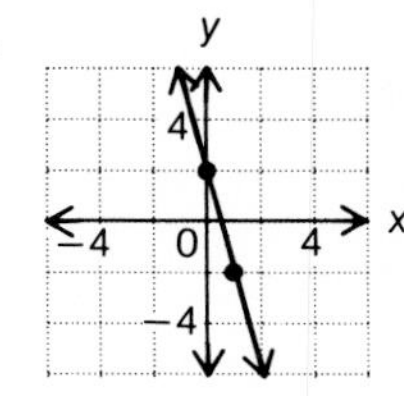

SECTION 4 Pages 333–334

Example 2

$$y = \frac{3}{2}x + b$$
$$-2 = \frac{3}{2}(4) + b$$
$$-2 = 6 + b$$
$$-8 = b$$
$$y = \frac{3}{2}x - 8$$

Example 4 $m = \frac{3}{4}$ $(x_1, y_1) = (4, -2)$

$$y - y_1 = m(x - x_1)$$
$$y - (-2) = \frac{3}{4}(x - 4)$$
$$y + 2 = \frac{3}{4}x - 3$$
$$y = \frac{3}{4}x - 5$$

The equation of the line is $y = \frac{3}{4}x - 5$.

Pages 335–336

1. The equation of the line is $y = 2x + 2$. **3.** The equation of the line is $y = -3x - 1$. **5.** The equation of the line is $y = \frac{1}{3}x$. **7.** The equation of the line is $y = \frac{3}{4}x - 5$. **9.** The equation of the line is $y = -\frac{3}{5}x$. **11.** The equation of the line is $y = \frac{1}{4}x + \frac{5}{2}$. **13.** The equation of the line is $y = 2x - 3$. **15.** The equation of the line is $y = -2x - 3$. **17.** The equation of the line is $y = \frac{2}{3}x$. **19.** The equation of the line is $y = \frac{1}{2}x + 2$. **21.** The equation of the line is $y = -\frac{3}{4}x - 2$. **23.** The equation of the line is $y = \frac{3}{4}x + \frac{5}{2}$.

REVIEW/TEST Pages 339–340

1.1
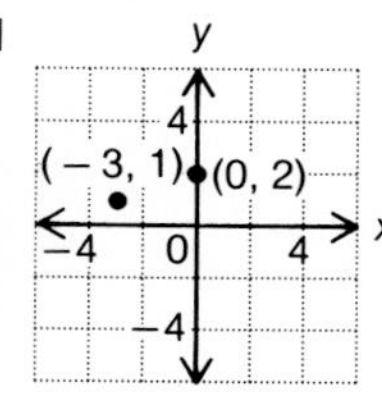

1.2 The ordered pair solution is (3,0).

1.3
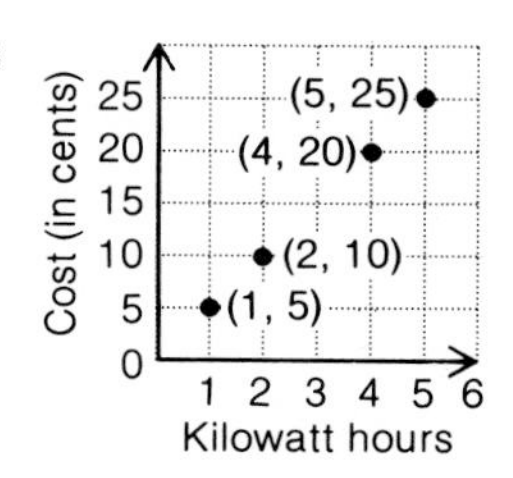

2.1a
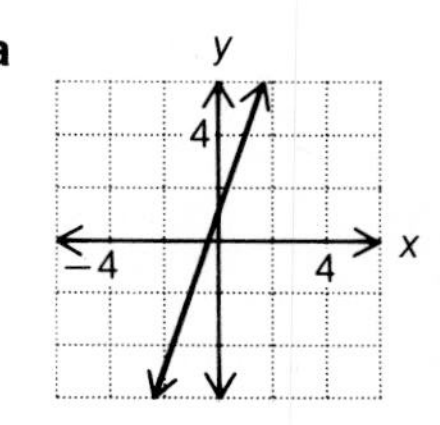

8

2.1b 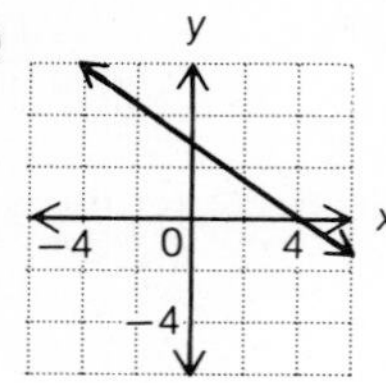

2.2a 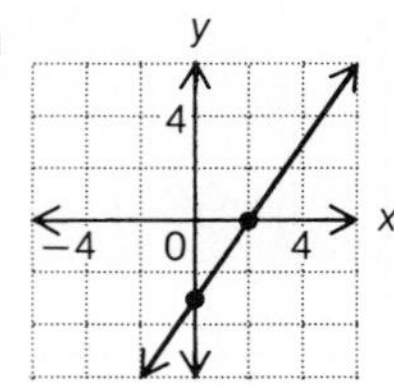

2.2b 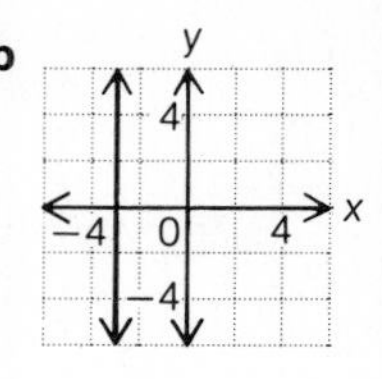

3.1a The x-intercept is (2,0), and the y-intercept is (0,−3). **3.1b** The x-intercept is (−2,0), and the y-intercept is (0,1). **3.2a** The slope is 2. **3.2b** The line has zero slope. **3.3a** 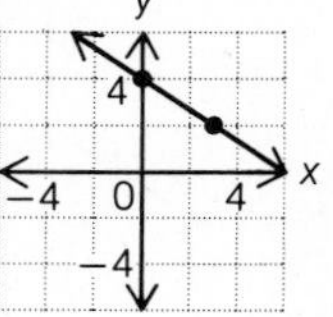

3.3b 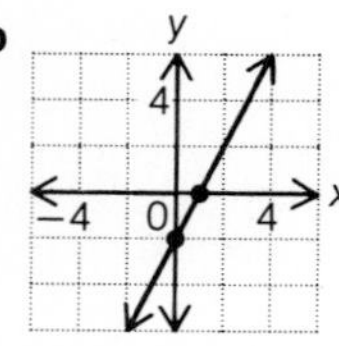

4.1a The equation of the line is $y = 3x - 1$. **4.1b** The equation of the line is $y = \frac{2}{3}x + 3$.

4.2a The equation of the line is $y = \frac{1}{2}x + 2$. **4.2b** The equation of the line is $y = -\frac{2}{3}x + \frac{4}{3}$.

CUMULATIVE REVIEW/TEST Pages 341–342

1. -12 (1.5.2) **2.** $-\frac{5}{8}$ (2.1.1) **3.** $-17x + 28$ (2.2.4) **4.** $x = \frac{3}{2}$ (3.2.1) **5.** $x = \frac{19}{18}$ (3.3.2) **6.** $\frac{1}{15}$ (4.1.1) **7.** $-32x^8y^7$ (5.2.1) **8.** $-3x^2$ (5.4.1) **9.** $x + 3$ (5.4.3) **10.** $5(x + 2)(x + 1)$ (6.2.2) **11.** $(a + 2)(x + y)$ (6.4.3) **12.** The solutions are 4 and −2. (6.5.1) **13.** $\frac{x^3(x + 3)}{y(x + 2)}$ (7.1.2) **14.** $\frac{3}{x + 8}$ (7.3.1) **15.** $x = 2$ (7.5.1) **16.** $y = -3 + \frac{4}{5}x$ (7.7.1) **17.** The ordered pair solution is (−2,−5). (8.1.2) **18.** The line has zero slope. (8.3.2) **19.** The equation of the line is $y = \frac{1}{2}x - 2$. (8.4.1) **20.** The equation of the line is $y = -3x + 2$. (8.4.1) **21.** The equation of the line is $y = 2x + 2$. (8.4.2) **22.** The equation of the line is $y = \frac{2}{3}x - 3$. (8.4.2) **23.** The sale price is \$62.30. (4.3.2) **24.** The present age of the gold coin is 135 years and the present age of the silver coin is 75 years. (4.8.3) **25.** The value of the home is \$110,000. (7.6.2) **26.** It would take $3\frac{3}{4}$ hrs with both the electrician and the apprentice working. (7.8.1)

27. 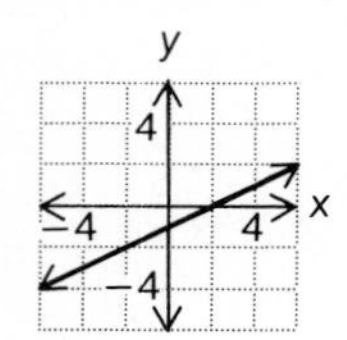

(8.2.1) **28.** 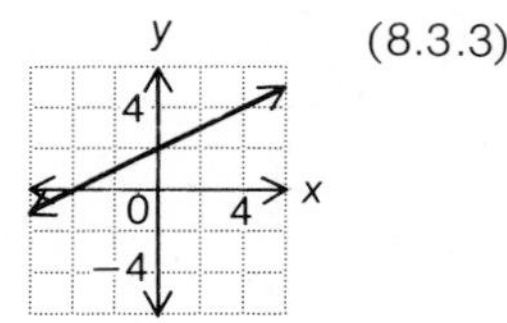

(8.3.3)

Answers to Chapter 9

SECTION 1 Pages 345–346

Example 2

$2x - 5y = 8$	
$2(-1) - 5(-2)$	8
$-2 + 10$	8
$8 = 8$	

$-x + 3y = -5$	
$-(-1) + 3(-2)$	-5
$1 + (-6)$	-5
$-5 = -5$	

Yes, $(-1,-2)$ is a solution of the system of equations.

Example 4

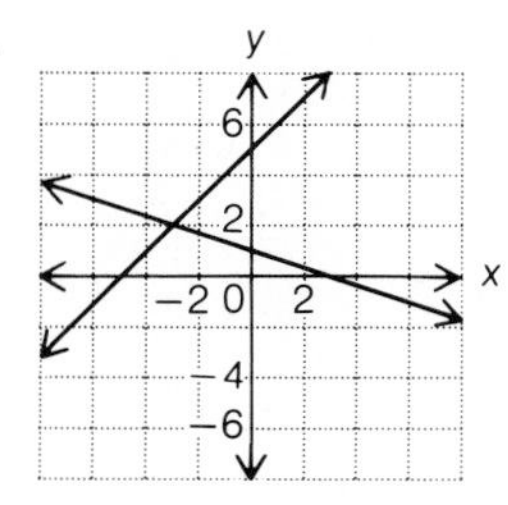

The solution is $(-3,2)$.

Example 6

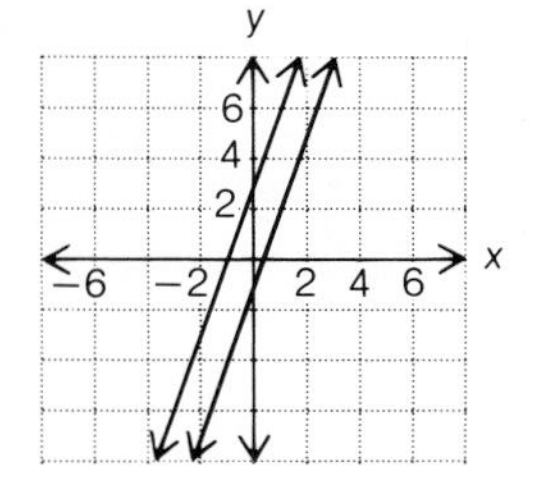

The lines are parallel and therefore do not intersect. The system of equations has no solution.

Pages 347–350

1. Yes, (2,3) is a solution of the system of equations. **3.** Yes, $(1,-2)$ is a solution of the system of equations. **5.** No, (4,3) is not a solution of the system of equations. **7.** No, $(-1,3)$ is not a solution of the system of equations. **9.** No, (0,0) is not a solution of the system of equations. **11.** Yes, $(2,-3)$ is a solution of the system of equations. **13.** Yes, (5,2) is a solution of the system of equations. **15.** Yes, $(-2,-3)$ is a solution of the system of equations. **17.** No, $(0,-3)$ is not a solution of the system of equations.

19.

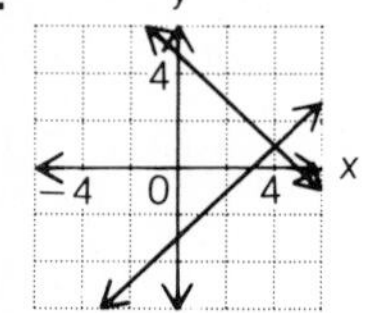

The solution is (4,1).

21.

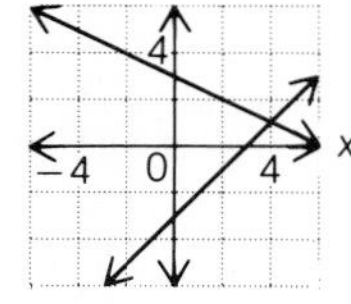

The solution is (4,1).

23.

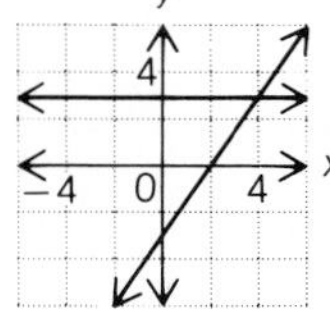

The solution is (4,3).

25.

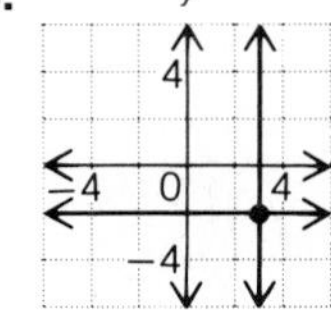

The solution is $(3,-2)$.

27.

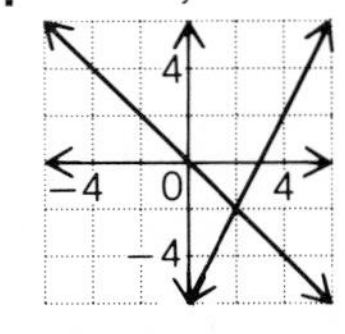

The solution is $(2,-2)$.

29.

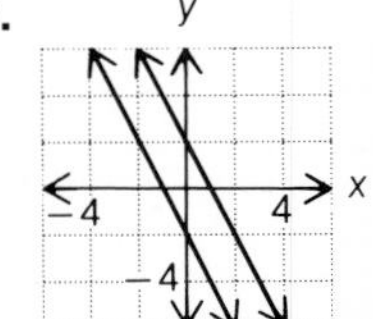

The system of equations has no solution.

31.

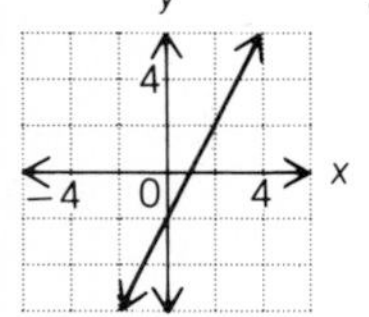

Any solution of one equation is also a solution of the other equation.

33.

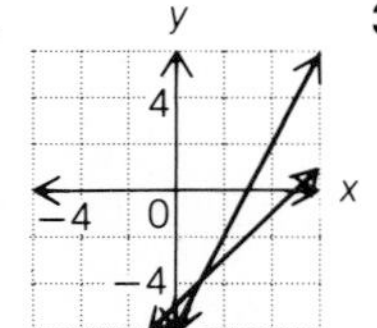

The solution is $(1,-4)$.

35.

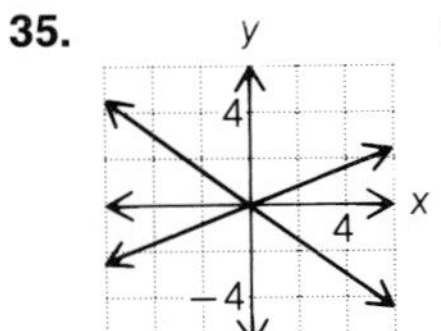

The solution is (0,0).

37.

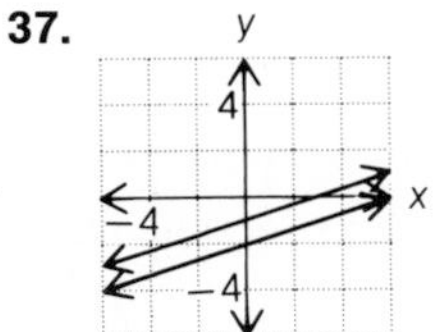

The system of equation has no solution.

39.

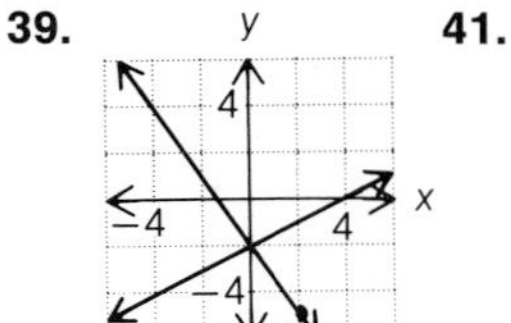

The solution is $(0,-2)$.

41.

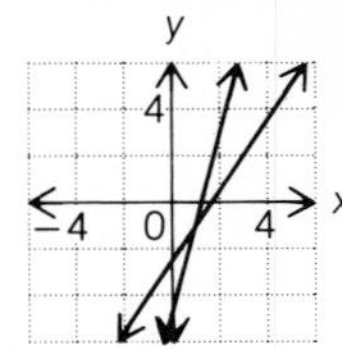

The solution is $(1,-1)$.

SECTION 2

Pages 351–352

Example 2

(1) $7x - y = 4$
(2) $3x + 2y = 9$

Solve equation (1) for y.

$$7x - y = 4$$
$$-y = -7x + 4$$
$$y = 7x - 4$$

Substitute in equation (2).

$$3x + 2y = 9$$
$$3x + 2(7x - 4) = 9$$
$$3x + 14x - 8 = 9$$
$$17x - 8 = 9$$
$$17x = 17$$
$$x = 1$$

Substitute in equation (1).

$$7x - y = 4$$
$$7(1) - y = 4$$
$$7 - y = 4$$
$$-y = -3$$
$$y = 3$$

The solution is (1,3).

Example 4

(1) $3x - y = 4$
(2) $y = 3x + 2$

$$3x - y = 4$$
$$3x - (3x + 2) = 4$$
$$3x - 3x - 2 = 4$$
$$-2 = 4$$

The lines are parallel. The system has no solution.

Example 6

(1) $y = -2x + 1$
(2) $6x + 3y = 3$

$$6x + 3y = 3$$
$$6x + 3(-2x + 1) = 3$$
$$6x - 6x + 3 = 3$$
$$3 = 3$$

The two equations represent the same line. Any ordered pair that is a solution of one equation is also a solution of the other equation.

Pages 353–354

1. The solution is (2,1). **3.** The solution is (4,1). **5.** The solution is (−1,1). **7.** The solution is (3,1). **9.** The solution is (1,1). **11.** The solution is (−1,1). **13.** The lines are parallel. The system of equations has no solution. **15.** The lines are parallel. The system of equations has no solution. **17.** The solution is $\left(-\frac{3}{4}, -\frac{3}{4}\right)$. **19.** The solution is (5,7). **21.** The solution is (1,7). **23.** The solution is $\left(\frac{17}{5}, -\frac{7}{5}\right)$. **25.** The solution is $\left(-\frac{6}{11}, \frac{31}{11}\right)$. **27.** The solution is (2,3). **29.** The solution is (0,0). **31.** The two equations represent the same line. Any ordered pair that is a solution of one equation is also a solution of the other equation. **33.** The solution is $\left(\frac{20}{17}, -\frac{15}{17}\right)$. **35.** The solution is (5,2). **37.** The solution is (−17,−8). **39.** The solution is $\left(-\frac{5}{7}, \frac{13}{7}\right)$. **41.** The solution is (3,−2).

SECTION 3 Pages 355–358

Example 2

(1) $x - 2y = 1$
(2) $2x + 4y = 0$

Eliminate y.

$$2(x - 2y) = 2 \cdot 1$$
$$2x + 4y = 0$$

$$2x - 4y = 2$$
$$2x + 4y = 0$$

Add the equations.

$$4x = 2$$
$$x = \frac{2}{4} = \frac{1}{2}$$

Replace x in equation (2).

$$2\left(\frac{1}{2}\right) + 4y = 0$$
$$1 + 4y = 0$$
$$4y = -1$$
$$y = -\frac{1}{4}$$

The solution is $\left(\frac{1}{2}, -\frac{1}{4}\right)$.

Example 6

(1) $4x + 5y = 11$
(2) $3y = x + 10$

Write equation (2) in the form $Ax + By = C$.

$$3y = x + 10$$
$$-x + 3y = 10$$

Eliminate x.

$$4x + 5y = 11$$
$$4(-x + 3y) = 4 \cdot 10$$

$$4x + 5y = 11$$
$$-4x + 12y = 40$$

Add the equations.

$$17y = 51$$
$$y = 3$$

Replace y in equation (1).

$$4x + 5y = 11$$
$$4x + 5 \cdot 3 = 11$$
$$4x + 15 = 11$$
$$4x = -4$$
$$x = -1$$

The solution is $(-1,3)$.

Example 4

(1) $2x - 3y = 4$
(2) $-4x + 6y = -8$

Eliminate y.

$$2(2x - 3y) = 2 \cdot 4$$
$$-4x + 6y = -8$$

$$4x - 6y = 8$$
$$-4x + 6y = -8$$

Add the equations.

$$0 + 0 = 0$$
$$0 = 0$$

The two equations represent the same line. Any ordered pair that is a solution of one equation is also a solution of the other equation.

Pages 359–360

1. The solution is $(5,-1)$. **3.** The solution is $(1,3)$. **5.** The solution is $(1,1)$. **7.** The solution is $(3,-2)$. **9.** The two equations represent the same line. Any ordered pair that is a solution of one equation is also a solution of the other equation. **11.** The solution is $(3,1)$. **13.** The two equations represent the same line. Any ordered pair that is a solution of one equation is also a solution of the other equation. **15.** The solution is $\left(-\frac{13}{17},-\frac{24}{17}\right)$. **17.** The solution is $(2,0)$. **19.** The solution is $(0,0)$. **21.** The solution is $(5,-2)$. **23.** The solution is $\left(\frac{32}{19},-\frac{9}{19}\right)$. **25.** The solution is $(3,4)$. **27.** The solution is $(1,-1)$. **29.** The two equations represent the same line. Any ordered pair that is a solution of one equation is also a solution of the other. **31.** The solution is $(3,1)$. **33.** The solution is $(-1,2)$. **35.** The solution is $(1,1)$. **37.** The solution is $\left(\frac{1}{2},-\frac{1}{2}\right)$. **39.** The solution is $\left(\frac{2}{3},\frac{1}{9}\right)$. **41.** The solution is $\left(\frac{7}{25},-\frac{1}{25}\right)$.

SECTION 4 Pages 361–364

Example 2

Strategy

▷ Rate of the current: c
Rate of the canoeist in calm water: r

	Rate	Time	Distance
With current	$r + c$	3	$3(r + c)$
Against current	$r - c$	5	$5(r - c)$

▷ The distance traveled with the current is 15 mi.
The distance traveled against the current is 15 mi.

Solution

$3(r + c) = 15$ $\qquad \frac{1}{3}\cdot 3(r + c) = \frac{1}{3}\cdot 15$

$5(r - c) = 15$ $\qquad \frac{1}{5}\cdot 5(r - c) = \frac{1}{5}\cdot 15$

$$r + c = 5$$
$$r - c = 3$$

$$2r = 8$$
$$r = 4$$

$$r + c = 5$$
$$4 + c = 5$$
$$c = 1$$

The rate of the current is 1 mph.
The rate of the canoeist in calm water is 4 mph.

Example 4

Strategy

▷ The number of dimes in the first bank: d
The number of quarters in the first bank: q

First bank:

	Number	Value	Total Value
Dimes	d	10	$10d$
Quarters	q	25	$25q$

Second bank:

	Number	Value	Total Value
Dimes	$\frac{1}{2}d$	10	$5d$
Quarters	$2q$	25	$50q$

▷ The total value of the coins in the first bank is \$4.80.
The total value of the coins in the second bank is \$8.40.

Solution

$10d + 25q = 480$ $\qquad 10d + 25q = 480$

$5d + 50q = 840$ $\qquad -2(5d + 50q) = -2\cdot 840$

$$10d + 25q = 480$$
$$-10d - 100q = -1680$$
$$-75q = -1200$$
$$q = 16$$

$$10d + 25q = 480$$
$$10d + 25(16) = 480$$
$$10d + 400 = 480$$
$$10d = 80$$
$$d = 8$$

There are 8 dimes and 16 quarters in the first bank.

Pages 365–366

1. The rate of the canoeist in calm water is 7 mph. The rate of the current is 1 mph. **3.** The rate of the motorboat in still water is 22.5 or $22\frac{1}{2}$ mph. The rate of the current is 2.5 or $2\frac{1}{2}$ mph. **5.** The rate of the rowing team in calm water is 6 mph. The rate of the current is 2 mph. **7.** The rate of the plane in calm air is 110 mph. The rate of the wind is 20 mph. **9.** The crew can row 14 km/h in calm water. The rate of the current is 6 km/h. **11.** The cost per copy for a black-and-white page is \$.04. The cost per copy for a color page is \$.10. **13.** The cost per kilogram of the tin alloy is \$3. The cost per kilogram of the aluminum alloy is \$2. **15.** There are 12 quarters and 6 nickels in the first bank. **17.** There are 12 dimes and 10 quarters in the bank. **19.** The present age of the adult is 31. The present age of the child is 7.

REVIEW/TEST Pages 369–370

1.1a Yes, $(-2,3)$ is a solution of the system of equations. **1.1b** Yes, $(1,-3)$ is a solution of the system of equations.
1.2

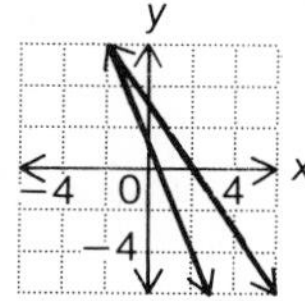

The solution is $(-2,6)$.
2.1a The solution is (3,1). **2.1b** The solution is $(1,-1)$. **2.1c** The solution is $(2,-1)$. **2.1d** The solution is $\left(\frac{22}{7},-\frac{5}{7}\right)$. **3.1a** The solution is (2,1). **3.1b** The solution is $\left(\frac{1}{2},-1\right)$. **3.1c** The solution is $(2,-1)$. **3.1d** The solution is $(1,-2)$. **4.1** The rate of the plane in calm air is 100 mph. The rate of the wind is 20 mph. **4.2** The price of a reserved-seat ticket is \$10. The price of a general-admission ticket is \$6.

CUMULATIVE REVIEW/TEST Pages 371–372

1. $\frac{3}{2}$ (2.1.1) **2.** $x = -\frac{3}{2}$ (3.1.3) **3.** $x = -\frac{7}{2}$ (3.3.2) **4.** $-6a^3 + 13a^2 - 9a + 2$ (5.3.2) **5.** $-2x^5y^2$ (5.4.1) **6.** $2b - 1 + \frac{1}{2b-3}$ (5.4.3) **7.** $-\frac{4y}{x^3}$ (5.5.1) **8.** $4y^2(xy - 4)(xy + 4)$ (6.4.4) **9.** The solutions are 4 and -1. (6.5.1) **10.** $x - 2$ (7.1.3) **11.** $\frac{x^2+2}{x^2+x-2}$ (7.3.2) **12.** $\frac{x-3}{x+1}$ (7.4.1) **13.** $x = -\frac{1}{5}$ (7.5.1) **14.** $r = \frac{A-P}{Pt}$ (7.7.1) **15.** The x-intercept is (6,0), and the y-intercept is $(0,-4)$. (8.3.1) **16.** The slope is $-\frac{7}{5}$. (8.3.2) **17.** The equation of the line is $y = -\frac{3}{2}x$. (8.4.1) **18.** Yes, (2,0) is a solution of the system of equations. (9.1.1) **19.** The solution is $(-6,1)$. (9.2.1) **20.** The solution is $(4,-3)$. (9.3.1) **21.** The amount invested at 9.6% is \$3750. The amount invested at 7.2% is \$5000. (4.4.1) **22.** The rate of the freight train is 48 mph. The rate of the passenger train is 56 mph. (4.6.1) **23.** A side of the original square measures 8 in. (6.5.2) **24.** The rate of the wind is 30 mph. (7.8.2)

25.

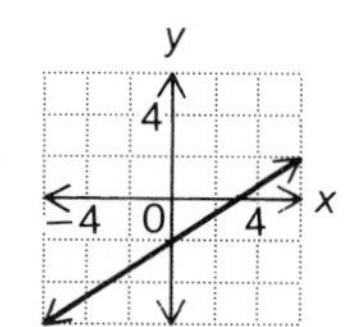

(8.2.2)

26.

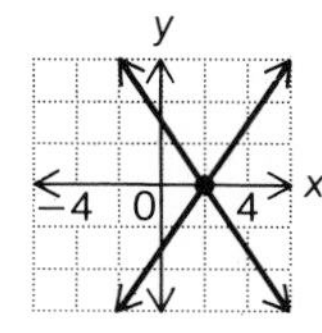

(9.1.2)

The solution is (2,0).

27. The rate of the boat in calm water is 14 mph. (9.4.1) **28.** There are 40 dimes in the first bank. (9.4.2)

9

Answers to Chapter 10

SECTION 1 Pages 375–378

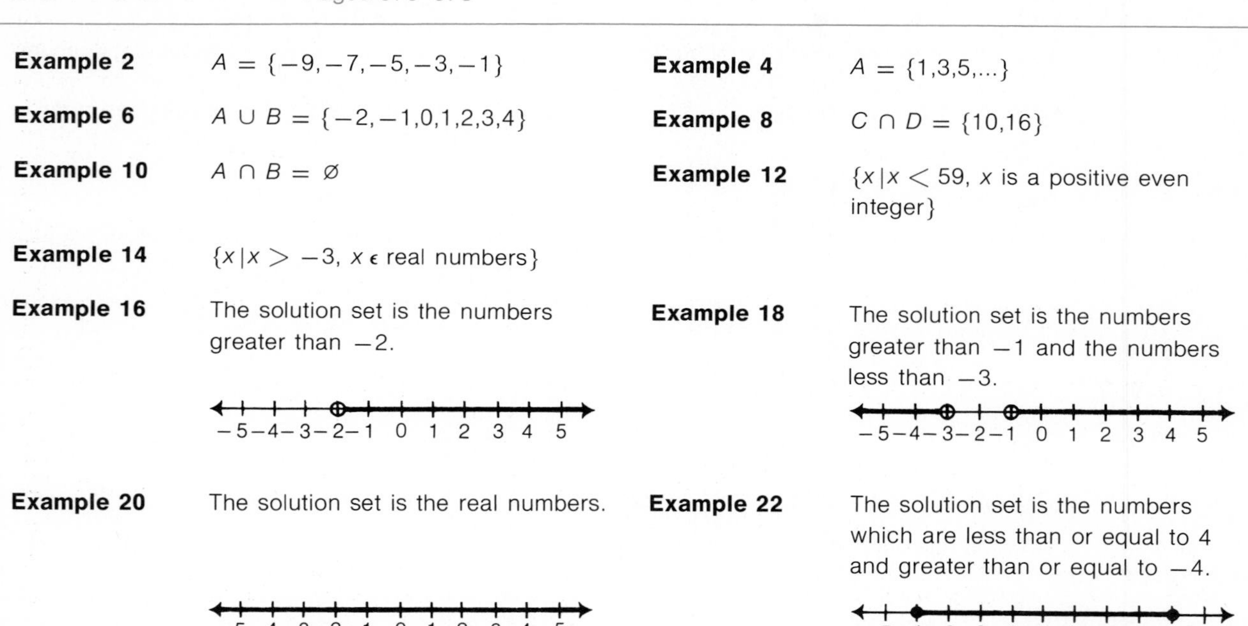

Example 2 $A = \{-9,-7,-5,-3,-1\}$

Example 4 $A = \{1,3,5,\ldots\}$

Example 6 $A \cup B = \{-2,-1,0,1,2,3,4\}$

Example 8 $C \cap D = \{10,16\}$

Example 10 $A \cap B = \varnothing$

Example 12 $\{x \mid x < 59, x \text{ is a positive even integer}\}$

Example 14 $\{x \mid x > -3, x \in \text{real numbers}\}$

Example 16 The solution set is the numbers greater than -2.

Example 18 The solution set is the numbers greater than -1 and the numbers less than -3.

Example 20 The solution set is the real numbers.

Example 22 The solution set is the numbers which are less than or equal to 4 and greater than or equal to -4.

Pages 379–380

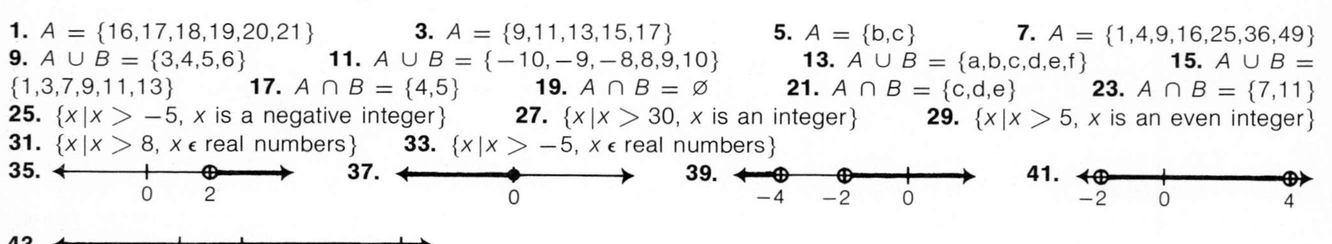

1. $A = \{16,17,18,19,20,21\}$ **3.** $A = \{9,11,13,15,17\}$ **5.** $A = \{b,c\}$ **7.** $A = \{1,4,9,16,25,36,49\}$ **9.** $A \cup B = \{3,4,5,6\}$ **11.** $A \cup B = \{-10,-9,-8,8,9,10\}$ **13.** $A \cup B = \{a,b,c,d,e,f\}$ **15.** $A \cup B = \{1,3,7,9,11,13\}$ **17.** $A \cap B = \{4,5\}$ **19.** $A \cap B = \varnothing$ **21.** $A \cap B = \{c,d,e\}$ **23.** $A \cap B = \{7,11\}$ **25.** $\{x \mid x > -5, x \text{ is a negative integer}\}$ **27.** $\{x \mid x > 30, x \text{ is an integer}\}$ **29.** $\{x \mid x > 5, x \text{ is an even integer}\}$ **31.** $\{x \mid x > 8, x \in \text{real numbers}\}$ **33.** $\{x \mid x > -5, x \in \text{real numbers}\}$ **35.** **37.** **39.** **41.** **43.**

SECTION 2 Pages 381–384

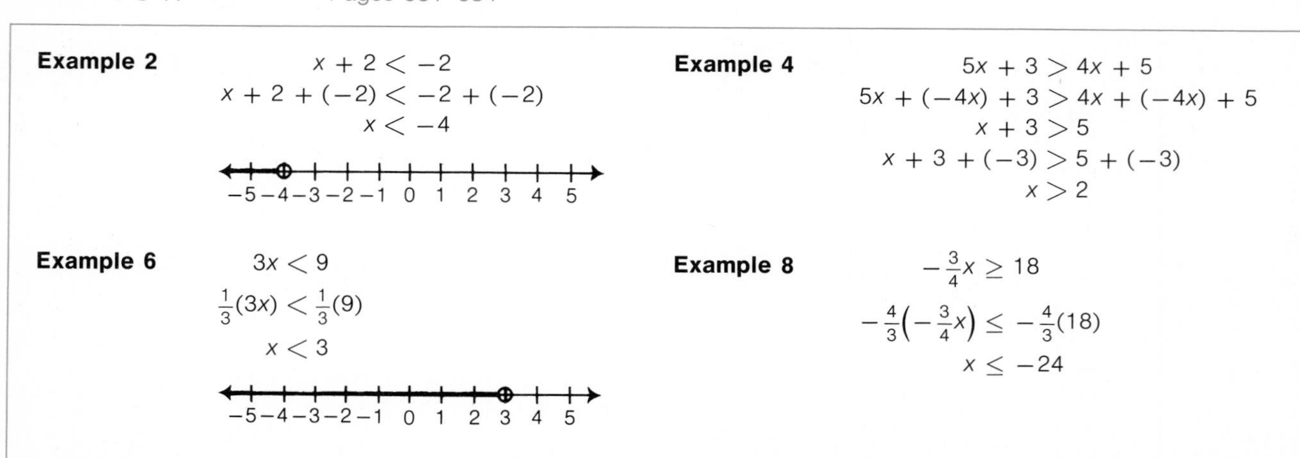

Example 2

$$x + 2 < -2$$
$$x + 2 + (-2) < -2 + (-2)$$
$$x < -4$$

Example 4

$$5x + 3 > 4x + 5$$
$$5x + (-4x) + 3 > 4x + (-4x) + 5$$
$$x + 3 > 5$$
$$x + 3 + (-3) > 5 + (-3)$$
$$x > 2$$

Example 6

$$3x < 9$$
$$\tfrac{1}{3}(3x) < \tfrac{1}{3}(9)$$
$$x < 3$$

Example 8

$$-\tfrac{3}{4}x \geq 18$$
$$-\tfrac{4}{3}\left(-\tfrac{3}{4}x\right) \leq -\tfrac{4}{3}(18)$$
$$x \leq -24$$

Example 10

Strategy To find the minimum selling price, write and solve an inequality using S to represent the selling price.

Solution

$$0.70S > 314$$
$$\frac{1}{0.70}(0.70S) > \frac{1}{0.70}(314)$$
$$S > 448.57142$$

The minimum selling price is $448.58.

Pages 385–387

1. $x < 2$ **3.** $x > 3$ **5.** $n \geq 3$

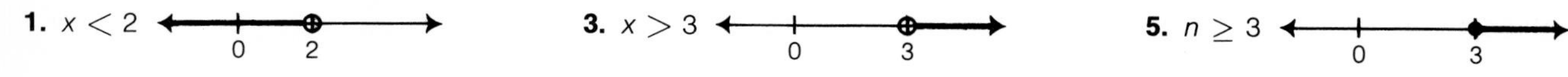

7. $x \leq -4$ **9.** $x \geq -1$ **11.** $y \geq -9$ **13.** $x < 12$

15. $x \geq 5$ **17.** $x < -11$ **19.** $x \leq 10$ **21.** $x \geq -6$ **23.** $x > 2$ **25.** $d < -\frac{1}{6}$ **27.** $x \geq -\frac{31}{24}$ **29.** $x < \frac{5}{8}$

31. $x < \frac{5}{4}$ **33.** $x > \frac{5}{24}$ **35.** $x < -3.8$ **37.** $x \leq -1.2$ **39.** $x < 5.6$

41. $x > -1.48$

43. $x < 4$ **45.** $y \geq 3$

47. $x \leq 1$ **49.** $x < -1$

51. $b < -4$ **53.** $y \leq -4$ **55.** $x > \frac{2}{7}$ **57.** $x \leq -\frac{5}{2}$ **59.** $x < 16$ **61.** $x \geq 16$

63. $x \geq -14$ **65.** $x \leq 21$ **67.** $x < \frac{7}{6}$ **69.** $x \leq -\frac{12}{7}$ **71.** $x > \frac{2}{3}$

73. $x \leq \frac{2}{3}$ **75.** $x \leq 2.3$ **77.** $x < -3.2$ **79.** $x \leq 5$ **81.** $x < -5.4$ **83.** $y \geq -3.15$ **85.** $x > -8.22$

87. $x \leq 7$

Page 388

1. $x > 18$. The smallest integer that satisfies the inequality is 19. **3.** $x \leq \frac{3}{2}$. The largest number that will satisfy the inequality is $\frac{3}{2}$. **5.** $N \geq 73$. The student's score on the last test must be equal to or greater than 73. **7.** $C \geq 1190$. The minimum commission during the fourth month must be $1190. **9.** $P > 613.63636$. The minimum selling price is $613.64.

SECTION 3

Pages 389–390

Example 2

$$\begin{aligned} 5 - 4x &> 9 - 8x \\ 5 - 4x + 8x &> 9 - 8x + 8x \\ 5 + 4x &> 9 \\ 5 + (-5) + 4x &> 9 + (-5) \\ 4x &> 4 \\ x &> 1 \end{aligned}$$

Example 4

$$\begin{aligned} 8 - 4(3x + 5) &\le 6(x - 8) \\ 8 - 12x - 20 &\le 6x - 48 \\ -12 - 12x &\le 6x - 48 \\ -12 - 12x + (-6x) &\le 6x + (-6x) - 48 \\ -12 - 18x &\le -48 \\ -12 + 12 - 18x &\le -48 + 12 \\ -18x &\le -36 \\ -\tfrac{1}{18}(-18x) &\ge -\tfrac{1}{18}(-36) \\ x &\ge 2 \end{aligned}$$

Example 6

Strategy To find the maximum number of miles:

▷ Write an expression for the cost of each car, using x to represent the number of miles driven during the week.

▷ Write and solve an inequality.

Solution

Cost of a Company A car	is less than	Cost of a Company B car

$$\begin{aligned} 8(7) + 0.10x &< 10(7) + 0.08x \\ 56 + 0.10x &< 70 + 0.08x \\ 56 + 0.10x + (-0.08x) &< 70 + 0.08x + (-0.08x) \\ 56 + 0.02x &< 70 \\ 56 + (-56) + 0.02x &< 70 + (-56) \\ 0.02x &< 14 \\ \tfrac{1}{0.02}(0.02x) &< \tfrac{1}{0.02}(14) \\ x &< 700 \end{aligned}$$

The maximum number of miles is 699.

Page 391

1. $x < 4$ **3.** $x < -4$ **5.** $x \ge 1$ **7.** $x \le 5$ **9.** $x < 0$ **11.** $x < 20$ **13.** $x > 500$ **15.** $x > 2$ **17.** $x \le -5$ **19.** $y \le \frac{5}{2}$ **21.** $x < \frac{25}{11}$ **23.** $x > 11$ **25.** $n \le \frac{11}{18}$ **27.** $x \ge 6$

Page 392

1. $x < -10$. The largest integer that will satisfy the inequality is -11. **3.** $x > 5$. The minimum length of the rectangle is 14 ft. **5.** $x < 150$. The maximum number of miles is 149. **7.** $A \le 50$. The maximum amount of fat is 50 lb. **9.** $x < 350$. The maximum number of miles is 349.

SECTION 4 Pages 393–394

Example 2

$$x - 3y < 2$$
$$x + (-x) - 3y < -x + 2$$
$$-3y < -x + 2$$
$$-\tfrac{1}{3}(-3y) > -\tfrac{1}{3}(-x + 2)$$
$$y > \tfrac{1}{3}x - \tfrac{2}{3}$$

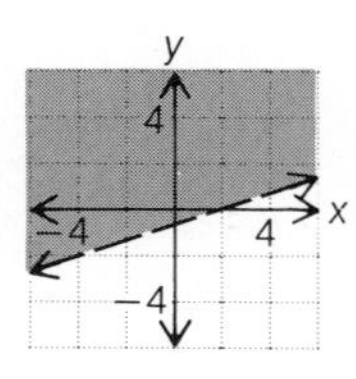

Example 4

$$2x - 4y \leq 8$$
$$2x + (-2x) - 4y \leq -2x + 8$$
$$-4y \leq -2x + 8$$
$$-\tfrac{1}{4}(-4y) \geq -\tfrac{1}{4}(-2x + 8)$$
$$y \geq \tfrac{1}{2}x - 2$$

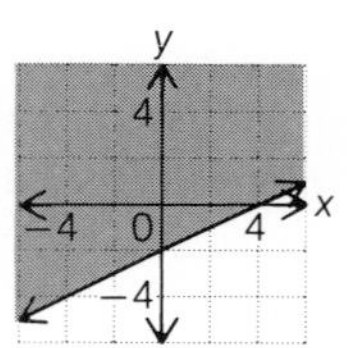

Example 6 $x < 3$

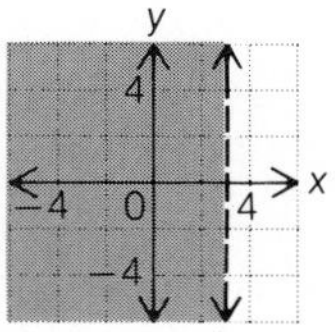

Pages 395–396

1. 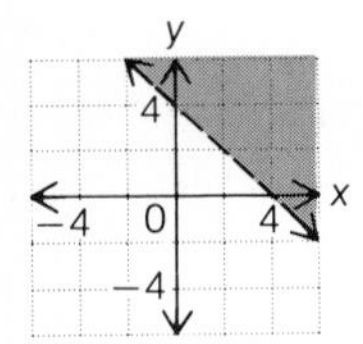

3. 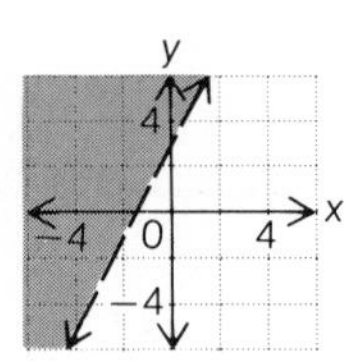

5. 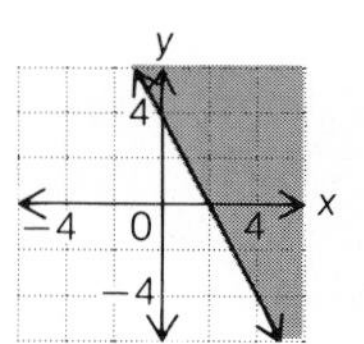

7. 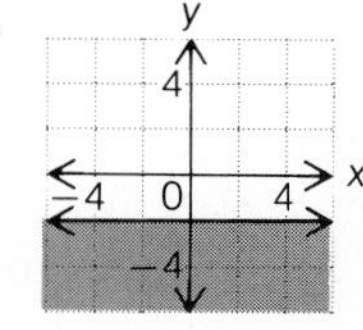

9.

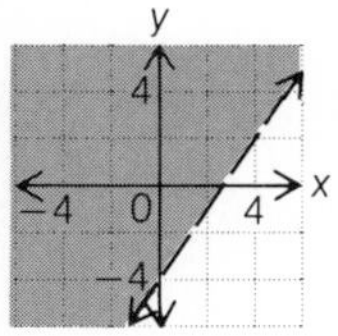

11. 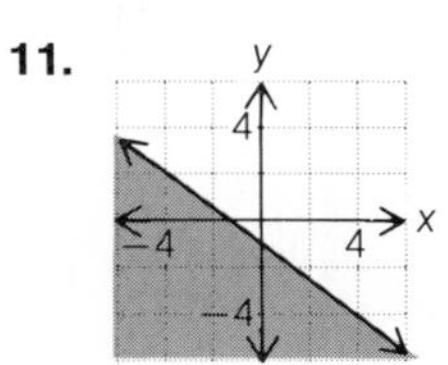

13. 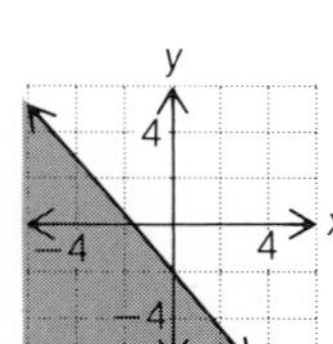

15. 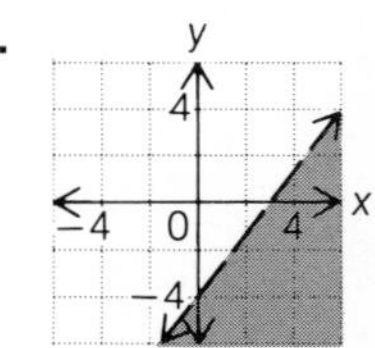

10

REVIEW/TEST Pages 399–400

1.1a $A = \{4,6,8\}$ **1.1b** $A \cap B = \{12\}$ **1.2a** $\{x \mid x < 50,\ x \text{ is a positive integer}\}$ **1.2b** $\{x \mid x > -23,\ x \in \text{real numbers}\}$ **1.3a**

−2 0

1.3b

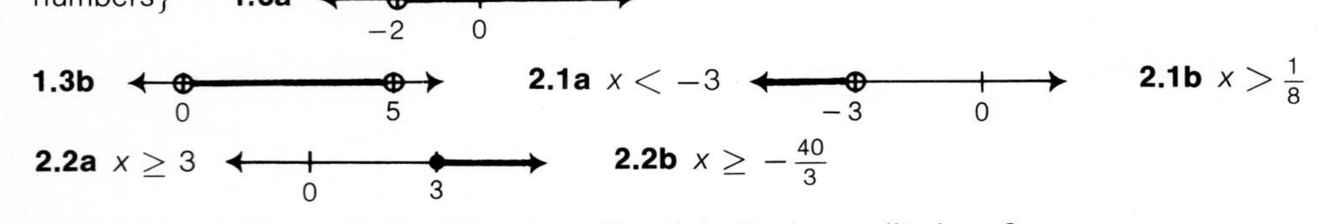

2.1a $x < -3$ **2.1b** $x > \frac{1}{8}$

2.2a $x \geq 3$

0 3

2.2b $x \geq -\frac{40}{3}$

2.3 $x < -8$. The largest integer that will satisfy the inequality is -9.

3.1a $x \leq -3$. **3.1b** $x < -\frac{22}{7}$ **3.2** The maximum width is 11 ft.

4.1a

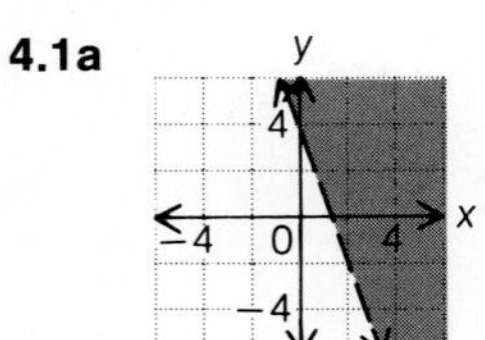

4.1b

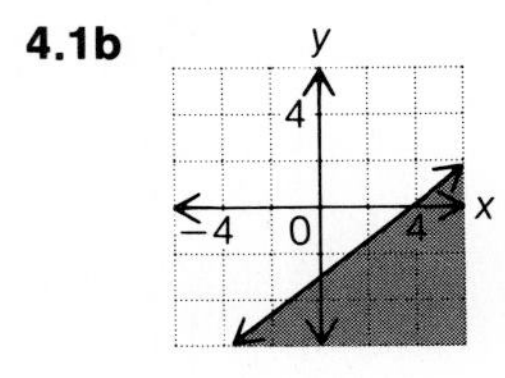

CUMULATIVE REVIEW/TEST Pages 401–402

1. $40a - 28$ (2.2.4) **2.** $x = \frac{1}{8}$ (3.1.3) **3.** $x = 4$ (3.3.2) **4.** $-12a^7b^4$ (5.2.2) **5.** $-\frac{1}{b^4}$ (5.4.1)

6. $4x - 2 - \frac{4}{4x-1}$ (5.4.3) **7.** $(4x - 1)(x - 5)$ (6.3.1) **8.** $3a^2(3x - 1)(3x + 1)$ (6.4.4) **9.** $\frac{1}{x+2}$ (7.1.2)

10. $\frac{18a}{2a^2 + 3a - 9}$ (7.3.2) **11.** $y = -\frac{5}{9}$ (7.5.1) **12.** $C = S + Rt$ (7.7.2) **13.** The slope is $-\frac{7}{3}$. (8.3.2)

14. The equation of the line is $y = -\frac{3}{2}x - \frac{3}{2}$. (8.4.2) **15.** The solution is (4,1). (8.2.1) **16.** The solution is (1,−4). (9.3.1) **17.** $A \cup B = \{-10,-2,0,1,2\}$ (10.1.1) **18.** $\{x \mid x < 48,\ x \in \text{real numbers}\}$ (10.1.2)

19.

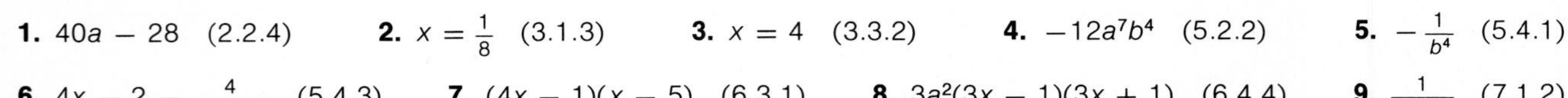

(10.1.3) **20.** $x > -2$ (10.2.2) **21.** $x < -15$ (10.2.2)

22. $x > 2$ (10.3.1) **23.** The largest integer that will satisfy the inequality is −26. (10.2.3) **24.** The maximum number of miles is 359. (10.3.2) **25.** There are 5000 fish in the lake. (7.6.2) **26.** There are seven 13¢ stamps in the drawer. (4.8.2) **27.**

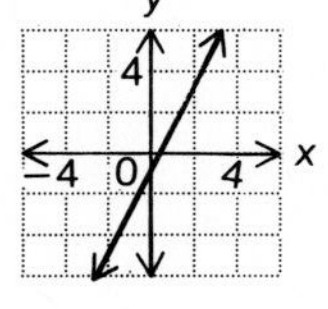

(8.2.2) **28.**

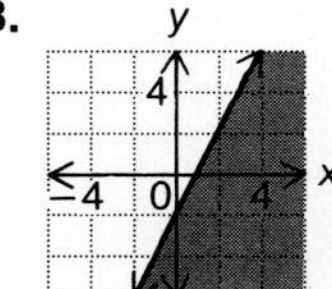

(10.4.1)

Answers to Chapter 11

SECTION 1 Pages 405–408

Example 2

$-5\sqrt{32} = -5\sqrt{2^5} = -5\sqrt{2^4 \cdot 2} = -5\sqrt{2^4}\sqrt{2} = -5 \cdot 2^2\sqrt{2} = -20\sqrt{2}$

Example 4

$\sqrt{216} = \sqrt{2^3 \cdot 3^3} = \sqrt{2^2 \cdot 3^2(2 \cdot 3)} = \sqrt{2^2 \cdot 3^2}\sqrt{2 \cdot 3} = 2 \cdot 3\sqrt{2 \cdot 3} = 6\sqrt{6} \approx 6(2.499) \approx 14.694$

Example 6

$\sqrt{y^{19}} = \sqrt{y^{18} \cdot y} = \sqrt{y^{18}}\sqrt{y} = y^9\sqrt{y}$

Example 8

$\sqrt{45b^7} = \sqrt{3^2 \cdot 5 \cdot b^7} = \sqrt{3^2b^6(5 \cdot b)} = \sqrt{3^2b^6}\sqrt{5b} = 3b^3\sqrt{5b}$

Example 10

$3a\sqrt{28a^9b^{18}} = 3a\sqrt{2^2 \cdot 7 \cdot a^9 \cdot b^{18}}$
$3a\sqrt{2^2a^8b^{18}(7a)} = 3a\sqrt{2^2a^8b^{18}}\sqrt{7a} =$
$3a \cdot 2 \cdot a^4 \cdot b^9\sqrt{7a} = 6a^5b^9\sqrt{7a}$

Example 12

$\sqrt{25(a-3)^2} = \sqrt{5^2(a-3)^2} = 5(a-3) = 5a - 15$

Example 14

$\sqrt{x^2 - 14x + 49} = \sqrt{(x-7)^2} = x - 7$

Pages 409–410

1. 4 **3.** 7 **5.** $4\sqrt{2}$ **7.** $2\sqrt{2}$ **9.** $18\sqrt{2}$ **11.** $10\sqrt{10}$ **13.** $\sqrt{15}$ **15.** $\sqrt{29}$ **17.** $-54\sqrt{2}$ **19.** $3\sqrt{5}$ **21.** 0 **23.** $48\sqrt{2}$ **25.** $\sqrt{105}$ **27.** 30 **29.** $30\sqrt{5}$ **31.** $5\sqrt{10}$ **33.** $4\sqrt{6}$ **35.** 18 **37.** 15.492 **39.** 16.968 **41.** 16 **43.** 16.585 **45.** 15.652 **47.** 18.76 **49.** x^3 **51.** $y^7\sqrt{y}$ **53.** a^{10} **55.** x^2y^2 **57.** $2x^2$ **59.** $2x\sqrt{6}$ **61.** $xy^3\sqrt{xy}$ **63.** $ab^5\sqrt{ab}$ **65.** $2x^2\sqrt{15x}$ **67.** $7a^2b^4$ **69.** $3x^2y^3\sqrt{2xy}$ **71.** $2x^5y^3\sqrt{10xy}$ **73.** $4a^4b^5\sqrt{5a}$ **75.** $8ab\sqrt{b}$ **77.** x^3y **79.** $8a^2b^3\sqrt{5b}$ **81.** $6x^2y^3\sqrt{3y}$ **83.** $4x^3y\sqrt{2y}$ **85.** $5a + 20$ **87.** $2x^2 - 8x + 8$ **89.** $x + 2$ **91.** $y - 1$ **93.** $x - 4$

SECTION 2 Pages 411–412

Example 2

$9\sqrt{3} + 3\sqrt{3} - 18\sqrt{3} = -6\sqrt{3}$

Example 4

$2\sqrt{50} - 5\sqrt{32} = 2\sqrt{2 \cdot 5^2} - 5\sqrt{2^5} =$
$2\sqrt{5^2}\sqrt{2} - 5\sqrt{2^4}\sqrt{2} = 2 \cdot 5\sqrt{2} - 5 \cdot 2^2\sqrt{2} =$
$10\sqrt{2} - 20\sqrt{2} = -10\sqrt{2}$

Example 6

$y\sqrt{28y} + 7\sqrt{63y^3} = y\sqrt{2^2 \cdot 7y} + 7\sqrt{3^2 \cdot 7 \cdot y^3} =$
$y\sqrt{2^2}\sqrt{7y} + 7\sqrt{3^2 \cdot y^2}\sqrt{7y} =$
$y \cdot 2\sqrt{7y} + 7 \cdot 3 \cdot y\sqrt{7y} = 2y\sqrt{7y} + 21y\sqrt{7y} =$
$23y\sqrt{7y}$

Example 8

$2\sqrt{27a^5} - 4a\sqrt{12a^3} + a^2\sqrt{75a} =$
$2\sqrt{3^3 \cdot a^5} - 4a\sqrt{2^2 \cdot 3 \cdot a^3} + a^2\sqrt{3 \cdot 5^2 \cdot a} =$
$2\sqrt{3^2 \cdot a^4}\sqrt{3a} - 4a\sqrt{2^2 \cdot a^2}\sqrt{3a} + a^2\sqrt{5^2}\sqrt{3a} =$
$2 \cdot 3 \cdot a^2\sqrt{3a} - 4a \cdot 2 \cdot a\sqrt{3a} + a^2 \cdot 5\sqrt{3a} =$
$6a^2\sqrt{3a} - 8a^2\sqrt{3a} + 5a^2\sqrt{3a} = 3a^2\sqrt{3a}$

Pages 413–414

1. $3\sqrt{2}$ **3.** $-\sqrt{7}$ **5.** $-11\sqrt{11}$ **7.** $10\sqrt{x}$ **9.** $-2\sqrt{y}$ **11.** $-11\sqrt{3b}$ **13.** $2x\sqrt{2}$ **15.** $-3a\sqrt{3a}$ **17.** $-5\sqrt{xy}$ **19.** $8\sqrt{5}$ **21.** $8\sqrt{2}$ **23.** $15\sqrt{2}-10\sqrt{3}$ **25.** $\sqrt{x}$ **27.** $-12x\sqrt{3}$ **29.** $2xy\sqrt{x}-3xy\sqrt{y}$ **31.** $-9x\sqrt{3x}$ **33.** $-13y^2\sqrt{2y}$ **35.** $4a^2b^2\sqrt{ab}$ **37.** $7\sqrt{2}$ **39.** $6\sqrt{x}$ **41.** $-3\sqrt{y}$ **43.** $-45\sqrt{2}$ **45.** $13\sqrt{3}-12\sqrt{5}$ **47.** $32\sqrt{3}-3\sqrt{11}$ **49.** $6\sqrt{x}$ **51.** $-34\sqrt{3x}$ **53.** $10a\sqrt{3b}+10a\sqrt{5b}$ **55.** $-2xy\sqrt{3}$ **57.** $-7b\sqrt{ab}+4a\sqrt{ab}$ **59.** $3ab\sqrt{2a}-ab+4ab\sqrt{3b}$

SECTION 3 Pages 415–418

Example 2

$\sqrt{5a}\sqrt{15a^3b^4}\sqrt{3b^5}=\sqrt{225a^4b^9}=\sqrt{3^25^2a^4b^9}=$
$\sqrt{3^25^2a^4b^8}\sqrt{b}=3\cdot 5a^2b^4\sqrt{b}=15a^2b^4\sqrt{b}$

Example 4

$\sqrt{5x}(\sqrt{5x}-\sqrt{25y})=\sqrt{5^2x^2}-\sqrt{5^3xy}=$
$\sqrt{5^2x^2}-\sqrt{5^2}\sqrt{5xy}=5x-5\sqrt{5xy}$

Example 6

$(2\sqrt{x}+7)(2\sqrt{x}-7)=4\sqrt{x^2}-7^2=4x-49$

Example 8

$(3\sqrt{x}-\sqrt{y})(5\sqrt{x}-2\sqrt{y})=$
$15\sqrt{x^2}-6\sqrt{xy}-5\sqrt{xy}+2\sqrt{y^2}=$
$15\sqrt{x^2}-11\sqrt{xy}+2\sqrt{y^2}=$
$15x-11\sqrt{xy}+2y$

Example 10

$\dfrac{\sqrt{15x^6y^7}}{\sqrt{3x^7y^9}}=\sqrt{\dfrac{15x^6y^7}{3x^7y^9}}=\sqrt{\dfrac{5}{xy^2}}=$
$\dfrac{\sqrt{5}}{y\sqrt{x}}=\dfrac{\sqrt{5}}{y\sqrt{x}}\cdot\dfrac{\sqrt{x}}{\sqrt{x}}=\dfrac{\sqrt{5x}}{xy}$

Example 12

$\dfrac{\sqrt{y}}{\sqrt{y}+3}=\dfrac{\sqrt{y}}{\sqrt{y}+3}\cdot\dfrac{\sqrt{y}-3}{\sqrt{y}-3}=\dfrac{y-3\sqrt{y}}{y-9}$

Example 14

$\dfrac{\sqrt{27x^3}-3\sqrt{12x}}{\sqrt{3x}}=\dfrac{\sqrt{27x^3}}{\sqrt{3x}}-\dfrac{3\sqrt{12x}}{\sqrt{3x}}=\sqrt{\dfrac{27x^3}{3x}}-3\sqrt{\dfrac{12x}{3x}}=$
$\sqrt{9x^2}-3\sqrt{4}=\sqrt{3^2x^2}-3\sqrt{2^2}=3x-3\cdot 2=3x-6$

Pages 419–420

1. 5 **3.** 6 **5.** x **7.** x^3y^2 **9.** $3ab^6\sqrt{2a}$ **11.** $12a^4b\sqrt{b}$ **13.** $2-\sqrt{6}$ **15.** $x-\sqrt{xy}$ **17.** $5\sqrt{2}-\sqrt{5x}$ **19.** $4-2\sqrt{10}$ **21.** $x-6\sqrt{x}+9$ **23.** $3a-3\sqrt{ab}$ **25.** $10abc$ **27.** $15x-22y\sqrt{x}+8y^2$ **29.** $x-y$ **31.** $10x+13\sqrt{xy}+4y$ **33.** 4 **35.** 7 **37.** 3 **39.** $x\sqrt{5}$ **41.** $\frac{a^2}{7}$ **43.** $\frac{\sqrt{3}}{3}$ **45.** $\frac{3\sqrt{x}}{x}$ **47.** $\frac{2\sqrt{y}}{xy}$ **49.** $\frac{2\sqrt{3x}}{3y}$ **51.** $-\frac{\sqrt{2}+3}{7}$ **53.** $\frac{15-3\sqrt{5}}{20}$ **55.** $\frac{x\sqrt{y}+y\sqrt{x}}{x-y}$ **57.** -5 **59.** $\frac{7}{4}$ **61.** $-x^2$

SECTION 4 Pages 421–424

Example 2

$\sqrt{4x}+3=7$
$\sqrt{4x}=4$
$(\sqrt{4x})^2=4^2$
$4x=16$
$x=4$

Check:
$\sqrt{4x}+3=7$
$\sqrt{4\cdot 4}+3=7$
$\sqrt{4^2}+3=7$
$4+3=7$
$7=7$

The solution is 4.

Example 4

$\sqrt{3x-2}-5=0$
$\sqrt{3x-2}=5$
$(\sqrt{3x-2})^2=5^2$
$3x-2=25$
$3x=27$
$x=9$

Check:
$\sqrt{3x-2}-5=0$
$\sqrt{3\cdot 9-2}-5=0$
$\sqrt{27-2}-5=0$
$\sqrt{25}-5=0$
$\sqrt{5^2}-5=0$
$5-5=0$
$0=0$

The solution is 9.

Example 6

$$\begin{aligned}\sqrt{4x+3} &= \sqrt{x+12}\\ (\sqrt{4x+3})^2 &= (\sqrt{x+12})^2\\ 4x+3 &= x+12\\ 4x &= x+9\\ 3x &= 9\\ x &= 3\end{aligned}$$

Check:

$$\begin{aligned}\sqrt{4x+3} &= \sqrt{x+12}\\ \sqrt{4\cdot 3+3} &= \sqrt{3+12}\\ \sqrt{12+3} &= \sqrt{15}\\ \sqrt{15} &= \sqrt{15}\end{aligned}$$

The solution is 3.

Example 8

Strategy

To find the distance, use the Pythagorean Theorem. The hypotenuse is the length of the ladder. One leg is the distance from the bottom of the ladder to the base of the building. The distance along the building from the ground to the top of the ladder is the unknown leg.

Solution

$$\begin{aligned}a^2 &= \sqrt{c^2 - b^2}\\ a^2 &= \sqrt{(8)^2 - (3)^2}\\ a^2 &= \sqrt{64 - 9}\\ a^2 &= \sqrt{55}\\ a^2 &\approx 7.416\end{aligned}$$

The distance is 7.416 ft.

Example 10

Strategy

To find the length of the pendulum, replace T in the equation with the given value and solve for L.

Solution

$$\begin{aligned}T &= 2\pi\sqrt{\frac{L}{32}}\\ 2.5 &= 2(3.14)\sqrt{\frac{L}{32}}\\ 2.5 &= 6.28\sqrt{\frac{L}{32}}\\ \frac{2.5}{6.28} &= \sqrt{\frac{L}{32}}\\ \left(\frac{2.5}{6.28}\right)^2 &= \left(\sqrt{\frac{L}{32}}\right)^2\\ \frac{6.25}{39.4384} &= \frac{L}{32}\\ (32)\left(\frac{6.25}{39.4384}\right) &= (32)\left(\frac{L}{32}\right)\\ \frac{200}{39.4384} &= L\\ 5.07 &\approx L\end{aligned}$$

The length of the pendulum is 5.07 ft.

Page 425

1. The rope is 4.47 ft. **3.** The solution is 144. **5.** The solution is 5. **7.** The solution is 16. **9.** The solution is 8. **11.** The equation has no solution. **13.** The solution is 6. **15.** The solution is 24. **17.** The solution is -1. **19.** The solution is $-\frac{2}{5}$. **21.** The solution is $\frac{4}{3}$. **23.** The solution is 15. **25.** The solution is 5. **27.** The solution is 1. **29.** The solution is 1.

Page 426

1. The number is 5. **3.** The height of the periscope must be 12.76 ft above the water. **5.** The object fell 64 ft. **7.** The height of the bridge is 36 ft. **9.** The length of the pendulum is 3.25 ft.

REVIEW/TEST Pages 429–430

1.1a $3\sqrt{5}$ **1.1b** $5\sqrt{3}$ **1.1c** 11.180 **1.2a** $11x^4y$ **1.2b** $6x^3y\sqrt{2x}$ **1.2c** $4a^2b^5\sqrt{2ab}$ **2.1a** $5\sqrt{y}$ **2.1b** $-5\sqrt{2}$ **2.1c** $21\sqrt{2y} - 12\sqrt{2x}$ **2.1d** $-2xy\sqrt{3xy} - 3xy\sqrt{xy}$ **3.1a** $4x^2y^2\sqrt{5y}$ **3.1b** $6x^2y\sqrt{y}$ **3.1c** $a - \sqrt{ab}$ **3.1d** $y + 2\sqrt{y} - 15$ **3.2a** 9 **3.2b** $7ab\sqrt{a}$ **3.2c** $\sqrt{3} + 1$ **3.2d** $x - 4$ **4.1a** The solution is 11. **4.1b** The solution is 25. **4.2a** The larger integer is 51. **4.2b** The length of the pendulum is 7.30 ft.

CUMULATIVE REVIEW/TEST Pages 431–432

1. $-\frac{1}{12}$ (1.5.2) **2.** $2x + 18$ (2.2.4) **3.** $x = \frac{1}{13}$ (3.3.2) **4.** $6x^5y^5$ (5.2.1) **5.** $-2b^2 + 1 - \frac{1}{3b^2}$ (5.4.2) **6.** $3x^2y^2(4x - 3y)$ (6.1.2) **7.** $2a(a - 5)(a - 3)$ (6.3.2) **8.** $\frac{1}{4(x + 1)}$ (7.1.3) **9.** $\frac{x + 3}{x - 3}$ (7.3.1) **10.** $x = \frac{5}{3}$ (7.5.1) **11.** The equation of the line is $y = \frac{1}{2}x - 2$. (8.4.1) **12.** The solution is (1,1). (9.2.1) **13.** The solution is (3,−2). (9.3.1) **14.** $x \le -\frac{9}{2}$ (10.3.1) **15.** $6\sqrt{3}$ (11.1.1) **16.** $-4\sqrt{2}$ (11.2.1) **17.** $4ab\sqrt{2ab} - 5ab\sqrt{ab}$ (11.2.1) **18.** $14a^5b^2\sqrt{2a}$ (11.3.1) **19.** $3\sqrt{2} - x\sqrt{3}$ (11.3.1) **20.** 8 (11.3.2) **21.** $-6 - 3\sqrt{5}$ (11.3.2) **22.** $x = 6$ (11.4.1) **23.** The book costs $24.50. (4.3.1) **24.** 56 oz of pure water must be added. (4.5.2) **25.** The numbers are 13 and 8. (6.5.2) **26.** Working alone, it would take the small pipe 48 h to fill the tank. (7.8.1)

27. (9.1.2)

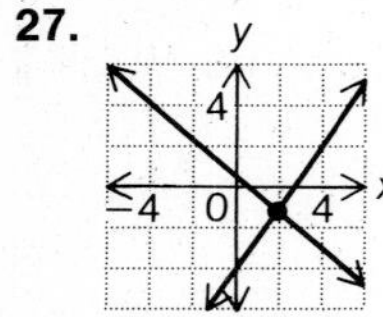

The solution is (2,−1).

28. (10.4.1)

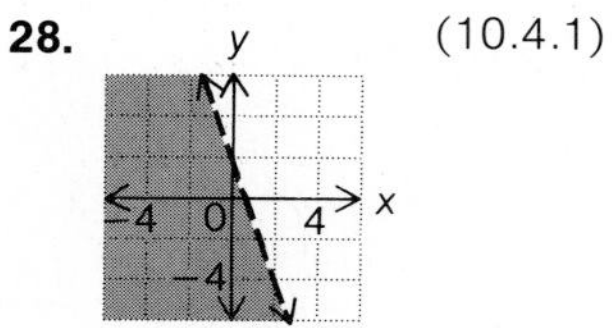

29. The smaller integer is 40. (11.4.2) **30.** The building is 400 ft high. (11.4.2)

Answers to Chapter 12

SECTION 1 Pages 435–436

Example 2

$$2x^2 = (x + 2)(x + 3)$$
$$2x^2 = x^2 + 5x + 6$$
$$x^2 - 5x - 6 = 0$$
$$(x + 1)(x - 6) = 0$$
$$x + 1 = 0 \qquad x - 6 = 0$$
$$x = -1 \qquad x = 6$$

The solutions are -1 and 6.

Example 4

$$x^2 + 81 = 0$$
$$x^2 = -81$$
$$\sqrt{x^2} = \sqrt{-81}$$

$\sqrt{-81}$ is not a real number.

The equation has no real number solution.

Pages 437–438

1. The solutions are -5 and 3. **3.** The solutions are 1 and 3. **5.** The solutions are -1 and -2. **7.** The solution is 3. **9.** The solutions are 0 and $-\frac{2}{3}$. **11.** The solutions are -2 and 5. **13.** The solutions are $\frac{2}{3}$ and 1. **15.** The solutions are -3 and $\frac{1}{3}$. **17.** The solution is $\frac{2}{3}$. **19.** The solutions are $-\frac{1}{2}$ and $\frac{3}{2}$. **21.** The solution is $\frac{1}{2}$. **23.** The solutions are -3 and 3. **25.** The solutions are $-\frac{1}{2}$ and $\frac{1}{2}$. **27.** The solutions are -3 and 5. **29.** The solutions are 1 and 5. **31.** The solutions are -1 and $\frac{13}{2}$. **33.** The solutions are 7 and -7. **35.** The solutions are 8 and -8. **37.** The solutions are $\frac{8}{3}$ and $-\frac{8}{3}$. **39.** The solutions are $\frac{5}{2}$ and $-\frac{5}{2}$. **41.** The solutions are $\frac{8}{5}$ and $-\frac{8}{5}$. **43.** The equation has no real number solution. **45.** The solutions are $4\sqrt{3}$ and $-4\sqrt{3}$. **47.** The solutions are 5 and -9. **49.** The solutions are 8 and -2. **51.** The solutions are $\frac{3}{2}$ and $-\frac{15}{2}$. **53.** The solutions are $\frac{26}{9}$ and $\frac{10}{9}$. **55.** The solutions are $-5 + 5\sqrt{2}$ and $-5 - 5\sqrt{2}$. **57.** The equation has no real number solution. **59.** The solutions are $-\frac{3}{4} + 2\sqrt{3}$ and $-\frac{3}{4} - 2\sqrt{3}$.

SECTION 2

Pages 439–442

Example 2

$$3x^2 - 6x - 2 = 0$$
$$3x^2 - 6x = 2$$
$$\frac{1}{3}(3x^2 - 6x) = \frac{1}{3}\cdot 2$$
$$x^2 - 2x = \frac{2}{3}$$

Complete the square.

$$x^2 - 2x + 1 = \frac{2}{3} + 1$$
$$(x - 1)^2 = \frac{5}{3}$$
$$\sqrt{(x - 1)^2} = \sqrt{\frac{5}{3}}$$
$$x - 1 = \pm\sqrt{\frac{5}{3}} = \pm\frac{\sqrt{15}}{3}$$
$$x - 1 = \frac{\sqrt{15}}{3} \qquad x - 1 = -\frac{\sqrt{15}}{3}$$
$$x = 1 + \frac{\sqrt{15}}{3} \qquad x = 1 - \frac{\sqrt{15}}{3}$$
$$= \frac{3 + \sqrt{15}}{3} \qquad = \frac{3 - \sqrt{15}}{3}$$

The solutions are $\frac{3 + \sqrt{15}}{3}$ and $\frac{3 - \sqrt{15}}{3}$.

Example 4

$$x^2 + 6x + 12 = 0$$
$$x^2 + 6x = -12$$
$$x^2 + 6x + 9 = -12 + 9$$
$$(x + 3)^2 = -3$$
$$\sqrt{(x + 3)^2} = \sqrt{-3}$$

$\sqrt{-3}$ is not a real number.

The quadratic equation has no real number solution.

Example 6

$$x^2 + 8x + 8 = 0$$
$$x^2 + 8x = -8$$
$$x^2 + 8x + 16 = -8 + 16$$
$$(x + 4)^2 = 8$$
$$\sqrt{(x + 4)^2} = \sqrt{8}$$
$$x + 4 = \pm\sqrt{8} = \pm 2\sqrt{2}$$
$$x + 4 = 2\sqrt{2} \qquad x + 4 = -2\sqrt{2}$$
$$x = -4 + 2\sqrt{2} \qquad x = -4 - 2\sqrt{2}$$
$$= -4 + 2(1.414) \qquad = -4 - 2(1.414)$$
$$= -4 + 2.828 \qquad = -4 - 2.828$$
$$= -1.172 \qquad = -6.828$$

The solutions are approximately -1.172 and -6.828.

Pages 443–444

1. The solutions are 1 and -3. **3.** The solutions are 8 and -2. **5.** The solution is 2. **7.** The quadratic equation has no real number solution. **9.** The solutions are -1 and -4. **11.** The solutions are -8 and 1. **13.** The solutions are $-2 + \sqrt{3}$ and $-2 - \sqrt{3}$. **15.** The solutions are $-3 + \sqrt{14}$ and $-3 - \sqrt{14}$. **17.** The solutions are $1 + \sqrt{2}$ and $1 - \sqrt{2}$. **19.** The solutions are $\frac{-3 + \sqrt{13}}{2}$ and $\frac{-3 - \sqrt{13}}{2}$. **21.** The solutions are 2 and 1. **23.** The solutions are $\frac{-1 + \sqrt{13}}{2}$ and $\frac{-1 - \sqrt{13}}{2}$. **25.** The solutions are $-5 + 4\sqrt{2}$ and $-5 - 4\sqrt{2}$. **27.** The solutions are $\frac{3 + \sqrt{29}}{2}$ and $\frac{3 - \sqrt{29}}{2}$. **29.** The solutions are $\frac{1 + \sqrt{17}}{2}$ and $\frac{1 - \sqrt{17}}{2}$. **31.** The quadratic equation has no real number solution. **33.** The solutions are 1 and $\frac{1}{2}$. **35.** The solutions are -3 and $\frac{1}{2}$. **37.** The solutions are 2 and $\frac{3}{2}$. **39.** The solutions are 1 and $-\frac{1}{2}$. **41.** The solutions are -2 and $\frac{1}{3}$. **43.** The solutions are -2 and $-\frac{2}{3}$. **45.** The solutions are $\frac{1}{2}$ and $-\frac{3}{2}$. **47.** The solutions are $\frac{1}{3}$ and $-\frac{3}{2}$. **49.** The solutions are $-\frac{1}{2}$ and $\frac{4}{3}$. **51.** The solutions are $\frac{1 + \sqrt{2}}{2}$ and

$\frac{1-\sqrt{2}}{2}$. **53.** The solutions are $\frac{2+\sqrt{5}}{2}$ and $\frac{2-\sqrt{5}}{2}$. **55.** The solutions are 2 and 4. **57.** The solutions are $1+\sqrt{6}$ and $1-\sqrt{6}$. **59.** The solutions are approximately -4.193 and 1.193. **61.** The solutions are approximately 2.766 and -1.266. **63.** The solutions are approximately -1.652 and 0.152.

SECTION 3 Pages 445–446

Example 2

$$3x^2 + 4x - 4 = 0$$
$$a = 3,\ b = 4,\ c = -4$$
$$x = \frac{-(4) \pm \sqrt{(4)^2 - 4(3)(-4)}}{2 \cdot 3}$$
$$= \frac{-4 \pm \sqrt{16 + 48}}{6}$$
$$= \frac{-4 \pm \sqrt{64}}{6} = \frac{-4 \pm 8}{6}$$
$$x = \frac{-4+8}{6} = \frac{4}{6} = \frac{2}{3} \qquad x = \frac{-4-8}{6} = \frac{-12}{6} = -2$$

The solutions are $\frac{2}{3}$ and -2.

Example 4

$$x^2 + 2x = 1$$
$$x^2 + 2x - 1 = 0$$
$$a = 1,\ b = 2,\ c = -1$$
$$x = \frac{-(2) \pm \sqrt{(2)^2 - 4(1)(-1)}}{2 \cdot 1}$$
$$= \frac{-2 \pm \sqrt{4+4}}{2} = \frac{-2 \pm \sqrt{8}}{2}$$
$$= \frac{-2 \pm 2\sqrt{2}}{2} = -1 \pm \sqrt{2}$$

The solutions are $-1+\sqrt{2}$ and $-1-\sqrt{2}$.

Pages 447–448

1. The solutions are 5 and -1. **3.** The solutions are -3 and 5. **5.** The solutions are -7 and 1. **7.** The solutions are 2 and -3. **9.** The solutions are 3 and -1. **11.** The solutions are -5 and 1. **13.** The solutions are $-\frac{1}{2}$ and 1. **15.** The quadratic equation has no real number solution. **17.** The solutions are 0 and 1. **19.** The solutions are $\frac{3}{2}$ and $-\frac{3}{2}$. **21.** The solutions are $\frac{3}{2}$ and $-\frac{5}{2}$. **23.** The solutions are 3 and $-\frac{2}{3}$. **25.** The solutions are -3 and $\frac{4}{5}$. **27.** The solutions are $-\frac{1}{2}$ and $\frac{2}{3}$. **29.** The quadratic equation has no real number solution. **31.** The solutions are $1+\sqrt{6}$ and $1-\sqrt{6}$. **33.** The solutions are $-3+\sqrt{10}$ and $-3-\sqrt{10}$. **35.** The solutions are $2+\sqrt{13}$ and $2-\sqrt{13}$. **37.** The solutions are $\frac{1+\sqrt{2}}{2}$ and $\frac{1-\sqrt{2}}{2}$. **39.** The solutions are $-3+2\sqrt{2}$ and $-3-2\sqrt{2}$. **41.** The solutions are $\frac{3+2\sqrt{6}}{2}$ and $\frac{3-2\sqrt{6}}{2}$. **43.** The solutions are $\frac{-1+\sqrt{2}}{3}$ and $\frac{-1-\sqrt{2}}{3}$. **45.** The solution is $-\frac{1}{2}$. **47.** The quadratic equation has no real number solution. **49.** The solutions are $\frac{-4+\sqrt{5}}{2}$ and $\frac{-4-\sqrt{5}}{2}$. **51.** The solutions are $\frac{1+2\sqrt{3}}{2}$ and $\frac{1-2\sqrt{3}}{2}$. **53.** The solutions are $\frac{-5+\sqrt{2}}{3}$ and $\frac{-5-\sqrt{2}}{3}$. **55.** The solutions are approximately 5.690 and -3.690. **57.** The solutions are approximately 7.690 and -1.690. **59.** The solutions are approximately 4.590 and -1.090. **61.** The solutions are approximately -2.118 and 0.118. **63.** The solutions are approximately 1.105 and -0.905.

SECTION 4 Pages 449–450

Example 2

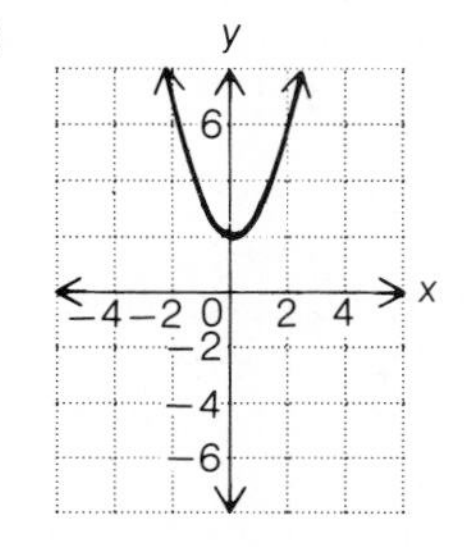

Example 4

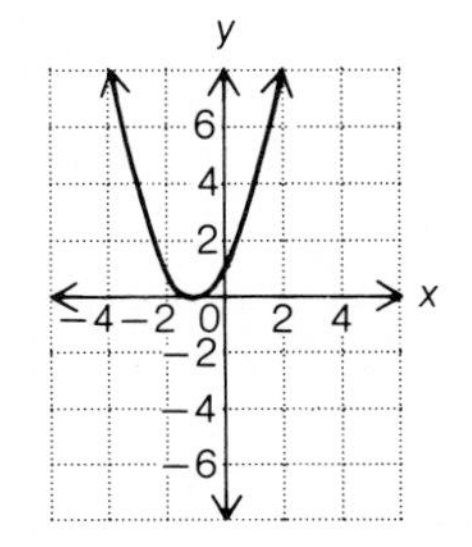

Pages 451–452

1.

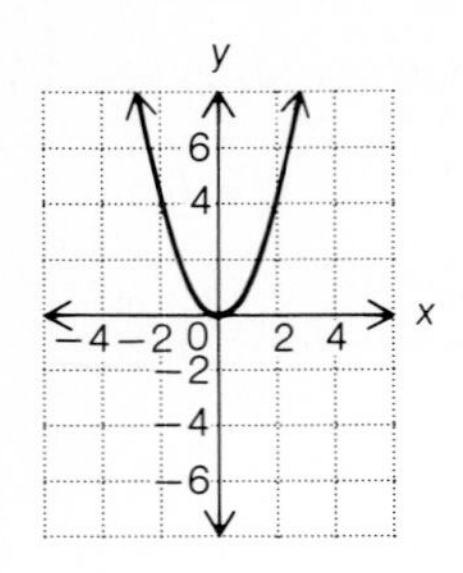

3.

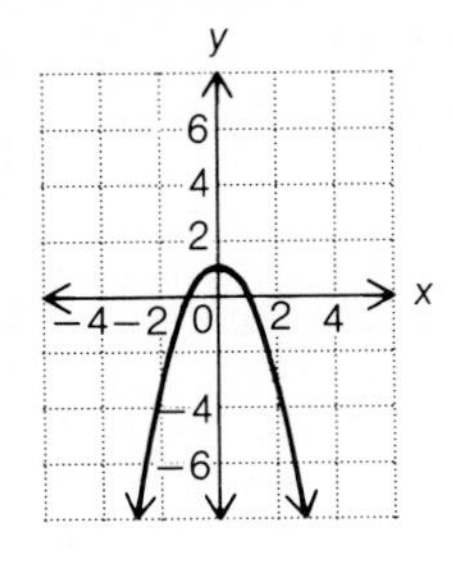

5.

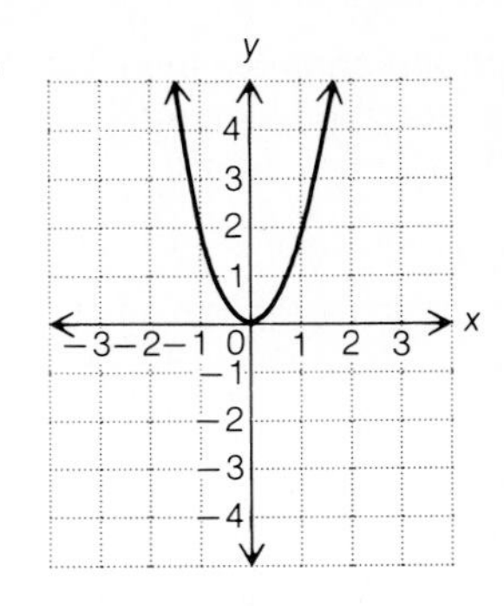

7.

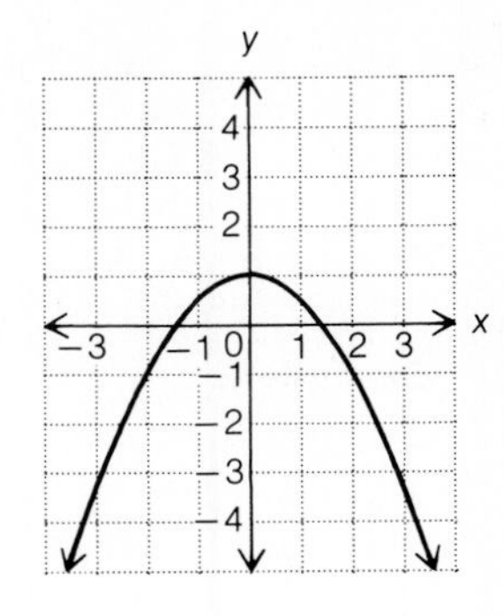

9.

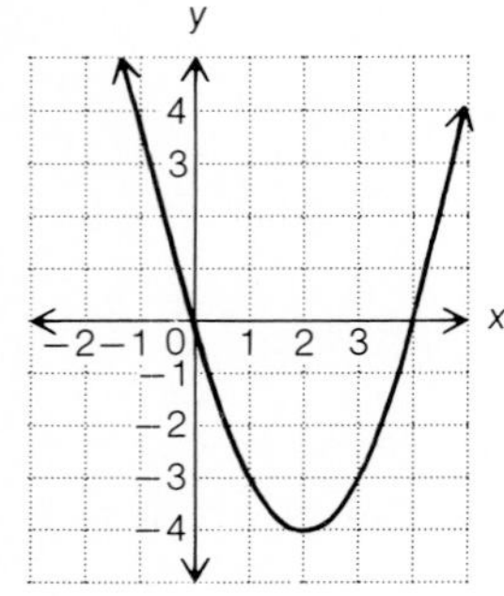

11.

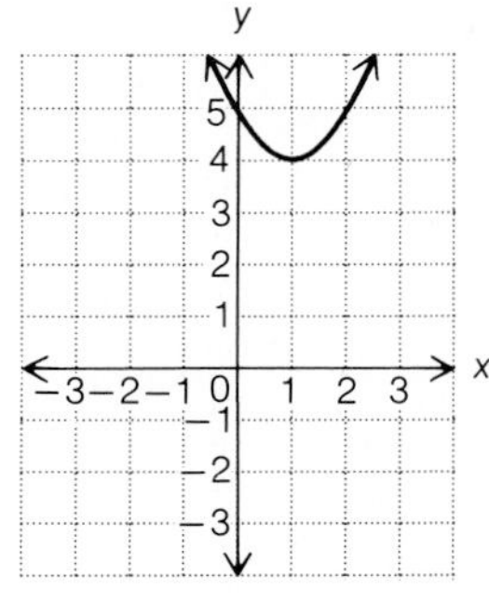

13.

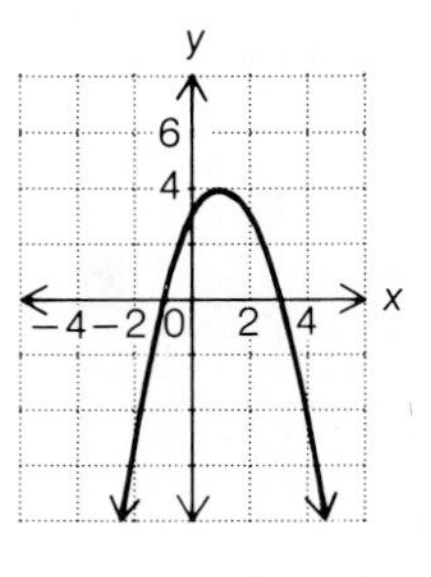

15.

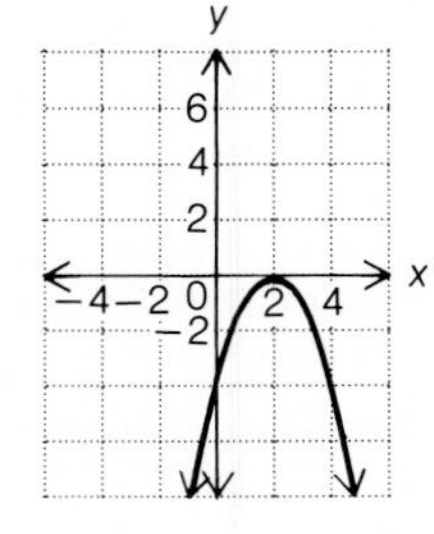

SECTION 5

Pages 453–454

Example 2

Strategy

▷ This is a geometry problem.

▷ Width of the rectangle: w

Length of the rectangle: $w + 2$

▷ Use the equation $A = l \cdot w$.

Solution

$$A = l \cdot w$$
$$15 = (w + 2)w$$
$$15 = w^2 + 2w$$
$$0 = w^2 + 2w - 15$$
$$0 = (w + 5)(w - 3)$$

$w + 5 = 0$ $\qquad$ $w - 3 = 0$

$w = -5$ $\qquad$ $w = 3$

The solution -5 is not possible.
The width is 3 m.

Pages 455–456

1. The width is 4 ft. The length is 8 ft. **3.** The base is 4 m. The height is 10 m. **5.** The integers are 3 and 5. **7.** The integers are -1, 0, and 1. **9.** The integer is 3. **11.** The first car is 8 years old and the second car is 6 years old. **13.** The first stamp is 27 years old and the second stamp is 9 years old. **15.** It would take the larger pipe 4 h to fill the tank. It would take the smaller pipe 12 h to fill the tank. **17.** The first machine can reproduce the report in 24 min. The second machine can reproduce the report in 8 min. **19.** The rate of the boat during the first 24 mi was 12 mph. **21.** The rate of the boat in calm water is 14 mph.

REVIEW/TEST Pages 459–460

1.1 The solutions are -4 and $\frac{5}{3}$. **1.2** The solutions are $-4 + 2\sqrt{5}$ and $-4 - 2\sqrt{5}$. **2.1a** The solutions are $-2 + 2\sqrt{5}$ and $-2 - 2\sqrt{5}$. **2.1b** The solutions are $\frac{-3+\sqrt{41}}{2}$ and $\frac{-3-\sqrt{41}}{2}$. **2.1c** The solutions are $\frac{3+\sqrt{7}}{2}$ and $\frac{3-\sqrt{7}}{2}$. **2.1d** The solutions are $\frac{-4+\sqrt{22}}{2}$ and $\frac{-4-\sqrt{22}}{2}$. **3.1a** The solutions are $-2 + \sqrt{2}$ and $-2 - \sqrt{2}$. **3.1b** The solutions are $\frac{3+\sqrt{33}}{2}$ and $\frac{3-\sqrt{33}}{2}$. **3.1c** The solutions are 3 and $-\frac{1}{2}$. **3.1d** The solutions are $\frac{1+\sqrt{13}}{6}$ and $\frac{1-\sqrt{13}}{6}$. **4.1**

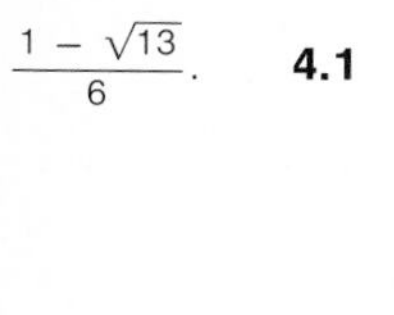

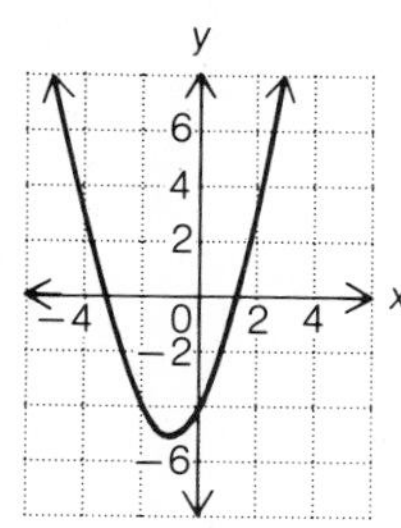

5.1a The width is 5 ft. The length is 8 ft. **5.1b** The rate of the boat in calm water is 11 mph.

CUMULATIVE REVIEW/TEST Pages 461–462

1. $-28x + 27$ (2.2.4) **2.** $x = \frac{3}{2}$ (3.1.3) **3.** $x = 3$ (3.3.2) **4.** $-12a^8b^4$ (5.2.2) **5.** $x + 2 - \frac{4}{x-2}$ (5.4.3) **6.** $x(3x - 4)(x + 2)$ (6.3.2) **7.** $\frac{9x^2(x-2)^2}{(2x-3)^2}$ (7.1.3) **8.** $\frac{x+2}{2x+2}$ (7.3.2) **9.** $\frac{x-4}{2x+5}$ (7.4.1) **10.** The x-intercept is (3,0), and the y-intercept is $(0,-4)$. (8.3.1) **11.** The equation of the line is $y = -\frac{4}{3}x - 2$. (8.4.2) **12.** The solution is (2,1). (9.2.1) **13.** The solution is $(2,-2)$. (9.3.1) **14.** $x > \frac{1}{9}$ (10.3.1) **15.** $a - 2$ (11.3.1) **16.** $6ab\sqrt{a}$ (11.3.2) **17.** $2x - 3$ (11.3.2) **18.** $x = 5$ (11.4.1) **19.** The solutions are $\frac{5}{2}$ and $\frac{1}{3}$. (12.1.1) **20.** The solutions are $5 + 3\sqrt{2}$ and $5 - 3\sqrt{2}$. (12.1.2) **21.** The solutions are $\frac{-7+\sqrt{13}}{6}$ and $\frac{-7-\sqrt{13}}{6}$. (12.2.1) **22.** The solutions are 2 and $-\frac{1}{2}$. (12.3.1) **23.** The selling price of the mixture is \$2.25/lb. (4.5.1) **24.** 250 additional shares are required. (7.6.2) **25.** The rate of the plane in still air is 200 mph. The rate of the wind is 40 mph. (9.4.1) **26.** The student must receive a score of 77 or above. (10.2.3) **27.** The integer is -5 or 5. (12.5.1) **28.** The rate for the last 8 miles was 4 mph. (12.5.1)

29. (10.4.1)

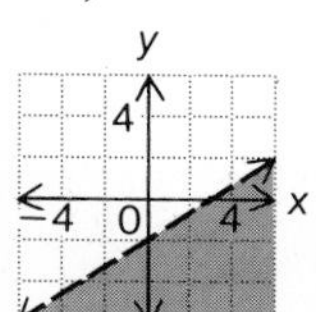

30. (12.4.1)

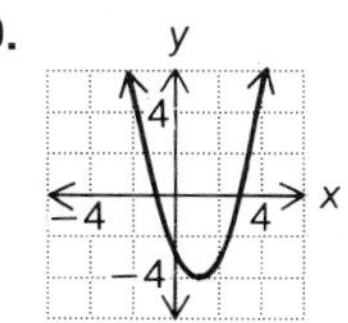

CUMULATIVE REVIEW/FINAL Pages 463–466

1. 10 (1.1.2) **2.** 14 (1.2.2) **3.** −72 (1.4.2) **4.** 9 (1.5.2) **5.** $-\frac{1}{5}$ (2.1.1) **6.** $-6a + 3b + 6$ (2.2.1) **7.** $6y$ (2.2.2) **8.** $9x - 27$ (2.2.4) **9.** $x = -48$ (3.1.3) **10.** $x = \frac{1}{3}$ (3.3.2) **11.** 37.5% (4.1.2) **12.** 45 (4.2.1) **13.** $-2x^2 + 2x + 12$ (5.1.2) **14.** $-32x^{10}y^5$ (5.2.2) **15.** $-2x^3 + 8x^2 - 7x + 3$ (5.3.2) **16.** $-\frac{8y^4}{3x}$ (5.4.1) **17.** $-8 + \frac{4}{x} - \frac{1}{x^2}$ (5.4.2) **18.** $3x + 3 + \frac{10}{2x - 3}$ (5.4.3) **19.** −6 (5.5.1) **20.** $-3y^2(y^2 - 2y + 7)$ (6.1.2) **21.** $(x - 9)(x + 1)$ (6.2.1) **22.** $(3x + 4)(2x - 3)$ (6.3.1) **23.** $-3x(2x + 3)(x + 2)$ (6.3.2) **24.** $(7y - 1)(7y + 1)$ (6.4.1) **25.** $(y - 1)(x - 2)$ (6.4.3) **26.** $3x^2(2 - 3y)(2 + 3y)$ (6.4.4) **27.** The solutions are −4 and 3. (6.5.1) **28.** 1 (7.1.3) **29.** $\frac{-4x^2 + 9x + 12}{(2x - 3)(x + 4)}$ (7.3.2) **30.** $y + 1$ (7.4.1) **31.** $x = 2$ (7.5.1) **32.** $t = \frac{L - a}{ac}$ (7.7.1) **33.** The slope is −3. (8.3.2) **34.** The equation of the line is $y = -\frac{3}{4}x - 2$. (8.4.1) **35.** The solution is (2, −1). (9.2.1) **36.** The solution is (−3,2). (9.3.1) **37.** $x > \frac{2}{3}$ (10.2.1) **38.** $y < \frac{1}{3}$ (10.3.1) **39.** $9a^2$ (11.1.2) **40.** $38\sqrt{3a} - 35\sqrt{b}$ (11.2.1) **41.** $3\sqrt{2} + 3$ (11.3.2) **42.** $x = 3$ (11.4.1) **43.** The solutions are $\frac{3}{2}$ and −1. (12.1.1) **44.** The solutions are $\frac{1 + \sqrt{10}}{3}$ and $\frac{1 - \sqrt{10}}{3}$. (12.3.1) **45.** $2n - 4(n + 2)$; $-2n - 8$ (2.3.3) **46.** The original value of the car was $7296. (3.4.2) **47.** The markup rate is 56%. (4.3.1) **48.** $3000 more must be invested at 9.5%. (4.4.1) **49.** The coffee mixture costs $4.50/lb. (4.5.1) **50.** The resulting mixture is 25% sugar. (4.5.2) **51.** For $1\frac{1}{2}$ hrs the plane flew at 180 mph, and for $2\frac{1}{2}$ hrs the plane flew at 140 mph. (4.6.1) **52.** The length is 25 ft. The width is 15 ft. (4.7.1) **53.** There are 17 dimes and 11 quarters in the bank. (4.8.2) **54.** The length is 20 in. The width is 16 in. (6.5.2) **55.** 76,000 people would be expected to vote. (7.6.2) **56.** With both computers operating, it will take $17\frac{1}{2}$ min to print the report. (7.8.1) **57.** The rate of the plane in calm air is 240 mph. The rate of the wind is 40 mph. (9.4.1) **58.** The rate of the plane in calm air is 240 mph. (12.5.1)

59. (9.1.2)

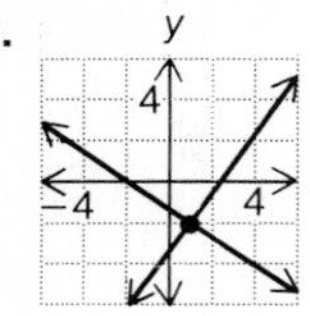

The solution is (1, −2).

60. (12.4.1)

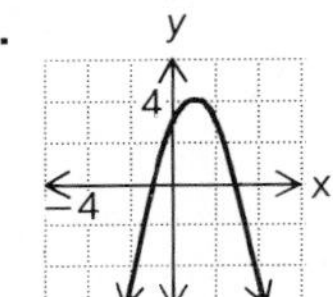

Index

Student Evaluation

To the student:

Houghton Mifflin, as a publisher of fine textbooks in mathematics, is constantly working to make good materials even better. We believe that the person best qualified to comment on how to improve a book is the student who has been learning from it. You can help us make better the learning experiences of future students if you will take a moment to share with us your impressions of INTRODUCTORY ALGEBRA: AN APPLIED APPROACH, SECOND EDITION.

The Publisher

About Your College

What college do you attend? __________

In what state is your college located? __________

What type of college? (2-year, 4-year, public, private, college or university) __________

About Your Course

Name the course in which INTRODUCTORY ALGEBRA is used: __________

How long is the course? (1-semester, 2-semester, etc.) __________

Does your college give a placement exam in math to entering students? __________

About Your Class

How many times does your class meet? __________

	Per Week	Per Term
For Lecture		
For Lab		
For Testing		
Other (Specify)		

How many students are in your class section? __________

Who is teaching your class? (Professor, Instructor, Graduate Student, etc.) __________

About Yourself

How many years of high school mathematics did you have? __________

How long has it been since your last math course? __________

How would you rate yourself as a math student? (1.0 Fair to 4.0 Top) __________

Are you working? __________ Full or part time? __________

What is your career goal? __________

About Your Book

How do you rate this text in these areas:

	Low	1	2	3	4	5	High
Organization							
Useful Content							
Explanations							
Overall Rating							

Identify (by page number) any parts of the book that you found especially difficult: ______

Do you know of another text that you like better? (Author/Title) ______

Will you keep this book for future reference? ______

About Your Ideas

What changes would you recommend be made to improve this text for future students?

From: ______
Name (optional)

______
Name of School

______
Address of School

Mail to: Houghton Mifflin
Attention: Editor-in-Chief
College Mathematics
One Beacon Street
Boston, MA 02108

15

$$\frac{13}{15} = \frac{P}{100}$$

$$\frac{15P}{15} = \frac{1300}{15}$$

$$P = 86\% = B$$

$$15\overline{)1300}$$ 86%

120

100

90

10